Logic Colloquium '95

Since their inception, the Perspectives in Logic and Lecture Notes in Logic series have published seminal works by leading logicians. Many of the original books in the series have been unavailable for years, but they are now in print once again.

This volume, the 11th publication in the Lecture Notes in Logic series, collects the proceedings of the Annual European Summer Meeting of the Association for Symbolic Logic, held in 1995. It includes papers in the core areas of set theory, model theory, proof theory and recursion theory, as well as the more recent topics of finite model theory and non-monotonic logic. It also includes a tutorial on interactive proofs, zero-knowledge and computationally sound proofs that reported on recent developments in theoretical computer science, and three plenary lectures dedicated to the foundational and technical evolution of set theory over the past 100 years.

JOHANN A. MAKOWSKY works in the Department of Computer Science at Technion – Israel Institute of Technology, Haifa.

ELENA V. RAVVE works in the Department of Computer Science at Technion – Israel Institute of Technology, Haifa.

LECTURE NOTES IN LOGIC

A Publication of The Association for Symbolic Logic

This series serves researchers, teachers, and students in the field of symbolic logic, broadly interpreted. The aim of the series is to bring publications to the logic community with the least possible delay and to provide rapid dissemination of the latest research. Scientific quality is the overriding criterion by which submissions are evaluated.

More information, including a list of the books in the series, can be found at http://www.aslonline.org/books-lnl.html

LECTURE NOTES IN LOGIC 11

Logic Colloquium '95

Proceedings of the Annual European Summer Meeting of the Association of Symbolic Logic, held in Haifa, Israel, August 9–18, 1995

Edited by

JOHANN A. MAKOWSKY

Technion – Israel Institute of Technology, Haifa

ELENA V. RAVVE

Technion – Israel Institute of Technology, Haifa

ASSOCIATION FOR SYMBOLIC LOGIC

CAMBRIDGE UNIVERSITY PRESS

CAMBRIDGE
UNIVERSITY PRESS

University Printing House, Cambridge CB2 8BS, United Kingdom

One Liberty Plaza, 20th Floor, New York, NY 10006, USA

477 Williamstown Road, Port Melbourne, VIC 3207, Australia

4843/24, 2nd Floor, Ansari Road, Daryaganj, Delhi – 110002, India

79 Anson Road, #06–04/06, Singapore 079906

Cambridge University Press is part of the University of Cambridge.

It furthers the University's mission by disseminating knowledge in the pursuit of education, learning, and research at the highest international levels of excellence.

www.cambridge.org
Information on this title: www.cambridge.org/9781107167902
10.1017/9781316716830

Association for Symbolic Logic
Richard A. Shore, Publisher
Department of Mathematics, Cornell University, Ithaca, NY 14853
http://www.aslonline.org

A catalogue record for this publication is available from the British Library.

ISBN 978-1-107-16790-2 Hardback

Preface

The Logic Colloquium '95 is the Annual European Summer Meeting of the *Association of Symbolic Logic (ASL)*. The ASL is the International Body of Logicians, both mathematical and philosophical. Its organizational center is in the US but its scope is truly international. The ASL was founded in the thirties by Alonzo Church. Since the sixties the ASL holds an annual symposium in Europe. *It is the first time that this meeting was held in Israel.* On behalf of the Organizing Committee I wish to thank the ASL, in particular Y. Moschovakis (former president of the ASL) and D. Lascar, for choosing Israel as the location of the Logic Colloquium '95 and for all its support in organizing it.

The Topics

The topics of this conference were the four classical topics of Mathematical Logic, Set Theory, Model Theory, Proof Theory and Recursion Theory, together with two topics which evolved more recently under the influence of Computer Science: Finite Model Theory and Non-monotonic Logic. A tutorial on Interactive Proofs, Zero Knowledge and Computationally Sound Proofs reported on recent developments in Theoretical Computer Science which shed new light on the nature of feasible computations.

We have added to this three plenary lectures dedicated to the foundational and technical evolution of Set Theory over the last 100 years. In this way we commemorated the 100th anniversary of the appearance in print of Cantor's fundamental paper which marks the beginning of Set Theory as a discipline.

The talks were divided into plenary lectures, special session lectures and contributed papers. The abstracts of the invited talks and contributed talks (of members of the ASL) were published in the *Bulletin of Symbolic Logic, vol. 3.1 (1997) pp. 73-147*. The contributions in this volume consists of papers and extended abstracts of the invited speakers. The papers do not necessarily coincide with the talks given. They were all submitted after the conference and were thoroughly refereed.

Logic in Israel

The fact that Israel was awarded the honour of hosting this Colloquium for the first time outside Europe is a result of the leading role that Israeli logicians play. Mathematical Logic has been a strong point in Israeli science since the beginning of modern academic activities, even before the foundation of the state. A.Fränkel, a founding father of the Hebrew University, was a true pioneer in Axiomatic Set Theory. Y.Bar–Hillel was a leading figure in the field of Philosophy of Science. A. Robinson, who together with A.Tarski initiated Model Theory, spent in the late fifties four years at the Hebrew University. H.Gaifman,

A.Levy and M.Rabin of the next generation of Israeli logicians continued this school which flourishes already in the fifth generation. In the last 40 years Mathematical Logic and the Foundations of Mathematics and Computer Science have become a focal point of research in Israel. M. Rabin left his imprint on the Theory of Computation and Theoretical Computer Science. S. Shelah has dramatically changed our perception and understanding of Model Theory and Set Theory. The program of this colloquium also bore witness to Shelah's overall impact on Mathematical Logic and we celebrated his 50th birthday together with the completion of his 600th publication with a special banquet. In the late seventies, the eminent Soviet logician B. Trakhtenbrot joined Tel Aviv University, where he founded a group which actively carries out research in Logic in Computer Science. P.Erdös, the most prolific contributor to Set Theoretic Combinatorics and laureate of the Wolf Prize, held a permanent visiting position in the Mathematics Department of the Technion. The tutorial given by S.Micali was dedicated to Interactive Proofs, an aspect of Computation theory in which Israeli researchers play a leading role. L.Babai, S.Micali C.Rackoff and two Israelis, S. Goldwasser (The Weizmann Institute) and S.Moran (Technion), received the Gödel Prize of the ACM and EATCS for their two seminal papers initiating this topic. A. Wigderson, a Technion graduate and Professor at the Hebrew University, was awarded the prestigious Nevanlinna Prize in 1994 for his overall contribution to Computation Theory.

Research in Logic is pursued very actively in all the seven institutions of higher learning in Israel. Prof. M. Rabin, laureate of the Turing Award, received the prestigious Israel Prize for his scientific and academic achievements in 1995. I have singled out only the most prominent Israeli researchers related to this conference but many others, among them the Israeli invited speakers, are worthy of similar praise. I am proud of being part of the Israeli Logic community, and I am sure all the active Israeli logicians share this feeling with me.

The Participants

This conference was truly international with over 230 participants (and over 20 contributions by title) from 34 countries and five continents:

Europe: Austria, Croatia, Czechia, Estonia, Finland, France, Germany, Great Britain, Hungary, Italy, Latvia, Lithuania, Norway, Poland, Romania, Russia, Slovakia, Spain, Sweden, Switzerland, Ukraine, Yugoslavia;
America: Brazil, Canada, Chile, USA;
Asia: China, Israel, Japan, Kasachstan, Singapore;
Oceania: Australia, New Zealand;
Africa: South Africa.

The four countries with delegations consisting of more than 10 participants are the USA, Germany, Israel, Russia and France.

The Hosting Institutions

Technion-Israel Institute of Technology is the oldest institution of higher learning in Israel and the first which adapted Hebrew as its language of instruction. Originally modeled after the German Technical Universities, it was founded in 1924. In the fifties it was expanded considerably and reorganized after the model of the American Institutes of Technology. The Mathematics Department was founded in the fifties, and the Institute of Advanced Studies in Mathematics (IASM) is attached to it. The Computer Science Departement was founded in the seventies. Logic is taught both at the Mathematics and the Computer Science Departments. Research in Mathematical Logic is more concetrated in Computer Science, but research in Set Theoretic Combinatorics is active in the Mathematics Department.

The University of Haifa was founded in the late sixties and plays an increasingly important role as the University of the North of Israel. Mathematics and Computer Science are part of the Faculty of Social and Mathematical Sciences. Logic is taught and research is active in Model Theory.

I would like to thank the Technion professors A.Solan (Senior Vice-President), N.Liron (Vice-President of Academic Affairs), M.Cwikel (Chairman IAMS), Y.Benjamini (Dean of Mathematics) and Z.Berk (former Dean of Computer Science). and the University of Haifa professors V.Harnik of the Organizing Committee, M.Schechter (Rector of the University of Haifa). and S.Waterman (Dean of the Faculty of Social and Mathematical Sciences). Their material and moral support was invaluable in the organization of the conference. In particular, I wish to thank Prof. Harnik for his part in the preparation for this conference.

Financial Support

This conference was financed through a joint financial effort of most of the Israeli Universities, the Israeli Academy of Sciences and the Ministry of Science and Arts and the Ministry of Tourism. Special thanks to the Hebrew University and the two hosting institutions. Substantial support was given by the Kurt Gödel Society and FoLLI, the European Association for Logic, Language and Information and the North Holland Publishing Company. Additionally, some of the invitations of the invited and special session speakers were supported by binational (BNSF, GIF, France-Israel) grants of Israeli researchers. All this money was used to support the invited speakers, and to subsidize the attendance of many outstanding logicians who currently suffer from the extremely poor financial conditions of their academic environment.

Furthermore, the ASL has contributed 10'000 US$ to be used for support of promising graduate students. 7500 US$ were used to support 13 applicants from America, Germany, Great Britain, Finland, France, Russia and Ukraine. 2500 US$ were used to support 10 Israeli students.

On behalf of the Organizing Committee I wish to thank all the sponsors for their support.

I wish to thank *all the participants*. The number of registrations surpassed our wildest dreams. Without them the conference would have ended in a financial disaster. I wish to thank the invited speakers who agreed to come here at their own cost and made the scientific program truly spectacular, all the special session speakers, and all those, who contributed their papers, for making the conference a truly memorable event.

Teamwork

I wish to thank the members of the Program Committee and especially Gerhard Jaeger, Menachem Magidor (chairman) and Richard Shore, who helped efficiently in choosing the invited and special session speakers. It was a pleasure to work with them.

I wish to thank the members of the Organizing Committee, who helped in raising the necessary funds. My special thanks are to V. Harnik and A. Levy.

I wish to thank the Local Organizing Staff for the enormous work they have invested in setting up the necessary framework. M.Kaminski and R.Shabalin helped in handling all the e-mail. G. Agranov provided graphics support. E.Ravve edited and compiled the abstracts distributed during the conference, Y.Sagi handled the registrations, Z.Keisari and her team from the External Studies Department of the Technion and the TANDU Travel Agency were in charge of organizing accommodation, meals and daily transportation. Last but not least, I wish to thank my wife M.N.Yelenevskaya, for her understanding and her enduring all kinds of deprivations while I was pursuing the preparations of this event.

János Makowsky

Since the holding of the
Logic Colloquium'95 four
logicians have passed away
to whom the colloquium was
indebted in one way or another:

Alonzo Church, 1903-1995, the founder of the ASL,
George Boolos, 1940-1996, then acting president of the ASL,
Paul Erdös, 1913-1996, who had a second home in Haifa, and
Laura Mayer, 1957-1997 who did attend and read a paper.

We shall remember them as colleagues and through their work.

Plenary sessions

Tutorials

- K. Compton (University of Michigan, USA)
  0-1 laws and finite model theory
- D. Marker (University of Illinois at Chicago, USA)
  Strongly minimal sets and geometry
- S. Micali (MIT, Cambridge Mass)
  Interactive proofs, zero knowledge and computationally sound proofs
- T. Slaman (University of Chicago, USA)
  Recent developments in recursion theory

Plenary addresses

- H.Becker (University of S. Carolina, USA)
  Path-connectedness, simple connectedness and descriptive set theory
- Z. Chatzidakis (Universite Paris 7, France)
  Model theory of finite fields
  Published as: Z. Chatzidakis, Model theory of finite and pseudo-finite fields, to appear in the Annals of Pure and Applied Logic, special issue of the Proceedings of the AILA-KGS conference.
- B. Cooper (University of Leeds, England)
  Beyond Goedel's theorem - the failure to capture information content
  Published as: S. Barry Cooper, 'Beyond Godel's Theorem: The failure to capture information content' in 'Complexity, Logic and Recursion Theory' (ed. A. Sorbi), Lecture Notes in Pure and Applied Mathematics no. 187, Marcel Dekker, 1997, pp. 93-122.
- A. Dawar (University of Swansea, Wales)
  Model theoretic methods for complexity theory
- R. Downey (Victoria University, New Zealand)
  On the low_2 recursively enumerable degrees
- I. Herzog (University of Illinois at Chicago, USA)
  Model theory of modules
- E. Hrushovski (Hebrew University, Israel)
  Stability and its heirs
- A. Kanamori (Boston University, USA)
  The evolution of techniques in set theory
- J. Krajicek (Mathematical Institute of CAS, Czechia)
  Propositional proofs, proofs of membership in polynomial ideals, and their complexity
- J. Lynch (Clarkson University, USA)
  On beyond zero-one laws
- P. Maddy (University of California at Irvine, USA)
  Justifying the axioms (V=L and maximize)

- M. Makkai (McGill University, Canada)
  Results in categorical model theory
- D.A. Martin (University of California at Los Angeles, USA)
  Recent work on large cardinals and determinacy
- A. Nies (Heidelberg University, Germany)
  Coding in distributive structures
- M. Pentus (Moscow University, Russia)
  Lambek calculus and formal languages
- A. Pillay (Notre Dame University, USA)
  Lecture 1: Superstable differential fields
  Lecture 2: S. Shelah's work in model theory
- Y. Peterzil (Haifa University, Israel)
  Zilber's trichotomy and o-minimal structures
- W. Pohlers (Universitaet Muenster, Germany)
  Subsystems of set theory vs. subsystems of second order number theory.
- M. Rathjen (Universitaet Muenster, Germany)
  The higher infinite in proof theory
- S. Shelah (Hebrew University, Israel)
  An iterated forcing for the continuum
- L. Soukup (Mathematical Institute HAS, Hungary)
  Combinatorial principles from adding Cohen reals
- A. Stolbushkin (UCLA, USA),
  Monotonicity in feasible models
- S. Wainer (University of Leeds, England)
  The proof theoretic complexity of recursion

Abstracts of all the lectures are published in the Bulletin of Symbolic Logic, vol. 3.1 (1997), pp. 73-147

Special sessions

Finite model theory and computer science

Coordinated by E. Grädel and J.A. Makowsky.

- J. Baldwin (University of Illinois at Chicago, USA)
- B. Courcelle (Bordeaux University, France)
- R. Downey (Victoria University, New Zealand)
- S. Lifsches (Hebrew University, Israel)
- M. Otto (Technical University Aachen, Germany)
- L. Pacholski (Wroclaw, Poland)
- E. Rosen (University of Pennsylvania, USA)
- M. Taitslin (Tver University, Russia)

A special issue of the *Archive for Mathematical Logic* will be dedicated to papers resulting from this session. It should contain the following papers:

- E. Graedel and A. Malmstroem, 0-1 Laws for Recursive Structures
- M. Otto, Bounded Variable Logic: Two, three, more,
- A. Stolbushkin and M. Taitslin, Normailzable orders and generic computations in finite models
- R. Downey and M. Fellows, Index sets and parametric reductions
- Eric Rosen, An existential fragment of second order logic,
- E. Graedel, M. Otto and E. Rosen, Undecidability results on two-variable logics,
- S. Lifsches and S. Shelah Random Graphs in the Monadic Theory of Order
- J. Nurmonen and L. Hella Vectorization hierarchies for some graph quantifiers

Model theory

Coordinated by E. Bouscaren and E. Hrushovski.

- O. Belegradek (Kemerovo, Russia)
- B. Herwig (Freiburg University, Germany)
- M. Junker (Universite de Paris 7, France)
- A. Khelif (Universite Paris 7, France)
- L. Newelski (Mathematical Institute of the Polish Academy of Sciences, Wroclav, Poland)
- E. Rabinovich (Ben Gurion University, Israel)
- Z. Sokolovic (McMaster University, Canada)

Three papers of this session will appear in the *Israel Journal of Mathematics*:

- B. Herwig, Extending partial isomorphisms for the small index property for many ω-categorical structures.
- L. Newelski, M-gap conjecture and m-normal theories.
- M. Junker, Completeness of Zariski groups.

Set theory

Coordinated by M. Magidor and J.A. Makowsky

- N. Brunner (Vienna University, Austria)
- L. Bukovsky (University of Kosice, Slovakia)
- J. Cummings (Hebrew University, Jerusalem, Israel)
- M. Foreman (UCI, Irvine, USA)
- S. Friedman (MIT, USA)
- M. Goldstern (TU Wien, Austria)
- T. Jech (Penn State University, USA)
- I. Juhasz (Math.Inst. of the Academy of Sciences, Budapest, Hungary)
- V. Kanovei (Moscow Transport Institute, Russia)
- P. Koepke (Universitaet Bonn, Germany)
- P. Komjath (Math.Inst. of the Academy of Sciences, Budapest, Hungary)
- L. Stanley (Lehigh University, Betlehem PA, USA)
- P. Vojtas (Slovac Academy of Sciences, Kosice, Slovakia)

A special issue of *Fundamenta Mathematicae* was dedicated to papers originating in this special session, edited by Akihiro Kanamori (Guest Editor), Alexander Kechris and Leszek Pacholski. The contents of this issue, volume 154, Number 2 (1997), consists of

- N. Brunner, P.Howard and J.E.Rubin, Choice principles in W/eglorz' models
- A.Dow and I.Juhász, Are initially ω_1-compact separable regular spaces compact?
- S.D.Friedman, Σ^* the approach to the fine structure of L
- S.Fuchino and L.Soukup, More set-theory around the weak Freese-Nation property
- W.Just and P.Vojtáš, On matrix rapid filters
- V.Kanovei, Two dichotomy theorems on colourability of non-analytic graphs
- P.Komjáth, A strongly non-Ramsey uncountable graph 2

Recursion theory

- M. Arslanov (Kazan, Russia)
- P. Cholak (University of Notre Dame, USA)
- C.T. Chong (Singapore)
- P. Fejer (University of Mass., Boston, USA)
- S. Goncharov (Novosibirsk, Russia)
- M. Groszek (Dartmouth College, Hannover, USA)
- V. Harizanov (George Washington University, USA)
- E. Herrman (Humboldt University Berlin, Germany)
- J. Knight, (Notre Dame University, USA)
- M. Kummer (University of Karlsruhe, Germany)
- M. Lerman (University of Connecticut, USA)

A special issue of *Annals of Pure and Applied Logic* was dedicated to papers originating in this special session, edited by R. Shore (Guest Editor). The contents of this issue, volume 87.2, (1997) consists of

- C.T. Chong and Y. Yang, Σ_2 induction and infinite injury arguments, Part II: Tame Σ_2 coding and the jump operator
- M.J. Groszek and T.A. Slaman, Π^0_1 classes and minimal degrees
- E. Herrmann, Boolean pairs formed by Δ^0_n sets
- C.J. Ash and J.F. Knight, Possible degrees and recursive copies II
- S. Lempp and M. lerman, A finite lattice without critical triple that cannot be embedded into the enumerable Turing degrees.

Proof theory

Coordinated by G. Jäger and S. Wainer.

- A. Avron (Tel Aviv University, Israel)
- L. Beklemishev (Steklov Mathematical Institute, Moscow, Russia)
- G. Bellin (Universite Paris 7, Paris, France)
- V. Danos (Universite Paris 7, Paris, France)
- W. Dekkers (University of Nijmegen, Holland)
- U. Kohlenbach (Frankfurt University, Germany)
- G. Mints (Stanford University, USA)
- W. Pohlers (Universitaet Muenster, Germany)
- T. Strahm (Universitaet Bern, Switzerland))
- A. Weiermann (Universitaet Muenster, Germany)

A special issue of the *Archive for Mathematical Logic* is being prepared with papers resulting from this session.

Non-monotonic and non-classical logic

Coordinated by D. Lehmann and W. Marek.

- G. Amati (Fondazione Ugo Bordoni, Roma, Italy)
- R. Ben-Eliyahu (Techion-Israel Institute of Technology, Israel)
- Tim Fernando (University of Stuttgart, Germany)
- M. Gelfond (University of Texas, USA)
- G. Gottlob (Technical University of Vienna, Austria)
- H. Herre (Universitaet Leipzig, Germany)
- W. Marek (University of Kentucky, USA)
- V.V. Rybakov (University of Krasnoyarsk, Russia)
- K. Schlechta (Universite Marseille, France)

Organizing and program committees

Program Committee

J. Barwise (USA), E. Bouscaren (France), E. Graedel (Germany), A. Hajnal (Hungary and USA), G. Jaeger (Switzerland), M. Magidor (Chairman, Israel), J.A. Makowsky (Coordinator, Israel), D. Mundici (Italy), R. Shore (USA), M.A. Taitslin (Russia), M. Ziegler (Germany).

Special Session Advisors

(Besides some of the Program Committee members)
M. Kaminski (Israel), E. Hrushovski (Israel), D. Lehmann (Israel), W. Marek (USA), L. Moss (USA), S. Wainer (Great Britain)

Organizing Committee

U. Abraham (Ben Gurion University). R. Aharoni (Technion), A. Avron (Tel Aviv University), N. Francez (Technion), V. Harnik (Haifa University), M. Kaminski (Technion), M. Koppel (Bar Ilan University), D. Lascar (Paris 7), A. Levy (Hebrew University), J.A. Makowsky (Chairman, Technion), M. Rubin (Ben Gurion University).

Local Organizing Staff

M. Kaminski (Technion, Department of Computer Science), Z. Keisari (Technion, Department of External Studies), S. Landmann (Haifa University (Faculty of Social and Mathematical Sciences), E. Ravve (Technion, Department of Computer Science), U. Rotics (Technion, Department of Computer Science), Y. Sagi (Technion, Department of Computer Science), S. Shabalin (Technion, Department of Computer Science), H. Shklarsky, (Technion, Department of External Studies), S. Schur (Technion, Institute of Advanced Studies in Mathematics), M. Yelenevskaya (Technion, Department of General Studies), S. Yarkoni (Haifa University, Public Relations Office), N. Zonshine (Technion, Department of External Studies).

Technical Support

Dr. Eli Biham (WWW advisor), Dr. G. Agranov, Dr. H. Gotsman, Arnon Amir (Computer Graphics) Y. Dubinsky, T. Uziel, D. Gutzeit (Computer Support).

Table of Contents

The Number of Path-Components of a Compact Subset of $\mathbb{R}^n$

Howard Becker *

Department of Mathematics
University of South Carolina
Columbia, SC 29208 USA
becker@math.sc.edu

§0. Introduction

This paper is concerned with the following question. Assume $\neg CH$; does there exist a compact set $K \subset \mathbb{R}^n$ such that K has exactly $\aleph_1$ path-components? For $\mathbb{R}^3$, the answer is yes. For $\mathbb{R}^2$, the answer is no, assuming a weak large cardinal axiom (which may or may not be necessary).

The proof of both results is descriptive set theoretic. Indeed, the motivation for asking the question is descriptive set theoretic. The same question for components, rather than path-components, would be a silly question; it is obvious (at least to descriptive set theorists) that the answer is no. It is also obvious that it is not possible that $2^{\aleph_0} \geq \aleph_3$ and that there is a compact $K \subset \mathbb{R}^n$ with $\aleph_2$ path-components. But the question as posed above does not seem to be a silly question. One of the purposes of this paper is to present the descriptive set theoretic point of view, and hopefully convince the reader that these "obvious" facts really are obvious. Two references for descriptive set theory are Kechris [13] and Moschovakis [17], and we follow their notation and terminology.

In both the $\mathbb{R}^3$ and $\mathbb{R}^2$ cases, we have results that are stronger than those stated above. In both cases, the size of the continuum is irrelevant and the theorem – properly stated – is nontrivial even if CH is true. These theorems will be given in §2. For $\mathbb{R}^3$, there is a more general theorem, a precise version of the following: Any $\boldsymbol{\Sigma}^1_1$ equivalence relation can be coded up as the equivalence relation of being in the same path-component of K, for some compact $K \subset \mathbb{R}^3$. From this it easily follows that there is a $K \subset \mathbb{R}^3$ with $\aleph_1$ path-components. That general theorem has other applications as well, one of which answers a question of Kunen-Starbird [14]. This paper is largely an explanation of the statement of these stronger theorems, and of the larger mathematical theory of which they are a part, that is, the descriptive set theory of equivalence relations. In the $\mathbb{R}^3$ case we say virtually nothing about the proof. In the $\mathbb{R}^2$ case we give an outline of the proof (§§6,7), containing several gaps, and using a stronger large cardinal axiom than required.

The author plans to some day write a long paper about path-connectedness, simple connectedness and descriptive set theory (Becker [3]). The results announced here will appear there with complete proofs. Most of Becker [3] will be concerned with calculating the complexity, with respect to the projective

* Partially supported by NSF Grant DMS–9505505

hierarchy, of the following pointsets in the space $\mathcal{K}(\mathbb{R}^n)$ of compact subsets of $\mathbb{R}^n$:

$$PC_n = \{K \in \mathcal{K}(\mathbb{R}^n) : K \text{ is path-connected}\},$$
$$SC_n = \{K \in \mathcal{K}(\mathbb{R}^n) : K \text{ is simply connected}\}.$$

Several theorems of this sort were announced in Becker [2, Example 16 ff.], and proofs of some of them have appeared in Kechris [13, Theorems 33.17 and 37.11]. (*Remark.* There has been one new result since these publications appeared. Darji [7] and Just [10], independently, proved that PC_2 is not $\boldsymbol{\Sigma}^1_1$.) This topic is related to the results in this paper. The proof that there is a $K \in \mathcal{K}(\mathbb{R}^3)$ with $\aleph_1$ path-components has much in common with the proof that PC_2 *is not* $\boldsymbol{\Pi}^1_1$. The proof that, assuming large cardinals and $\neg CH$, there is no such K in $\mathcal{K}(\mathbb{R}^2)$, has much in common with the proof that SC_2 *is* $\boldsymbol{\Pi}^1_1$.

We work in ZFC. When anything more is used in a theorem it will be explicitly stated in the hypothesis.

§1. Path-components

Our basic reference for topological matters is Kuratowski [15]. Our terminology is standard, and mostly consistent with that reference.

Set theorists have a habit of calling practically anything "the reals". But here, topology actually matters, so the *reals* always means the reals. It is denoted by $\mathbb{R}$. The letter K will always denote a compact subset of $\mathbb{R}^n$ for some n. While our main interest is in such a space K, we give the definitions in more generality.

Definition 1. Definition Let X be a topological space and let $\mathbf{p}$, $\mathbf{q}$ be points in X. A *path from* $\mathbf{p}$ *to* $\mathbf{q}$ *in* X is a continuous function $\gamma : [0,1] \to X$ such that $\gamma(0) = \mathbf{p}, \gamma(\mathbf{1}) = \mathbf{q}$. An *arc* is a one-to-one path.

We sometimes abuse the language and refer to the pointset $Im(\gamma)$ as "the path γ" or "the arc γ".

For any topological space X, let $\approx_X$ denote the following equivalence relation on X:

$$\mathbf{p} \approx_{\mathbf{X}} \mathbf{q} \Longleftrightarrow \text{there exists a path from } \mathbf{p} \text{ to } \mathbf{q} \text{ in } \mathbf{X}.$$

The $\approx_X$-equivalence classes are called the *path-components* of X. X is *path-connected* if it has only one path-component.

Path-connectedness and path-components should not be confused with a different notion: connectedness and components. (*Connected* means no nontrivial clopen sets, and a *component* is a maximal connected subset.) While path-connectedness implies connectedness, the converse is false, even for compact subsets of $\mathbb{R}^2$. The standard counterexample is $K^* = A_1 \cup A_2$, where

$$A_1 = \{(x,y) : -1 \le x < 0 \text{ and } y = sin(1/x)\},$$
$$A_2 = \{(x,y) : x = 0 \text{ and } -1 \le y \le 1\}$$

(see Figure 1). K^* is connected. But K^* is not path-connected; it has exactly two path-components, A_1 and A_2.

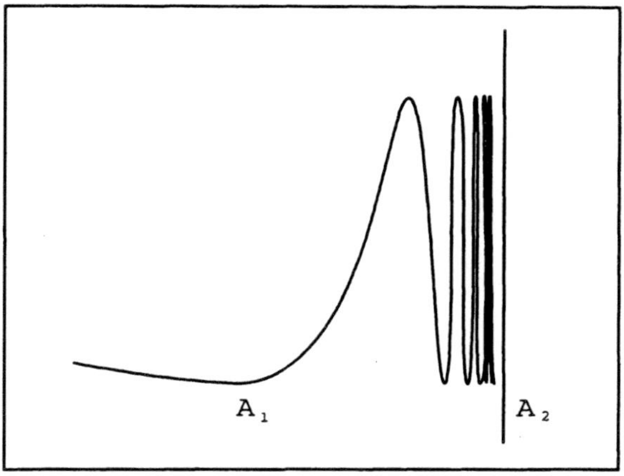

Figure 1

Theorem 1.1 Let $\mathbf{p}, \mathbf{q} \in \mathbb{R}^n$ and let $\gamma : [0,1] \longrightarrow \mathbb{R}^n$ be a path from $\mathbf{p}$ to $\mathbf{q}$. If $\mathbf{p} \neq \mathbf{q}$ then there is an arc γ' from $\mathbf{p}$ to $\mathbf{q}$ such that $Im(\gamma') \subset Im(\gamma)$.
Proof. See Kuratowski [15, §50, I, Theorem 2 and II, Theorem 1]. □

By 1.1, for any $X \subset \mathbb{R}^n$, path-components are the same thing as *arc-components* and path-connectedness the same as *arc-connectedness*. (In fact, for any Hausdorff space, the two concepts coincide.)

§2. Statement of theorems

We have two theorems, 2.1 and 2.2, below, which answer the question posed at the beginning of this paper.
Theorem 2.1 There is a compact set $K \subset \mathbb{R}^3$ with the following properties.

(a) K has exactly $\aleph_1$ path-components.
(b) There does not exist a nonempty perfect set $P \subset K$ such that any two distinct points of P are in different path-components of K.

The above theorem is proved in ZFC. The next theorem is not quite proved in ZFC, but rather in $ZFC + \epsilon$. (A precise description of ϵ is given below.)
Theorem 2.2 Assume ϵ. For any compact set $K \subset \mathbb{R}^2$, one of the following holds:

(i) K has only countably many path-components;
(ii) There is a nonempty perfect set $P \subset K$ such that any two distinct points of P are in different path-components of K. (Hence K has $2^{\aleph_0}$ path-components.)

The axiom ϵ is the following statement:

Every uncountable $\boldsymbol{\Sigma}^1_2$ set of reals contains a nonempty perfect subset.

By a theorem of Solovay (see Kanamori [11, Theorem 14.10]) ϵ is equivalent to:

$$\text{For all } a \subset \omega,\ \aleph_1^{L[a]} < \aleph_1.$$

The axiom ϵ is equiconsistent with the existence of an inaccessible cardinal (see Kanamori [11, Theorem 11.6]), and thus it is a "large cardinal axiom" by virtue of its consistency strength, although it does not, of course, imply the actual existence of large cardinals. Serious large cardinal axioms, e.g., the existence of a measurable cardinal, imply that ϵ is *true* (as opposed to merely consistent). Hence these large cardinal axioms imply that the conclusion of 2.2 is true. For more information on large cardinal axioms, see Kanamori [11].

This axiom has been around for a long time, and has been explicitly considered as a hypothesis of theorems, but does not seem to have ever been given a name. To rectify that oversight, I have decided to call it ϵ. Compared to the large cardinal axioms commonly used in set theory these days, this axiom is a *very* weak assumption – the name ϵ is entirely appropriate.

Theorem 2.2 leads to an interesting open question in reverse mathematics: Is 2.2 provable in weak subsystems of $ZFC+\epsilon$, such as ZFC? It is possible that it is provable in ZFC. But I would conjecture that it is not, and that, in fact, the following is provable in ZFC: There exists a compact $K \subset \mathbb{R}^2$ and a bijection between the path-components of K and $\aleph_1^L$. If this is the case, then in all models where $\aleph_1^L = \aleph_1 < 2^{\aleph_0}$, the answer to the question posed at the beginning of this paper would be yes, even for $\mathbb{R}^2$; hence a large cardinal axiom really would be necessary to get a no answer.

§3. Descriptive set theory and equivalence relations, I: Theorems of Silver and Burgess

If E is an equivalence relation on X and $Y \subset X$, Y is called *E-invariant* if for all $y, y' \in X$:

$$y \in Y \text{ and } yEy' \Longrightarrow y' \in Y.$$

Definition 2. Definition Let X be a Polish space, let E be an equivalence relation on X, and let $Y \subset X$ be E-invariant. We say that Y has *perfectly many* E-equivalence classes if there is a nonempty perfect set $P \subset Y$ such that no two distinct points of P are E-equivalent.

Clearly perfectly many equivalence classes implies $2^{\aleph_0}$ equivalence classes. In fact, "perfectly many" is, in some sense, an effectivized version of "continuum many": Y has continuum many equivalence classes iff there is some (arbitrary) function f from the Cantor set $\mathcal{C}$ into Y, such that for $x, y \in \mathcal{C}$, if $x \neq y$ then $f(x) \not\!\!E f(y)$; Y has perfectly many equivalence classes iff there is a *continuous* f as above. "Perfectly many", unlike "continuum many", is absolute whenever E and Y are absolutely - Δ^1_2 (which is the only situation we consider in this paper). Therefore, the size of $2^{\aleph_0}$ is irrelevant to the question of whether there are perfectly many equivalence classes.

In this terminology, Theorem 2.1(b) (respectively, Theorem 2.2(ii)) states that K does not have (respectively, does have) perfectly many path-components.

In the next two theorems, we consider this property in the case $Y = X$, when E is $\boldsymbol{\Pi}^1_1$ (coanalytic) and when E is $\boldsymbol{\Sigma}^1_1$ (analytic), where E is regarded as a pointset in the space $X \times X$.

Theorem 3.1 (Silver). Let X be a Polish space and let E be a $\boldsymbol{\Pi}^1_1$ equivalence relation on X. One of the following two cases holds:

(i) X has countably many E-equivalence classes;
(ii) X has perfectly many E-equivalence classes.

This dichotomy theorem is not true for $\boldsymbol{\Sigma}^1_1$ equivalence relations. The following equivalence relation E^* on $\mathcal{C}$ is a counterexample:

$$xE^*y \iff [(x \notin WO \text{ and } y \notin WO) \text{ or } |x| = |y|],$$

where WO denotes the set of ordinal codes and $|x|$ denotes the ordinal encoded by x. Clearly there are exactly $\aleph_1$ E^*-equivalence classes, and the Boundedness Theorem implies that (even if CH is true) there are not perfectly many classes.

Theorem 3.2 (Burgess). Let X be a Polish space and let E be a $\boldsymbol{\Sigma}^1_1$ equivalence relation on X. One of the following three cases holds:

(i) X has countably many E-equivalence classes;
(ii) X has $\aleph_1$ and not perfectly many E-equivalence classes;
(iii) X has perfectly many E-equivalence classes.

As shown above, case (ii) of 3.2 can occur. Thus $\boldsymbol{\Sigma}^1_1$ equivalence relations come in three types. Assuming $\neg CH$, the three types are just three cardinalities for the set of equivalence classes: $\aleph_0, \aleph_1, 2^{\aleph_0}$. But if CH is true, we need a different way of distinguishing case (ii) from case (iii), and that is where the concept "perfectly many" comes in.

The original proof of Theorem 3.1 appeared in Silver [18]. A simpler proof, essentially due to Harrington, can be found in Martin-Kechris [16]. The original proof of Theorem 3.2 is in Burgess [5]. Shelah later discovered an extremely general theorem, of which both 3.1 and 3.2 are special cases – this can be found in Harrington-Shelah [9].

Although case (iii) is absolute, the distinction between cases (i) and (ii) of Theorem 3.2 is not, in general, absolute. For it is provable in ZFC that there is a $\boldsymbol{\Sigma}^1_1$ equivalence relation E^{**} on $\mathcal{C}$ and a bijection between the equivalence classes of E^{**} and $\aleph_1^L$. (*Proof.* Let $C_1 = \{x : x \in L_{\omega_1^x}\}$ be the largest thin Π^1_1 set – see Kechris [12] for details. Then define

$$xE^{**}y \iff [(x \notin C_1 \text{ and } y \notin C_1) \text{ or } x = y].$$

Since $card(C_1) = card(\aleph_1^L)$, this works.) On the other hand, for some $\boldsymbol{\Sigma}^1_1$ equivalence relations, such as E^*, case (ii) holds in *every* model.

Now consider those Polish spaces K which are compact subsets of $\mathbb{R}^n$, and the equivalence relation $\approx_K$ on K of being in the same path-component. Clearly

$\approx_K$ is $\boldsymbol{\Sigma}^1_1$, since

$$\text{(3.3)} \qquad \mathbf{p} \approx_K \mathbf{q} \iff (\exists \gamma \in (C[0,1])^n) F(\mathbf{p}, \mathbf{q}, \gamma),$$

where F is the following closed subspace of the Polish space $K \times K \times (C[0,1])^n$:

$$F(\mathbf{p}, \mathbf{q}, \gamma) \iff [Im(\gamma) \subset K \text{ and } \gamma(0) = \mathbf{p} \text{ and } \gamma(1) = \mathbf{q}].$$

Therefore Burgess's Theorem is applicable to $\approx_K$, and so, as pointed out in the introduction, it is not possible that $2^{\aleph_0} \geq \aleph_3$ and K has $\aleph_2$ path-components.

The equivalence relation of being in the same *component* of K is closed, hence $\boldsymbol{\Pi}^1_1$, and therefore Silver's Theorem is applicable. That is, for any K,

(3.4) K has either countably many or perfectly many components.

So, as was also pointed out in the introduction, assuming $\neg CH$, K cannot have $\aleph_1$ components. These facts about components can be proved directly, without going through Silver's Theorem.

But is the $\boldsymbol{\Sigma}^1_1$ equivalence relation $\approx_K$ also $\boldsymbol{\Pi}^1_1$? Note that by Suslin's Theorem, it is $\boldsymbol{\Pi}^1_1$ iff it is Borel.

It has been known since the work of Kunen-Starbird [14] in 1982 that there exists a compact $K \subset \mathbb{R}^3$ for which $\approx_K$ is not Borel (and that therefore Silver's Thorem is not, in general, applicable to the equivalence relation $\approx_K$). It is still an open question whether or not for every compact $K \subset \mathbb{R}^2$, $\approx_K$ is Borel. While it is possible that for all $K \in \mathcal{K}(\mathbb{R}^2)$, $\approx_K$ is Borel, it is not the case that $\approx_K$ is Borel *uniformly* in K. For if it was, PC_2 would be a $\boldsymbol{\Pi}^1_1$ set, which is not true (see Becker [2, Theorem 2.2]).

This is the background which motivated the question posed at the beginning of this paper. (That question was asked by the author in 1984 in several talks and in the circulated notes Becker [1], but never asked in print.) To summarize: We have a collection $\mathcal{E} = \{\approx_K : K \in \mathcal{K}(\mathbb{R}^n)\}$ of $\boldsymbol{\Sigma}^1_1$, generally non-Borel, equivalence relations; Theorem 3.2 classifies $\boldsymbol{\Sigma}^1_1$ equivalence relations into three types, all of which can occur; the question is whether type (ii) ($\aleph_1$, not perfectly many) can occur for equivalence relations in $\mathcal{E}$. There are many interesting questions (some solved, some open) of precisely this form: Given a proper subclass of the class of all $\boldsymbol{\Sigma}^1_1$ equivalence relations, can type (ii) occur in this subclass? For example, Vaught's Conjecture is such a question, since isomorphism for countable structures – restricted to the Borel set of models of a first-order theory – is a $\boldsymbol{\Sigma}^1_1$ equivalence relation.

Remark. For the equivalence relation of isomorphism, the distinction between cases (i) and (ii) of Theorem 3.2 is absolute. Thus if there is a counterexample to Vaught's Conjecture in L it remains a counterexample in V (even if $\aleph_1^L < \aleph_1$). See Becker-Kechris [4, §7.2]. In this respect, there is a descriptive set theoretic difference between Vaught's Conjecture and the analogous conjecture for path-components with which this paper is concerned.

§4. Path-components in compact subsets of $\mathbb{R}^3$

The question, as posed in §3, was whether case (ii) of Theorem 3.2 – which does occur among arbitrary $\boldsymbol{\Sigma}^1_1$ equivalence relations – can occur for a special sort of $\boldsymbol{\Sigma}^1_1$ equivalence relation, those of the form $\approx_K$. Of course, Theorem 2.1 says that it does. The way 2.1 is proved is to show that equivalence relations of the form $\approx_K$ are really not all that special; *any* $\boldsymbol{\Sigma}^1_1$ equivalence relation can be coded up as $\approx_K$ for some $K \in \mathcal{K}(\mathbb{R}^3)$. This is made precise in Theorem 4.1, below.

Let $\mathcal{C}$ denote the Cantor middle third set in $[0, 1]$.

Theorem 4.1 Let E be a $\boldsymbol{\Sigma}^1_1$ equivalence relation on $\mathcal{C}$. There exists a compact set $K_E \subset \mathbb{R}^3$ satisfying the following three properties.

(a) For all $x \in \mathbb{R}$, $(x, 0, 0) \in K_E$ iff $x \in \mathcal{C}$.
(b) For all $\mathbf{p} \in \mathbf{K_E}$ there exists an $x \in \mathcal{C}$ such that $(x, 0, 0) \approx_{K_E} \mathbf{p}$.
(c) For all $x, y \in \mathcal{C}$, xEy iff $(x, 0, 0) \approx_{K_E} (y, 0, 0)$.

Both a proof of Theorem 4.1 and a magnificent 3-dimensional picture of K_E will appear in Becker [3].

Note that if the word "compact" was removed from 4.1, the proof would be quite easy. For each pair (x, y) such that xEy, we could pick a path $\gamma^{(x,y)}$ connecting x and y, and since we are in 3-dimensional space, there is enough room to pick these paths so that no two intersect except at the endpoints; then let K_E be the union of all these paths. However a K_E constructed in this naive manner will not even be a Borel set. The trick is to get it to be compact. The construction of K_E is similar to the constructions in Kechris [13, Theorems 33.17 and 37.11].

Theorem 2.1 is a corollary of Theorem 4.1. To see this, just consider a $\boldsymbol{\Sigma}^1_1$ equivalence relation E on $\mathcal{C}$ with $\aleph_1$ and not perfect many equivalence classes, and let K_E be as in Theorem 4.1, for this particular E. It is not hard to show that K_E satisfies 2.1.

Kunen-Starbird [14] proved that there is a $K \in \mathcal{K}(\mathbb{R}^3)$ which has a non-Borel path-component, and asked: Does there exist a $K \in \mathcal{K}(\mathbb{R}^3)$ such that no path-component of K is Borel?

Corollary 4.2 There is a compact set $K \subset \mathbb{R}^3$ such that no path-component of K is Borel.

Proof. It is well known (but apparently unpublished) that there is a $\boldsymbol{\Sigma}^1_1$ equivalence relation E on $\mathcal{C}$ such that no E-equivalence class is Borel. (*Proof.* It will suffice to find such a $\boldsymbol{\Sigma}^1_1$ equivalence relation E' on a standard Borel space. Let S be a $\boldsymbol{\Sigma}^1_1$ non-Borel subset of $\mathbb{R}$, and let $F(\mathbb{R})$ and $F(S)$ be the free groups generated by $\mathbb{R}$ and S, respectively. Let E' be the equivalence relation on $F(\mathbb{R})$ given by the coset decomposition $F(\mathbb{R})/F(S)$.) Let K_E be as in Theorem 4.1, for this particular E. By 4.1(c), if any path-component of K_E was Borel, the corresponding E-equivalence class would be Borel.

Remark. In both 2.1 and 4.2, the K's can be taken to be connected (that is, to be *continua*). This is so because the components of the original K are compact and connected, so in 4.2, we can pass from K to any component, and in 2.1, to any component which consists of $\aleph_1$ path-components. Such a component must exist, by 3.4.

There are some very complicated $\boldsymbol{\Sigma}^1_1$ equivalence relations – complicated in both the intuitive sense, and in the precise sense of *definable cardinality*, as explained in Becker-Kechris [4, §8]. One example of a complicated $\boldsymbol{\Sigma}^1_1$ equivalence relation is Turing-equivalence. By 4.1, all this complexity exists in the path-component equivalence relation for compact subsets of $\mathbb{R}^3$.

All of the above results trivially transfer from $\mathbb{R}^3$ to $\mathbb{R}^n$, for $n \geq 3$. What about $n = 2$? Of course, the analog of Theorem 2.1 is false for $\mathbb{R}^2$ (assuming ϵ). The analog of Corollary 4.2 is also false for $\mathbb{R}^2$ (in ZFC); that is, for any compact $K \subset \mathbb{R}^2$, at least one path-component of K is a Borel set. These facts seem to mean that it is not possible to code up arbitrary $\boldsymbol{\Sigma}^1_1$ equivalence relations as the path-component equivalence relation for some $K \in \mathcal{K}(\mathbb{R}^2)$, under any conceivable meaning of "code up". This still leaves open the question of whether $\approx_K$ can ever be "complicated" for $K \in \mathcal{K}(\mathbb{R}^2)$, e.g., can it be as complicated as Turing-equivalence? There are no known examples (from any axioms) of a $K \in \mathcal{K}(\mathbb{R}^2)$ such that $\approx_K$ is not *smooth*, i.e., such that $\approx_K$ is more complicated than the equality relation on $\mathcal{C}$ (see Becker-Kechris [4, §3.4] for definitions and details).

§5. Descriptive set theory and equivalence relations, II: Stern's Theorem

In this section, we consider Borel equivalence relations, which are much better behaved than arbitrary $\boldsymbol{\Pi}^1_1$ equivalence relations. At first glance, Silver's Theorem (3.1) would seem to say that nothing could be better behaved than $\boldsymbol{\Pi}^1_1$ equivalence relations. The problem is that the Silver dichotomy for $\boldsymbol{\Pi}^1_1$ equivalence relations applies only to the *entire Polish space* X. If E is a $\boldsymbol{\Pi}^1_1$ equivalence relation on X, there may well be a simply definable – in fact, $\boldsymbol{\Pi}^1_1$ – E-invariant set $Y \subset X$ such that $E|(Y \times Y)$ does not have either countably many or perfectly many equivalence classes. For example, let E^{***} be the following $\boldsymbol{\Pi}^1_1$ equivalence relation on $\mathcal{C}$:

$$xE^{***}y \iff [(x \in WO \text{ and } y \in WO \text{ and } |x| = |y|) \text{ or } x = y].$$

Clearly WO is $\boldsymbol{\Pi}^1_1$ and E^{***}-invariant, and $E^{***}|(WO \times WO)$ violates the dichotomy. For Borel equivalence relations, this situation does not occur.

Theorem 5.1 (Stern) Assume ϵ. Let X be a Polish space, let E be a Borel equivalence relation on X and let $Y \subset X$ be an E-invariant $\boldsymbol{\Sigma}^1_2$ set. One of the following two cases holds:

(i) Y has countably many E-equivalence classes;
(ii) Y has perfectly many E-equivalence classes.

Proof. See Stern [19].□

To put Stern's Theorem in its proper context, the following two remarks may be helpful. First, fix a Borel equivalence relation E on X with perfectly many equivalence classes. Assuming the full axiom of determinacy (which contradicts the axiom of choice), *every* E-invariant set $Y \subset X$ has either countably many or perfectly many E-equivalence classes. This follows from Stern's Theorem together with a result of Harrington-Sami [8, Theorem 2]. Obviously, using the axiom of choice, we can pick out a set of $\aleph_1$ E-equivalence classes; and, in fact, even if CH is true, using choice we can get an E-invariant set $Y \subset X$ with uncountably many but not perfectly many equivalence classes. But such a Y will not be definable. Thus E-invariant sets $Y \subset X$ which violate the dichotomy are like sets of real numbers which are not Lebesgue measurable: Such pathological sets do exist, but one cannot explicitly define an example. That's not provable in ZFC, but all right-thinking people know it is true. Regarding provability, the analogy between sets $Y \subset X$ violating the dichotomy and nonmeasurable sets of reals still holds: Stronger and stronger large cardinal axioms imply larger and larger classes of sets are nonpathological. Stern's Theorem is that the axiom ϵ is sufficient to prove that $\boldsymbol{\Sigma}^1_2$ sets Y are nonpathological.

Second, consider the case where X is the reals and E is equality. For this special case, the conclusion of Theorem 5.1 is that for any $\boldsymbol{\Sigma}^1_2$ set $Y \subset \mathbb{R}$, either Y is countable or Y has a perfect subset. That is, the conclusion of 5.1 is literally the axiom ϵ. So clearly this assumption is necessary. Stern's Theorem says that *if* equality has this property, then every Borel equivalence relation has this property. And as shown by the examples E^* and E^{***}, above, "Borel" is best possible.

§6. Theta-curves

Definition. A *theta-curve* (in $\mathbb{R}^2$) is a 5-tuple $(\mathbf{u}, \mathbf{v}, \gamma_1, \gamma_2, \gamma_3)$ such that $\mathbf{u}, \mathbf{v} \in \mathbb{R}^2$, each γ_i is an arc from $\mathbf{u}$ to $\mathbf{v}$ in $\mathbb{R}^2$, and if $i \neq j$ then $\gamma_i \cap \gamma_j = \{\mathbf{u}, \mathbf{v}\}$.

We sometimes abuse the language and refer to the pointset $Im(\gamma_1) \cup Im(\gamma_2) \cup Im(\gamma_3)$ in $\mathbb{R}^2$ as the "theta-curve". Figure 2 is a picture of a theta-curve in this latter sense.

We need a theorem about the topology of the plane – the theorem says that the picture in Figure 2 is correct. It is actually a very deep theorem, and to motivate it one should first consider the famous Jordan Curve Theorem. A *circle* always means a topological circle. The Jordan Curve Theorem states: If C is any circle embedded in $\mathbb{R}^2$, then $\mathbb{R}^2 \backslash C$ has exactly two components; and furthermore, the boundary of each of the two components is C. There is a similar theorem for theta-curves.

Theorem 6.1 Let $(\mathbf{u}, \mathbf{v}, \gamma_1, \gamma_2, \gamma_3)$ be a theta-curve, and let $\tilde{\gamma}_i = Im(\gamma_i)$. $\mathbb{R}^2 \backslash (\tilde{\gamma}_1 \cup \tilde{\gamma}_2 \cup \tilde{\gamma}_3)$ has exactly three components. The boundary of one component is $\tilde{\gamma}_1 \cup \tilde{\gamma}_2$. The boundary of another component is $\tilde{\gamma}_2 \cup \tilde{\gamma}_3$. And the boundary of the third component is $\tilde{\gamma}_3 \cup \tilde{\gamma}_1$.

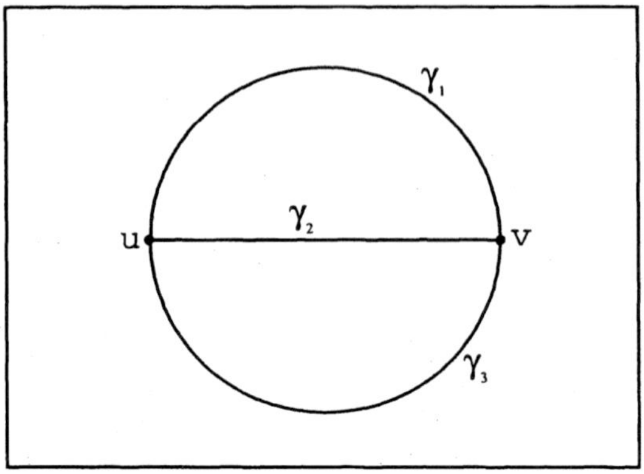

Figure 2

Proof. See Kuratowski [15, §61, II, Theorem 2]. □

Theorem 6.2 Let K be a compact subset of $\mathbb{R}^2$. If there is no theta-curve lying in K, then the equivalence relation $\approx_K$ is Borel.

Corollary 6.3 Assume ϵ. Let K be a compact subset of $\mathbb{R}^2$. If there is no theta-curve lying in K, then for any $\approx_K$-invariant $\boldsymbol{\Sigma}^1_2$ set $Y \subset K$, one of the following two cases holds:

(i) Y has countably many path-components;
(ii) Y has perfectly many path-components.

Proof. This follows from Theorems 5.1 and 6.2. Note that since Y is $\approx_K$-invariant, $\approx_Y$ is $\approx_K |(Y \times Y)$, i.e., every path-component of Y is also a path-component of K. □ In §7, we give a proof of Theorem 2.2 (from a stronger large cardinal axiom than ϵ). That proof uses both Theorem 6.1 and Corollary 6.3. We remark that one could also consider theta-curves in $\mathbb{R}^n$, for any n, and that both 6.2 and 6.3 would still be valid in the n-dimensional case. But the 3-dimensional analog of Theorem 6.1 is obviously false. Theorem 6.1 is the one and only place in the proof of Theorem 2.2 where the hypothesis that $K \subset \mathbb{R}^2$ is used.

The rest of §6 consists of a sketch of the proof of Theorem 6.2. This proof involves *effective* descriptive set theory, that is, recursion theoretic methods. Moschovakis [17] is the reference for this subject.

We work with recursively presented Polish spaces (as defined in Moschovakis [17, Page 128]). The Polish spaces $\mathbb{R}^2, \mathcal{K}(\mathbb{R}^2)$ and $(C[0,1])^2$ are all recursively presented, hence so are all finite products of these spaces. We regard compact subsets of $\mathbb{R}^2$ as points in the space $\mathcal{K}(\mathbb{R}^2)$, and we regard paths in $\mathbb{R}^2$ as points in the space $(C[0,1])^2$. For any recursively presented Polish spaces X and Y, and any points $x \in X$ and $y \in Y$, $x \leq_h y$ means that x is hyperarithmetic-in-y, or equivalently, that x is $\Delta^1_1(y)$. This is defined in Moschovakis [17, Pages 151 and 157].

The key step in the proof that SC_2 is $\boldsymbol{\Pi}^1_1$ is the following theorem, announced in Becker [2, Theorem 2.5]: If K is a compact simply connected subset of $\mathbb{R}^2$, then for any points $\mathbf{p}, \mathbf{q} \in \mathbf{K}$, there exists a path γ from $\mathbf{p}$ to $\mathbf{q}$, lying in K, such that $\gamma \leq_h (K, \mathbf{p}, \mathbf{q})$. A trivial special case of this theorem is the following fact: If K is a compact path-connected subset of $\mathbb{R}^2$, and there is no circle in K, then for any points $\mathbf{p}, \mathbf{q} \in \mathbf{K}$, there exists a path γ from $\mathbf{p}$ to $\mathbf{q}$, lying in K, such that $\gamma \leq_h (K, \mathbf{p}, \mathbf{q})$. We generalize this latter fact here (but generalize it in a different – and easier – way than the above theorem about simply connected sets).

Lemma 6.4 If K is a compact subset of $\mathbb{R}^2$, and there is no theta-curve in K, then for any points $\mathbf{p}, \mathbf{q} \in \mathbf{K}$ such that $\mathbf{p} \approx_{\mathbf{K}} \mathbf{q}$, there exists a path γ from $\mathbf{p}$ to $\mathbf{q}$, lying in K, such that $\gamma \leq_h (K, \mathbf{p}, \mathbf{q})$.

Sketch of proof As $\mathbf{p} \approx_{\mathbf{K}} \mathbf{q}$, by Theorem 1.1, there exists an *arc* from $\mathbf{p}$ to $\mathbf{q}$, lying in K. Throughout this proof, we consider only arcs, not arbitrary paths. Suppose Y is a recursively presented Polish space, $y \in Y$, $\mathbf{r}, \mathbf{s} \in \mathbb{R}^2$, and δ is any arc from $\mathbf{r}$ to $\mathbf{s}$ in $\mathbb{R}^2$; it can be shown that if the pointset $Im(\delta)$ is a $\Delta^1_1(y, \mathbf{r}, \mathbf{s})$ set in $\mathbb{R}^2$, then there is an arc $\delta' \in (C[0,1])^2$ such that $Im(\delta') = Im(\delta)$ and $\delta' \leq_h (y, \mathbf{r}, \mathbf{s})$. Therefore, to prove 6.4, it will suffice to show that there is an arc γ from $\mathbf{p}$ to $\mathbf{q}$, lying in K, such that $Im(\gamma)$ is a $\Delta^1_1(K, \mathbf{p}, \mathbf{q})$ subset of $\mathbb{R}^2$. The image of the arc from $\mathbf{p}$ to $\mathbf{q}$ lying in K is almost unique. It fails to be unique only because there is a countable set $\{C_i : i \in I\}$ of circles lying along the arc, and there are two distinct points $\mathbf{a_i}$, $\mathbf{b_i}$ on each circle C_i, such that $\mathbf{a_i}, \mathbf{b_i}$ lie on the arc (with $\mathbf{a_i}$ occurring before $\mathbf{b_i}$), and at each circle C_i one can go around the circle from $\mathbf{a_i}$ to $\mathbf{b_i}$ in either of two ways. That is, K must look pretty much like the pointset in Figure 3 (except that there may be a countably infinite set of circles lying along the arc, ordered in any countable order-type).

The above description is not very precise, and in this sketch we will not make it precise, let alone prove it. We merely point out that there are a number of different things that have to be proved, and for all of them the proof is straightforward and has the same structure: If K was in any way different from the above description, it would contain a theta-curve.

We now choose a canonical arc (more precisely, image of an arc) from $\mathbf{p}$ to $\mathbf{q}$ in K. This amounts to choosing which of the two ways to go around each of the circles C_i. Note that we cannot make an arbitrary choice. If we chose, say, to always go clockwise, we would not, in general, end up with an arc, since we could get a $sin(1/x)$-type situation (see Figure 1). We therefore choose to go around each circle the "short way", where "short" does not refer to arc length, but rather the *short* arc of C_i from $\mathbf{a_i}$ to $\mathbf{b_i}$ is the one which can be covered by a disc of smaller radius. Our hypothesis is that *there exists* an arc from $\mathbf{p}$ to $\mathbf{q}$ in K, and assuming this, it can be shown that if we always go around the circles the short way, that, too, will be an arc. We call that the *canonical arc* (actually the canonical image of an arc).

Let $D \subset \mathbb{R}^2$ be the canonical arc. All that remains to be proved is that D is $\Delta^1_1(K, \mathbf{p}, \mathbf{q})$.

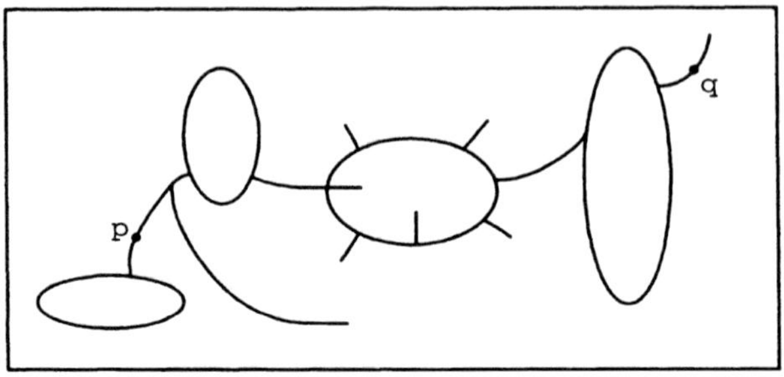

Figure 3

Let

$\mathcal{J} =$

$\{C \in \mathcal{K}(\mathbb{R}^2) : C$ is a circle and $C \subset K$ and $(\exists \gamma \in (C[0,1])^2)(\gamma$ is an arc from $\mathbf{p}$ to $\mathbf{q}$, lying in $\mathbf{K}$, and $\gamma \cap \mathbf{C}$ contains at least two points$)\}$.

Then $\mathcal{J}$ is a $\Sigma_1^1(K, \mathbf{p}, \mathbf{q})$ set (in the space $\mathcal{K}(\mathbb{R}^2)$). And $\mathcal{J}$ is countable. By the Effective Perfect Set Theorem, there is a $\Delta_1^1(K, \mathbf{p}, \mathbf{q})$ enumeration $\{C_j : j \in \omega\}$ of $\mathcal{J}$. Now for any point $\mathbf{r}$ in $\mathbb{R}^2$:

$$\begin{aligned} \mathbf{r} \in \mathbf{D} \iff & (\exists \gamma \in (C[0,1])^2)[\gamma \text{ is an arc from } \mathbf{p} \text{ to } \mathbf{q}, \text{ lying in } \mathbf{K}, \\ & \quad \text{and } (\forall j \in \omega)(\gamma \text{ goes around } C_j \text{ the short way}) \text{ and } \mathbf{r} \text{ lies on } \gamma] \\ \iff & (\forall \gamma \in (C[0,1])^2)[\text{if } (\gamma \text{ is an arc from } \mathbf{p} \text{ to } \mathbf{q}, \text{ lying in } \mathbf{K}, \\ & \quad \text{and } (\forall j \in \omega)(\gamma \text{ goes around } C_j \text{ the short way })) \text{ then } \mathbf{r} \text{ lies on } \gamma]. \end{aligned}$$

The first formula shows that D is $\Sigma_1^1(K, \mathbf{p}, \mathbf{q})$ and the second formula shows that D is $\Pi_1^1(K, \mathbf{p}, \mathbf{q})$. Therefore, D is $\Delta_1^1(K, \mathbf{p}, \mathbf{q})$. □

Remark. In Lemma 6.4, "hyperarithmetic" is best possible. For any countable ordinal α, there exists a $K, \mathbf{p}, \mathbf{q}$ satisfying the hypothesis of 6.4 such that no $\Delta_\alpha^0(K, \mathbf{p}, \mathbf{q})$ path γ satisfies the conclusion of 6.4.

Proof of 6.2 For any K (whether K contains a theta-curve or not), $\approx_K$ is $\boldsymbol{\Sigma}_1^1$; see 3.3. Since this K does not contain a theta-curve, Lemma 6.4 implies that for all $\mathbf{p}, \mathbf{q} \in \mathbf{K}$,

$$\mathbf{p} \approx_{\mathbf{K}} \mathbf{q} \iff (\exists \gamma \leq_{\mathbf{h}} (\mathbf{K}, \mathbf{p}, \mathbf{q}))\mathbf{F}(\mathbf{p}, \mathbf{q}, \gamma),$$

where F is as in 3.3. By a theorem of Kleene (see Moschovakis [17, Theorem 4D.3]), the pointclass $\boldsymbol{\Pi}_1^1$ is closed under quantification of the form "$\exists x \leq_h y$". Hence the above formula shows that $\approx_K$ is $\boldsymbol{\Pi}_1^1$. Since $\approx_K$ is both $\boldsymbol{\Sigma}_1^1$ and $\boldsymbol{\Pi}_1^1$, by Suslin's Theorem, it is Borel. □

§7. Path-components in compact subsets of $\mathbb{R}^2$

While Theorem 2.2 can be proved in $ZFC + \epsilon$, its proof is *much* easier if one assumes a stronger large cardinal axiom. We give this easier proof here in §7.

The axiom which will be used is denoted #, and is the following statement:

$$\text{For all } a \subset \omega,\ a^{\#} \text{ exists.}$$

The axiom # is equivalent to $\boldsymbol{\Pi}^1_1$-determinacy, and also equivalent to several other interesting propositions. It implies ϵ, but not conversely; in fact, # has greater consistency strength than ϵ. The existence of a measurable cardinal implies # (but is far stronger). We again refer the reader to Kanamori [11] for details.

For any topological space X, let $X/\approx_X$ denote the set of path-components of X. Let us again note that if $Y \subset X$ is $\approx_X$-invariant then $\approx_Y$ is $\approx_X |(Y \times Y)$.

Lemma 7.1 Assume #. Let K be a compact subset of $\mathbb{R}^n$ which has uncountably many but not perfectly many path-components. There exists a non-trivial countably additive two-valued measure, defined on a σ-algebra $\mathcal{S}$ of subsets of $K/\approx_K$, such that for any $\approx_K$-invariant $\boldsymbol{\Sigma}^1_2$ set $Y \subset K$, $Y/\approx_Y$ is in $\mathcal{S}$.

To understand 7.1, recall that the axiom of determinacy (AD) implies that $\aleph_1$ is a measurable cardinal, and that by Theorem 3.2, the set K of 7.1 has $\aleph_1$ path-components. So assuming AD, we could take $\mathcal{S}$ to be the full power set of $K/\approx_K$. But to prove Theorem 2.2 it is not necessary to measure *arbitrary* sets of path-components; it will suffice to measure $\boldsymbol{\Sigma}^1_2$ (in fact, $\boldsymbol{\Pi}^1_1$) sets of path-components. So we clearly do not need the full force of AD – some weak version of it will do. Although it is not obvious, $\boldsymbol{\Pi}^1_1$-determinacy is sufficient to measure the sets we need to measure; since $\boldsymbol{\Pi}^1_1$-determinacy is equivalent to #, that is the content of Lemma 7.1. I do not know whether or not 7.1 – or 7.1 with $\boldsymbol{\Sigma}^1_2$ replaced by $\boldsymbol{\Pi}^1_1$ – is provable in $ZFC + \epsilon$.

The proof of 7.1 breaks into two parts. First, any $\boldsymbol{\Sigma}^1_2$ set of countable ordinals either contains or is disjoint from a closed-unbounded set. This much is provable assuming only ϵ. But we do not want to measure sets of ordinals, we want to measure sets of path-components. So we also need a second fact: The path-components of K can be paired up with the countable ordinals in a $\boldsymbol{\Delta}^1_2$ way. This is a special case of a theorem of Burgess [6], and the proof seems to require #.

Proof of 2.2, assuming # First note that there does not exist an ω_1-sequence $\langle K_\alpha : \alpha < \omega_1 \rangle$ of compact subsets of $\mathbb{R}^2$ with the property that if $\alpha > \beta$ then K_α is a proper subset of K_β. We prove Theorem 2.2 by assuming that 2.2 is false and showing that such an ω_1-sequence does exist. So assume # and assume 2.2 is false. Fix a compact set $K \subset \mathbb{R}^2$ which has uncountably many but not perfectly many path-components. Fix a measure μ on $K/\approx_K$ satisfying Lemma 7.1. We construct, by induction, an ω_1-sequence $\langle (K_\alpha, A_\alpha) : \alpha < \omega_1 \rangle$ satisfying the following six properties.

(1) $K_0 = K$.

(2) K_α is a compact set.

(3) If $\alpha > \beta$ then K_α is a proper subset of K_β.
(4) For μ-a.e. path-component A of K, $A \subset K_\alpha$.
(5) A_α is a path-component of K.
(6) Let $\mathbf{p} \in \mathbf{K}_\alpha \backslash [\cup\{\mathbf{A}_\beta : \beta \leq \alpha\}]$. Let $\mathbf{q} \approx_{\mathbf{K}} \mathbf{p}$. Then $\mathbf{q} \in \mathbf{K}_\alpha$.

Property (6) requires some explanation. While K_α is a subset of K, it is *not* a $\approx_K$-invariant subset. Therefore it is possible that two points in K_α are in the same path-component of K but in different path-components of K_α. So a single path-component of K could break up into two path-components of K_α, or even into perfectly many path-components of K_α. The point of (6) is that while path-components of K can get broken apart, at any fixed stage α in the construction only countably many path-components of K get broken apart; we keep track of these countably many bad path-components; they are the A_β's. With these countably many exceptions, every path-component of K is either entirely in K_α or entirely out of K_α. (Of course, by (4), μ-a.e. path-component is entirely in.)

In the sequel we carefully distinguish $\approx_K$ from $\approx_{K_\alpha}$. Note that A_α is a path-component of K, not of K_α. Also note that the A_α's need not be distinct.

We now give the inductive construction of K_α and A_α, and thereby complete the proof.

$\alpha = 0$. Let $K_0 = K$ and let A_0 be any path-component of K.

α a successor ordinal. Let $\alpha = \alpha' + 1$, and suppose that the sequence has been constructed out to stage α' and that (1)-(6) hold.

Claim A. There is a theta-curve lying in $K_{\alpha'}$.

To prove Claim A, assume it is false. Let

$$Y = K_{\alpha'} \backslash [\cup\{A_\beta : \beta \leq \alpha'\}].$$

The A_β's are path-components of K, hence $\approx_{K_{\alpha'}}$-invariant. So clearly Y is a $\approx_{K_{\alpha'}}$-invariant subset of $K_{\alpha'}$. As path-components are $\boldsymbol{\Sigma}^1_1$ sets, Y is $\boldsymbol{\Pi}^1_1$ (hence $\boldsymbol{\Sigma}^1_2$). Applying Corollary 6.3, we see that one of the following two cases holds:

(i) Y contains only countably many path-components of $K_{\alpha'}$;
(ii) Y contains perfectly many path-components of $K_{\alpha'}$.

By (6), Y is $\approx_K$-invariant and $\approx_K |(Y \times Y)$ is the same as $\approx_{K_{\alpha'}} |(Y \times Y)$. So in case (ii), K has perfectly many path-components, contrary to assumption. Similarly, in case (i), the $\approx_K$-invariant set $Y \subset K$ contains only countably many path-components of K. Therefore, $K_{\alpha'}$ intersects only countably many path-components of K, namely those in Y, plus A_β for $\beta \leq \alpha'$. But μ is a non-trivial countably additive measure, hence (4) implies that $K_{\alpha'}$ contains uncountably many path-components of K. This proves Claim A.

Fix a theta-curve lying in $K_{\alpha'}$. Let I_1, I_2 and I_3 be the three components of the complement (in $\mathbb{R}^2$) of this theta-curve, as described in Theorem 6.1.

Consider the following four sets of path-components of K:

$$Q_0 = \{A \in K/\approx_K : A \text{ intersects the theta-curve}\};$$
$$Q_1 = \{A \in K/\approx_K : A \subset I_1\};$$
$$Q_2 = \{A \in K/\approx_K : A \subset I_2\};$$
$$Q_3 = \{A \in K/\approx_K : A \subset I_3\}.$$

Claim B. (a) Q_0 has only one element.

(b) Every path-component of K is in $Q_0 \cup Q_1 \cup Q_2 \cup Q_3$.

The proofs of both parts of Claim B are immediate. Since the theta-curve is a path-connected subset of K, (a) holds. To prove (b), note that if $A \in K/\approx_K$ does not intersect the theta-curve, then A must lie entirely inside one of the three components I_1, I_2, I_3; for if A intersected two of these components it would be disconnected.

Thus we have partitioned K into four $\approx_K$-invariant sets: $\cup Q_0, \cup Q_1, \cup Q_2, \cup Q_3$. The first of these is $\boldsymbol{\Sigma}^1_1$, the other three are $\boldsymbol{\Pi}^1_1$, hence all four Q_i's are μ-measurable. So one of the four must have measure 1. Let $j \in \{1,2,3\}$ be such that $\mu(Q_j) = 1$. Then let $K_\alpha = K_{\alpha'} \cap (closure(I_j))$ and let A_α be the unique member of Q_0.

With this choice of K_α and A_α, properties (2), (4) and (5) are obvious, and – toward proving (3) – it is also obvious that K_α is a subset of $K_{\alpha'}$. The reason that K_α must be a *proper* subset of $K_{\alpha'}$ is that one of the three arcs of the theta-curve is removed. This uses Theorem 6.1. Finally, to see that (6) holds, note that by Claim B, every path-component of K other than A_α must lie entirely inside I_j or entirely outside $closure(I_j)$. Therefore the only path-component of K to get broken apart in passing from $K_{\alpha'}$ to K_α is the path-component A_α, and so, by induction, (6) holds.

α a limit ordinal. Let $K_\alpha = \cap_{\beta<\alpha} K_\beta$ and let A_α be any path-component of K.

Since α is a countable ordinal and μ is countably additive, (4) holds. All the other properties are obvious. □

Remark. In the proof of 2.2, we emphasized the fact that $\approx_{K_\alpha}$ is not the same thing as $\approx_K |(K_\alpha \times K_\alpha)$. It is not hard to see that in the above proof, this phenomenon does not occur at the successor stages $\alpha = \alpha' + 1$; $\approx_{K_\alpha}$ is, indeed, the same as $\approx_{K_{\alpha'}} |(K_\alpha \times K_\alpha)$. That is, the procedure of intersecting the path-component of a theta-curve with the closure of one of the components of the complement of that theta-curve, does not destroy path-connectedness. The problem occurs at limit ordinals. If this procedure is done infinitely many times to the same path-component A (using different theta-curves), the part of A which is retained need not be path-connected.

References

[1] H. Becker, *Some theorems and questions about some natural examples in descriptive set theory* Xeroxed Notes, 1984

[2] H. Becker, *Descriptive set theoretic phenomena in analysis and topology* Set Theory of the Continuum (H.Judah, W. Just and H. Woodin, eds.) Springer-Verlag, 1992, pp. 1-25.

[3] H. Becker, *Path-connectedness, simple connectedness and descriptive set theory,* info in preparation.

[4] H. Becker and A.S. Kechris, *The Descriptive Set Theory of Polish Group Actions, London Mathematical Society Lecture Note Series* , Cambridge University Press, to appear

[5] J.P. Burgess, *Equivalences generated by families of Borel sets*, Proceedings of the American Mathematical Society **69** (1978), pp. 323-326.

[6] J.P. Burgess, *Effective enumeration of classes in a Σ^1_1 equivalence relation*, Indiana University Mathematics Journal **28** (1979), pp. 353-364.
[7] U.B. Darji, *Complexity of hereditarily decomposable continua*, preprint.
[8] L. Harrington and R. Sami, *Equivalence relations, projective and beyond*, Logic Colloquium 78 (M. Boffa, D. van Dalen and K. McAloon, eds.), North-Holland, 1979, pp. 247-264.
[9] L. Harrington and S. Shelah, *Counting equivalence classes for co-κ-Suslin equivalence relations*, Logic Colloquium '80 (D. van Dalen, D. Lascar and T.J. Smiley, eds.), North-Holland, 1982, pp. 147-152.
[10] W. Just, *Path-connectedness in $\mathbb{R}^2$ is Π^1_1-hard*, unpublished TeX file, 1993.
[11] A. Kanamori, *The Higher Infinite*, Springer-Verlag, 1994.
[12] A.S. Kechris, *The theory of countable analytical sets*, Transactions of the American Mathematical Society **202** (1975), pp. 259-297.
[13] A.S. Kechris, *Classical Descriptive Set Theory*, Springer-Verlag, 1995.
[14] K. Kunen and M. Starbird, *Arc components in metric continua*, Topology and its Applications **14** (1982), pp. 167-170.
[15] K. Kuratowski, *Topology, Volume II*, Academic Press, 1968.
[16] D.A. Martin and A.S. Kechris, *Infinite games and effective descriptive set theory*, Analytic Sets (C.A. Rogers et al., eds.), Academic Press, 1980, pp. 404-470.
[17] Y.N. Moschovakis, *Descriptive Set Theory*, North-Holland, 1980.
[18] J.H. Silver, *Counting the number of equivalence classes of Borel and co-analytic equivalence relations*, Annals of Mathematical Logic **18** (1980), pp. 1-28.
[19] J. Stern, *On Lusin's restricted continuum problem*, Annals of Mathematics **120** (1984), pp. 7-37.

Intervals Without Critical Triples

Peter Cholak

Department of Mathematics
University of Notre Dame
Notre Dame, IN 46556-5683, USA
cholak@turing.math.nd.edu

Rod Downey

Department of Mathematics
Victoria University of Wellington
Wellington, New Zealand
rod.downey@vuw.ac.nz

Richard Shore *

Department of Mathematics
White Hall, Cornell University
Ithaca, NY 14853, USA
shore@math.cornell.edu

Abstract. This paper is concerned with the construction of intervals of computably enumerable degrees in which the lattice M_5 (see Figure 1) cannot be embedded. Actually, we construct intervals $\mathcal{I}$ of computably enumerable degrees without any weak critical triples (this implies that M_5 cannot be embedded in $\mathcal{I}$, see Section 2) . Our strongest result is that there is a low$_2$ computably enumerable degree $\mathbf{e}$ such that there are no weak critical triples in either of the intervals $[\mathbf{0}, \mathbf{e}]$ or $[\mathbf{e}, \mathbf{0}']$.

1 Introduction

A set of natural numbers is computably (or recursively) enumerable if it is the range of a function computed by a Turing machine. We say one set of natural numbers, A, is Turing computable from another, B, if there is a Turing machine, which using an oracle for B, computes A. Equivalence classes under this

* Research partially supported by a Mathematical Sciences Postdoctoral Research Fellowship (Cholak); ARO through MSI, Cornell University, DAAL-03-C-0027 (Cholak, Downey, Shore); a U.S.-New Zealand Binational Cooperative Research grant NSF INT 90-20558 (Cholak, Downey and Shore); IGC of Victoria University, Wellington (Downey); New Zealand Marsden Fund for Basic Science under contract number VIC-509 (Downey) and NSF Grants DMS-9204308 and DMS-9503503 (Shore).

reduction are called Turing degrees or just degrees. In this paper we will restrict our attention to those degrees which contain a computably enumerable set; the computably enumerable degrees.

The computably enumerable degrees form an upper semilattice, $\mathcal{R}$. Despite the fact that $\mathcal{R}$ is not a lattice, there has been a long series of results each demonstrating larger and larger classes of lattices that could be embedded into $\mathcal{R}$. Examples include the results of [Lac72] and [ASL89]. On the other hand, in [LSo80] it is shown that not every lattice can be embedded into $\mathcal{R}$. Naturally, understanding precisely which lattices can be embedded into $\mathcal{R}$ is central to the question of the decidability of the $\exists\forall$ theory of $\mathcal{R}$. For instance, in the language of $\{\leq\}$, one needs to demonstrate an algorithm that will decide which lattices can be embedded into $\mathcal{R}$ (or the nonexistence of such an algorithm).

In the present paper, our concern is the lattice M_5 (see Figure 1), the modular 5 element lattice and its distribution in $\mathcal{R}$. In [Lac72] it is shown that M_5 is embeddable into $\mathcal{R}$. However, Lachlan's proof exhibited certain technical features such as "continuous tracing" which had not been necessary in previous lattice embedding theorems. Lachlan and Soare's proof that S_8 (see Figure 2) is not embeddable in some sense demonstrates that to embed M_5 requires such features and furthermore these features can be incompatible with other lattice properties such as simultaneously controlling the infima of a pair (or more) of degrees as with $\mathbf{a_1} \cap \mathbf{a_2} = \mathbf{a_3}$ in Figure 2).

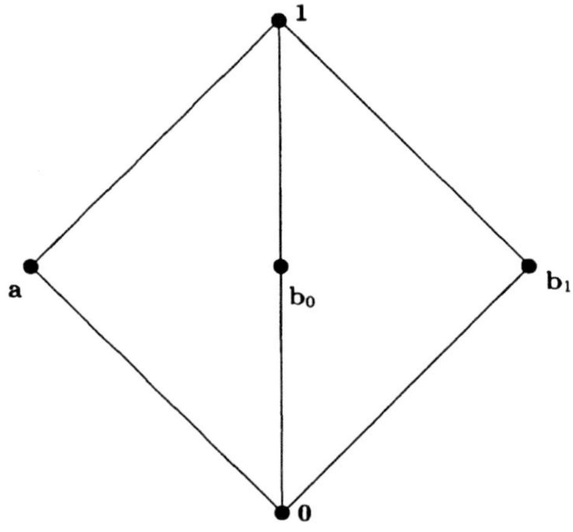

Fig. 1. The lattice M_5

These observations led Lerman to conjecture that essentially the "M_5 phenomenon" interacting with the infima was the only blockage to embeddability. The central role of the "M_5 phenomenon" was also demonstrated in [Dow90] and [Wei88] where it was proved that there are initial segments of $\mathcal{R}$ into which

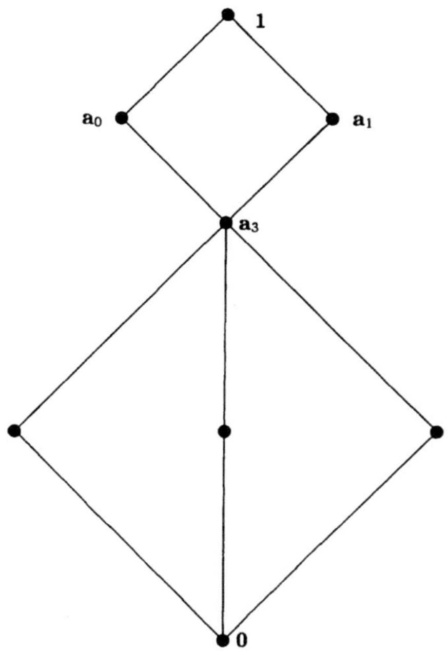

Fig. 2. The lattice S_8

M_5 cannot be embedded.

Actually, the Downey and Weinstein results were stated in terms of (weak) critical triples (see Section 2 for a definition). Critical triples is a lattice theoretic condition which reflects the need for "continuous tracing" in the construction of such a triple. These results as well as [CDo93] demonstrate that embedding of critical triples and the structure of $\mathcal{R}$ interact in very interesting ways.

Indeed, embeddings of nondistributive lattices is closely tied to the structure of $\mathcal{R}$. For instance, following the work in [DLe], it is shown in [ASF96] that the degrees which are tops of N_5 (see Figure 3) are precisely the "noncontiguous" degrees which are not "locally distributive" in $\mathcal{R}$.

We remark that recently in [LLe] it is shown that there are lattices without critical triples which are not embeddable into $\mathcal{R}$. But despite the fact that critical triples do not completely capture the nonembeddability picture, it is obvious that their definition plays a central role in our understanding of embeddings into $\mathcal{R}$.

Critical triples and nondistributive lattices also seem to have a connection with a natural operator on the Turing degrees called the jump operator. Informally the jump of a set of natural numbers X, X', is the set of numbers e such that the eth Turing machine (under some standard indexing of all Turing machines) with an oracle for X halts with input e. We use the jump operator to

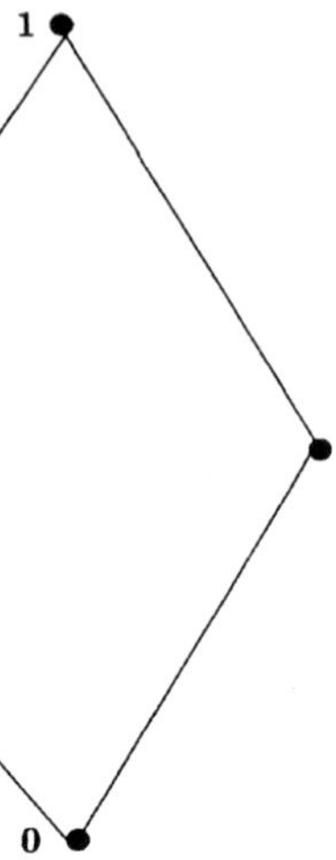

Fig. 3. The lattice N_5

define jump classes: sets of degrees which have the same nth jump for some n. For example, Low_2 is the collection of all computably enumerable Turing degrees $\mathbf{d}$ whose double jump is as low as possible, i.e. $\mathbf{d}'' \equiv \mathbf{0}''$.

Now in [DSh96], it is demonstrated that if $\mathbf{d}$ is nonlow_2 then one can embed M_5 below $\mathbf{d}$. The Ambos-Spies and Fejer result proves that if $\mathbf{d}$ is nonlow_2 then $\mathbf{d}$ is the top of an N_5. Because of these results one is led to believe that lattice embeddings and the jump operator are deeply related.

In the back of our minds when we began this project we hoped to show that the low_2 degrees are definable in the computably enumerable Turing degrees (in the language of partial orders) via embedding properties of M_5. There are techniques for working with low_2 computably enumerable degrees [SSl91, DSh95] and nonlow_2 computably enumerable degrees [DSh96]. Downey and Shore used these techniques to show that a computably enumerable tt degree $\mathbf{d}$ is low_2 iff $\mathbf{d}$ has a minimal cover in the computably enumerable tt degrees [DSh95] and hence the low_2 computably enumerable tt degrees are definable in the computably enumerable tt degrees. Our plan was to use these techniques to show $\mathbf{d}$ is low_2 iff there is a computably enumerable degree $\mathbf{e}$ such that the lattice M_5 cannot be embedded into either $[\mathbf{0}, \mathbf{e}]$ or $[\mathbf{e}, \mathbf{d}]$.

We had reason to hope that we might be successful: [DSh96] had recently shown that the lattice M_5 can be embedded below any nonlow_2 computably enumerable Turing degree and from [CDo93] we knew how to construct intervals of computably enumerable degrees where it is impossible to embed M_5. Hence, we needed to first extend the Downey and Shore result to the following: If $\mathbf{d}$ is nonlow_2 and $\mathbf{e}$ is a low_2 degree below $\mathbf{d}$ (and so $\mathbf{d}$ is nonlow_2 relative to $\mathbf{e}$), then the lattice M_5 can be embedded in the interval $[\mathbf{e}, \mathbf{d}]$. And second, improve the techniques of Cholak and Downey to show that for each low$_2$ degree $\mathbf{d}$ we could find an $\mathbf{e}$ such that the lattice M_5 cannot be embedded into the intervals $[\mathbf{0}, \mathbf{e}]$ and $[\mathbf{e}, \mathbf{d}]$. This would have given us a formula defining low$_2$.

However, one of our results implies that there is a computably enumerable degree $\mathbf{e}$ such that the lattice M_5 cannot be embedded into either $[\mathbf{0}, \mathbf{e}]$ or $[\mathbf{e}, \mathbf{0}']$. Thus the proposed formula does not define the low$_2$ computably enumerable Turing degrees.

Before we turn to our own work we should note that [NSS] have recently announced that the jump classes Low$_n$ and High$_m$ are definable, for $n \geq 2$ and $m \geq 1$. Their methods however, rely on coding models of arithmetic and analyzing the complexity of certain lattice like structures that can be coded below a given degree. We should also mention that there have been at least two other unsuccessful attempts to show that Low$_2$ is "naturally" definable: one by Leonhardi [Leoi 1994] and the other by Cooper and Yi [CYi]. For more of a discussion of definability and computably enumerable degrees, the reader is directed to Shore [Sho].

The rest of this paper is concerned with the construction of intervals of computably enumerable degrees in which M_5 cannot be embedded. Actually, we construct intervals $\mathcal{I}$ of computably enumerable degrees without any weak critical triples (this implies M_5 cannot be embedded in $\mathcal{I}$). The definitions of and the relationship between critical triples and weak critical triples (and the lattice M_5) are isolated in the next section.

The concept of a critical triple first arose implicitly in [ASL86] and [ASL89].In [ASL89] it is shown that a finite lattice $\mathcal{L}$ can be embedded into $\mathcal{R}$ if there is no (weak) critical triple $\mathbf{a}$, $\mathbf{b}_0$ and $\mathbf{b}_1$ and no pair $\mathbf{p}$ and $\mathbf{q}$ in $\mathcal{L}$ such that $\mathbf{b}_0 \leq \mathbf{p} \cap \mathbf{q} \leq \mathbf{b}_0 \cup \mathbf{a}$. The concept of a critical triple was first explicitly isolated in [Dow90]. In [Dow90], it is shown that there is a degree which does not bound a critical triple. In [CDo93], this work was extended to show that if $\mathbf{a} < \mathbf{b}$ are degrees then there is a degree $\mathbf{e}$ such that $\mathbf{a} < \mathbf{e} < \mathbf{b}$ and there is no critical triple in the interval $[\mathbf{a}, \mathbf{e}]$. The definition of a weak critical triple first appeared in [Wei88] under the name of "pre 1-3-1". He showed that there is a degree below which there is no weak critical triple. Our results further point out the importance of critical triples and weak critical triples in our quest to know what lattices can and cannot be embedded into intervals of computably enumerable degrees.

Our results fall into two groups. The first group of results concerns degrees which do not bound a weak critical triple. We show that every degree can be split into two degrees neither of which bounds a weak critical triple. Hence, the class of degrees which fail to bound a weak critical triple generates $\mathcal{R}$. We also show that there is a properly low$_2$ degree with no weak critical triple below it. Therefore, the result of Downey and Shore [DSh96] that the lattice M_5 can be embedded below any non-low$_2$ degree is the best possible in terms of jump classes. These results are presented in Section 4. The other group of results is presented in Section 6 and concerns degrees above which there is no weak critical triple. We show there is an incomplete degree above which there is no weak critical triple. In addition, we show there is a degree above and below which there is no weak critical triple. By the above result of Downey and Shore such a degree must be low_2.

There are two different types of requirements that reflect the grouping of results: There are the requirements $\mathcal{N}_\rceil$ which ensure that there is no weak critical triple in the desired lower cone and there are requirements $\mathcal{P}_\rangle$ which ensure that there is no weak critical triple in the desired upper cone. Each $\mathcal{N}_\rceil$ is a negative requirement in that it restrains elements from entering the constructed set. The requirements $\mathcal{N}_\rceil$ are presented in Section 3. Each $\mathcal{P}_\rangle$ is an infinite positive requirement; it may add infinitely many elements into the constructed set. The requirements $\mathcal{P}_\rangle$ are presented in Section 5. The two types of requirements and the strategies used to meet them are in a sense duals of each other.

Although demonstrating that there are no weak critical triples in some interval is stronger than the corresponding result for critical triples, it is actually easier to construct intervals of computably enumerable degrees without weak critical triples than it is to construct intervals of computably enumerable degrees without critical triples (at least, as this was done in [Dow90] and [CDo93]). Fix a triple of computably enumerable degrees, $\mathbf{a}$, $\mathbf{b}_0$ and $\mathbf{b}_1$ such that $\mathbf{a} \cup \mathbf{b}_0 = \mathbf{a} \cup \mathbf{b}_1$. To show, as in [Dow90] and [CDo93], that $\mathbf{a}$, $\mathbf{b}_0$ and $\mathbf{b}_1$ is not a critical triple a computably enumerable degree $\mathbf{d}$ is built such that $\mathbf{d} \leq \mathbf{b}_0, \mathbf{b}_1$ and if $\mathbf{a} \not\leq \mathbf{b}_0$ then $\mathbf{d} \not\leq \mathbf{a}$. Determining whether $\mathbf{a}$ is Turing reducible to $\mathbf{b}_0$ is a Σ_3 question and therefore it is not surprising that these arguments turn out to be $\mathbf{0}'''$ arguments. To show that $\mathbf{a}$, $\mathbf{b}_0$ and $\mathbf{b}_1$ is not a weak critical triple, we construct a computably enumerable degree $\mathbf{d}$ such that $\mathbf{d} \leq \mathbf{b}_0, \mathbf{b}_1$ and $\mathbf{b}_0 \leq \mathbf{d} \cup \mathbf{a}$. As it turns out, this can be done by a $\mathbf{0}''$ argument. In terms of lattice embeddings, however, the results are equivalent for, as we will see below, a lattice contains a critical triple if and only if it contains a weak critical triple.

A question we tried to answer but could not is whether the lattice M_5 can be embedded above every low degree. For more on this issue, the reader is directed to Section 6.3. It is also open whether one can extend our result that there is an incomplete degree which does not bound a weak critical triple to show such a degree must exist above every nonlow_2 degree.

Remark Notation. Our notation is standard and generally follows [Soa87] with the following important exceptions: The use of a computation $\Phi(X_s; x)$ is denoted by $\varphi_s(x)$ and similarly for other Greek letters. We assume the uses of all functionals not constructed by us to be nondecreasing in the stage, s, and increasing in the argument, x, for each stage. Furthermore, if the underlying set involved in a computation changes below the use of the computation at some stage s, we will assume that computation diverges at stage s. For example, if we are given Ψ and for some x and s, $\Psi_s(X_s; x)\downarrow$ and $X_s \upharpoonright (\psi_s(x)+1) \neq X_{s+1} \upharpoonright (\psi_s(x)+1)$ then $\Psi_{s+1}(X_{s+1}; x)\uparrow$. When the oracle of a functional is given as the join of sets we assume the use to be computed separately on each set. To make life easier, we will assume that $X \oplus Y \oplus Z$ is defined as

$$\{3x : x \in X\} \cup \{3y+1 : y \in Y\} \cup \{3z+2 : z \in Z\}.$$

All other joins are defined normally. When we choose a *large* number we mean a number larger than any other number mentioned or used so far. For any of the functionals (or parameters) which we are building, we will assume that if the underlying set changes at some stage on the use (for this stage) then this functional (or parameter) is undefined at this stage unless otherwise explicitly defined. For example, if we are building Γ and for some x and s, $\Gamma_s(X_s;x)\downarrow$ and $X_s \upharpoonright (\gamma_s(x)+1) \neq X_{s+1} \upharpoonright (\gamma_s(x)+1)$ then unless we otherwise explicitly define $\Gamma_{s+1}(X_{s+1};x)$, $\Gamma_{s+1}(X_{s+1};x)\uparrow$. Otherwise, all parameters remain the same from stage to stage unless explicitly redefined. We assume the reader is familiar with $\mathbf{0}''$ arguments as in [Soa87].

2 Definitions and Examples

Definition 1. Let $\mathbf{a}$, $\mathbf{b}_0$ and $\mathbf{b}_1$ be elements in any upper semilattice $\mathcal{L}$ (such as the Turing degrees or the computably enumerable Turing degrees). We say that $\mathbf{a}$, $\mathbf{b}_0$ and $\mathbf{b}_1$ form a *critical triple* if $\mathbf{a} \cup \mathbf{b}_0 = \mathbf{a} \cup \mathbf{b}_1$, $\mathbf{b}_0 \not\leq \mathbf{a}$ and for $\mathbf{d} \in \mathcal{L}$, if $\mathbf{d} \leq \mathbf{b}_0, \mathbf{b}_1$ then $\mathbf{d} \leq \mathbf{a}$.

Definition 2. Let $\mathbf{a}$, $\mathbf{b}_0$ and $\mathbf{b}_1$ be elements in any upper semilattice $\mathcal{L}$. We say that $\mathbf{a}$, $\mathbf{b}_0$ and $\mathbf{b}_1$ form a *weak critical triple* if $\mathbf{a} \cup \mathbf{b}_0 = \mathbf{a} \cup \mathbf{b}_1$, $\mathbf{b}_0 \not\leq \mathbf{a}$ and for $\mathbf{d} \in \mathcal{L}$, if $\mathbf{d} \leq \mathbf{b}_0, \mathbf{b}_1$ then $\mathbf{b}_0 \not\leq \mathbf{d} \cup \mathbf{a}$.

Any upper semilattice in which the lattice M_5 (for a diagram of the lattice M_5 see Figure 1) can be embedded (as a lattice) contains a critical triple. If $\mathbf{a}$, $\mathbf{b}_0$ and $\mathbf{b}_1$ form a critical triple in some upper semilattice $\mathcal{L}$ then $\mathbf{a}$, $\mathbf{b}_0$ and $\mathbf{b}_1$ form a weak critical triple in $\mathcal{L}$. In Figure 4, a weak critical triple $\mathbf{a}$, $\mathbf{b}_0$ and $\mathbf{b}_1$ is identified within a lattice. Within this lattice, $\mathbf{a}$, $\mathbf{b}_0$ and $\mathbf{b}_1$ do not form a critical triple but the elements $\mathbf{a} \cup (\mathbf{b}_0 \cap \mathbf{b}_1)$, $\mathbf{b}_0$ and $\mathbf{b}_1$ do. In fact, if $\mathbf{a}$, $\mathbf{b}_0$ and $\mathbf{b}_1$ form a weak critical in an upper semilattice $\mathcal{L}$ and $\mathbf{b}_0 \cap \mathbf{b}_1$ exists (for example, this must occur if $\mathcal{L}$ is a lattice), then $\mathbf{a} \cup (\mathbf{b}_0 \cap \mathbf{b}_1)$, $\mathbf{b}_0$ and $\mathbf{b}_1$ form a critical triple in $\mathcal{L}$. On the other hand, it is possible to construct an infinite upper semilattice which contains a weak critical triple but no critical triples.

As for $\mathcal{R}$, it is unknown whether there exists an interval $\mathcal{I}$ such that there is a weak critical triple in $\mathcal{I}$ but $\mathcal{I}$ does not contain a critical triple or whether there exits an interval $\mathcal{I}$ such that there is a critical triple in $\mathcal{I}$ but M_5 cannot be embedded into $\mathcal{I}$.

3 The requirements $\mathcal{N}_\urcorner$

Our goal is to build a computably enumerable set E such that the following requirements are met:

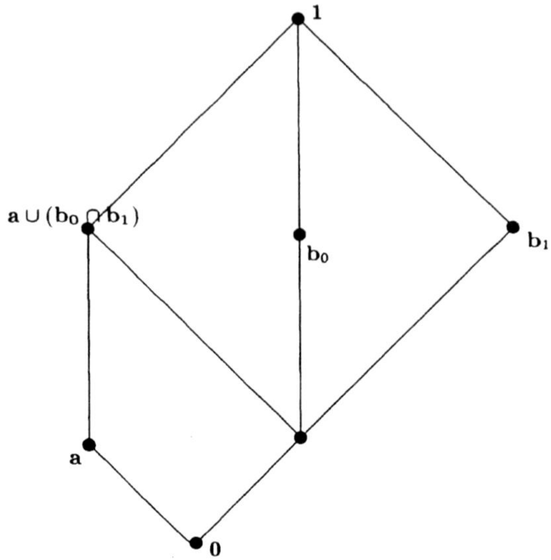

Fig. 4. A weak critical triple which is not a critical triple

$$\begin{array}{c}\text{If } \Lambda(E) = A \oplus B_0 \oplus B_1 \text{ and } \Psi_i(A \oplus B_{\bar{i}}) = B_i \\ \text{then there exists a computably enumerable set } D \\ \text{and functionals } \Delta_i \text{ and } \Gamma \text{ such that} \\ \Delta_i(B_i) = D \text{ and } \Gamma(A \oplus D) = B_0\end{array} \qquad \mathcal{N}$$

where Λ and Ψ_i ($i = 0, 1$) are functionals and A, B_0 and B_1 are computably enumerable sets. These six items, the requirement, the set D and the three functionals Δ_i and Γ will later be indexed in some fashion by e; but for now, we will drop the e. If we meet $\mathcal{N}$ then either E does not bound the computably enumerable sets A, B_0 and B_1 or the degrees represented by these sets do not form a weak critical triple. (If $B_0 \leq_T A$ then the requirement can be easily met.)

We will need some auxiliary functions (at first it may seem that we are generating more notation than needed but we will use all these functions later in the construction):

$$\begin{aligned} L(s) = \max\{x :& (\forall y < x)(\forall z \in \{3y, 3y+1, 3y+2\}) \\ & [\Lambda_s(E_s; z) = A_s \oplus B_{0,s} \oplus B_{1,s}(z)]\} \end{aligned} \tag{3.1}$$

$L(s)$ is the length of agreement function between $\Lambda_s(E_s)$ and $A_s \oplus B_{0,s} \oplus B_{1,s}$. We use the convention that if $\Lambda_s(E_s; y) = A_s \oplus B_{0,s} \oplus B_{1,s}(y)$ and $E_s \restriction \lambda_s(y)+1$

does not change then no new numbers can enter $A_s \oplus B_{0,s} \oplus B_{1,s} \upharpoonright y+1$. Let $\{i, \bar{i}\} = \{0, 1\}$.

$$l^{\Psi_i}(s) = \max\{x : (\forall y < x)[\Psi_{i,s}(A_s \oplus B_{\bar{i},s}; y) = B_{i,s}(y) \text{ and } \psi_{i,s}(y) < L(s)]\} \tag{3.2}$$

$l^{\Psi_i}(s)$ is the E_s-correct length of agreement function between $\Psi_{i,s}(A_s \oplus B_{\bar{i},s})$ and $B_{i,s}$. We will use these length of agreement functions to define, by induction, the following length of agreement and use functions:

$$l^0(s) = l^{\Psi_0}(s) \tag{3.3}$$

$$\text{If } x < l^0(s) \text{ let } \rho^0(A, x, s) = \psi_{0,s}(x), \quad \rho^0(B_1, x, s) = \psi_{0,s}(x) \text{ and } \rho^0(B_0, x, s) = x. \tag{3.4}$$

$$l^{2i+1}(s) = \max\{x : (\forall y < x)[y < l^{2i}(s) \text{ and } \rho^{2i}(B_1, y, s) < l^{\Psi_1}(s)]\} \tag{3.5}$$

$$\text{If } x < l^{2i+1}(s) \text{ let } \rho^{2i+1}(A, x, s) = \psi_{1,s}(\rho^{2i}(B_1, x, s)), \quad \rho^{2i+1}(B_0, x, s) = \psi_{1,s}(\rho^{2i}(B_1, x, s)) \text{ and } \rho^{2i+1}(B_1, x, s) = \rho^{2i}(B_1, x, s). \tag{3.6}$$

$$l^{2i+2}(s) = \max\{x : (\forall y < x)[y < l^{2i+1}(s) \text{ and } \rho^{2i+1}(B_0, y, s) < l^{\Psi_0}(s)]\} \tag{3.7}$$

$$\text{If } x < l^{2i+2}(s) \text{ let } \rho^{2i+2}(A, x, s) = \psi_{0,s}(\rho^{2i+1}(B_0, x, s)), \quad \rho^{2i+2}(B_1, x, s) = \psi_{0,s}(\rho^{2i+1}(B_0, x, s)) \text{ and } \rho^{2i+2}(B_0, x, s) = \rho^{2i+1}(B_0, x, s). \tag{3.8}$$

$$\text{If } x < l^i(s), \text{ let } \rho^i(E, x, s) = \max\{\lambda_s(3\rho^i(A, x, s)), \lambda_s(3\rho^i(B_0, x, s) + 1), \lambda_s(3\rho^i(B_1, x, s) + 2)\}. \tag{3.9}$$

$$\text{If } x > l^i(s) \text{ then for all } X \in \{A, B_0, B_1, E\}, \rho^i(X, x, s)\uparrow. \tag{3.10}$$

$$\text{If for almost all stages } s, \rho^i(X, x, s)\downarrow \text{ then let } \rho^i(X, s) = \lim \rho^i(X, x, s), \text{ for } X \in \{A, B_0, B_1, E\}. \tag{3.11}$$

If $x < l^i(s)$ then we have an E_s-correct computation involving $i+1$ layers, that is $\Psi_{0,s}(A_s \oplus B_{1,s}) \upharpoonright x+1 = B_{0,s} \upharpoonright x+1, \Psi_{1,s}(A_s \oplus B_{0,s}) \upharpoonright \rho^0(B_1, x, s)+1 = B_{1,s} \upharpoonright \rho^0(B_1, x, s)+1$, $\Psi_{0,s}(A_s \oplus B_{1,s}) \upharpoonright \rho^1(B_0, x, s)+1 = B_{1,s} \upharpoonright \rho^1(B_0, x, s)+1, \ldots$ and $\rho^i(X, x, s)$ is the initial segment of the set X used in this computation, for all $X \in \{A, B_0, B_1, E\}$. Diagram 5 may be helpful at this point. If we preserve E_s below $\rho^i(E, x, s)+1$ then the computations used in defining $l^i(s) > x$ will not change and we will preserve X_s below $\rho^i(X, x, s)+1$, for all $X \in \{A, B_0, B_1\}$.

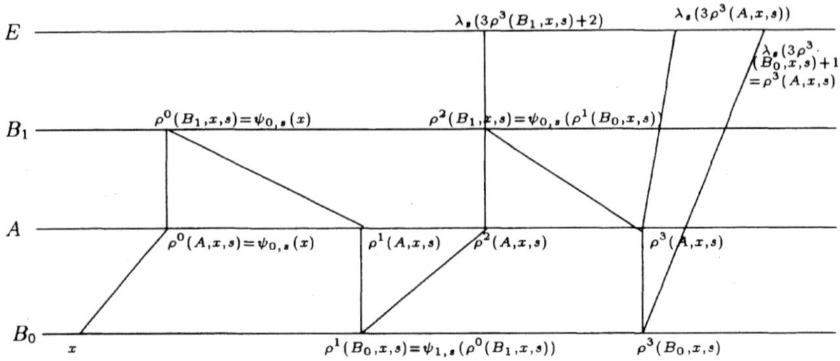

Fig. 5. 4 layers

Associated with each number x will be another number $i(x)$. We should view $i(x)$ as the number of injuries $\Gamma_s(A_s \oplus D_s; x) = B_{0,s}(x)$ can sustain and still ensure that $\Gamma(A \oplus D; x) = B_0(x)$ (more on this later). The value of $i(x)$ will be determined *before* the construction. In fact, determining the value of $i(x)$ will play a large role in all of our proofs. We will assume that $i(x)$ is a non-decreasing function. We will define the length of agreement $l(s)$ as

$$l(s) = \max\{x : (\forall y < x)[y < l^{i(y)}(s)]\}. \tag{3.12}$$

If $x < l(s)$ then let $\rho(X, x, s) = \rho^{i(x)}(X, x, s)$ and $\rho(X, x) = \rho^{i(x)}(X, x)$. Think of $\rho(X, x, s)$ as the complete use function for x on X. We say a stage s is expansionary iff $s = 0$ or $t < s$ is the last expansionary stage and $l(s) > l(t)$.

If $\Lambda(E) = A \oplus B_0 \oplus B_1$ and $\Psi_i(A \oplus B_{\bar{i}}) = B_i$ then there will be infinitely many expansionary stages. Hence, to meet $\mathcal{N}$ it is enough to build a computably enumerable set D and the functionals Δ_i and Γ such that if there are infinitely many expansionary stages then $\Delta_i(B_i) = D$ and $\Gamma(A \oplus D) = B_1$. We will do this by splitting $\mathcal{N}$ into infinitely many subrequirements, $\mathcal{N}_\S$:

> If there are infinitely many stages s such that $l(s) > x$ then $\Gamma(A \oplus D; x) = B_0(x)$ and $\Delta_i(B_i) \restriction (\gamma(x) + 1) = D \restriction (\gamma(x) + 1)$. $\qquad \mathcal{N}_\S$

Informally, the strategy to meet $\mathcal{N}_\S$ is first to redefine $\Gamma_s(A_s \oplus D_s; x)$ to reflect any change in $B_{0,s}(x)$, if we can. If $\Gamma_s(A_s \oplus D_s; x)\downarrow \neq B_{0,s}(x)$ and for all $i \in \{0, 1\}$, $\Delta_{i,s}(B_{i,s}; \gamma_s(x))\uparrow$ then add $\gamma_s(x)$ to D_{s+1} which will allow us to later redefine $\Gamma(A \oplus D; x)$. Our goal is to prove that this works.

We will now present the formal details. We will have two additional parameters $e^x(s)$ and $p^x(s)$; they will be used to keep track of which layer has been "peeled away" with $e^x(s) \in \{0, 1\}$ indicating which side has changed and $p^x(s)$ at which layer.

Action for $\mathcal{N}_\S$ at stage $s+1$. Assume $l(s+1) > x$ (otherwise do nothing). Let $t < s+1$ be the last stage such that $\Gamma_t(A_t \oplus D_t; x)\downarrow$ and $l(t) > x$, if such a stage exists. Do the first of the following cases which applies:

Case 1 t does not exist.
Action Let d be large, define $\Gamma_{s+1}(A_{s+1} \oplus D_{s+1}; x) = B_{0,s+1}(x)$ with use $\gamma_{s+1}(x) = d > \rho(A, x, s)$, and define $\Delta_{i,s+1}(B_{i,s+1}; d) = 0$ with use $\delta_{i,s}(d) = \rho(B_i, x, s)$. For all $\gamma_{s+1}(x-1) < y < d$, if $\Delta_{i,s}(B_{i,s}; y)\uparrow$ then define $\Delta_{i,s+1}(B_{i,s+1}; y) = D_{s+1}(y)$ with use 0. If $i(x)$ is even let $e^x(s+1) = 1$, otherwise let $e^x(s+1) = 0$. Let $p^x(s+1) = i(x) + 1$.
Case 2 There is a stage t' such that $t < t' < s+1$ and either Case 3, 4 or 5 applies at stage t' for some $\mathcal{N}_\dagger$, where $y \leq x$.
Action Same as in Case 1.
Case 3 $A_{s+1} \restriction \rho(A, x, t) \neq A_t \restriction \rho(A, x, t)$.
Action Do nothing.
Case 4 For $i \in \{0, 1\}$, and $l = p^x(t) - 1$, $B_{i,s+1} \restriction \rho^l(B_i, x, t) \neq B_{i,t} \restriction \rho^l(B_i, x, t)$ (hence $\Delta_{i,s+1}(B_{i,s+1}; \gamma_t(x))$ will diverge).
Action Add $\gamma_t(x)$ to D_{s+1} (by our convention this will cause $\Gamma_{s+1}(A_{s+1} \oplus D_{s+1}; x)\uparrow$ and will cause Case 2 to apply at the next possible stage).
Case 5 Currently never applies. Will be used in Section 5.
Case 6 For $i = e^x(t)$ and $l = p^x(t)$, $B_{i,s+1} \restriction \rho^l(B_i, x, t) \neq B_{i,t} \restriction \rho^l(B_i, x, t)$ (hence, by Equations (3.6) and (3.8), $B_{i,s+1} \restriction \rho^{l-1}(B_i, x, t) \neq B_{i,t} \restriction \rho^{l-1}$ (B_i, x, t) and therefore $\Delta_{i,s+1}(B_{i,s+1}; \gamma_t(x))$ would diverge except that we now redefine it).
Action Let $e^x(s+1) = \bar{i}$ and $p^x(s+1) = p^x(s) - 1$. If $\Gamma_s(A_s \oplus D_s; x)\uparrow$ then define $\Gamma_{s+1}(A_{s+1} \oplus D_{s+1}; x) = B_{0,s+1}(x)$ with use $\gamma_t(x)$. Define $\Delta_{i,s+1}(B_{i,s+1}; \gamma_t(x)) = 0$ with large use. For $\gamma_t(x-1) < y < \gamma_t(x)$ and $j \in \{0, 1\}$, let $\Delta_{j,s+1}(B_{j,s+1}; y) = \Delta_{j,t}(B_{j,t}; y)$ with use $\delta_{j,t}(y)$. For $j = \bar{i}$, let $\Delta_{j,s+1}(B_{j,s+1}; \gamma_t(x)) = \Delta_{j,t}(B_{j,t}; \gamma_t(x))$ with use $\delta_{j,t}(\gamma_t(x))$.
Case 7 Otherwise.
Action If $\Gamma_s(A_s \oplus D_s; x)\uparrow$ then define $\Gamma_{s+1}(A_{s+1} \oplus D_{s+1}; x) = B_{0,s+1}(x)$ with use $\gamma_t(x)$. For $\gamma_t(x-1) < y \leq \gamma_t(x)$ and $j \in \{0, 1\}$, let $\Delta_{j,s+1}(B_{j,s+1}; y) = \Delta_{j,t}(B_{j,t}; y)$ with use $\delta_{j,t}(y)$.

Remark Coordinating the action for different $\mathcal{N}_\S$. We will assume that at any stage s we take the action needed for $\mathcal{N}_\S$, for $x < s$, in increasing order. Hence since $i(x)$ is a nondecreasing function, $\gamma_s(x)$ is nondecreasing as a function of s and increasing as a function of x. Similarly for the use functions $\gamma_{i,s}$.

Definition 3. If Cases 1 or 2 apply or $s+1 = 0$ then we call $s+1$ a *free-clear* stage. If Cases 3, 4 or 5 apply for some $\mathcal{N}_\dagger$, where $y \leq x$, we call $s+1$ an *almost free-clear* stage (assuming that there are infinitely many stages where $l(s) > x$, the next such stage will be a free-clear stage).

Definition 4. Let t be a free-clear stage. Let t' be the next almost free-clear stage or ∞ if there is none. We let $s \in I_x^t$ if either $s = t$ or $t < s < t'$ and $p^x(s) < p^x(s-1)$ (i.e. Case 6 holds).

Lemma 5. *Let t be a free-clear stage. Suppose for all $s \in I_x^t$, $p^x(s) \geq 0$. Then (i) For all $l < p^x(s)$ and for all $X \in \{A, B_0, B_1, E\}$, $X_s \upharpoonright (\rho^l(X,x,t)+1) = X_t \upharpoonright (\rho^l(X,x,t)+1)$ (by induction on s, this implies that $\rho^l(X,x,s) = \rho^l(X,x,t)$). (ii) For $l = p^x(s)$ and $i = e^x(s)$, $B_{i,s} \upharpoonright (\rho^l(B_i,x,t)+1) = B_{i,t} \upharpoonright (\rho^l(B_i,x,t)+1)$ (again, by induction, this implies that $\rho^l(B_i,x,s) = \rho^l(B_i,x,t)$).*

Proof. By induction. Clearly the lemma holds for $s = t$. Let $j = e^x(s-1) = \bar{i}$. Since s is in I_x^t and $\rho^l(B_j,x,t) = \rho^{l+1}(B_j,x,t)$ (see Equations (3.6) and (3.8)), for some $k \leq l$, $B_{j,s-1} \upharpoonright (\rho^k(B_j,x,t)+1) \neq B_{j,t} \upharpoonright (\rho^k(B_j,x,t)+1)$. Let k be as small as possible. Suppose $k < l$. Then, by the layering and the induction hypothesis, for some $X \in \{A, B_i\}$, $X_s \upharpoonright (\rho^k(X,x,t)+1) \neq X_t \upharpoonright (\rho^k(X,x,t)+1)$. So either Case 3 or 4 applies at stage s and hence $s \notin I_x^t$. Hence $k = l$ and (i) holds. $B_{i,s} \upharpoonright (\rho^l(B_i,x,t)+1) = B_{i,t} \upharpoonright (\rho^l(B_i,x,t)+1)$, otherwise s is an almost free-clear stage.

We have just shown that, between (almost) free-clear stages, we only peel back the layers one at a time with changes alternating between B_0 and B_1. The idea of peeling back the layers is called the "top-down approach" in [Wei88]. We say that it is possible to *peel back* (or away) the $p^x(s)^{\text{th}}$ layer of x at stage s.

It may be helpful to refer to Diagram 6 this point. This Diagram is a subdiagram of Diagram 5. The numbered areas show where the changes in these sets must occur to peel back all 4 layers. The numbers refer to the sequence of events; i.e. there must be a change in first region followed by a change in the second, etc.

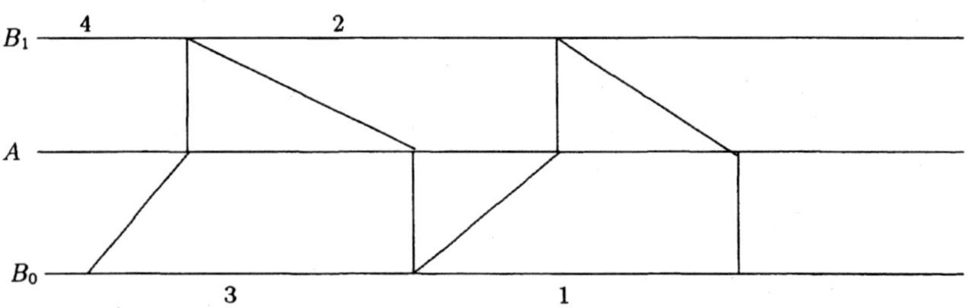

Fig. 6. Where the changes must occur, in the required order, to peel back all 4 layers

Lemma 6. *Either $\Gamma(A \oplus D; x)\downarrow$ and for all $y \leq \gamma(x)$, $\Delta_i(B_i; y)\downarrow$ (i.e. the functionals are well-defined) or it is not the case that $\Lambda(E) = A \oplus B_0 \oplus B_1$ and $\Psi_i(A \oplus B_{\bar{i}}) = B_i$.*

Proof. By induction on x. We can assume that there are infinitely many stages s where $l(s) > x$ (otherwise we are done). Case 1 can only apply once. If Case 2 applies infinitely often then it is not the case that $\Lambda(E) = A \oplus B_0 \oplus B_1$ and

$\Psi_i(A \oplus B_{\bar{i}}) = B_i$. Case 6 can apply at most $i(x) - 2$ times between stages where Case 1 or 2 apply. Hence we can assume Case 7 applies almost always. Therefore $\lim_s \gamma_s(x)$ exists and $\Gamma(A \oplus D; x)\downarrow$. Furthermore, for all $\gamma(x-1) < y \leq \gamma(x)$, $\delta_i(y) = \lim_s \delta_{i,s}(y)$ exists and $\Delta_i(B_i; y)\downarrow$.

Lemma 7. *Suppose $\Lambda(E) = A \oplus B_0 \oplus B_1$ and $\Psi_i(A \oplus B_{\bar{i}}) = B_i$, there is a* last *free-clear stage t and $|I_x^t| \leq i(x) + 1$. Then $\mathcal{N}_\S$ is met.*

Proof. By induction on x. By the above lemma, $\Gamma(A \oplus D; x)\downarrow$ and for all $y \leq \gamma(x)$, $\Delta_i(B_i; y)\downarrow$. Since t is the last free-clear stage, $\Gamma(A \oplus D; x) = \Gamma_t(A_t \oplus D_t; x)$ and for all $y \leq \gamma(x)$, $\Delta_i(B_i; y) = \Delta_{i,t}(B_{i,t}; y)$.

Let s be the greatest stage in I_x^t. By Lemma 5 and the above hypothesis, $B_{0,s} \restriction (\rho^0(B_0, x, t) + 1) = B_{0,t} \restriction (\rho^0(B_0, x, t) + 1)$. $\rho^0(B_0, x, s) = \rho^0(B_0, x, t) = x$. Therefore $B_{0,s}(x) = B_{0,t}(x) = B_0(x)$ (otherwise s is not the greatest stage in I_x^t). Hence $\Gamma(A \oplus D; x) = \Gamma_t(A_t \oplus D_t; x) = B_{0,t}(x) = B_0(x)$.

Let $\gamma(x-1) < y \leq \gamma(x)$ (by Remark 3, we know $\gamma(x)$ is an increasing function). By the induction hypothesis, y can only enter D at some stage s iff $y = \gamma_{t'}(x')$, $l(s) > x$, and for both i, $\Delta_{i,s+1}(B_{i,s+1}; y)\uparrow$, for some x' and t', where $t' \leq t$ (see Case 4 or 5). By stage t we will have to redefine Δ_i to reflect this change in D with use 0. Hence $\Delta_i(B_i) \restriction (\gamma(x) + 1) = D \restriction (\gamma(x) + 1)$.

Corollary 8. *Suppose for every free-clear stage t, $|I_x^t| \leq i(x) + 1$. Then $\mathcal{N}_\S$ is met.*

Definition 9. Let s be a stage such that we act for $\mathcal{N}_\S$ as above at stage s and let t be the greatest stage less than s such that we act for $\mathcal{N}_\S$ as above at stage t (see Remark 3). Then $s \in I_x$ iff $E_t \restriction (\rho(E, x, t)\downarrow + 1) \neq E_s \restriction (\rho(E, x, t)\downarrow + 1)$ (if t does not exist then $s \notin I_x$).

Corollary 10. *If $|I_x| \leq i(x) + 1$ then $\mathcal{N}_\S$ is met.*

Proof. If $s \in I_x^t$ then $s \in I_x$. (Note that the converse of the last statement does not hold.)

Hence the easy way to met $\mathcal{N}_\S$ is to ensure that $|I_x| \leq i(x) + 1$. In this case $\mathcal{N}_\S$ applies finite restraint. We will close this section with a few remarks:

Remark The priority ordering. We will *always* assume that $\mathcal{N}_\S$ has higher priority than $\mathcal{N}_\dagger$ if and only if $x < y$.

Remark Initializing the strategy used for $\mathcal{N}_\S$. We cannot initialize $\mathcal{N}_\S$ without initializing $\mathcal{N}_\dagger$ for all y. The strategies used by these requirements are all helping construct the same sets and functionals. Hence we will only allow initialization of all $\mathcal{N}_\S$, in which case we can restart the construction of the needed sets and functionals and redefine the function $i(x)$. A requirement $\mathcal{R}$ will be allowed to initialize (cancel) the $\mathcal{N}_\S$'s iff $\mathcal{R}$ has higher priority than $\mathcal{N}_\prime$ (and hence all the $\mathcal{N}_\S$'s, by the above remark).

Remark Outcomes of $\mathcal{N}_{\S}$. $\mathcal{N}_{\S}$ has $i(x) + 2$ outcomes which we will use when needed. If Case 1, 3, 4 or 5 applies at stage $s+1$ then this strategy has outcome 0 at stage $s+1$. Otherwise this strategy has outcome $p^x(s+1)+1$ at stage $s+1$. The final outcome of the strategy is the lim inf of the outcomes. Since our hope is to meet $\mathcal{N}_{\S}$, by Corollary 8, we can assume that $p^x(s) \geq 0$, for all s. So if this strategy has outcome 0, then $\Gamma(A \oplus D; x)\uparrow$ (the converse is not true).

Remark Stages. In one of the following proofs we will need to restrict the stages that action for $\mathcal{N}_{\S}$ can take place to *accessible* stages (see Section 4.3). In another we will need to restrict the action for $\mathcal{N}_{\S}$ to *expansionary* stages (see Section 6.2). The above lemmas and discussion still hold true as long as we restrict all stages used above to appropriate stages.

Remark Indexing. All the sets A, B_0 and B_1 and all the functionals Λ, Ψ_0 and Ψ_1 should be indexed in some uniform fashion. Let e be an index for the sets A, B_0 and B_1 and the functionals Λ, Ψ_0 and Ψ_1. The above requirement and everything associated with it should be indexed by e. In particular, the subrequirement $\mathcal{N}_{\S}$ should be $\mathcal{N}_{\rceil,\S}$ and $i(x)$ should be $i_e(x)$.

4 Lower Cones

In this section we present three theorems. All of these theorems involve the construction of a degree(s) which does not bound a weak critical triple. This section is split into three subsection each containing a theorem and its proof.

4.1 Nonbounding

Theorem 11 [Wei88, Dow90]. *There is a noncomputable, computably enumerable degree* $\mathbf{e}$ *which does not bound a weak critical triple in the computably enumerable degrees.*

It is enough to build a computably enumerable set E which meets all the requirements $\mathcal{N}_{\rceil,\S}$ and the requirements

$$\bar{E} \neq W_e. \qquad \mathcal{R}_{\rceil}$$

We meet $\mathcal{R}_{\rceil}$ by using the following procedure:

Action for $\mathcal{R}_{\rceil}$ at stage $s+1$. Do the first of the following cases which applies:

Case 1 If there is no witness w for $\mathcal{R}_{\rceil}$ and $E_s \cap W_{e,s} = \emptyset$, choose a large witness w.

Case 2 If a witness w for $\mathcal{R}_{\rceil}$ exists, $E_s \cap W_{e,s} = \emptyset$ and $w \in W_{e,s}$, add w to E_{s+1}. (If this case occurs then we say $\mathcal{R}_{\rceil}$ acts.)

This requirement is positive; it wants to add elements to E. $\mathcal{R}_\rceil$ acts at most once. To initialize this strategy means to discard the current witness. This strategy will met $\mathcal{R}_\rceil$ as long as it is not initialized infinitely often.

Choose some appropriate computable ω-ordering of all the requirements. Given a requirement $\mathcal{N}_{\rceil,\S}$ we will let $i_e(x)$ be the number of positive requirements of higher priority.

The Construction at stage $s+1$. For the first s requirements take the action described above (in order of increasing priority). For all e and x, if $\rho^s_e(E,x,s)\downarrow\neq\rho^s_e(E,x,s-1)$ then initialize all positive requirements which have lower priority than $\mathcal{N}_{\rceil,\S}$.

Lemma 12. *(i)* $|I_{e,x}|\leq i_e(x)$. *Hence* $\mathcal{N}_{\rceil,\S}$ *is met. Furthermore, either* $\rho_e(E,x)$ *exists or for almost all* s, $\rho_e(E,x,s)\uparrow$.

(ii) $\mathcal{R}_\rceil$ *is only initialized finitely often. Hence* $\mathcal{R}_\rceil$ *is met.*

Proof. By induction on priority.

(i) By the above initialization, only those $\mathcal{R}_{\rceil'}$ with higher priority can put s into $I_{e,x}$ and each one of these $i_e(x)$ requirements can act at most once. By the induction hypothesis, there is a stage t by which all higher priority $\mathcal{R}_{\rceil'}$ which ever act have acted. If there is a stage $s\geq t$ such that $\rho_e(E,x,s)\downarrow$ then $\rho_e(E,x)=\rho_e(E,x,s)$.

(ii) Wait for a stage s where all higher priority $\mathcal{R}_{\rceil'}$ which are going to act have acted and for all higher priority $\mathcal{N}_{\rceil,\S}$ if $\rho_e(E,x)$ exists then $\rho_e(E,x)=\rho_e(E,x,s)$. $\mathcal{R}_\rceil$ is never initialized after stage s.

4.2 Splitting

Clearly there are computably enumerable degrees which bound a weak critical triple in the computably enumerable degrees. The next theorem implies that the set of computably enumerable degrees which do not bound a weak critical triple generates the computably enumerable degrees (under join) and so does not form an ideal in the computably enumerable degrees.

Theorem 13. *Every degree* $\mathbf{g}$ *is the join of two degrees* $\mathbf{e}_0$ *and* $\mathbf{e}_1$ *neither of which bound a weak critical triple in the computably enumerable degrees.*

It is enough to build a pair of computably enumerable sets E_0 and E_1 and a functional Θ such that the requirements $\mathcal{N}^{|}_{\rceil,\S}$ (replace "E" with "E_j", for $j\in\{0,1\}$, and index everything with an additional superscript "j" in the discussion in Section 3) are met and the requirements

$$\Theta(E_0\oplus E_1;y)=G(y) \qquad \Theta_y$$

are met. To meet Θ_y, we will use the following simple scheme: For all y, if $\Theta_s(E_{0,s}\oplus E_{1,s};y)\downarrow\neq G_s(y)$ then add some $z\leq\theta_s(y)$ to one of the E_i's. Θ_y is a positive requirement.

Choose some appropriate computable ω-ordering of all the requirements such that Θ_y has higher priority than Θ_z iff $y < z$. Assume $\mathcal{N}^{|\|}_{1\|,\S\|}$ is the kth negative requirement. Let z_k^* be the least number such that $\Theta_{z_k^*}$ has lower priority than $\mathcal{N}^{|\|-\infty}_{1\|-\infty,\S\|-\infty}$. We define two functions p and $\tilde{p}$ by induction as follows: $p(1) = \tilde{p}(1) = 1$, $p(k+1) = p(k) + \tilde{p}(k+1)$, and

$$\tilde{p}(k+1) = z_{k+1}^* + \sum_{k' \leq k} \tilde{p}(k'), \tag{4.1}$$

Let $i_{e_k}^{j_k}(x_k) = p(k)$. We make no claims that this function is a tight bound on the number of injuries. In the future we will drop the k from z^* since the requirement to which the z_k^* refers will always be clear from the context.

The Construction at stage $s+1$. For the first s requirements take the action described above (i.e for $\mathcal{N}^{\rangle}_{1,\S}$) or below (i.e. for Θ_y) in order of increasing priority.

Action for $\Theta_\dagger$ **at stage** $s+1$. Do the first of the following cases which applies, if any:

Case 1 Suppose $\Theta_s(E_{0,s} \oplus E_{1,s}; y)\uparrow$. Then define $\Theta_{s+1}(E_{0,s+1} \oplus E_{1,s+1}; y) = G_{s+1}(y)$ with large use.

Case 2 Suppose $\Theta_s(E_{0,s} \oplus E_{1,s}; y)\downarrow \neq G_s(y)$. Let $\mathcal{N}^{|}_{1,\S}$ be the highest priority negative requirement such that if $\rho^l(E_j, x, s)\downarrow$ then $\rho^l(E_j, x, s) \geq \theta_s(y)$, where $l = p_e^{j,x}(s) - 1$. If $\mathcal{N}^{|}_{1,\S}$ exists, then add $\theta_s(z^*)$ and $\theta_s(y)$ to E_{s+1}^j (in this case we say that $\mathcal{N}^{|}_{1,\S}$ takes action for Θ). If $\mathcal{N}^{|}_{1,\S}$ does not exist, then add $\theta_s(y)$ to E_{s+1}^0.

Clearly all the positive requirements are met. The following lemma shows that the negative requirements are also met.

Lemma 14. *$\mathcal{N}^{|}_{1,\S}$ only takes action for Θ at most $\tilde{p}(k)$ times, where $\mathcal{N}^{|}_{1,\S}$ is the kth negative requirement. $|I_{e,x}^j| \leq i_e^j(x)$. Hence $\mathcal{N}^{|}_{1,\S}$ is met.*

Proof. By induction on priority order. Assume $\rho_e^l(E_j, x, s)\downarrow$, where $l = p_e^{j,x}(x) - 1$ and $\theta_s(z) \leq \rho_e^l(E_j, x, s)$ enters E_j at stage s. Then for all $z' \geq z$ and all stage $t > s$, if $\theta(z')\downarrow$, $\theta(z') > \rho_e^{l-1}(E_j, x, s)$. Unless some $\theta_t(z')$ later enters E_j, where $z' < z$, $\mathcal{N}^{|}_{1,\S}$ will never take action for Θ again. By induction, the higher priority negative requirements can only cause such a z' to later enter E_j, $\sum_{k'<k} \tilde{p}(k')$ many times. There are $z^* - 1$ many positive requirements which also could cause such a z' to later enter E_j. Hence Equation 4.1 bounds the number of times $\mathcal{N}^{|}_{1,\S}$ can take action.

4.3 Nonlow nonbounding

Theorem 15. *There is a nonlow computably enumerable degree* e *which does not bound a weak critical triple in the computably enumerable degrees.*

To show E is not low it is enough to build E and a set, W^E, which is computably enumerable in E, such that the following requirements are met:

$$w \in W^E \text{ iff } \lim_l \varphi_e(w,l) = 0, \text{ for some witness } w. \qquad \mathcal{R}_\rceil$$

(If E is low then every set which is computably enumerable in E can be computably approximated. Hence W^E witnesses the fact that E is not low.) We meet $\mathcal{R}_\rceil$ by using the following procedure:

Action for $\mathcal{R}_\rceil$ at stage $s+1$. Do the first of the following cases which applies, if any:

Case 1 A witness w does not exist. Choose a large witness w and a large use $u_{e,s+1}(w)$ and let $w \in W_{s+1}^{E_{s+1}}$ iff $\varphi_{e,s}(w,s) \neq 0$. If $\varphi_{e,s}(w,s)\downarrow$ then let $l(s+1) = s+1$, otherwise let $l(s+1) = s$.
Case 2 A witness w exists, $\varphi_{e,s}(w, l(s)) = 0$ and $w \notin W_s^{E_s}$. Add $u_s(w)$ to E_{s+1}, add w to $W_{s+1}^{E_{s+1}}$, let $u_{e,s+1}(w)$ be large and let $l(s+1) = l(s)+1$.
Case 3 A witness w exists, $\varphi_{e,s}(w,l(s)) = 1$ and $w \in W_s^{E_s}$. Add $u_s(w)$ to E_{s+1}, remove w from $W_{s+1}^{E_{s+1}}$, let $u_{e,s+1}(w)$ be large and let $l(s+1) = l(s)+1$.
Case 4 A witness w exists and $\varphi_{e,s}(w,l(s))\downarrow$. Let $l(s+1) = l(s)+1$.

This is a positive requirement; it wants to add elements to E. $\mathcal{R}_\rceil$ can act infinitely many times. We say this strategy has outcome 0 if Case 2 or 3 applies infinitely often and outcome 1, otherwise. To initialize this strategy means to discard the current witness. This strategy will meet $\mathcal{R}_\rceil$ as long as it is not initialized infinitely often.

We will meet all the negative requirements $\mathcal{N}_{\rceil,\S}$ and all the positive requirements $\mathcal{R}_\rceil$ by using a priority tree. Let $T = \{\omega\}^{<\omega}$ (the subtree of nodes which are accessible at some stage will be a finitely branching tree). Choose some appropriate computable ω-ordering of all the requirements $\mathcal{N}_{\rceil,\S}$. Assume $\mathcal{N}_{\rceil_\|,\S_\|}$ is the kth negative requirement. Let $i_{e_k}(x_k) = 2^{2k-1}$ (as always we make no claims this bound is tight). *All* the nodes of length $2k$ will work on meeting $\mathcal{N}_{\rceil_\|,\S_\|}$ using the *same* strategy as described in Section 3. Each node, α, of length $2e+1$ will work on meeting $\mathcal{R}_\rceil$ using the above procedure on stages where α is accessible with its *own* witness w_α.

The Construction of E at stage $s+1$. By induction on k. Let $\beta_{s+1,0} = \lambda$. Let $\beta = \beta_{s+1,k}$. There are two cases:

Case 1 Suppose $|\beta| = 2m$. Use the above procedure for $\mathcal{N}_{\rceil_{\mathfrak{a}},\S_{\mathfrak{a}}}$. Let o be the outcome of this strategy as determined in Remark 3. If $k < s+1$ then let $\beta_{s+1,k+1} = \beta_{s+1,k}\widehat{\ }o$.

Case 2 Suppose $|\beta| = 2m+1$. Use the above procedure for $\mathcal{R}_{\Updownarrow}$ (using the witness w_β). If $k < s+1$ then if either Case 2 or Case 3 applies (for the strategy based at the node β), let $\beta_{s+1,k+1} = \beta_{s+1,k}{}^\frown 0$, otherwise let $\beta_{s+1,k+1} = \beta_{s+1,k}{}^\frown 1$.

Let $\beta_{s+1} = \beta_{s+1,s+1}$. Initialize all nodes of odd length which are to the right of β_{s+1}.

Lemma 16. *Let $f = \liminf \beta_s$. For all k,*
(i) For all stages t, $|I^t_{e_k,x_k}| \leq i_{e_k}(x_k)$. Hence $\mathcal{N}_{\urcorner_{\|},\S_{\|}}$ is met.
(ii) Let $\alpha = f \restriction (2k+1)$. α is only initialized finitely often. Hence $\mathcal{R}_{\|}$ is met.

Proof. By induction on k.

(i) Let $e = e_k$ and $x = x_k$. We say α injures $\mathcal{N}_{\urcorner,\S}$ at stage $s+1$ if $s+1$ is not an almost free-clear stage, $s+1$ is not a free-clear stage, $u_s(w_\alpha) < \rho^l_e(E, x, s)$ and $u_s(w_\alpha)$ enters E at stage $s+1$, where $l = p^x_e(s) - 1$. Let $t < s+1$ be the last free-clear stage and $t' < t$ be the last almost free-clear stage (let $t' = t$, if such a stage does not exist). At stage t' all α such that $|\alpha| > 2k$, $|\alpha|$ is odd and $\alpha(2k) \neq 0$ are initialized. Hence no such α can injure $\mathcal{N}_{\urcorner,\S}$. If α is such that $|\alpha| > 2k$, $|\alpha|$ is odd and $\alpha(2k) = 0$ then α can only act at almost free-clear stages for $\mathcal{N}_{\urcorner,\S}$. Hence no such α can injure $\mathcal{N}_{\urcorner,\S}$. Therefore if α injures $\mathcal{N}_{\urcorner,\S}$ at stage $s+1$, $|\alpha| < 2k$ and α cannot injure again $\mathcal{N}_{\urcorner,\S}$ until after the next almost free-clear stage (its use is too large). There are less than 2^{2k-1} such nodes.

(ii) Clear.

5 The requirement $\mathcal{P}_{\rangle}$

For $X \in \{A, B_0, B_1\}$, let $\widehat{X} = X \oplus E$. Let $\widehat{E} = K$. Our goal is to build a computably enumerable set E such that the following requirements are met:

$$\begin{gathered}\text{If } \widehat{\Lambda}(\widehat{E}) = \widehat{A} \oplus \widehat{B}_0 \oplus \widehat{B}_1 \text{ and } \widehat{\Psi}_i(\widehat{A} \oplus \widehat{B}_{\bar{i}}) = \widehat{B}_i \\ \text{then there exists a computably enumerable set } \widehat{D} \\ \text{and functionals } \widehat{\Delta}_i \text{ and } \widehat{\Gamma} \text{ such that} \\ \widehat{\Delta}_i(\widehat{B}_i) = \widehat{D} \text{ and } \widehat{\Gamma}(\widehat{A} \oplus \widehat{D}) = \widehat{B}_0\end{gathered} \qquad \mathcal{P}$$

where $\widehat{\Lambda}$ and $\widehat{\Psi}_i$ are functionals and $\widehat{A}$, $\widehat{B}_0$ and $\widehat{B}_1$ are computably enumerable sets. If we meet $\mathcal{P}$ then the degrees represented by these three sets do not form a weak critical triple.

$\mathcal{P}$ is the dual of $\mathcal{N}$ under the operation of hatting. Hence we can use the dual of the strategy used for $\mathcal{N}$ to meet $\mathcal{P}$ and the dual of everything in Section 3 applies to the requirement $\mathcal{P}$. As in Section 3, we split $\mathcal{P}$ into infinitely many subrequirements:

If there are infinitely many stages s such that $\widehat{l}(s) > x$ then
$\widehat{\Gamma}(\widehat{A} \oplus \widehat{D}; x) = \widehat{B}_0(x)$ and $\widehat{\Delta}_i(\widehat{B}_i) \upharpoonright (\widehat{\gamma}(x)+1) = \widehat{D} \upharpoonright (\widehat{\gamma}(x)+1)$. $\mathcal{P}_\S$

The strategy for $\mathcal{P}_\S$ is the dual of the strategy used for $\mathcal{N}_\S$ with two additional features. First, there is a restraint function $\widehat{r}_x(s)$. $\widehat{r}_x(s)$ will be controlled by the negative requirements and at each stage will be determined *before* the strategy for $\mathcal{P}_\S$ acts. $\widehat{r}_x(s)$ will be a nondecreasing function (in s). (Initially, $\widehat{r}_x(0) = 0$.) Second, Case 5 now reads:

Case 5 For $i = \widehat{e}^x(t)$ and $l = \widehat{p}^x(t)$, $\widehat{B}_{i,s+1} \upharpoonright \widehat{\rho}^l(\widehat{B}_i, x, t) \neq \widehat{B}_{i,t} \upharpoonright \widehat{\rho}^l(\widehat{B}_i, x, t)$ and $\widehat{\delta}_{\bar{i},s}(x) > \widehat{r}_x(s+1)$.

Action Add $\widehat{\delta}_{\bar{i},s}(x)$ to E_{s+1} and $\widehat{\gamma}_t(x)$ to $\widehat{D}_{s+1}$ (by our convention this will make $\widehat{\Delta}_{\bar{i},s}(\widehat{B}_{\bar{i}}, x)\uparrow$, $\widehat{\Gamma}_{s+1}(\widehat{A}_{s+1} \oplus \widehat{D}_{s+1}; x)\uparrow$ and will cause Case 2 to apply at the next possible stage).

By Definition 3, we know if Case 5 applies at stage $s+1$, then $s+1$ is an almost free-clear stage. Case 5 adds numbers to E. In fact, this strategy may add infinitely many numbers into E. Hence this strategy is an infinite positive strategy (this makes sense since the dual of a negative strategy should be a positive strategy). Using the dual of the arguments in Section 3, one can prove:

Lemma 17 The Dual of Corollary 8. *Suppose for every free-clear stage t, $|\widehat{I}^t_x| \leq \widehat{i}(x)+1$. Then $\mathcal{P}_\S$ is met.*

A word of caution: Since $\widehat{E} = K$, we cannot control $|\widehat{I}^t_x|$ ($|\widehat{I}_x|$). Hence while Lemma 17 (the dual of Lemma 10) holds we cannot ensure that the hypothesis holds. We need some other way to ensure that $\mathcal{P}_\S$ is met.

Definition 18. Let t be a free-clear stage. Let t' be the next almost free-clear stage or ∞ if there is no such stage. We let $s \in \widehat{R}^t_x$ iff either $s = t$ or $t < s < t'$ and $\widehat{r}_x(s) > \widehat{r}_x(s-1)$ (since $\widehat{r}_x(s)$ is nondecreasing in s, these are the only stages at which the value of $\widehat{r}_x(s)$ changes).

Lemma 19. *Suppose for every free-clear stage t, $|\widehat{R}^t_x| \leq \frac{1}{2}\widehat{i}(x) - 1$. Then $|\widehat{I}^t_x| \leq \widehat{i}(x)$ and so $\mathcal{P}_\S$ is met.*

Proof. First note there cannot be three stages, s_0, s_1 and s_2, such that $t \leq s_0 \leq s_1 < s_2 < t'$, $s_0 \in \widehat{R}^t_x$, $s_1 \in \widehat{I}^t_x$, $s_2 \in \widehat{I}^t_x$ and for all s if $s_0 < s \leq s_2$, then $s \notin \widehat{R}^t_x$. (Assume otherwise then at stage s_2 Case 5 will apply rather than Case 6 and hence s_2 is an almost free-clear stage.) So two layers cannot be peeled away between stages when the restraint increases. Therefore $|\widehat{I}^t_x| \leq 2(|\widehat{R}^t_x|+1) \leq \widehat{i}(x)$.

One should think of $\frac{1}{2}\widehat{i}(x) - 1$ as a bound on the number of times the restraint can increase between a free-clear stage and the next almost free-clear stage and still meet $\mathcal{P}_\S$.

6 Upper Cones

In this section we present two theorems each of which is presented in its own subsection. Both of these theorems involve the construction of a degree above which there is no weak critical triple (in the computably enumerable degrees). We end this section with a subsection which concerns mixing the requirements $\mathcal{P}_{\S}$ with the standard lowness requirements.

6.1 Upward nonbounding

The following theorem follows from Theorem 11 and the pseudo-jump theorem in [JSh83] (see [CDo93] for more details). A direct proof of the theorem using Harrington's "levels" method appears in [Wei88].

Theorem 20 [Wei88]. *There is an incomplete computably enumerable degree* e *above which there is no weak critical triple in the computably enumerable degrees.*

To show E is not complete it is enough to build E and a computably enumerable set, W such that the following requirements are meet:

$$w \in W \text{ iff } \Phi_e(E; w) = 0, \text{ for some witnesses } w. \qquad \mathcal{R}_{\rceil}$$

(If E is complete then every set which is computably enumerable is computable in E. Hence W witnesses that E is not complete.) We meet $\mathcal{R}_{\rceil}$ by using the following procedure:

Action for $\mathcal{R}_{\rceil}$ at stage $s+1$. Do the first of the following cases which applies, if any:

Case 1 If a witness w does not exist, then choose a large witness w.
Case 2 If a witness w exists and $\Phi_{e,s}(E_s; w) = 0$, add w to W and restrain E below $\varphi_{e,s}(w) + 1$.

This is a negative requirement; it wants to stop elements from entering E. This strategy is injured if $w \in W_s$ and some x later enters E below $\varphi_{e,s}(w) + 1$. When this occurs we will initialize the strategy. To initialize this strategy means to discard the current witness. $\mathcal{R}_{\rceil}$ can act at most once unless initialized. This strategy will meet $\mathcal{R}_{\rceil}$ as long as it is not initialized infinitely often.

We will meet all the negative requirements $\mathcal{R}_{\rceil}$ and all the positive requirements $\mathcal{P}_{\rceil,\S}$ by using a tree. Let $T = \{\omega\}^{<\omega}$ (again, the subtree of nodes which are accessible at some stage will be a finitely branching tree). Let f be the true path and β_s be the approximation to the true path at stage s.

Choose some appropriate computable ω-ordering of all the requirements $\mathcal{P}_{\rceil,\S}$. Assume $\mathcal{P}_{\rceil_{\|},\S_{\|}}$ is the kth positive requirement. *All* the nodes of length $2k$ will work on meeting $\mathcal{P}_{\rceil_{\|},\S_{\|}}$ using the *same* strategy as described in Section 5 and the *same* restraint $\widehat{r}_{e_k,x_k}$. For shorthand, we will let $\widehat{r}_{e_k,x_k} = \widehat{r}_{2k}$.

A node β of length $2e+1$ will work on meeting $\mathcal{R}_{\rceil}$ using the above procedure at stages when β is accessible with its *own* witness w_β. It would be nice if no node below β could injure the strategy used at β but this is not possible. We can *only* restore all needed layers of a positive requirement when there is an almost free-clear stage. Hence we cannot ensure the needed layers are available when the computation converges. These layers may have been used on a previous computation which was later injured or initialized.

To determine $\widehat{\imath}_{e_k}(x_k)$ we need the following functions:

$$\text{If } k \le e \text{ then } g(2e+1,k) = 0 \tag{6.1}$$

$$\text{If } k+1 > e \text{ then } g(2e+1,k+1) = 1 + 2g(2e+1,k) \tag{6.2}$$

$$h(k) = \sum_{j<k} (g(2j+1,k)+1) \tag{6.3}$$

$$\widehat{\imath}_{e_k}(x_k) = 2(h(k)+1) \tag{6.4}$$

(We make no claims that our bounds are tight.)

The Construction of E at stage $s+1$. By induction on k. Let $\beta_{s+1,0} = \lambda$. Let $\beta = \beta_{s+1,k}$. There are two cases:

Case 1 Suppose $|\beta| = 2m$. Use the above procedure for $\mathcal{P}_{\rceil_{\Updownarrow},\S_{\Updownarrow}}$. We say β injures α at stage $s+1$ if $|\alpha| = 2e+1$, $\varphi_{e,s}(w_\alpha)\downarrow$ and Case 5 applies and adds a number less than or equal to $\varphi_{e,s}(w_\alpha)$ into E at stage $s+1$. Let o be the outcome of this strategy as determined by the dual of Remark 3. If $k < s+1$ then let $\beta_{s+1,k+1} = \beta^\frown o$.

Case 2 Suppose $|\beta| = 2m+1$. Use the above procedure for $\mathcal{R}_{\Updownarrow}$ (using the witness w_β). If $k < s+1$ then let $\beta_{s+1,k+1} = \beta^\frown 0$. Let $t > 2m$ be the last stage such that $\beta_t <_{\mathrm{L}} \beta$ (if such a stage does not exist then let $t = 2m+1$). If $|\alpha| > t$ then increase $\widehat{r}_{|\alpha|}(s+1)$ to $s+1$. For $|\alpha| \le t$, let t_α be the last stage such that α injured β at stage t_α. Suppose that t_α exists, $t < t_\alpha$ and for all γ if $|\gamma| > |\alpha|$ then either t_γ does not exist or $t_\gamma < t_\alpha$. Then increase $\widehat{r}_{|\alpha|}(s+1)$ to $s+1$.

Let $\beta_{s+1} = \beta_{s+1,s+1}$. Initialize all nodes of odd length which are to the right of β_{s+1}.

Lemma 21. *Let $\beta \subset f$ (the true path on T) and $|\beta| = 2e+1$. Fix t, t' and k. Assume that for all stages s if $t \le s < t'$, $\beta \le \beta_s$. Assume no node of length $2k+1$ or greater injures β during those stages s where $t \le s < t'$. Then β can be injured at most $g(2e+1,k)$ times during those stages s where $t \le s < t'$.*

Proof. By induction on k. Clearly, the lemma holds for $k \le e$ (otherwise β is initialized). All the nodes of length $2k$ are using the same strategy. If one of these nodes injures β, they all are later restrained from injuring β. Hence, collectively, these nodes can injure β at most once. But this injury allows all the nodes of length less than $2k$ to reinjure β, if needed.

Lemma 22. *Let $\beta \subset f$ and $|\beta| = 2e+1$. Let $2k > |\beta|$. Then β can only increase the restraint function $\widehat{r}_{2k}$ $(g(2e+1,k)+1)$ times between a free-clear stage (for $\mathcal{P}_{\rceil_{\|},\S_{\|}}$) and the next almost free-clear stage (for $\mathcal{P}_{\rceil_{\|},\S_{\|}}$).*

Proof. Let t be a free-clear stage and t'' be the next almost free-clear stage. Let t' be the least stage such that $t < t' \leq t''$ and either $\beta_{t'} <_{\mathrm{L}} \beta$ or some node γ of length greater than $2k$ injures β. After stage t', β will not increase the restraint function $\widehat{r}_{2k}$ until after stage t''. If such a t' does not exist let $t' = t''$. Now by the above lemma, β can be injured at most $g(2e+1, k)$ times during those stages s where $t \leq s < t'$.

Lemma 23. *Let* $2k > |\beta|$.

(i) The restraint function $\widehat{r}_{2k}$ *only increases* $h(k)$ *times between a free-clear stage (for* $\mathcal{P}_{\rceil_{\|},\S_{\|}}$*) and the next almost free-clear stage (for* $\mathcal{P}_{\rceil_{\|},\S_{\|}}$*).*

(ii) $|\widehat{R}^t_{e_k,x_k}| \leq \frac{1}{2}\widehat{i}_{e_k}(x_k) - 1$. *Hence* $\mathcal{P}_{\rceil_{\|},\S_{\|}}$ *is met.*

Proof. (i) Follows from Lemma 22 and the fact that if $|\beta| > 2k$ then β can never increase $\widehat{r}_{2k}$.

(ii) Follows from (i) and Lemma 19.

Lemma 24. *Let* $\beta \subset f$ *and* $|\beta| = 2e + 1$. β *is only injured finitely often. Hence* $\mathcal{R}_\rceil$ *is met.*

Proof. Let $2t \geq 2e+1$ be the least stage such that for all stages $s \geq 2t$, $\beta \leq_{\mathrm{L}} \beta_s$. No node of length $2e$ or less ever injures β after stage t. No node of length $2t+1$ or greater ever injures β after stage t. Now, by Lemma 21, β is injured at most $g(2e+1, t)$ more times.

6.2 Upward and downward nonbounding

Theorem 25. *There is a computably enumerable degree* **e** *above which there is no weak critical triple* and *below which there is no weak critical triple.*

By [DSh96] we know that e must be low_2 since they show that below every nonlow_2 degree one can embed a copy of M_5.

We will meet all the negative requirements $\mathcal{N}_{\rceil,\S}$ and all the positive requirements $\mathcal{P}_{\rceil,\S}$ by using a tree. Let $T = \{\omega\}^{<\omega}$ (again, the subtree of nodes which are accessible at some stage will be a finitely branching tree). Let f be the true path and β_s be the approximation to the true path at stage s.

Choose some appropriate computable ω-ordering of all the requirements $\mathcal{P}_{\rceil,\S}$. Assume $\mathcal{P}_{\rceil_{\|},\S_{\|}}$ is the kth positive requirement. If $x_k = 0$ then all nodes of length $2k$ will work on meeting $\mathcal{P}_{\rceil_{\|},\S_{\|}}$; each using a *different* strategy. Such a node is called a *parent* node. If $|\alpha| = 2k$, α's parent node is the substring of length $2k'$, where $e_{k'} = e_k$ and $x_{k'} = 0$ (by the dual of Remark 3 such a k' always exists). All nodes α of length $2k$ which share the same parent node will work on meeting $\mathcal{P}_{\rceil_{\|},\S_{\|}}$ using the *same* strategy as described in Section 5 and the *same* restraint $\widehat{r}_\alpha$. Such a node is called a *child* node of its parent. Hence if α and α' are using the same strategy to meet a positive requirement, $\widehat{r}_\alpha = \widehat{r}_{\alpha'}$.

Choose some appropriate computable ω-ordering of all the requirements $\mathcal{N}_{\rceil,\S}$. Assume $\mathcal{N}_{\rceil_{\|},\S_{\|}}$ is the kth negative requirement. If $x_k = 0$ then all nodes of the nodes of length $2k+1$ will work on meeting $\mathcal{N}_{\rceil_{\|},\S_{\|}}$; each using a *different* strategy. Such a node is called a *parent* node. If γ is a parent node and $|\gamma|$ is odd then let l_γ be the length of agreement function as defined in Equation 3.12. If $|\alpha| = 2k+1$, α's parent node is the substring of length $2k'+1$, where $e_{k'} = e_k$ and $x_{k'} = 0$ (by Remark 3 such a k' always exists). All nodes α of length $2k+1$ which share the same parent node will work on meeting $\mathcal{N}_{\rceil_{\|},\S_{\|}}$ using the *same* strategy as described in Section 3. Such a node is called a *child* node of its parent (parents are children of themselves). We will let $\rho_\alpha(E,s) = \rho_{e_k}(E, x_k, s)$ if α is working on $\mathcal{N}_{\rceil_{\|},\S_{\|}}$. Hence if α and α' are using the same strategy to meet a negative requirement, $\rho_\alpha(E,s) = \rho_{\alpha'}(E,s)$.

We can initialize parent nodes γ (see Remark 3) and all of their children. At which point we can redefine the function $i_{e_k}(x_k)$ for those children which are initialized.

We will meet $\mathcal{N}_{\rceil_{\|},\S_{\|}}$ by using Corollary 10. Hence, we must be able to count the number of possible injuries. Similarly, since we will use Lemma 19, we must also count the number of times the restraint for $\mathcal{P}_{\rceil_{\|},\S_{\|}}$ will increase between free-clear stages.

Informally, the key idea is: Assume β is working on $\mathcal{N}_{\rceil_{\|},\S_{\|}}$. Let γ be β's parent node. Let α be a node of even length. Let $2t^* + j < t$ (for $j \in \{0,1\}$) be the last stage at which γ was initialized (if such a stage does not exist let $t^* = 0$). If $|\alpha| > t^* + |\beta|$ then α can never injure β. Assume $|\alpha| \leq t^* + |\beta|$. If α is started before β, α may injure β once. After this initial injury, α cannot injure β unless some α', where $|\alpha'| > |\alpha|$ injures β first.

Actions and accessibility at stage $s+1$: Assume β is working on $\mathcal{N}_{\rceil_{\mathfrak{a}},\S_{\mathfrak{a}}}$. Fix a stage $s \geq |\beta|$. Let γ be β's parent node. $s+1$ is γ-expansionary iff $s = 0$ or $l_\gamma(s) > l_\gamma(t)$, for all $t < s$. β will only *act* at accessible γ-expansionary stages $s+1$ where $l_\gamma(s) > x_m$. We say β is *started* by stage $s+1$ if some β' working on $\mathcal{N}_{\rceil_{\mathfrak{a}},\S_{\mathfrak{a}}}$ using the same strategy as β acts (for the first time) by stage $s+1$.

Injuries at stage $s+1$: Assume that α is working on $\mathcal{P}_{\rceil_{\mathfrak{a}},\S_{\mathfrak{a}}}$ and that α acts by adding a number n into E at stage $s+1$. Assume β is working on $\mathcal{N}_{\rceil_{\|},\S_{\|}}$, γ is β's parent node, $\gamma \subset \alpha$ and β has not been injured since the last stage t when it acted (defined below). If $n < \rho_\beta(E,t)$ then we say α *injures* β at stage $s+1$. If α injures β at stage $s+1$ and α' is also working on $\mathcal{P}_{\rceil_{\mathfrak{a}},\S_{\mathfrak{a}}}$ using the same strategy as α, then we say α' injures β. The same thing occurs when β and β' are both using the same strategy to meet $\mathcal{N}_{\rceil_{\|},\S_{\|}}$. Hence a child of γ can be injured at most once between γ-expansionary stages. Let $\mathcal{C} = \{\beta_\rangle\}$ be a finite collection of γ's children. We say this finite collection is injured at stage $s+1$ if the collection has not been injured since the last γ-expansionary stage and some β_i is injured at stage $s+1$.

Restraint at stage $s+1$: Assume that β is working on $\mathcal{N}_{\rceil_{\mathfrak{a}},\S_{\mathfrak{a}}}$ and that β acts at stage $s+1$. Let γ be β's parent. Let $2t^* + j < s+1$ be the last stage at which γ was initialized and $j \in \{0,1\}$ (if such a stage does not exist let $t^* = 0$). Let $t+1 \leq s$ be the last γ-stage. If $\rho_\beta(E,t) \neq \rho_\beta(E,s)$ or β is started by stage $s+1$ then do the following: Assume α is working on $\mathcal{P}_{\rceil_{\|},\S_{\|}}$ and $\gamma \subset \alpha$. If $|\alpha| > 2t^* + |\beta|$ then increase $\widehat{r}_\alpha(s+1)$ to $s+1$. Let t' be the last stage when α

injured some child of γ (if t' does not exist let $t' = 0$). Suppose that $2t^* + j < t'$ and β was started by stage t'. For $|\alpha'| > |\alpha|$ let $t_{\alpha'}$ be the last stage such that α' injured some child of γ at stage $t_{\alpha'}$. Suppose for all α', if $|\alpha| < |\alpha'|$ then either $t_{\alpha'}$ does not exist or $t_{\alpha'} < t$. Then increase $\widehat{r}_\alpha(s+1)$ to $s+1$.

Determining the stages $s+1$ at which β restrains α only depends on whether $s+1$ is γ-expansionary, the last stage where γ was initialized, $l_\gamma(s) > x_m$ and $\rho_\beta(E,s)$ has increased since the last γ-expansionary stage. Hence this only depends on γ and the stage. We will assume that α's restraint at stage $s+1$ is determined before α acts at stage $s+1$.

To determine $i_{e_k}(x_k)$ and $\widehat{i}_{e_k}(x_k)$ we need the following functions: Given k, let $p(k)$ be the length of the parent node for all nodes of length k (this is a well-defined function).

$$\text{If } k \leq m \text{ then } g(2m+1,k) = 0 \tag{6.5}$$

$$\begin{gathered}\text{If } m < k+1 \text{ then}\\ g(2m+1,k+1) = 2^{2k} + (2^{2k}+1)g(2m+1,k)\end{gathered} \tag{6.6}$$

$$h(k) = \sum_{m<k} 2g(2m+1,k-1) + (k-m) + 2 \tag{6.7}$$

$$\widehat{i}_{e_k}(x_k) = 2(h(k)+1) \tag{6.8}$$

If β is working on $\mathcal{N}_{\rceil_{\|},\S_{\|}}$ and β is initialized at stage $2t^* + j$, where $j \in \{0,1\}$, then define $i_{e_k}(x_k) = g(p(2k+1), t^* + k)$ at stage $2t^* + j$. (We make no claim that our bounds are tight.)

The Construction of E at stage $s+1$. By induction on k. Let $\beta_{s+1,0} = \lambda$. Let $\beta = \beta_{s+1,k}$. There are two cases:

Case 1 Suppose $|\beta| = 2m$. Use the above procedure for $\mathcal{P}_{\rceil_{\mathfrak{o}},\S_{\mathfrak{o}}}$ (in Section 5). Let o be the outcome of this strategy as determined by the dual of Remark 3. If $k < s+1$ then let $\beta_{s+1,k+1} = \beta^\frown o$.

Case 2 Suppose $|\beta| = 2m+1$. Let γ be β's parent node. If s is β-expansionary and $l_\gamma(x) > x_m$, use the above procedure for $\mathcal{N}_{\rceil_{\mathfrak{o}},\S_{\mathfrak{o}}}$ (in Section 3) and impose restraint as described above. If $k < s+1$ then let $\beta_{s+1,k+1} = \beta^\frown 0$ (we do not care about the outcome in the sense of Remark 3).

Let $\beta_{s+1} = \beta_{s+1,s+1}$. Initialize all nodes whose parent node is to the right of β_{s+1}.

Lemma 26. *Let γ be a parent node of length $2m+1$. Fix k and some stages $t < t'$. Assume for all stages s, if $t \leq s \leq t'$, $\gamma \leq \beta_s$. Let $2t^* + j < t$ (for $j \in \{0,1\}$) be the last stage at which γ was initialized (if such a stage does not exist let $t^* = 0$). Let $C = \{\beta_\rangle\}$ be a finite collection of γ's children such that for all i either β_i has been started by stage t or $2t^* + |\beta_i| < 2k$. Assume no node of length $2k+1$ or greater injures any β_i during those stages s where $t \leq s < t'$. Then the collection $\{\beta_i\}$ can be injured at most $g(2m+1,k)$ times by nodes of length $2k$ or smaller during stages s where $t \leq s < t'$.*

Proof. By induction on k. Clearly, the lemma holds for $k \leq m$ (otherwise γ is initialized). Suppose a node α of length $2k$ injures β_i at some stage between t and t'. Hence α last injured a child of γ before β_i was started and $2t^* + |\beta_i| > 2k$. At the next γ-expansionary stage, α is restrained from injuring any β_i. There are 2^{2k} such nodes and they may act independently. But each one of these injuries allows all the nodes of length less than k to reinjure the collection $\mathcal{C}$, if needed.

Lemma 27. *Let $\gamma \subset f$ be a parent node of length $2m+1$. Let $2t^* + j < t$ (for $j \in \{0,1\}$) be the last stage at which γ was initialized (if such a stage does not exist let $t^* = 0$). Let β be a child of γ working on $\mathcal{N}_{\rceil_{\|},\S_{\|}}$ of length $2k+1$. Then β can be injured at most $i_{e_k}(x_k) = g(2m+1, t^*+k)$ times after stage $2t^*+j$. Hence for all e and x, $\mathcal{N}_{\rceil,\S}$ is meet.*

Proof. First note no node of length greater than $2t^* + 2k + 1 = 2(t^* + k) + 1$ can ever injure β. Now apply Lemma 26 to $\mathcal{C} = \{\beta\}$.

Lemma 28. *Let $\gamma \subset f$ be a parent node of length $2m+1$. Fix $k > m$. Let α be a node of length $2k$. Let t be a free-clear stage for $\mathcal{P}_{\rceil_{\|},\S_{\|}}$ and t' be the next almost free-clear stage t' (for $\mathcal{P}_{\rceil_{\|},\S_{\|}}$). Let $2t^* + j < t$ (for $j \in \{0,1\}$) be the last stage at which γ was initialized (if such a stage does not exist let $t^* = 0$).*

(i) After stage s, the children β of γ such that $2t^* + |\beta| > 2k$ can collectively increase α's restraint at most $g(2m+1, k-1) + 1$ times between t and t'.

(ii) After stage s, the children of γ such that $2t^* + |\beta| < 2k$ can collectively increase α's restraint at most $g(2m+1, k-1) + (k-m) + 1$ times between t and t'.

(iii) After stage s, the children of γ of any length can collectively increase α's restraint at most $2g(2m+1, k-1) + (k-m) + 2$ times between t and t'.

Proof. (i) Let s_i be the stages such that $t \leq s_0 < s_1 \ldots < s_j \leq t'$, $\widehat{r}_\alpha(s_i) > \widehat{r}_\alpha(s_i - 1)$ and this increase was caused by a child β of γ such that $2t^* + |\beta| > 2k$. Let β_i be a child of γ such that β_i restrained α at stage s_i and $2t^* + |\beta_i| > 2k$. β_i was started by stage t. No α' such that $|\alpha| \leq |\alpha'|$ can injure any child of γ between stages t and s_j (otherwise β_j cannot restrain α at stage s_j). γ cannot be initialized between stages t and s_j (otherwise β_j cannot restrain α at stage s_j). Let $i < i'$. Since s_i is a γ-expansionary stage and $\beta_{i'}$ was started by stage t, $\rho_{\beta_{i'}}(E, s_i)\downarrow < s_i$. So once β_i restrains α, $\beta_{i'}$ cannot restrain α unless it is later injured. Now Lemma 26 applies to $\mathcal{C} = \{\beta_\rangle\}$. Hence $j \leq g(2m+1, k-1) + 1$.

(ii) Let s_i be the stages such that $t \leq s_0 < s_1 \ldots < s_j \leq t'$, $\widehat{r}_\alpha(s_i) > \widehat{r}_\alpha(s_i - 1)$ and this increase was caused by a child β of γ such that $2t^* + |\beta| < 2k$. Let β_i be a child of γ of length such that β_i restrained α at stage s_i and $2t^* + |\beta_i| < 2k$. No node of length $2k$ or greater can ever injure any β_i. γ cannot be initialized between stages t and s_j (otherwise β_j cannot restrain α at stage s_j). Let $i < i'$. Since s_i is a γ-expansionary stage, if $\beta_{i'}$ was started by stage s_i, $\rho_{\beta_{i'}}(E, s_i)\downarrow < s_i$. Hence once β_i restrains α, $\beta_{i'}$ cannot restrain α unless it is later started or injured. After $(k-m)$ γ-expansionary stages, all the β_i are started. Lemma 26 applies to $\mathcal{C} = \{\beta_\rangle\}$. Hence $j \leq g(2m+1, k-1) + (k-m) + 1$.

(iii) Since γ has no child of even length this follows from (i) and (ii).

Corollary 29. *Suppose α is working on $\mathcal{P}_{\rceil_{\|},\S_{\|}}$ of length $2k$. Let t be a free-clear stage for $\mathcal{P}_{\rceil_{\|},\S_{\|}}$ and t' be the next almost free-clear stage t' (for $\mathcal{P}_{\rceil_{\|},\S_{\|}}$).*

(i) α's restraint can increase at most $h(k)$ times between t and t'.

(ii) $|\widehat{R}^t_{e_k,x_k}| \leq \frac{1}{2}\widehat{i}_{e_k}(x_k) - 1$. Hence $\mathcal{P}_{\rceil_{\|},\S_{\|}}$ is met.

Proof. (i) Only those β whose parent node γ is a substring of α can restrain α. Hence the above lemma applies at most k times.

(ii) Follows from (i) and Lemma 19.

6.3 $\mathcal{P}_{\rceil,\S}$ and lowness requirements

The astute reader might wonder why one could not combine the requirements $\mathcal{P}_{\rceil,\S}$ and lowness requirements in the following fashion: First note that one meets the standard lowness requirements by finite restraint. All the negative requirements in this section were met by finite restraint. However one cannot meet the standard lowness requirements by acting independently at different nodes on a tree with infinitary positive requirements. Now turn the infinite positive strategy used for $\mathcal{P}_{\rceil,\S}$ into a finite positive strategy by only allowing Case 5 to act when x actually enters B_0 and ignoring the outcomes. Hence a negative requirement can be injured at most once by each positive requirement of higher priority. So hopefully we can have enough layers in the positive requirements of lower priority to handle this fixed number of increases in the restraint.

However, the problem with this approach, is that a positive requirement $\mathcal{P}_{\rceil,\S}$ may in fact injure a negative requirement of higher priority: The strategy used for $\mathcal{P}_{\rceil,\S}$ may have peeled away all of its layers and be in a position that it *must* add some number n into E iff x enters B_0. Now the computation for the negative requirement converges and wishes to restrain n. Then x enters B_0 which forces n to enter E. This may be repeated infinitely many times for the same negative requirement. This remains true even if we restrict the positive requirements to a single fixed e. Thus the question of whether M_5 can be embedded above every low computably enumerable degree remains open.

References

[ASF96] Ambos-Spies, K. , Fejer, P. A. 1996. Personal communication.

[ASL86] Ambos-Spies, K. , Lerman, M. 1986. Lattice embeddings into the recursively enumerable degrees, *J. Symbolic Logic* **51**: 257–272.

[ASL89] Ambos-Spies, K. , Lerman, M. 1989. Lattice embeddings into the recursively enumerable degrees II, *J. Symbolic Logic* **54**: 735–759.

[CDo93] Cholak, P. , Downey, R. 1993. Lattice nonembeddings and intervals of the recursively enumerable degrees, *Ann. Pure Appl. Logic* **61**: 195–221.

[CYi] Cooper, B. , Yi, X. n.d. . Non-splitting and the high/low hierarchy. In preparation.

[Dow90] Downey, R. G. 1990. Lattice nonembedding and initial segments of the r. e. degrees, *Ann. Pure Appl. Logic* **49**: 97–119.

[DLe] Downey, R. G. , Lempp, S. n.d. . Contiguity and distributivity in the computably enumerable turing degrees, *J. Symbolic Logic* . To appear.

[DSh95] Downey, R. G. , Shore, R. A. 1995. Degree theoretic definitions of low_2 recursively enumerable sets, *J. Symbolic Logic* **60**(3): 727–756.

[DSh96] Downey, R. G. , Shore, R. A. 1996. Lattice embedding below a $nonlow_2$ recursively enumerable degree, *Israel J. Math.* **94**: 221–246.

[JSh83] Jockusch, Jr., C. G. , Shore, R. A. 1983. Pseudo-jump operators I: the r.e. case, *Trans. Amer. Math. Soc.* **275**: 599–609.

[Lac72] Lachlan, A. H. 1972. Embedding nondistributive lattices in the recursively enumerable degrees, *in* W. Hodges (ed.), *Conference in Mathematical Logic, London, 1970*, Vol. 255 of *Lecture Notes in Mathematics*, Springer–Verlag, Heidelberg, pp. 149–177.

[LSo80] Lachlan, A. H. , Soare, R. I. 1980. Not every finite lattice is embeddable in the recursively enumerable degrees, *Adv. in Math.* **37**: 74–82.

[LLe] Lempp, S. , Lerman, M. n.d. . A finite lattice without a critical triple that cannot be embedded into the enumerable turing degree, *Ann. Pure Appl. Logic* . To appear.

[Leoi 1994] Leonhardi, S. 1994. *Generalized Nonsplitting in the Recursively Enumerable Degrees*, PhD thesis, University of Wisconsin.

[NSS] Nies, A., Shore, R. A. , Slaman, T. A. n.d. . Definability in the recursively enumerable degrees. In preparation.

[Sho] Shore, R. A. n.d. . The recursively enumerable degrees, *in* E. Griffor (ed.), *The Handbook of Recursion Theory*, North-Holland. To appear.

[SSl91] Shore, R. A. , Slaman, T. A. 1991. Working below a low_2 recursively enumerable degree, *Ann. Pure Appl. Logic* **52**: 1–25.

[Soa87] Soare, R. I. 1987. *Recursively Enumerable Sets and Degrees*, Perspectives in Mathematical Logic, Omega Series, Springer–Verlag, Heidelberg.

[Wei88] Weinstein, B. 1988. *On Embedding of the Lattice 1-3-1 into the Recursively Enumerable Degrees*, PhD thesis, University of California, Berkeley.

Beyond Gödel's Theorem: Turing Nonrigidity Revisited

S. Barry Cooper*

School of Mathematics
University of Leeds
Leeds, LS2 9JT, England
pmt6sbc@@sun.leeds.ac.uk

In [4] it was argued that incomputability in nature dictates a mathematical model (due to Turing [39]) based on mechanical processes relative to appropriate abstractions of incomputable phenomena. (One should refer to that paper for background historical and technical detail.) For instance, it is Turing definability rather than the more familiar notions of provability, and completeness of axiomatic theories, which are more relevant to an analysis of the scope of scientific understanding in the real world; while recent results concerning Turing invariance and nonrigidity have both negative and positive consequences for science as a means to knowledge. Turing nonrigidity (see [6]) may reinforce scepticism about a narrow perspective based on scientific observation, as modelled by Turing computable processes: but the proliferation of invariant substructures of the Turing universe (see, for example, Cooper [5], Nies, Shore and Slaman [27] or Odifreddi [28]) can be viewed as reflecting negatively on the more radical postmodernist and (post-) structuralist views of the roles of culture and language in relation to science (cf. Gross and Levitt [17]) – objective reality does exist. Such comments can be framed in terms of qualifications to the Duhem-Quine thesis ("Any statement can be held true come what may, if we make drastic enough adjustments elsewhere in the system.", Quine [33], p. 43), and have obvious consequences for related empiricist and pragmatist views of the world (cf. again Quine, p. 44, "Physical objects are conceptually imported into the situation as convenient intermediaries – not by definition in terms of experience, but simply as irreducible points comparable, epistemologically, to the gods of Homer.") Of course, the theory itself does indicate difficulties in substantiating the Turing model, but, if not overstretched (viz. the ubiquitous Gödel's [15], [16] Theorem) such asymptotic representations can be useful and productive adjuncts to subjective intuition. For instance, unlike in mathematics where

* Partially supported by E.P.S.R.C. Research Grants nos. GR/J28018, GR/H91213 ('Leeds Recursion Theory Year'), GR/H02165 ('Rigidity and Definability in the Noncomputable Universe') and E.C. Human Capital and Mobility network 'Complexity, Logic and Recursion Theory'.

small variations in axioms can lead to fundamentally different theories, Turing nonrigidity and known countable automorphism bases indicate that although diverse basic assumptions about the real world, related to culture or religion, for example, are inevitable (perhaps even necessary), relative to the Turing model there is a convergence at higher levels of the informational structure suggested by relative rigidity of substructures.

The purpose of this note is to describe how, at a more basic level, the *material* Universe can be modelled according to the underlying structure of its information content. Modern information theory since Shannon [35], linking physical interactions and particular kinds of information transference, provides the background. See [8] for a more detailed discussion. *Some* such mathematical framework is essential to attempts to reduce the ad hoc arbitrariness of cosmological theory by situating it within a logically coherent Universe (see for instance Davies [9], p. 68: "On the philosophical side, there is an urgent need for these speculations to be placed in the context of a theory of mathematics"). In developing such a model, the concept of *immanence* potentially extends not just to the actions of natural laws, but to the process of formation of those laws. (Cf. Penrose's description of his 'strong determinism', [30], pp. 106–107, and his comment that "Like Einstein and his hidden-variable followers, I believe strongly that it is the purpose of physics to provide an *objective* description of reality.")

If one abstracts from the Universe its information content, structured via the basic (computable, at least in the sense of Solomonoff [38]) fundamental laws of nature, one obtains a particular (partial) manifestation of the Turing universe (usually investigated via the Turing degree structure), within which vague questions attain a precise analogue of quite plausible validity. Of course, the usefulness of such a model will depend on a clearer idea of *which* substructure of the Turing universe is being materialised. One can then get theoretical analogues to such questions as:

- How is the Universe capable of determining observed reality?

- How is the observed ambiguity at the quantum level reconciled with classical reality? And how can one explain nonlocal quantum phenomena (such as that that implied by Bell's [1] Theorem and the Einstein-Podolsky-Rosen [10] paradox)?

- Can one add to theoretical projections concerning the Universe under extreme conditions, such as near the Big Bang 'singularity'?

Basic to this is a discussion (see [8]) of how incomputability in nature arises in finite time from ostensibly computable structures, as evidenced by chaotic natural phenomena, graphic examples such as the Julia and Mandelbrot sets of how how incomputability arises mathematically from the iteration of simple rules, and, perhaps most significantly, by the discovery by Pour-El and Richards [31], [32] of a general class of differential equations giving rise to noncomputable solutions from computable initial data. It must be the case that a Universe low in information content, in which the essential character of its laws is defined according to an immanently determined logic, is incapable of defining anything unambiguously. Of course, the Turing model obviously requires some assumption such as the well-foundedness of the hierarchy of subatomic particles, but not necessarily discreteness of time and space.

The theoretical key to the first of the above questions is then the automorphism group of the Turing universe, encapsulating the range and interrelationships of the parallel worlds allowed by the underlying model. Given that information content in nature must first evolve via the quantum level, with the local theory of the Turing model immediately relevant, one needs to know what the global context dictates as regards local invariance, relative rigidity, automorphism bases, and structure of the computably enumerable Turing degrees. It may well be that a deeper understanding of the arcane mysteries of the Turing structure of computably enumerable objects (see, for example, Soare [37] or Lerman [25]) has the potential to dissipate the impression of arbitrariness attaching to the details of subatomic structure, although the possibility that other noncanonical levels of the arithmetical hierarchy may be relevant cannot be excluded.

By Slaman and Woodin [36], it is known that the computably enumerable Turing degrees form an automorphism base for the global structure. The latter has a countably infinite automorphism group (see [6]), providing a theoretical explanation for the observed quantum ambiguity. The form of the noninvariance attached to subatomic individual states suggested by the Heisenberg [21] Uncertainty Principle (but not necessarily to non-unary relations, such as those derived from the weak and the strong nuclear forces) would lead one to expect many local relations, but not singletons, rigid relative to the global model. The former is borne out by Cooper [5], Jockusch and Shore [23], Nerode and Shore [26] and Nies, Shore and Slaman [27], for instance, providing a rich source of subatomic structure, but there appear to be no invariant computably enumerable singletons other than $\mathbf{0}$ and $\mathbf{0}'$. It is worth noting that the relationship between invariance and definability is not well understood, so it may be that there are well-defined elements of the quantum environment which cannot be theoretically captured in a framework derived (inductively or otherwise) from scientific observation.

The consolidation of quantum entities into classically observable objects has an immediate parallel in the actions of the algebraic operations on the Turing degrees and the invariance of $\mathbf{0}'$ (see [5]). The world we observe already encapsulates a high level of concealed information content. It is Slaman and Woodin's proof of the relative rigidity of the Turing universe above $\mathbf{0}''$ which provides the theoretical counterpart of the classically observable universe. The phase transition from quantum ambiguity to an immanently defined reality will involve an accretion of information content concerning the way in which contingencies retrospectively select a consistent invariant history, which will lift the phenomena in question to the cone above $\mathbf{0}''$. It is this process which also provides a theoretical explanation of EPR in terms of the higher order logical structure of the Universe, and hence gives substance to Bohm's (see [2]) concept of globally originating laws of cause and effect, the explanations in terms of decoherence due to Gell-Mann and Hartle [14], Omnès [29] and others, and the orderly retreat from

Laplacian determinism. In this context (providing not just 'explanation' but the 'understanding' looked for in Feynman [12]), the unsatisfactory elements of the various ontological interpretations of quantum theory (including the 'many worlds' interpretation of Everett [11], and its variants), and their disagreements, originating with their attempts (or lack of them) to place classical reality within a universe of quantum uncertainty, are potentially removable. It is worth mentioning here that although the cone above $\mathbf{0}'$ is not an automorphism base (in the proof [6] of nonrigidity, it is relatively straightforward to make the presentation $*$ of the nontrivial automorphism low, so that the automorphism itself will leave unchanged everything above $\mathbf{0}'$), one can construct a Turing automorphism which witnesses the existence of a noninvariant atomic jump class, so that relative rigidity of the cone above $\mathbf{0}'$ fails.

Finally, depending on the extent to which the Turing model is found to be helpful, one may use it to obtain mathematical projections of the behaviour of the Universe near the Hawking and Penrose [19] 'initial singularity'. In doing this, consideration of the extent to which the global theory is realised at the lower levels of the Turing model appropriate to the Universe at different stages of its development becomes crucial. The mathematical singularity within the Big Bang (and any final 'Big Crunch') can only be extrapolated from a context in which gravity and the laws of general relatively derive from an extra-Universal source, whereas in the Turing model such relations derive their authority from the invariance of (first order) arithmetically defined relations on the cone above $\mathbf{0}''$ (Slaman and Woodin [36]). But it is known that at increasingly local levels globally definable relations lose their Turing invariance (for instance, Cooper [5], [4], lowness is defined globally but is noninvariant within the structure of the computably enumerable degrees). The scenario suggested close to the Big Bang, and tacitly accepted by Hawking (see Hartle and Hawking [18]) in more recent work concerning the 'No-Boundary' model, is one in which, travelling backwards in time, there occurs a great homogenisation of information content which is likely to be incapable of uniquely defining the familiar features of the material universe. Not only do quantum effects come into play, they increasingly predominate, and eventually change out of all recognition. This is in a theoretical framework in close agreement with Hawking's assumption of a combination of the many worlds interpretation of quantum theory (as opposed to the Copenhagen Interpretation) and (see [13]) Feynman's 'sum-over-histories' as the preferred approach to a quantum description of the early Universe. Information content is not necessarily destroyed, but is increasingly randomised, necessitating appropriate theoretical counterparts. (Note, Kucera [24], that the 2-random Turing degrees avoid the cone above any nonzero degree below $\mathbf{0}'$.) The fact that the random degrees form an automorphism base for the Turing universe suggests that despite the huge apparent chaos near the Big Bang, and lack of structured information content, the latter may still be sufficient to deterministically decide the form of the Universe at other points in time. Hence, assuming Penrose's *cosmic censorship conjecture* ('Nature abhors a naked singularity', [20], p. 21), one obtains a theoretical framework capable of reinstating meaningful cyclic models of the Universe, in which the force of any rebound associated with the 'singularities' is directly related to the proximity of the dissolution of invariant gravity (presumably not within the Planck interval of 10^{-43} seconds), which in turn depends on the size of the Universe and hence how much information content is involved. [2]

The resulting answer to the question: "If the laws of physics could break down at the beginning of the universe, why couldn't they break down anywhere?" ([20], p. 76), while not being precisely that envisaged by Hawking himself in his article, is a particularly satisfying one.

References

1. J.S. Bell. On the Einstein-Podolsky-Rosen paradox. Physics **1** (1964), 195–200; reprinted in J.S. Bell, *Speakable and Unspeakable in Quantum Mechanics*, Cambridge University Press, 1987, pp. 14–21.
2. D. Bohm and B.J. Hiley. *The Undivided Universe: An ontological interpretation of quantum theory.* Routledge, London, New York, 1993.
3. S.B. Cooper. Definability and global degree theory. In J. Oikkonen and J. Väänänen, eds.,*Logic Colloquium '90*, Lecture Notes in Logic vol. 2, Springer-Verlag, Berlin, Heidelberg, New York, pp. 25–45.
4. S.B. Cooper. Beyond Gödel's Theorem: The failure to capture information content. In A. Sorbi, editor, *Complexity, Logic and Recursion Theory*, Lecture Notes in Pure and Applied Mathematics, vol. 187, Marcel Dekker, 1997, pp. 93–122.
5. S.B. Cooper. On a conjecture of Kleene and Post. To appear.
6. S.B. Cooper. The Turing universe is not rigid. To appear.
7. S.B. Cooper. Observation, understanding and Turing definability. In preparation.
8. S.B. Cooper. Turing nonrigidity and quantum theory. In preparation.
9. P.C.W. Davies. Why is the physical world so comprehensible? In W.H. Zurek, ed., *Complexity, Entropy, and the Physics of Information*, Santa Fe Inst. Studies in the Sciences of Complexity, vol. 8, Addison-Wesley, Reading, Mass., 1990, pp. 61–70.
10. A. Einstein, B. Podolsky and N. Rosen. Phys. Rev., **47** (1935), 777–780.
11. H.Everett, III. "Relative state" formulation of quantum mechanics. Rev. Mod. Phys., **29** (1957), 454–462.
12. R.P. Feynman, R.B. Leighton, M. Sands. *The Feynman Lectures on Physics, Vol. III.* Addison-Wesley, Reading, Mass., 1965.
13. R.P. Feynman and A.R. Hibbs. *Quantum Mechanics and Path Integrals.* McGraw-Hill, New York, London, Sydney, 1965.
14. M. Gell-Mann and J.B. Hartle. Quantum mechanics in the light of quantum cosmology. In W.H. Zurek, ed., *Complexity, Entropy, and the Physics of Information*, Santa Fe Inst. Studies in the Sciences of Complexity, vol. 8, Addison-Wesley, Reading, Mass., 1990, pp. 425–458.

[2] The mystery of what lies at the base of this hierarchy of Turing computable and Turing definable relations remains. But strange attractors (see, for example [22]) provide concrete examples of Turing definability arising from simple computational relationships. There is no surprise in this primitive emergence of definability to one familiar with the role of Sacks' [34] Splitting Theorem in the proof (see [3]) of Turing definability of the various levels of the arithmetical hierarchy.

15. K. Gödel. Über formal unentscheidbare Sätze der Principia Mathematica und verwandter Systeme I. Monatsh. Math. Phys., **38** (1931), 173–198.
16. K. Gödel. On undecidable propositions of formal mathematical systems. Mimeographed notes, 1934; reprinted in M. Davis, editor, *The Undecidable. Basic Papers on Undecidable Propositions, Unsolvable Problems, and Computable Functions*, Raven Press, New York, 1965, pp. 39–71.
17. P.R. Gross and N. Levitt. *Higher Superstition: The academic left and its quarrels with science.* John Hopkins University Press, 1994.
18. J.B. Hartle and S.W. Hawking. Wave function of the universe. Phys. Rev., **D28** (1983), 2960–2975.
19. S.W. Hawking and R. Penrose. The singularities of gravitational collapse and cosmology. Proc. Roy. Soc. London, **A314** (1970), 529–48.
20. S.W. Hawking and R. Penrose. *The Nature of Space and Time.* Princeton University Press, Princeton, N.J., 1996.
21. W. Heisenberg. Über den anschaulichen Inhalt der quantentheoretischen Kinematik und Mechanik. Z. Phys., **43** (1927), 172–198.
22. M. Hénon and Y. Pomeau. Two strange attractors with a simple structure. In R. Temam, ed., *Turbulence and Navier Stokes Equations*, Lecture Notes in Mathematics, vol. 565, Springer-Verlag, Berlin, Heidelberg, New York, 1976, pp. 29–68.
23. C.G. Jockusch, Jr. and R.A. Shore. Pseudo jump operators II: Transfinite iterations, hierarchies, and minimal covers. J. Symbolic Logic, **49** (1984), 1205–1236.
24. A. Kučera. Randomness and generalizations of fixed point free functions. In K. Ambos-Spies, G. Müller and G.E. Sacks, eds., *Recursion Theory Week, Proceedings Oberwolfach 1989*, Springer, Berlin, 1990, pp. 245–254.
25. M. Lerman. Embedding partial lattices into the computably enumerable degrees. To appear.
26. A. Nerode and R.A. Shore. Reducibility orderings: theories, definability and automorphisms. Ann. MathLogic, **18** (1980), 61–89.
27. A. Nies, R.A. Shore and T.A. Slaman. Definability in the recursively enumerable degrees. To appear.
28. P. Odifreddi. *Classical Recursion Theory.* North-Holland, Amsterdam, New York, Oxford, 1989.
29. R. Omnès. *The Interpretation of Quantum Mechanics.* Princeton University Press, Princeton, N.J., 1994.
30. R. Penrose. Quantum physics and conscious thought. In B.J. Hiley and F.D. Peat, editors, *Quantum Implications: Essays in honour of David Bohm*, Routledge & Kegan Paul, London, New York, 1987, pp. 105–120.
31. M.B. Pour-El and I. Richards. The wave equation with computable initial data such that its unique solution is not computable. Advances in Math., **39** (1981), 215–239.
32. M.B. Pour-El and J.I. Richards. *Computability in Analysis and Physics.* Springer-Verlag, Berlin, Heidelberg, New York, London, Paris, Tokyo, 1989.
33. W.V.O. Quine. Two dogmas of empiricism. In W.V.O. Quine, *From A Logical Point Of View*, Harvard University Press, Cambridge, Mass. and London, 1953, pp. 20–46.
34. G.E. Sacks. On the degrees less than $\mathbf{0}'$. Ann. of Math. (2) **77** (1963), 211–231.
35. C.E. Shannon and W. Weaver. *The mathematical theory of communication.* University of Illinois Press, 1949.
36. T.A. Slaman and W.H. Woodin. *Definability in Degree Structures.* To appear.

37. R.I. Soare *Recursively Enumerable Sets and Degrees*. Springer-Verlag, Berlin, Heidelberg, London, New York, 1987.
38. R.J. Solomonoff. A formal theory of inductive inference. Part I. Infor. and Control **7** (1964), 1–22.
39. A.M. Turing. Systems of logic based on ordinals. Proc. London Math. Soc., **45** (1939), 161–228; reprinted in M. Davis, editor, *'The Undecidable. Basic Papers on Undecidable Propositions, Unsolvable Problems, and Computable Functions'*, Raven Press, New York, 1965, pp. 154–222.

Types and Indiscernibles in Finite Models

Anuj Dawar

Department of Computer Science,
University of Wales Swansea,
Singleton Park, Swansea SA2 8PP, England
a.dawar@swansea.ac.uk

Abstract. We consider L^k — first order logic restricted to k variables, and interpreted in finite structures. The study of classes of finite structures axiomatisable with finitely many variables has assumed importance through connections with computational complexity. In particular, we investigate the relationship between the size of a finite structure and the number of distinct types it realizes, with respect to L^k. Some open questions, formulated as finitary Löwenheim-Skolem properties, are presented regarding this relationship. This is also investigated through finitary versions of an Ehrenfeucht-Mostowski property.

1 Introduction

In this paper, we are concerned only with finite structures. That is, our logical formulas are interpreted in relational structures with finite domain. Interest in finite model theory has largely grown from the fact that there is a close connection between definability of classes of finite structures and their computational complexity. For instance, Fagin [10] showed that a class of finite structures is definable in existential second order logic if, and only if, it is decidable by a non-deterministic Turing machine in polynomial time (i.e. in the class NP). Similarly, many naturally arising complexity classes have been characterised as definability classes in appropriate logics.

These results raised the hope that model theoretic methods could be deployed to attack some of the notoriously open problems in complexity theory. Unfortunately, most methods and results developed in model theory fail to work when only finite structures are considered (see [12]). For instance, the compactness theorem for first order logic trivially fails, as do most of its consequences. Indeed, as we make the transition from arbitrary structures to finite structures, first order logic loses its central role. Two significant reasons can be discerned for this: on the one hand, first order logic is too strong; and on the other hand, first order logic is too weak.

First order logic is too strong in the sense that for every finite structure $\mathfrak{A}$, there is a first order sentence $\varphi_{\mathfrak{A}}$ that describes it up to isomorphism: that is, for any structure $\mathfrak{B}$, if $\mathfrak{B} \models \varphi_{\mathfrak{A}}$, then $\mathfrak{B}$ is isomorphic to $\mathfrak{A}$. This means that the relation $\equiv$ of elementary equivalence is trivial on finite structures — it coincides with the isomorphism relation. Since a large part of model theory can arguably be described as the study of the structure of the elementary equivalence relation and its variants, this raises a serious question about the applicability of model theoretic methods.

First order logic is too weak in the sense that the class of finite models of any first order sentence is of very low computational complexity. It can, in fact, be recognised by a deterministic Turing machine using only a logarithmic amount of workspace (in other words, it is in the complexity class DSPACE[$\log n$]). Moreover, there are easily computed properties in DSPACE[$\log n$] that are not first order definable.

The disparity in the above two statements can be accounted for by contrasting the expressive power of single sentences with that of theories, i.e. sets of sentences. In model theory, we are most often concerned with the limits on the expressive power of sets of sentences. Clearly, two elementarily equivalent structures cannot be distinguished by any such set. However, when we only consider finite structures, *any* isomorphism closed class S of structures is defined by the countable set of sentences[1] :

$$\Sigma = \{\neg\varphi_{\mathfrak{A}} \mid \mathfrak{A} \notin S\},$$

where $\varphi_{\mathfrak{A}}$ is as above. On the other hand, as already mentioned, the expressive power of first order sentences is extremely weak. Another way of stating the same thing is that, while every class of structures is *axiomatisable*, few interesting classes are *finitely axiomatisable.*

One conclusion one can draw from the above is that in order to analyse the expressive power of first order logic (i.e. of first order sentences) and, indeed, of more powerful logics, we need equivalence relations that are coarser than the relation of elementary equivalence. There are two main approaches to coarsening the elementary equivalence relation that have proved useful. The first is to stratify the relation according to quantifier rank. This is useful in analysing first order definability. Indeed, it can be shown that a class of structures is finitely axiomatisable if, and only if, it is closed under the elementary equivalence relation of a fixed quantifier rank.[2] The other approach is provided by a restriction on the number of variables considered in defining the equivalence relation. This way we obtain a notion of axiomatisability intermediate between finite and full axiomatisiability — namely the notion of a class of structures axiomatised with a finite number of variables. This turns out to be a valuable approach in studying several natural extensions of first order logic, particularly extensions by inductive and fixed point operators. For several such extensions, it can be shown that any definable class of structures is, in fact, axiomatisable with a finite number of variables. Moreover, it has been shown that many of the difficult open questions of computational complexity can be recast as questions on the expressive power of these logics.

In the rest of this paper, a signature σ is a finite relational signature, i.e. it is a finite sequence of relation symbols $(R_1, \ldots, R_m)$ and associated arities $(r_1, \ldots, r_m)$, where $r_i \in \omega$. A σ-structure $\mathfrak{A}$ is an interpretation of this signature

[1] Here, to say that Σ defines S means that S is exactly the class of finite models of Σ.

[2] Provided that the vocabulary is purely relational, i.e. it contains no function symbols.

in a finite domain A, i.e. $\mathfrak{A} = (A, R_1^{\mathfrak{A}}, \ldots, R_m^{\mathfrak{A}})$. The *width* of σ, denoted width(σ) is defined as $\max(r_1, \ldots, r_m)$. The domain A of a structure $\mathfrak{A}$ is often denoted $|\mathfrak{A}|$, while the cardinality of a set S is denoted card(S). For a sentence φ (or a set of sentences Σ), Mod(φ) (or Mod(Σ)) denotes the set of finite models of φ (or Σ).

2 Finite Variable Logics

Define L^k to be the fragment of first order logic in which we only use the variables $x_1, \ldots, x_k$. For any sentence φ of L^k, we denote by $qr(\varphi)$ the *quantifier rank* of φ, i.e. the depth of nesting of quantifiers in φ.

The following definition introduces some basic notation used throughout the rest of this paper.

Definition 1. For two structures $\mathfrak{A}$ and $\mathfrak{B}$,

- we write
$$\mathfrak{A} \equiv^k \mathfrak{B}$$
if the two structures agree on all sentences of L^k.
- If s is a tuple of elements in $\mathfrak{A}$ and t is a tuple of elements in $\mathfrak{B}$, with length(s) = length(t) $\leq k$, by abuse of notation, we also write
$$(\mathfrak{A}, s) \equiv^k (\mathfrak{B}, t)$$
to denote that for any formula φ of L^k,
$$\mathfrak{A} \models \varphi[s] \quad \text{if, and only if,} \quad \mathfrak{B} \models \varphi[t].$$
- k-size($\mathfrak{A}$) is the index (i.e. the number of equivalence classes) of the equivalence relation $\equiv^k$ on the set $\{(\mathfrak{A}, s) \mid s \in |\mathfrak{A}|^k\}$.

It can be shown that, if $\mathfrak{A}$ is a finite structure and $\mathfrak{A} \equiv^k \mathfrak{B}$, then
$$k\text{-size}(\mathfrak{A}) = k\text{-size}(\mathfrak{B})$$
(see, for instance, [8]).

The equivalence relation $\equiv^k$ has a natural characterisation in terms of two player pebble games in the style of Ehrenfeucht-Fraïssé games. This characterisation is essentially due to Barwise [4] (see also [13, 18]).

The game board consists of two structures $\mathfrak{A}$ and $\mathfrak{B}$ and a supply of k pairs of pebbles $(a_i, b_i), 1 \leq i \leq k$. The pebbles $a_1, \ldots, a_l$ are initially placed on the elements of an l-tuple s in $\mathfrak{A}$, and the pebbles $b_1, \ldots, b_l$ on a tuple t in $\mathfrak{B}$. There are two players, Spoiler and Duplicator. At each move of the game, Spoiler picks up a pebble (either an unused pebble or one that is already on the board) and places it on an element of the corresponding structure. For instance he might take pebble b_i and place it on an element of $\mathfrak{B}$. Duplicator must respond by placing the other pebble of the pair in the other structure. In the above example, she must place a_i on an element of $\mathfrak{A}$. If at the end of the move the partial map

$f : \mathfrak{A} \to \mathfrak{B}$ given by $a_i \mapsto b_i$ is not a partial isomorphism, then Spoiler has won the game, otherwise it can continue for another move. Duplicator has a strategy to avoid losing for q moves, starting with the initial position $(\mathfrak{A}, s)$ and $(\mathfrak{B}, t)$ if, and only if, $(\mathfrak{A}, s)$ and $(\mathfrak{B}, t)$ cannot be distinguished by any formula of L^k of quantifier rank q or less. Hence, if Duplicator has a strategy to play the game indefinitely without losing, then $(\mathfrak{A}, s) \equiv^k (\mathfrak{B}, t)$.

The following fact regarding the relations $\equiv^k$ was established in [8]:

Theorem 2. *For every finite structure $\mathfrak{A}$ and tuple s of elements of $\mathfrak{A}$, with* $\text{length}(s) \leq k$, *there is a formula φ of L^k such that $\mathfrak{B} \models \varphi[t]$ if, and only if,* $(\mathfrak{A}, s) \equiv^k (\mathfrak{B}, t)$. *Moreover,* $qr(\varphi) \leq k\text{-size}(\mathfrak{A}) + k$.

In particular, it follows from Theorem 2 that the set of L^k sentences true in any finite model is, in fact, finitely axiomatisable within L^k.

Another useful notion is that of the L^k-type of a tuple in a structure. The following definition is from [8].

Definition 3. Let s be an l-tuple of elements from a structure $\mathfrak{A}$ for $l \leq k$. The *L^k-type* of s in $\mathfrak{A}$, denoted $\text{Type}_k(\mathfrak{A}, s)$, is the set of formulas φ of L^k with free variables among $x_1, \ldots, x_l$, such that $\mathfrak{A} \models \varphi[s]$.

Thus, $(\mathfrak{A}, s) \equiv^k (\mathfrak{B}, t)$ can be seen as shorthand for $\text{Type}_k(\mathfrak{A}, s) = \text{Type}_k(\mathfrak{B}, t)$, and Theorem 2 states that every L^k-type is completely determined by one of its elements.

The interest in finite variable logics stems largely from the fact that a number of logics that have been considered in finite model theory — particularly in the characterisation of computational complexity classes — have the property that the definable classes of structures can all be axiomatised with a finite number of variables.

Thus, for instance, LFP is the closure of first order logic under an operation for forming the least fixed points of positive formulas. Similarly, PFP closes first order logic under an operation forming the iterative fixed points of arbitrary formulas [2]. It was independently proved by Immerman and Vardi [14, 20] that LFP exactly characterises polynomial time complexity (the class PTIME) on structures equipped with a linear order, while Abiteboul and Vianu [2] show that PFP similarly characterises polynomial space complexity (the class PSPACE) on ordered structures.

By results of Kolaitis and Vardi [15] and Dawar *et al.* [8], we know the following fact:

> For any sentence φ of LFP (or PFP), there is a k and a set Σ of sentences of L^k such that $\text{Mod}(\varphi) = \text{Mod}(\Sigma)$.[3]

The significance of this fact lies in the observation (see [15]) that most results showing that some property cannot be expressed in a logic such as LFP or PFP,

[3] Kolaitis and Vardi [15] show that each such φ is equivalent to a sentence of the infinitary logic $L^{\omega}_{\infty\omega}$, while in [8], it is shown that every sentence of $L^{k}_{\infty\omega}$ is equivalent to a single countable conjunction of sentences of L^k.

actually show that the property cannot be axiomatised in L^k, for any k. The essential tool for establishing this is the pebble game defined above.

What's more, Abiteboul and Vianu [3] show that, even without the restriction to ordered structures,

$$\text{LFP} = \text{PFP} \quad \text{if, and only if,} \quad \text{PTIME} = \text{PSPACE},$$

where, equality for the logics means that they have the same expressive power.

Recent work has shown that this result of Abiteboul and Vianu can be extended to a variety of complexity classes. That is, for a large number of such classes, logics have been found that characterise the class on ordered structures and for which the expressive power is bounded by finite variable axiomatisability (see [1, 6, 7]). Moreover, the inclusion relations between these logics exactly mirror the inclusions among the corresponding complexity classes. Abiteboul *et al.* [1] have shown that definability in these logics corresponds to complexity on a relational model of computation, in which complexity is measured not as a function of the size of the input, but as a function of the k-size of the input.

Translations of complexity theoretic questions into the framework of logics with limited discerning power has not (as yet, anyway) thrown new light on the separation of complexity classes. One reason for this was hinted at above, where it was stated that the main tool for establishing the inexpressibility of a class of structures in a logic such as LFP is the pebble game for L^k, which is used to show that the class is not closed under the relation $\equiv^k$ for any k. But then, the class is not definable in any of the logics we are considering, and this tool is not directly useful in separating the logics in question.

Nevertheless, some finer separation results have been obtained. For instance, in [8] it is shown that there is a polynomial time decidable property that is closed under $\equiv^k$ for a fixed k, and is still not definable in LFP and in [7] two different ways of restricting implicit first order definitions to finite variable axiomatisibility are separated. Both of these results rely on the construction of classes of structures in which, though the k-size is unbounded, it grows much more slowly than the cardinality of the structures themselves. This suggests that a greater understanding of the relationship between the size of a structure and its k-size could be very useful (see [5] for a discussion of some of these issues). Some questions regarding this relationship are investigated in the next section.

3 Finitary Löwenheim-Skolem Properties

We begin this section with a few straightforward remarks about the relationship between the size of a structure and its k-size. First of all, it is immediate from the definition that, if $\text{card}(|\mathfrak{A}|) = n$, then $k\text{-size}(\mathfrak{A}) \leq n^k$, since there are at most n^k tuples in $\mathfrak{A}$. This bound is actually achieved in the class of linear orders, for all $k \geq 2$.

If we turn the question around and ask how large a structure can we find with a given k-size, at first things appear equally simple. It is trivial to construct arbitrarily large such structures. Thus, if we consider the class of structures in the empty signature (sometimes called pure sets), it is clear that the k-size of structures in this class is bounded by a constant, and yet there are arbitrarily large structures in this class. On closer examination, we find that this class splits into only finitely many $\equiv^k$ equivalence classes of structures (indeed, this is necessarily true of any class of structures in which the k-size of structures is bounded — see [9]). So, the existence of arbitrarily large structures of a fixed k-size follows simply from the existence of infinite $\equiv^k$ equivalence classes. The more interesting question is: how large can the *smallest* or *largest* structure in a $\equiv^k$ equivalence class be, with respect to its k-size (recall that the k-size is invariant among structures in a given $\equiv^k$ equivalence class)? Are there any recursive bounds on either of these? This leads us to formulate the following two open questions, which can be seen as downward and upward Löwenheim-Skolem properties for finite structures (over some fixed signature σ, with

$$2 \leq \text{width}(\sigma) \leq k$$

— the case of unary signatures is straightforward).

Open Question 1 Is there a recursive function f_k, for every k, such that, for every finite structure $\mathfrak{A}$ with k-size$(\mathfrak{A}) \leq n$, there is a $\mathfrak{B}$ such that

$$\text{card}(|\mathfrak{B}|) \leq f_k(n) \text{ and } \mathfrak{A} \equiv^k \mathfrak{B}?$$

Open Question 2 Is there a recursive function g_k for every k, such that, if $\mathfrak{B}$ is a finite structure with k-size$(\mathfrak{B}) \leq n$ and card$(|\mathfrak{B}|) \geq g_k(n)$, then there are arbitrarily large finite $\mathfrak{A}$ such that $\mathfrak{A} \equiv^k \mathfrak{B}$?

We first observe that if we drop the word recursive from the above questions, then the answer is yes in both cases. This is based on the fact that for any fixed q, there are only finitely many sentences φ of L^k with $qr(\varphi) \leq q$.[4] Thus, for any n, the relation $\equiv^k$ has finite index on the class of structures $\mathfrak{A}$ such that k-size$(\mathfrak{A}) \leq n$ (by virtue of Theorem 2). We can then define the functions $d_k, e_k : \mathbf{N} \to \mathbf{N}$ by:

$$d_k(n) = \max\{\text{card}(\mathfrak{A}) \mid k\text{-size}(\mathfrak{A}) \leq n \text{ and } \mathfrak{A} \text{ is smallest in its } \equiv^k \text{ class.}\}$$

$$e_k(n) = \max\{\text{card}(\mathfrak{A}) \mid k\text{-size}(\mathfrak{A}) \leq n \text{ and } \mathfrak{A} \text{ belongs to a finite } \equiv^k \text{ class.}\}$$

Here, by finite $\equiv^k$ class, we mean a class that contains finitely many structures up to isomorphism. The functions d_k and e_k are well defined, because they involve taking, for each n, the maximum of a finite set of cardinalities.

The Open Questions 1 and 2 posed above can now be seen as asking if there are recursive upper bounds on the functions d_k and e_k. Moreover, since the relation $\equiv^k$ is decidable, if such upper bounds exist, then the functions d_k and e_k are themselves recursive.

It is also a worthwhile line of investigation to establish lower bounds on the functions d_k and e_k. At present, the best known lower bounds on these functions are exponential, as described in the next section. The following open question can thus be added to the two mentioned above:

---

[4] Once again, we use the fact that our signatures are purely relational.

Open Question 3 Are there super-exponential lower bounds on the functions d_k and e_k, for any k?

By super-exponential, we mean of the order of $2^{f(n)}$ for some super-polynomial function f. Even though the exponential lower bounds established for trees (and discussed in the next section) are of the form $2^{g(n)}$, where the function g is $O(n)$, there are constructions known that raise this to at least 2^{n^2}.

It should also be pointed out that the case of $k = 2$ is solved, as linear lower and upper bounds are essentially established in [17]. Thus, we adopt the following proviso for the rest of the paper:

Wherever k refers to the number of variables, we assume that $k \geq 3$.

4 Trees

The exponential lower bounds on the functions d_k and e_k can be established by considering the class of complete binary trees. We introduce this class through a series of definitions.

Definition 4. – A tree T is a structure $(V, \preceq)$ where,

- $\preceq$ is a partial order on the set of vertices V;
- there is a unique $\preceq$-minimal element of V, called the *root*; and
- every element $v \in V$ (other than the root) has a unique $\preceq$-predecessor, called the *parent* of v.

– Two vertices are *siblings* if they have the same parent. If u is the parent of v, we also call v a *child* of u.
– The *subtree* rooted at a vertex v is the substructure of T formed by the set $\{u \in V \mid v \preceq u\}$.
– The *height* of a tree is the length of its longest maximal chain.
– For any $c \in \omega$, a *c-ary tree* is a tree in which every vertex has at most c $\preceq$-successors.
– A *complete c-ary tree* is one in which every vertex has either 0 or c successors, and all maximal chains are of equal length.

Now, a lower bound on the functions d_k and e_k is obtained from the following three facts. Let T_h denote the (unique up to isomorphism) complete binary tree of height h.

1. T_h is characterised up to isomorphism by a sentence of L^3.
2. $\text{card}(T_h) = 2^h - 1$.
3. For any k, there is a k' such that $k\text{-size}(T_h) \leq h^{k'}$, for all h.

For details of the above, the reader is referred to [8]. Here, we only note that (1) above is based on the following lemma (stated here without proof, details can be found in [5], for instance), which will be of use later:

Lemma 5. *For every $d \in \omega$, there is a formula δ_d of L^2 such that, in any partial order $P = (V, \preceq)$, $P \models \delta_d[u]$ if, and only if, the length of the maximal chain below u is d.*

A natural first attempt to improve on the exponential lower bound is to consider trees with higher than binary branching. Clearly, facts analogous to the above hold for c-ary trees for any constant c, i.e. each complete c-ary tree is described up to isomorphism by a sentence with no more than $\max(c+1, 3)$ variables; the size of the complete c-ary tree of height h is exponential in h, but the k-size is polynomial in h. This, of course, does not improve on the exponential lower bound.

Another approach is to consider trees of uniform arity, whose arity is a function of their height (this approach is considered in [1]). However, this also fails as we can no longer describe such a tree up to isomorphism with a fixed (independent of the height) number of variables. In particular, if T is a complete c-ary tree, then

1. there is a T' such that $T \equiv^k T'$ and the arity of T' is at most k; and
2. if $c > k$, there are infinitely many non-isomorphic T'' such that $T \equiv^k T''$.

The above facts are consequences of the following general property of trees:

Lemma 6. *Let $v_1, \ldots, v_{k+1}$ be siblings in a tree T, let $T_1, \ldots, T_{k+1}$ be the respective subtrees rooted at these nodes, and let v be their common parent. If $T_i \equiv^k T_j$ for $1 \leq i, j \leq k+1$, then*

1. *$T \equiv^k T'$, where T' is obtained from T be removing T_{k+1}.*
2. *$T \equiv^k T''$, where T'' is obtained from T by adding a new subtree $T_{k+2} \equiv^k T_{k+1}$ above v.*

Proof: The proof is a straightforward pebble game. We describe Duplicator's winning strategy in the game played on T and T'. The game on T and T'' is analogous.

First of all, if Spoiler plays anywhere (in T or T') other than in the subtrees $T_1, \ldots, T_{k+1}$, Duplicator simply plays on the same node in the other tree. If Spoiler plays in one of the trees T_i (in either T or T') which does not contain a pebble, Duplicator responds by playing in a subtree T_j of the other tree not containing a pebble (not counting the pebble being currently moved). Such a subtree must exist, because there are at least k matching subtrees and only $k-1$ pebbles on the board. The vertex played by Duplicator is determined by her winning strategy in the k pebble game on structures T_i and T_j, given by the fact that $T_i \equiv^k T_j$. If Spoiler plays a pebble in a subtree T_i which already contains a pebble, Duplicator responds with a pebble in the subtree T_j containing the matching pebble. Again, she uses the winning strategy for the game on T_i and T_j to guarantee that the result is a legal position. It is not difficult to verify that this describes a winning strategy for Duplicator on T and T'. ■

It is possible to use Lemma 6 to show an exponential *upper* bound on the functions d_k and e_k on the class of trees. Thus, the lower bound based on complete binary trees cannot be improved as long as we confine ourselves to trees.

In order to prove this upper bound, it is useful to establish some more basic properties of the $\equiv^k$ relation on trees.

Lemma 7. *For any tree $T = (V, \preceq)$, any vertices $u, v \in V$ and any $k \geq 3$, if*

$$(T, u) \equiv^k (T, v)$$

then the following two conditions hold:

1. *$(T, u') \equiv^k (T, v')$, where u' and v' are the parents of u and v respectively; and*
2. *$T_u \equiv^k T_v$, where T_u and T_v are the subtrees of T rooted at u and v respectively.*

Proof: Since $(T, u) \equiv^k (T, v)$, Duplicator has a winning strategy in the k pebble game played on two copies of T that begins with a pebble on u in one copy and on v in the other. In the following we will refer to these two copies as T^1 and T^2 respectively. We will also use superscripts 1 and 2 to distinguish vertices and subtrees of T^1 from those of T^2.

If Spoiler's first move is to place a pebble on u' (in T^1), then Duplicator must respond with a pebble on v' (in T^2), as placing a pebble on any other vertex $v'' \prec v^2$ would allow Spoiler to pebble a vertex in between v^2 and v'' (for instance, v') resulting in a losing position for Duplicator. Now, from the assumption that Duplicator has a winning strategy, we conclude that Duplicator wins the game from the position (T^1, u, u'), (T^2, v, v') and this implies $(T, u') \equiv^k (T, v')$, establishing condition 1.

To establish condition 2, we argue that if, in the game starting from (T, u) and (T, v), Spoiler confines his moves to the subtrees T_u in T^1 and to T_v in T^2, then Duplicator has a winning strategy also confining her moves to these subtrees. Suppose this were not so and, for some move by Spoiler within the tree T_u^1 (without loss of generality), Duplicator is forced to respond with a vertex outside T_v^2. Note that, after the first such move by Duplicator, T^2 contains pebbles both within and outside T_v^2, since there was a pebble on v initially, while all pebbles in T^1 are within T_u. Spoiler now places a pebble on u^1, using a previously unused pebble if there are only two pairs of pebbles currently on the board, or possibly moving a pebble otherwise (though not moving the pebble used in the previous move). In either case, there remain pebbles within and outside T_v^2, not counting the pebble that Duplicator must now place. Duplicator must respond with a vertex v'' in T^2 that is a lower bound (in the partial order $\preceq$) of all the currently pebbled vertices in T^2. Since these pebbled vertices include vertices both within and outside of T_v^2, it must be the case that $v'' \prec v$. This means that u and v'' can be distinguished by a formula δ_d as in Lemma 5, giving Spoiler a winning strategy from this point on. ■

In fact, the converse of Lemma 7 holds as well, but we do not require it for our purposes. One immediate consequence of Lemma 7 that is of interest to us is the following:

Lemma 8. *If $(T, u) \not\equiv^k (T, v)$ and $u \preceq u'$, then there is no v', such that $v \preceq v'$ and $(T, u') \equiv^k (T, v')$.*

In other words, if $(T, u) \not\equiv^k (T, v)$, then the subtrees T_u and T_v realize completely disjoint sets of L^k-types in T.

Also, combining Lemma 6 with Lemma 7 gives us the following:

Lemma 9. *For every tree T there is a tree T', with $T \equiv^k T'$, such that there is no set $v_1, \ldots, v_{k+1}$ of distinct sibling nodes in T' with $(T', v_i) \equiv^k (T', v_j)$ for $1 \leq i, j \leq k+1$.*

Moreover, if T contains a set $v_1, \ldots, v_k$ of distinct sibling nodes such that $(T, v_i) \equiv^k (T, v_j)$ for $1 \leq i, j \leq k$, then there are arbitrarily large T'' such that $T \equiv^k T''$.

In addition, using Lemma 8 we can establish bounds on the size of the tree T' of Lemma 9, as in the following.

Lemma 10. *If T does not contain a set $v_1, \ldots, v_{k+1}$ of distinct sibling nodes such that $(T, v_i) \equiv^k (T, v_j)$ for $1 \leq i, j \leq k+1$ and k-size$(T) \leq n$, then* $\mathrm{card}(|T|) \leq k^n$.

Proof: We first note that, by virtue of Lemma 5, k-size(T) is at least as large as the height of T. Now, we establish the claim by induction on the height of the tree. The basis is trivial. For the induction step, we group the children of the root of T into equivalence classes according to their L^k-type in T. By the hypothesis of the lemma, there are at most k elements in each equivalence class. Let us say there are e equivalence classes: $E_1, \ldots, E_e$. Associate with each E_i, the number n_i of distinct L^k-types that are realized in the subtree rooted at one of the elements of E_i (this is, by Lemma 7, the same for all the elements of E_i).

By the induction hypothesis, the total number of elements in all the subtrees rooted at elements in E_i is at most $k \cdot k^{n_i}$, and thus

$$\mathrm{card}(|T|) \leq 1 + k \sum_{1 \leq i \leq e} k^{n_i}.$$

Since $k \geq 3$, for each i, $k^{n_i} \geq 3$, and therefore $\sum_{1 \leq i \leq e} k^{n_i} < k^{\sum_{1 \leq i \leq e} n_i}$. But, by Lemma 8, $\sum_{1 \leq i \leq e} n_i < n$, and we conclude that $\mathrm{card}(|T|) \leq k^n$. ■

Note that, in the proof of Lemma 10, it suffices to count the L^k-types of individual elements. The k-size would be larger than this number, and the bound obtained could be made tighter. However, we are only interested in establishing the exponential order of the bound, and the simple argument given above serves that purpose.

Now, Lemmas 9 and 10 immediately yield the upper bounds on the functions d_k and e_k on the class of trees:

Theorem 11. *If T is a tree with k-size$(T) \leq n$, then there is a tree T' with* $\mathrm{card}(|T'|) \leq k^n$ *such that $T \equiv^k T'$. Moreover, if* $\mathrm{card}(|T|) > k^n$, *then there are arbitrarily large T'' such that $T \equiv^k T''$.*

The question arises as to whether any of the techniques that work for trees can be generalised to other classes of structures, and perhaps even to obtain a positive resolution of Open Questions 1 and 2. We outline one such attempt at generalisation in the next section, though it is one that fails to resolve the questions, in the general case.

5 Indiscernibles

The following definitions introduce notions of sets and sequences of indiscernibles for finite variable logics, that are analogous to the standard definitions from model theory.

First, recall the following definition:

Definition 12. Given a structure $\mathfrak{A}$ and a sequence of elements $s = (a_1, \ldots, a_k)$ from $\mathfrak{A}$, the *basic equality type* of $(\mathfrak{A}, s)$ is the formula

$$\bigwedge_{(i,j)\in S} (x_i = x_j) \wedge \bigwedge_{(i,j)\in T} \neg(x_i = x_j),$$

where $S = \{(i,j) \mid i < j \text{ and } a_i = a_j\}$, and $T = \{(i,j) \mid i < j \text{ and } a_i \neq a_j\}$.

That is, the basic equality type of a tuple is its complete quantifier free description in the language of identity.

Now we can introduce the definitions of indiscernible sets and sequences.

Definition 13. A set $X \subseteq |\mathfrak{A}|$ is a *k-indiscernible set* in $\mathfrak{A}$, if whenever $s, t \in X^k$ are such that s and t have the same basic equality type, then $(\mathfrak{A}, s) \equiv^k (\mathfrak{A}, t)$.

$X \subseteq |\mathfrak{A}|$ is a *k-indiscernible sequence* in $\mathfrak{A}$, if there is some (not necessarily definable) linear order on X such that for any order increasing $s, t \in X^l$, $l \leq k$, $(\mathfrak{A}, s) \equiv^k (\mathfrak{A}, t)$.

Clearly, any k-indiscernible set X is a k-indiscernible sequence, as any order on X would satisfy the conditions of the definition. However, the converse may fail.

The next definition is also a straightforward analogue, with finitely many variables, of a standard definition from model theory.

Definition 14. A structure $\mathfrak{B}$ is a *k-elementary substructure* of $\mathfrak{A}$, denoted $\mathfrak{B} \preceq^k \mathfrak{A}$, if $|\mathfrak{B}| \subseteq |\mathfrak{A}|$ and for every $t \in |\mathfrak{B}|^k$ and every formula φ of L^k,

$$\mathfrak{B} \models \varphi[t] \quad \text{if, and only if,} \quad \mathfrak{A} \models \varphi[t].$$

By analogy with the Ehrenfeucht-Mostowski stretching technique, we make the following definition.

Definition 15. A class of structures C has the *k-Ehrenfeucht-Mostowski property* (or k-EM property, for short) if, for every structure $\mathfrak{A}$ in C, if $X \subseteq |\mathfrak{A}|$ is a k-indiscernible sequence in $\mathfrak{A}$, with $\text{card}(X) \geq k$, then

1. there is an $\mathfrak{A}' \in C$ with $\mathfrak{A}' \preceq^k \mathfrak{A}$ and $\text{card}(|\mathfrak{A}'| \cap X) \leq k$.
2. for any set Y disjoint from $|\mathfrak{A}|$, there is an $\mathfrak{A}'' \in C$ such that $\mathfrak{A} \preceq^k \mathfrak{A}''$ and $X \cup Y$ is a k-indiscernible sequence in $\mathfrak{A}''$.

We motivated the introduction of indiscernible sets as an attempt at generalising the techniques used in the previous section on trees. We now justify this through the following theorem.

Theorem 16. *The class of trees has the k-Ehrenfeucht-Mostowski property.*

Proof: We prove that if $X = \{a_1, \ldots, a_n\}$ is a k-indiscernible sequence in the tree T with the order $a_i < a_j$ for $i < j$, then there is a collection of disjoint subtrees $T_1, \ldots, T_n$ of T such that:

1. $a_i \in T_i$ for $1 \leq i \leq n$;
2. $T_i \equiv^k T_j$ for $1 \leq i, j \leq n$; and
3. the roots of T_i and T_j are siblings for $1 \leq i, j \leq n$.

The result then follows from Lemma 6.

In order to establish 1–3, we first note that a straightforward modification of the formula δ_d of Lemma 5 yields a formula γ_d of L^3, such that $T \models \gamma_d[a, b]$ if, and only if, the distance from the root to the greatest lower bound of a and b is d. Thus, from the assumption that X is a sequence of indiscernibles, we conclude that there is a single vertex g such that g is the greatest lower bound of any pair a_i, a_j. By taking T_i to be the subtree rooted at a successor of g which contains a_i, properties 1 and 3 are easily established. Furthermore, since the elements of X all have the same L^k-type, property 2 follows immediately from Lemmas 7 and 8. ■

Other classes of structures that have the k-EM property include the class of linear orders (trivially, since no linear order contains a large enough indiscernible sequence), and the class of pure sets, i.e. structures in the empty signature. We shall see below examples of classes of structures where the k-EM property fails.

We will now show that we can establish recursive upper bounds for the functions d_k and e_k for any class of structures having the k-EM property.

Lemma 17. *There is a recursive function ι_k, for every $k \geq 3$, such that for any structure $\mathfrak{A}$, with k-size$(\mathfrak{A}) \leq n$ and card$(|\mathfrak{A}|) \geq \iota_k(n, m)$, there is a set $X \subseteq |\mathfrak{A}|$ which forms a k-indiscernible sequence in $\mathfrak{A}$ with card$(X) \geq m$.*

Proof: The proof is a direct application of the finite Ramsey theorem (see [11]). This guarantees the existence of a function ι_k such that if we take any ordered set A of more than $\iota_k(n, m)$ elements, and partition the set of order-increasing tuples in A^k into at most n sets $P_1, \ldots, P_n$, then there must exist a homogeneous subset X of A with card$(X) \geq m$, i.e. there is an i such that $[X]^k \subseteq P_i$, where $[X]^k$ is the set of order increasing tuples of elements of X.

Now, taking A to be the universe of $\mathfrak{A}$, choosing an arbitrary ordering of the elements, and letting the sets P_i be the $\equiv^k$ equivalence classes on the order increasing tuples in $|\mathfrak{A}|^k$, the resulting set X is guaranteed to be a k-indiscernible sequence in $\mathfrak{A}$. ■

Now, the following theorem is immediate.

Theorem 18. *If a class of structures C has the k-EM property, then there are recursive functions f_k and g_k satisfying the conditions of Open Questions 1 and 2, on C.*

Proof: For the function g_k, just take $g_k(n) = \iota_k(n, k)$.
For f_k, let $f_k(n) = \iota_k(n, k+1) - 1$. Note that any structure $\mathfrak{A}$ with $k-size(\mathfrak{A}) \leq n$ and $\mathrm{card}(\mathfrak{A}) > f_k(n)$ m has an indiscernible sequence of $k+1$ elements, and therefore contains a proper k-elementary substructure. ■

Unfortunately, the k-EM property does not generalise from the class of trees to the class of all finite structures. It does not even generalise to the class of all partial orders, as the following example shows.

Let B_n be the (unique up to isomorphism) partial order induced by a Boolean algebra with n atoms. This structure is characterised up to isomorphism by a sentence of L^3, namely the sentence that is the conjunction of the axioms of Boolean algebras (stated in terms of the partial order relation) with $\exists x \delta_n \wedge (\neg \exists x \delta_{n+1})$, where the formulas δ are from Lemma 5. Moreover, the set A of atoms is a k-indiscernible set in B_n, for any k, since any permutation of A can be extended to an automorphism of B_n. Thus, we obtain structures with arbitrarily large sets of k-indiscernibles, yet which are, up to isomorphism, unique in their $\equiv^k$ equivalence class.

Though the class of Boolean algebras does not have the k-EM property, it certainly does have recursive (even exponential) bounds on the functions d_k and e_k. This suggests that the k-EM property is too strong a requirement. One way to weaken it is to require collapsing and stretching only for sets of k-indiscernibles of size much larger than k itself. In fact, as the above argument shows, we cannot replace k by any constant in this requirement, and get a property that is satisfied by the class of Boolean algebras. We must, instead, replace it by a function of the k-size of $\mathfrak{A}$.

Definition 19. A class of structures C has the *weak k-Ehrenfeucht-Mostowski property* (or weak k-EM property, for short), if there is a recursive function h such that for every structure $\mathfrak{A}$ in C, if k-size$(\mathfrak{A}) \leq n$ and $X \subseteq |\mathfrak{A}|$ is a k-indiscernible sequence in $\mathfrak{A}$, with $\mathrm{card}(X) \geq h(n)$, then

1. there is an $\mathfrak{A}' \preceq^k \mathfrak{A}$ with $\mathrm{card}(|\mathfrak{A}'| \cap X) \leq h(n)$.
2. for any set Y disjoint from $|\mathfrak{A}|$, there is an $\mathfrak{A}''$ such that $\mathfrak{A} \preceq^k \mathfrak{A}''$ and $X \cup Y$ forms a k-indiscernible sequence in $\mathfrak{A}''$.

With this definition, it can be verified that the class of Boolean algebras has the weak k-EM property.

Moreover, the weak k-EM property is still strong enough for the following:

Theorem 20. *If a class of structures C has the weak k-EM property, then there are recursive functions f_k and g_k satisfying the conditions of Open Questions 1 and 2, on C.*

Proof: Just take $g_k(n) = \iota_k(n, h(n))$ and $f_k(n) = \iota_k(n, h(n)+1) - 1$. ■

However, the weak k-EM property is still too strong to hold of the class of all finite structures. This is a consequence of a recent result due to Luczak [16] (see also [19] for a proof of this result). Luczak shows that the sentence θ_k which is the conjunction of all extension axioms with k variables (see [15] for the definition of θ_k) has infinitely many minimal models with respect to the substructure relation. Since Kolaitis and Vardi [15] showed that all models of θ_k are $\equiv^k$ equivalent, it follows that the class $\text{Mod}(\theta_k)$ does not have the weak k-EM property. Note that this does not negatively resolve Open Questions 1 and 2, but it does show that we cannot hope to replace $\mathfrak{A} \equiv^k \mathfrak{B}$ with the stronger requirement $\mathfrak{B} \preceq^k \mathfrak{A}$ (at least in Question 1).

We end with the observation that we can define a stronger notion of indiscernibility, that allows us to prove collapsing and stretching properties (*à la* Ehrenfeucht-Mostowski) over the class of all structures. This is given by the following.

Definition 21. A set $X \subseteq |\mathfrak{A}|$ is a set of *strong k-indiscernibles* in $\mathfrak{A}$, if whenever $s \in |\mathfrak{A}|^k$, $a, a' \in X$ and a' is not in s, then, $(\mathfrak{A}, s) \equiv^k (\mathfrak{A}, s[a/a'])$, where $s[a/a']$ is the tuple obtained from s by replacing all occurrences of a by a'.

We then have the following:

Theorem 22. *If* $X = \{a_1, \ldots, a_{k+1}\} \subseteq |\mathfrak{A}|$ *is a set of strong k-indiscernibles in* $\mathfrak{A}$*, then*

1. *there is an* $\mathfrak{A}' \preceq^k \mathfrak{A}$ *with* $|\mathfrak{A}'| = |\mathfrak{A}| \setminus \{a_1\}$
2. *for any set* Y *disjoint from* $|\mathfrak{A}|$*, there is an* $\mathfrak{A}''$*, with* $|\mathfrak{A}''| = |\mathfrak{A}| \cup Y$*, such that* $\mathfrak{A} \preceq^k \mathfrak{A}''$ *and* $X \cup Y$ *is a set of strong k-indiscernibles in* $\mathfrak{A}''$.

The proof of the stretching part (part 2) of Theorem 22 appears in [18]. The proof of the first part is similar, and we omit it.

Unfortunately, with this stronger notion of indiscernibility, we do not have an analogue of Lemma 17.

Acknowledgements: I am grateful to Martin Grohe, Henrik Imhof and Eric Rosen for carefully reading an unpolished version of this paper.

References

1. S. Abiteboul, M. Y. Vardi, and V. Vianu. Fixpoint logics, relational machines, and computational complexity. In *Proc. 7th IEEE Symp. on Structure in Complexity Theory*, 1992.
2. S. Abiteboul and V. Vianu. Datalog extensions for database queries and updates. *Journal of Computer and System Sciences*, 43:62–124, 1991.
3. S. Abiteboul and V. Vianu. Computing with first-order logic. *Journal of Computer and System Sciences*, 50(2):309–335, 1995.
4. J. Barwise. On Moschovakis closure ordinals. *Journal of Symbolic Logic*, 42:292–296, 1977.
5. A. Dawar. *Feasible Computation through Model Theory*. PhD thesis, University of Pennsylvania, 1993.

6. A. Dawar. A restricted second order logic for finite structures. In D. Leivant, editor, *Logic and Computational Complexity*, volume 960 of *LNCS*, pages 393–413. Springer-Verlag, 1995.
7. A. Dawar, L. Hella, and Ph. G. Kolaitis. Implicit definability and infinitary logic in finite model theory. In Z. Fülöp and F. Gécseg, editors, *ICALP95*, volume 944 of *LNCS*, pages 624–635. Springer-Verlag, 1995.
8. A. Dawar, S. Lindell, and S. Weinstein. Infinitary logic and inductive definability over finite structures. *Information and Computation*, 119(2):160–175, 1995.
9. A. Dawar, S. Lindell, and S. Weinstein. First order logic, fixed point logic and linear order. In H. Kleine-Büning, editor, *Computer Science Logic '95*, volume 1092 of *LNCS*, pages 161–177. Springer-Verlag, 1996.
10. R. Fagin. Generalized first-order spectra and polynomial-time recognizable sets. In R. M. Karp, editor, *Complexity of Computation, SIAM-AMS Proceedings, Vol 7*, pages 43–73, 1974.
11. R. L. Graham, B. L. Rothschild, and J. H. Spencer. *Ramsey Theory*. Wiley, 1980.
12. Y. Gurevich. Toward logic tailored for computational complexity. In M. Richter et al., editors, *Computation and Proof Theory*, pages 175–216. Springer Lecture Notes in Mathematics, 1984.
13. N. Immerman. Upper and lower bounds for first-order expressibility. *Journal of Computer and System Sciences*, 25:76–98, 1982.
14. N. Immerman. Relational queries computable in polynomial time. *Information and Control*, 68:86–104, 1986.
15. Ph. G. Kolaitis and M. Y. Vardi. Infinitary logics and 0-1 laws. *Information and Computation*, 98(2):258–294, 1992.
16. T. Łuczak, 1995. Personal Communication.
17. M. Otto. Ptime canonization for two variables with counting. In *Proc. 10th IEEE Symp. on Logic in Computer Science*, pages 342–352, 1995.
18. B. Poizat. Deux ou trois choses que je sais de L_n. *Journal of Symbolic Logic*, 47(3):641–658, 1982.
19. E. Rosen, S. Shelah, and S. Weinstein. k-Universal finite graphs. In R. Boppana and J. F. Lynch, editors, *Logic and Random Structures*, DIMACS Series in Deiscrete Mathematics and Theoretical Computer Science. American Mathematical Society, 1996. to appear.
20. M. Y. Vardi. The complexity of relational query languages. In *Proceedings of the 14th ACM Symposium on the Theory of Computing*, pages 137–146, 1982.

Model Theory of Modules

(Extended Abstract)

Ivo Herzog

Department of Mathematics
University of Notre Dame
Notre Dame, IN 46556-5683, USA
ivo.herzog.2@nd.edu

This note is meant to give a brief account of some themes in the model theory of modules as they have developed since the subject's beginning. We shall survey the situation from the perspective that has grown in recent years out of an effort to formulate the theory in a way that stresses the common ground shared by the model theory of modules with other areas of mathematics, most notably representation theory and abelian category theory. We will take this opportunity to state the classical theorems of the theory in a context more general and hopefully more accessible to those working in other areas. On the other hand, this exposition hopes to illustrate how the theory of left exact functors, developed by Gabriel [4], figures into model theory. The two standard references for this subject are [7] and [12].

The seminal results of what has turned out to be the model theory of modules were attained by Wanda Szmielew [10] in her work on abelian groups, that is, modules over the ring of integers. Among other things, Szmielew characterized the complete theories of abelian groups. Quite sometime later, Eklof and Fisher [3] exploited the theory of algebraically compact (= pure-injective) abelian groups that had developed in the meantime to present Szmielew's classification in a more conceptual framework. The following result is not the classification itself but a result that generalizes nicely.

Theorem. Every abelian goup is elementarily equivalent to a direct product of pure-injective indecomposable abelian groups.

The pure-injective indecomposable abelian groups were classified by Kaplansky [5] into four groups:

Torsion free, divisible. The group of rational numbers Q.
Torsion, divisible. The Prüfer groups $Z(p^\infty)$, p prime.
Torsion free, not divisible. The p-adic completions $\overline{Z_p}$ of the integers, p prime.
Torsion, not divisible. The cyclic groups of order a prime power $Z(p^n)$.

The theorem thus gives a complete list of abelian groups up to elementary equivalence. This list is however not without repetition so the problem that emerges is to describe how the pure-injective indecomposables relate to each other. More precisely, one must consider the following question.

Problem. Given pure-injective indecomposable abelian groups U and $\{V_i\}_i$, when does there exist a group A such that

$$U \mid A \equiv \prod_i V_i^{\alpha_i}$$

for some cardinals α_i. ($U \mid A$ means that U is a direct summand of A.)

In his breakthrough paper [12] on the model theory of modules, Ziegler generalized the theorem to R-modules (where a theory of pure-injectivity also exists) and he showed that if the answer to the problem is affirmative, we may take $\alpha_i = 1$ for all i. But most spectacularly, he introduced into the theory a topological space - known as the Ziegler spectrum of the ring R - which has ever since played a monolithic role in the model theory of modules. The points of the space are the pure-injective indecomposable left R-modules and a point U belongs to the closure of a subset $\{V_i\}_i$ if and only if the problem above has an affirmative answer. This definition of a closed subset is really an elementary analogue of the notion of a torsion free class of indecomposables.

Decidability.

Szmielew [10] proved that the theory $T(Z)$ of abelian groups is decidable. Ziegler [12] showed that if the ring R is recursive, an effective description of the Ziegler spectrum of R yields the decidability of $T(R)$, the theory of left R-modules. This has provided the main line of attack towards the following conjecture of Prest.

Conjecture. Let Λ be a recursive finite dimensional algebra over an algebraically closed field. The theory of left R-modules is decidable if and only if the algebra Λ is tame.

Such a characterization of tameness would certainly aid to dispell the general dissatisfaction among algebraists of present definitions of the notion. But in any case, to provide an effective description of the Ziegler spectrum is very much in line with present work of representation theorists a portion of which has turned its attention to infinite dimensional representations. The Ziegler spectrum and some variants of it have offered the best context for such an undertaking.

Purity.

The language $\mathcal{L}(R) = (+, 0, r)_{r \in R}$ for left R-modules consists of the binary function symbol $+$, a constant symbol 0 and for every $r \in R$, a unary function symbol intended to interpret the action of r on the abelian group interpreted by $+$ and 0. Clearly the set of axioms $T(R)$ for a left R-module is expressible in this language.

Definition. A formula $\varphi(\mathbf{x})$ of $\mathcal{L}(R)$ is *positive-primitive* if it is equivalent modulo $T(R)$ to an existentially quantified system of linear equations. More precisely, there are matrices A and B of appropriate size with entries in R such that

$$T(R) \models \varphi(\mathbf{x}) \leftrightarrow \exists \mathbf{y}\ (A, B) \begin{pmatrix} \mathbf{x} \\ \mathbf{y} \end{pmatrix} \doteq 0.$$

An inclusion $M \subseteq N$ of left R-modules is said to be *pure* if for every n and positive-primitive formula $\varphi(\mathbf{x})$ in n free variables,

$$\varphi(M) = M^n \cap \varphi(N).$$

A left R-module U is *pure-injective* if for every such inclusion, there is a retraction $f : N \to M$, $f\mid_M = 1_M$. The pure-injective modules U have a nice model-theoretic characterization. They are the pp-saturated modules, which means that every U-consistent type $p(x, A)$, $A \subseteq U$, of positive-primitive formulae is realized in U.

A left R-module K is called *pure-projective* if every pure inclusion $M \subseteq N$ with $M/N \cong K$ has a retraction. These modules are precisely the summands of direct sums of finitely presented modules [11]. The model theoretic characterization of such modules K is a bit trickier. But they do belong to the class of Mittag-Leffler modules [6], which are the pp-atomic modules. This means that for every tuple $\mathbf{a} \in M^n$, there is a positive-primitive formula $\varphi_{\mathbf{a}}(\mathbf{x}) \in \mathrm{tp}_M(\mathbf{a})$ such that

$$T(R) \models \varphi_{\mathbf{a}} \to \psi$$

for every positive-primitive formula $\psi \in \mathrm{tp}_M(\mathbf{a})$.
R^{eq}.

In this section, we shall describe the abelian category R^{eq}. This category appears in different guises in several areas of mathematics. It is the free abelian category over the ring R and it is also equivalent to the category introduced by Auslander [1] of coherent functors on the category of finitely presented left R-modules. While in the other areas it is treated more as a mathematical object of interest in its own right, the model theory of modules comes closest to comprising a "representation theory" of this category.

The category R^{eq} has as its objects the positive-primitive sorts of $\mathcal{L}(R)^{\mathrm{eq}}$, that is, those of the form φ/ψ where φ and ψ are positive-primitive formulae such that $T(R) \models \psi \to \varphi$. The morphisms of R^{eq} are the definable functions between sorts that are expressible by a positive-primitive formula. The category R^{eq} is abelian and to it we associate this positive-primitive reduct of $\mathcal{L}(R)^{\mathrm{eq}}$. This can be done with any small abelian category $\mathcal{A}$ to get a many-sorted abelian language $\mathcal{L}(\mathcal{A})$ defined as follows. The sorts of $\mathcal{L}(\mathcal{A})$ correspond to the objects of $\mathcal{A}$ and the function symbols to the morphisms. Each sort is equiped with a language for abelian groups. The intended interpretation of a structure for this language is that of an additive functor from the category $\mathcal{A}$ to the category Ab of abelian groups. This is clearly an elementary class denoted by $(\mathcal{A}, \mathrm{Ab})$.

Because the language $\mathcal{L}(R^{\mathrm{eq}})$ is a reduct of $\mathcal{L}(R)^{\mathrm{eq}}$, every left R-module $M \models T(R)$ has a unique expansion M^{eq} to $\mathcal{L}(R^{\mathrm{eq}})$ if the sorts are interpreted as intended. This is a first order property of M^{eq} and we denote the theory of such expansions by $T(R)^{\mathrm{eq}}$. In fact, it may be shown that this class is precisely the obviously elementary class of exact functors.

Theorem. $\mathrm{Mod}(T(R)^{\mathrm{eq}}) = \mathrm{Ex}(\mathcal{A}, \mathrm{Ab})$.

The category R^{eq} was also used in the original definition of the Ziegler spectrum by giving a basis of quasi-compact open subsets which have the form

$$\mathcal{O}(\varphi/\psi) := \{U : \varphi/\psi(U) \neq 0\}$$

where φ/ψ is a sort (object) of R^{eq}.

Elimination of quantifiers.

In 1976, Baur [2] proved the following elimination of quantifiers result. It generalizes the correponding result of Szmielew for abelian groups.

Theorem. Let $T \supseteq T(R)$ be a complete theory of left R-modules. Given a formula $\rho(\mathbf{x})$ of $\mathcal{L}(R)$, there is a boolean combination of positive-primitive formulae $\beta(\mathbf{x})$ such that

$$T \models \rho(\mathbf{x}) \leftrightarrow \beta(\mathbf{x}).$$

This result may also be thought of as generalizing the full elimination of quantifiers attained by Eklof and Sabbagh for the complete theory of an absolutely pure module over a left coherent ring. Now a ring R is left coherent if and only if the category R-Mod is a locally coherent Grothendieck category. By Mitchell's Representation Theorem, every such category may be represented up to equivalence as the category of left exact functors $\mathrm{Lex}(\mathcal{A}, \mathrm{Ab})$ on some small abelian category $\mathcal{A}$. This is clearly an elementary class of $\mathcal{L}(\mathcal{A})$-structures whose theory we denote by $T(\mathcal{A})$. The absolutely pure objects of $\mathrm{Lex}(R^{\mathrm{eq}}, \mathrm{Ab})$ are precisely the models of $T(R)^{\mathrm{eq}}$ and the complete theory of such a model admits full elimination of quantifiers in $\mathcal{L}(R^{\mathrm{eq}})$ by an abelian version of Baur's result.

The functor $M \mapsto M^{\mathrm{eq}}$ from R-Mod to $\mathrm{Lex}(R^{\mathrm{eq}}, \mathrm{Ab})$ is a left adjoint to the functor that associates to $A \in \mathrm{Lex}(R^{\mathrm{eq}}, \mathrm{Ab})$ the left R-module $H(A)$ that lives on the home sort of A:

$$\mathrm{Hom}_R(M, H(A)) \cong \mathrm{Hom}_{\mathrm{Lex}}(M^{\mathrm{eq}}, A).$$

Complete theories.

Let $\mathcal{A}$ be a small abelian category and $M : \mathcal{A} \to \mathrm{Ab}$ an exact functor. We may associate to M a character $\chi_M : \mathcal{A} \to \mathcal{N} \cup \{\infty\}$ defined by

$$\chi_M(S) := \mid S(M) \mid,$$

the cardinality of the sort $S(M)$ (modulo infinity). This is the Baur-Garavaglia-Monk character of the functor M. It was Garavaglia who made explicit use of the fact that two laeft exact functors M and N are elementarily equivalent if and only if their characters are the same. For left R-modules, this means that $M^{eq} \equiv N^{eq}$ if and only if for every pair of positive-primitive formulae $\psi \leq \varphi$, $|\varphi(M)/\psi(M)| = |\varphi(N)/\psi(N)|$ (modulo ∞).

A Nullstellensatz.

Sabbagh [9] showed, and it is corollary of Baur's elimination of quantifiers, that a monomorphism of left R-modules $f : M \to N$ is elementary if and only if it is pure and $M \equiv N$. That f is pure is equivalent to the condition that the Lex(R^{eq}, Ab)-morphism $f^{eq} : M^{eq} \to N^{eq}$ is a monomorphism, so that Sabbagh's result asserts that every complete extension of $T(R)^{eq}$ is model complete. To state the ensuing Nullstellensatz, one needs the notion of a Serre subcategory of R^{eq}.

Definition. A (full) subcategory $\mathcal{S} \subseteq R^{eq}$ is called *Serre* if for every short exact sequence of sorts

$$0 \to S_1 \to S \to S_2 \to 0$$

in R^{eq}, $S \in \mathcal{S}$ if and only if $S_1, S_2 \in \mathcal{S}$.

The Serre subcategories of R^{eq} play a role similar to that of the ideals of R. For example, if M is a left R-module, then the "annihilator" of M in R^{eq} is the Serre subcategory

$$\mathcal{S}(M) := \{S : S(M) = 0\} = \operatorname{Ker} \chi_M.$$

Theorem. There is a bijective correspondence between the Serre subcategories $\mathcal{S}$ of R^{eq} and the open subsets $\mathcal{O}$ of the Ziegler spectrum of R. The correspondence is given by the maps

$$\mathcal{S} \mapsto \mathcal{O}(\mathcal{S}) := \bigcup_{S \in \mathcal{S}} \mathcal{O}(S),$$

$$\mathcal{O} \mapsto \{S : \mathcal{O}(S) \subseteq \mathcal{O}\}$$

which are mutually inverse.

The theorem can be used to show that every Serre subcategory of R^{eq} is of the form $\mathcal{S}(M)$ for some left R-module M. In this way, one may associate to a left R-module M the closed subset of the Ziegler spectrum which is the complement of $\mathcal{O}(\mathcal{S}(M))$.

It is proved in [8] that the injective indecomposable left R-modules form a closed subset of the Ziegler spectrum of R if and only if R is left coherent. In the same paper, Prest, Rothmaler and Ziegler [8] gave an example of a ring R that is not von Neumann regular, but whose Ziegler spectrum is indiscrete.

Duality.

If the ring R is not commutative, then the opposite ring R^{op} is distinct from R and it is possible that the category of R^{op}-modules (= right R-modules) drastically differs from the category of left R-modules. But model theoretically, due to the limitations on the power of expression of the language $\mathcal{L}(R)$, left and right R-modules are dual in many ways. More precisely, there is the following categorical duality.

Theorem. For very ring R, $(R^{\text{op}})^{\text{eq}} \cong (R^{\text{eq}})^{\text{op}}$.

This has a some interesting consequences for the relationship between the model theory of left and right R-modules. The Serre subcategories of R^{eq} are in bijective correspondence with those of $(R^{\text{op}})^{\text{eq}}$ and this implies that the topologies (viewed as algebras of open subsets) of the respective Ziegler spectra of R and R^{op} are isomorphic. This will at times induce a homeomorphism between the two spectra, but even when the ring is commutative this self-homeomorphism need not be the identity. In the example of the ring of integers, the points Q and $Z(p^n)$ remain fixed, but as in Pontryagin duality, the p-adic completions of Z are interchanged with the Prüfer groups.

The Pontryagin duality is the functor $DM := \text{Hom}_Z(M, \mathcal{C}^*)$ where $\mathcal{C}^*$ denotes the group of nonzero complex numbers under multiplication. This is an exact functor $D : R\text{-Mod} \to \text{Mod-}R$ that takes on pure-injective values (because D takes discrete modules to compact modules and every compact topological module is algebraically compact and hence pure-injective). But moreover, the functor D respects elementary equivalence and gives a bijective correspondence between complete theories of left R-modules and complete theories of right R-modules such that

$$\chi_M(S) = \chi_{DM}(DS)$$

where DS denotes the value that S takes on under the duality of the theorem.

References

1. Auslander, M. Coherent functors, in *Proceedings of the Conference on Categorical Algebra* (La Jolla 1965), S. Eilenberg, D.K. Harrison, S. MacLane, H. Röhrl, eds.
2. Baur, W. Elimination of quantifiers for modules, Israel Journal of Mathematics **25** (1976) 64-70.
3. Eklof, P. and Fisher, E. The elementary theory of abelian groups, Annals of Mathematical Logic 4(2) (1972) 115-171.
4. Gabriel, P. Des Catégories Abéliennes, Bull. Soc. Math. France **90** (1962) 323-448.
5. Kaplansky, I. *Infinite Abelian Groups*, University of Michigan Press, Ann Arbor, 1969.
6. Rothmaler, Ph. Mittag-Leffler modules. Habilitationschrift, 1995, Kiel.
7. Prest, M. *Model Theory and Modules*, London Mathematical Society Lecture Notes Series **130**, Cambridge University Press, Cambridge, 1988.

8. Prest, M., Rothmaler, Ph. and Ziegler, M. Absolutely pure and flat modules and "indiscrete" rings, Journal of Algebra, **174** (1995) 349-372.
9. Sabbagh, G. Sur la pureté dans les modules, C.R Acad. Sci. Paris, **271** (1970) 865-867.
10. Szmielew, W. Elementary properties of abelian groups, Fundamenta Mathematicae **41** (1955) 203-271.
11. Warfield, R. Purity and algebraic compactness for modules, Pacific Journal of Mathematics **28** (1969) 699-719.
12. Ziegler, M. Model theory of modules, Annals of Pure and Applied Logic **26** (1984) 149-213.

Noninterpretability of Infinite Linear Orders

Wilfrid Hodges

School of Mathematical Science
Queen Mary and Westfield College
Mile End Rd., London E1 4NS, England
w.hodges@qmw.ac.uk

André Nies *

Department of Mathematics
University of Chicago
5734 S. University Ave., Chicago, IL 60657, USA
nies@math.uchicago.edu

Abstract. We prove that no infinite linear order can be interpreted without parameters in any structure of the form $B \times B$, and describe applications of this result in recursion theory.

1 Squares

We are interested structures **A** which *fail* to satisfy the following property.

(ILOP) It is possible to interpret an infinite linear order in **A** without parameters.

The abbreviation stands for Infinite Linear Order Property. Interpretations of a structure **C** in a structure **A** are studied in [Ho 93]. In brief, an interpretation is a way to encode **C** into **A** using a finite collection of first–order formulas in the language of **A** as a decoding key. Elements of **C** are represented by tuples of a fixed length m of elements of B, modulo some definable equivalence relation. For the special case of an interpretation of a reflexive linear order Q in **A**, it is enough to consider a decoding key consisting of a single formula $\varphi(\overline{x}, \overline{y})$, also written as $\overline{x} \leq_Q \overline{y}$. Here $\overline{x}, \overline{y}$ are tuples of variables of length m, and, in **A**, the formula defines a linear pre–ordering on the domain $\{\overline{x} : \overline{x} \leq_Q \overline{x}\}$. For instance, $(\mathbb{Q}, \leq)$ can be interpreted in the ring $\mathbb{Z}$ using the formula

$$(z, w) \leq_Q (z', w') \equiv \psi_{\leq}(1, w) \wedge \psi_{\leq}(1, w') \wedge \psi_{\leq}(z \cdot w', z' \cdot w),$$

* The second author was supported under NSF-grant DMS–9500983

where $\psi_{\leq}(x, y)$ expresses in the language of rings that $x \leq y$ (which is possible using Lagranges Theorem: an integer is nonnegative if and only if it is the sum of 4 squared integers).

We write $\overline{x} <_Q \overline{y}$ for $\overline{x} \leq_Q \overline{y} \wedge \overline{y} \not\leq_Q \overline{x}$ and $\equiv_Q$ for the equivalence relation on m-tuples

$$\{\overline{x}, \overline{y} : \overline{x} \leq_Q \overline{y} \leq_Q \overline{x}\}.$$

The fact that (ILOP) holds via a fixed decoding key can be expressed by an infinite collection of first-order sentences saying for each n that in the interpreted linear order there is a chain of length n. Therefore, (ILOP) only depends on the theory of **A**.

Recall that we are interested in structures where (ILOP) fails. A weaker property than (ILOP) is the strict order property ([Ho 93], p. 317) of a structure **A**: some formula $\varphi(\overline{x}, \overline{y})$ defines a *partial* order with arbitrarily long finite chains in **A**. Even the weaker strict order property implies that the theory of **A** is unstable. However, the converse implication fails to hold: The random graph for instance has an unstable theory, but does not satisfy the strict order property. The random partial order (i.e. the Fraissé limit in the sense of [Ho 94] of the class of finite partial orders) or the random distributive lattice obviously have the strict order property, but still fail to satisfy (ILOP).

In asking whether (ILOP) holds via some given decoding key, it is safe to assume that

$$\{\overline{x} : \overline{x} \leq_Q \overline{x}\} = A^m$$

because else one can form a new equivalence class $\{\overline{x} : \overline{x} \not\leq_Q \overline{x}\}$ and put it at the beginning of an extended definable linear order, modifying $\leq_Q$ in the appropriate way.

The product $\mathbf{C} \times \mathbf{D}$ of structures **C**, **D** for the same language is defined in the expected way ([Ho 93]). We some times call structures of the form $\mathbf{C} \times \mathbf{C}$ *squares*.

Theorem 1. *It is not possible to interpret an infinite linear order Q in a structure of the form $\mathbf{A} \times \mathbf{A}$ without parameters.*

Proof. Let $\mathbf{B} = \mathbf{A} \times \mathbf{A}$. The proof is by contradiction. Suppose that an interpretation of an infinite linear order in **B** is given so that $\{\overline{x} : \overline{x} \leq_Q \overline{x}\} = B^m$. Since a decoding key uses only finitely many symbols, we can assume that the language of **A** has a finite signature. To make the argument clearer, we will first assume that $m = 1$. We make use of the following fact. Suppose $\leq_Q$ is a definable linear order on a structure **C** such that the domain $\{x : x \leq_Q x\}$ equals C. If an automorphism π of **C** exchanges u and v, then $u \equiv_Q v$. This is because

$$u \leq_Q v \Rightarrow \pi(u) \leq_Q \pi(v) \Rightarrow v \leq_Q u. \tag{1}$$

Clearly **B** can be interpreted in **A**. Thus "$\leq_Q$" can be viewed both as a definable binary relation on **B** and as a definable 4–ary relation on **A**. Moreover the statement "$\leq_Q$ defines an infinite linear order" is a statement about Th(**A**).

We claim that the following 6-type in the language of $\mathbf{A}$ is consistent with $\mathrm{Th}(\mathbf{A})$:

(a) $(x_0, y_0) <_Q (x_1, y_1) <_Q (x_2, y_2)$
(b) The collection of first order statements expressing that, for each i, j, $0 \leq i < j \leq 2$, (x_i, y_i, x_j, y_j) realize the same 4–type.

The statements in (b) are of the form

$$\varphi(x_0, y_0, x_1, y_1) \Leftrightarrow \varphi(x_0, y_0, x_2, y_2) \Leftrightarrow \varphi(x_1, y_1, x_2, y_2),$$

for any formula φ with four free variables. They also imply that the same 2-type is realized by any pair (x_i, y_i), $0 \leq i \leq 2$.

To prove the consistency of the 6–type with $\mathrm{Th}(\mathbf{A})$, we apply Ramseys Theorem as in the construction of Ehrenfeucht–Mostowski models. Since Q is infinite, we can suppose that there is an infinite ascending chain

$$(a_0, b_0) <_Q (a_1, b_1) <_Q \ldots$$

in Q (if there is an infinite descending chain, one can argue similarly). Given finitely many formulas φ_k $(k = 0, \ldots, n-1)$ in 4 free variables, we assign one of 2^n possible colours to an unordered pair $\{(a_i, b_i), (a_j, b_j)\}$ $(i < j)$ depending on which formulas $\varphi_k(a_i, b_i, a_j, b_j)$ hold. Then any homogeneous (in the sense of Ramsey) set of cardinality 3 shows the consistency of $\mathrm{Th}(\mathbf{A})$,

$$(x_0, y_0) <_Q (x_1, y_1) <_Q (x_2, y_2)$$

and the statements in (b) determined by a formula φ_k $(k = 0, \ldots, n-1)$. Thus the 6–type is consistent with $\mathrm{Th}(\mathbf{A})$.

Now, by standard results from model theory we can choose a countable ω–homogeneous model $\widetilde{\mathbf{A}}$ of $\mathrm{Th}(\mathbf{A})$ such that elements

$$(a_0, b_0) <_Q (a_1, b_1) <_Q (a_2, b_2)$$

realize the 6–type above. We will derive a contradiction. We need the following fact: If two pairs (a, b) and (a', b') realize the same 2–type in $\widetilde{\mathbf{A}}$, then

$$(a, b') \equiv_Q (a', b). \tag{2}$$

For let α be an automorphism of $\widetilde{\mathbf{A}}$ mapping a to a' and b to b'. Let S be the automorphism of $\widetilde{\mathbf{B}} := \widetilde{\mathbf{A}} \times \widetilde{\mathbf{A}}$ mapping a pair (x, y) to (y, x). Then, the automorphism of $\widetilde{\mathbf{B}}$

$$S \circ (\alpha, \alpha^{-1}),$$

exchanges (a, b') and (b, a').

Applying fact (1), we can conclude that $(a, b') \equiv_Q (b, a')$, and applying the same fact to S alone, $(a', b) \equiv_Q (b, a')$. So, by the transitivity of $\equiv_Q$,

$$(a, b') \equiv_Q (a', b).$$

We apply (2) two times to obtain the desired contradiction. First, a_0, b_1 and a_0, b_2 realize the same 2–type in $\widetilde{\mathbf{A}}$ since they are the projections to the first and fourth components of the 4–types of a_0, b_0, a_1, b_1 and of a_0, b_0, a_2, b_2, respectively. Thus $(a_0, b_1) \equiv_Q (a_0, b_2)$. Also, a_0, b_2 and a_1, b_2 realize the same 2–type, so $(a_0, b_2) \equiv_Q (a_1, b_2)$. Now, by transitivity,

$$(a_0, b_1) \equiv_Q (a_1, b_2).$$

However, since $\widetilde{\mathbf{A}}$ is ω–homogeneous, again by our definition of the 6–type, we can choose an automorphism β of $\widetilde{\mathbf{A}}$ such that $\beta(b_i) = b_{i+1}$ $(i = 0, 1)$. Then, since $(\mathrm{Id}_{\widetilde{\mathbf{A}}}, \beta)$ is an automorphism of $\widetilde{\mathbf{B}}$, $(a_0, b_0) <_Q (a_1, b_1)$ implies that

$$(a_0, b_1) <_Q (a_1, b_2),$$

a contradiction.

We now adapt the argument to the case that $m = 2$. Then, by obvious notational changes the general case can be obtained. Note that $\leq_Q$ now is a 4–ary relation on $\mathbf{B}$, which can be viewed as an 8–ary relation on $\mathbf{A}$ via

$$(x_0, y_0, z_0, w_0) \leq_Q (x_1, y_1, z_1, w_1) \text{ in } \mathbf{A}$$
$$\Leftrightarrow ((x_0, y_0), (z_0, w_0)) \leq_Q ((x_1, y_1), (z_1, w_1)) \text{ in } \mathbf{B}$$

Fact (1) is used in a modified form: If an automorphism π of $\mathbf{C}$ (a structure like $\mathbf{B}$) exchanges u with v and p with q, then $(u, p) \equiv_Q (v, q)$.

The following 12–type in the language of $\mathbf{A}$ is consistent with $\mathrm{Th}(\mathbf{A})$:

(a') $(x_0, y_0, z_0, w_0) <_Q (x_1, y_1, z_1, w_1) <_Q (x_2, y_2, z_2, w_2)$
(b') The collection of first order–statements expressing that, for each i, j, $0 \leq i < j \leq 2$, $(x_i, y_i, z_i, w_i, x_j, y_j, z_j, w_j)$ realize the same 8–type.

The consistency of the 12–type with $\mathrm{Th}(\mathbf{A})$ is proved as before. Now let $\widetilde{\mathbf{A}}$ be a countable ω–homogeneous model of $\mathrm{Th}(\mathbf{A})$ such that the elements

$$(a_0, b_0, c_0, d_0) <_Q (a_1, b_1, c_1, d_1) <_Q (a_2, b_2, c_2, d_2)$$

realize the 12–type. We derive a contradiction by applying the previous argument simultaneously to the first and the second half of the quadruples. If two pairs (a, b, c, d) and (a', b', c', d') realize the same 4–type in $\widetilde{\mathbf{A}}$, then

$$(a, b', c, d') \equiv_Q (a', b, c', d).$$

For if α is an automorphism of $\widetilde{\mathbf{A}}$ mapping (a, b, c, d) to (a', b', c', d'), and S is as before, then $S \circ (\alpha, \alpha^{-1})$ exchanges the pairs of elements of $\widetilde{\mathbf{B}}$ $((a, b'), (c, d'))$ and $((b, a'), (d, c'))$, and S exchanges the last pair of pairs with $((a', b), (c', d))$.

Now, by similar arguments as before,

$$(a_0, b_1, c_0, d_1) \equiv_Q (a_0, b_2, c_0, d_2) \equiv_Q (a_1, b_2, c_1, d_2).$$

But we can also choose an automorphism β of $\widetilde{\mathbf{A}}$ such that $\beta(b_i) = b_{i+1}$ and $\beta(d_i) = d_{i+1} (i = 0, 1)$. Then $(a_0, b_0, c_0, d_0) <_Q (a_1, b_1, c_1, d_1)$ implies that $(a_0, b_1, c_0, d_1) <_Q (a_1, b_2, c_1, d_2)$, contradiction. ◇

The statement of the theorem becomes false if one allows interpretations with parameters. For instance, if $B = \mathbb{N} \times \mathbb{N}$, consider $a = (1,0)$ as a parameter. The set of elements u in B which satisfy $a \cdot u = u$ is definable from a and equals the set of elements $\{(x,0) : x \in \mathbb{N}\}$, which is infinite and linearly ordered by $\leq^{\mathbf{B}}$. The same example shows that the statement is false in general for structures $\mathbf{C} \times \mathbf{D}$, even if $\mathbf{C}, \mathbf{D}$ are infinite: let $\mathbf{C} = (\mathbb{N}, 0)$ and $\mathbf{D} = (\mathbb{N}, 1)$.

2 Applications in recursion theory

In this section we give examples of structures arising naturally in recursion theory which are squares and therefore fail to satisfy (ILOP). First we discuss the relevance of (ILOP) in recursion theory. Examples of structures arising naturally are the lattice $\mathcal{E}$ of r.e. sets under inclusion and the structure $\mathcal{R}_\mathcal{T}$ of r.e. Turing degrees. The use of coding methods has been a very successful method of analyzing those structures. For all structures, coding methods were first used to prove the undecidability of the elementary theory. Stronger results were obtained by giving a (natural) many-one reduction of true arithmetic, i.e. Th($\mathbb{N}$), to the theory of the structure. Such a reduction shows that the theory is as complex as possible, provided that the structure itself is arithmetical, which is the case for $\mathcal{E}$, $\mathcal{R}_\mathcal{T}$ and many other structures occuring in recursion theory. However, the existence of such a reduction does not mean that an interpretation of the model $\mathbb{N}$ in the recursion theoretic structure can be given. Instead, to obtain the reduction, some coding of copies of $\mathbb{N}$ with parameters $\overline{p}$ is developed. As a next step, one finds a first-order correctness condition $\psi(\overline{p})$ on parameter lists which holds for some list, and always implies that the model coded using $\overline{p}$ is a standard model. Then, existentially quantifying over correct lists $\overline{p}$, one obtains an interpretation of true arithmetic.

All theories we consider in this section have the same computational complexity. If one looks at the structures themselves rather than just at their theories, finer distinctions emerge. In $\mathcal{R}_{\Updownarrow}$ (the structure of r.e. many-one degrees) and in $\mathcal{R}_\mathcal{T}$, a further development of the coding methods allows one to find an interpretation of $\mathbb{N}$ without parameters ([N95] and [NSSlta], respectively). Certainly this implies (ILOP). However, for $\mathcal{E}$, L. Harrington showed that even (ILOP) fails. Harrington used $\mathcal{E}$-specific methods, thereby paving the way for further noncoding and nondefinability results (see [HaNta]). Here we obtain Harringtons result in a purely model theoretic way, since $\mathcal{E}$ is a square. To see this, let $S \subseteq \omega$ be any recursive infinite and coinfinite set. Then the collection of subsets of S forms a lattice isomorphic to $\mathcal{E}$. Thus the map

$$X \mapsto (X \cap R, X \cap (\omega - R))$$

gives an isomorphism as required.

We now consider other structures from recursion theory where (ILOP) fails. The major subset relation is defined as follows: for $A, B \in \mathcal{E}$,

$$A \subset_m B \Leftrightarrow A \subset_\infty B \wedge (\forall W)[B \cup W = \omega \Rightarrow A \cup W =^* \omega].$$

Maass and Stob [MSt83] proved that the interval $[A, B]_{\mathcal{E}}$ has the same structure for any pair A, B such that $A \subset_m B$. This structure is denoted by $\mathcal{M}$. Thus, $\mathcal{M} = [A, B]$, where $A \subset_m B$. From the Maas–Stob result, it follows that $\mathcal{M}$ is a distributive lattice with strong homogeneity properties: all nontrivial closed intervals are isomorphic to the whole structure, and all nontrivial complemented elements are automorphic within $\mathcal{M}$. The undecidability of Th($\mathcal{M}$) is proved in [Nta1].

Theorem 2. *$\mathcal{M}$ fails to satisfy (ILOP).*

Proof. Choose $A \subset_m B$. Clearly, $\overline{B} \cup A$ is non–r.e. Then, by the Owings Splitting Theorem (see [So87]), there is a splitting $B = B_1 \cup B_2$ into disjoint sets such that $\overline{B}_i \cup A$ is non–r.e. for $i = 1, 2$. If $C_i = A \cup B_i$, then $A \subset_m C_i$ and hence, by the Maass–Stob result, $[A, C_i] \cong \mathcal{M}$. Now the map

$$X \mapsto (X \cap C_1, X \cap C_2)$$

($X \in [A, B]$) gives an isomorphism $\mathcal{M} \cong \mathcal{M} \times \mathcal{M}$. So $\mathcal{M}$ fails to satisfy (ILOP). $\Diamond$

Note that the arguments above also work for the quotient lattices modulo finite differences $\mathcal{E}^*$ and $\mathcal{M}^*$.

Question. As mentioned in the introduction, the random graph, random partial order, and the random distributive lattice fail to satisfy (ILOP). On the other hand, $(\mathbb{Q}, \leq)$ certainly satisfies (ILOP). It would be interesting to find a criterion when a countable structure **A** such that Th(**A**) is ω–categorical and has quantifier elimination satisfies (ILOP) (see also [Ho93], p. 350). m

Acknoledgement. The question answered by Theorem 1.1 arose in discussions between L. Harrington and the second author.

References.

[HaN ta] L. Harrington, A. Nies. Coding in the lattice of enumerable sets. Submitted.

[Ho93] W. Hodges. Model Theory. Encyclopedia of mathematics and its applications 42, Cambridge University Press, 1993.

[MaSt83] W. Maass, M. Stob. The interval of the lattice of r.e. sets determined by major subsets. Annals of Pure and applied Logic, 24 (1983), 189-212.

[N94] A. Nies. The last question on recursively enumerable many one degrees. Algebra Logika 33 (5), transl. July 1995 550-563.

[Nta1] A. Nies. Intervals of $\mathcal{E}$. To appear.

[Nta2] A. Nies. Flexible coding of standard models of arithmetic. To appear.

[NSSlta] A. Nies, R. Shore and T. Slaman. Definability in the r.e. Turing degrees. To appear.

[So87] R. Soare, "R.e. sets and degrees". Springer 1987.

Combinatorial Principles from Adding Cohen Reals *

I. Juhász

Mathematical Institute of the Hungarian Academy of Sciences
Budapest, Hungary
juhasz@math-inst.hu

L. Soukup **

Mathematical Institute of the Hungarian Academy of Sciences
Budapest, Hungary
soukup@math-inst.hu

Z. Szentmiklóssy

Department of Analysis
Eötvös Loránd University
Budapest, Hungary
szentmiklossy@math-inst.hu

Abstract. We first formulate several "combinatorial principles" concerning $\kappa \times \omega$ matrices of subsets of ω and prove that they are valid in the generic extension obtained by adding any number of Cohen reals to any ground model V, provided that the parameter κ is an ω-inaccessible regular cardinal in V.
Then in section 4 we present a large number of applications of these principles, mainly to topology. Some of these consequences had been established earlier in generic extensions obtained by adding ω_2 Cohen reals to ground models satisfying CH, mostly for the case $\kappa = \omega_2$.

1 Introduction

The last 25 years have seen a furious activity in proving results that are independent of the usual axioms of set theory, that is ZFC. As the methods of these independence proofs (e.g. forcing or the fine structure theory of the constructible universe) are often rather sophisticated, while the results themselves are usually of interest to "ordinary" mathematicians (e.g. topologists or analysts), it has

* The preparation of this paper was supported by the Hungarian National Foundation for Scientific Research grant no. 16391.
** The second author was supported by DFG (grant Ko 490/7-1)

been natural to try to isolate a relatively small number of principles, i.e. independent statements that a) are *simple* to formulate and b) are ***useful*** in the sense that they have many interesting consequences. Most of these statements, we think by necessity, are of combinatorial nature, hence they have been called combinatorial principles.

In this paper we propose to present several new combinatorial principles that are all statements about $\mathcal{P}(\omega)$, the power set of the natural numbers. In fact, they all concern matrices of the form $\langle A(\alpha,n) : \langle \alpha, n\rangle \in \kappa \times \omega\rangle$, where $A(\alpha,n) \subset \omega$ for each $\langle \alpha, n\rangle \in \kappa \times \omega$, and, in the interesting cases, κ is a regular cardinal with $c = 2^{\omega} \geq \kappa > \omega_1$.

We show that these statements are valid in the generic extensions obtained by adding any number of Cohen reals to any ground model V, assuming that the parameter κ is a regular and ω-inaccessible cardinal in V (i.e. $\lambda < \kappa$ implies $\lambda^{\omega} < \kappa$).

Then we present a large number of consequences of these principles, some of them combinatorial but most of them topological, mainly concerning separable and/or countably tight topological spaces. (This, of course, is not surprising because these are objects whose structure depends basically on $\mathcal{P}(\omega)$.)

The above formulated criteria a) and b) as to what constitutes a combinatorial principle are often contrary to each other: for more usefulness one often has to sacrifice some simplicity. It is not clear whether an ideal balance exists between them. It is up to the reader to judge if we have come close to this balance.

We would like to express our thanks to the referees for the careful reading of the manuscript and fot the numerous valuable suggestions.

2 The combinatorial principles

The principles we formulate here are all statements on $\kappa \times \omega$ matrices of subsets of ω claiming - roughly speaking - that all these matrices contain large "submatrices" satisfying certain homogeneity properties.

To simplify the formulation of our results we introduce the following pieces of notation. If S is an arbitrary set and k is a natural number then let

$$(S)^k = \{s \in S^k : |\operatorname{ran} s| = k\}$$

and

$$(S)^{<\omega} = \bigcup_{k<\omega} (S)^k.$$

For $D_0, \ldots, D_{k-1} \subset S$ we let

$$(D_0, \ldots, D_{k-1}) = \{s \in (S)^k : \forall i \in k \ (s(i) \in D_i)\}.$$

Definition 1. If S is a set of ordinals denote by $\mathcal{M}(S)$ the family of all $S \times \omega$-matrices of subsets of ω, that is, $\mathcal{A} \in \mathcal{M}(S)$ if and only if

$$\mathcal{A} = \langle A(\alpha, i) : \alpha \in S, i < \omega\rangle,$$

where $A(\alpha, i) \subset \omega$ for each $\alpha \in S$ and $i < \omega$. If

$$\mathcal{A} = \langle A(\alpha, i) : \alpha \in S, i < \omega \rangle \in \mathcal{M}(S)$$

and $R \subset S$ we define the restriction of $\mathcal{A}$ to R, $\mathcal{A} \restriction R$ in the straightforward way: $\mathcal{A} \restriction R = \langle A(\alpha, i) : \alpha \in R, i < \omega \rangle$. If

$$\mathcal{A} = \langle A(\alpha, i) : \alpha \in S, i < \omega \rangle \in \mathcal{M}(S),$$

$t \in \omega^{<\omega}$ and $s \in (S)^{|t|}$ then we let

$$A(s, t) = \bigcap_{i<|t|} A(s(i), t(i)).$$

Now we formulate our first and probably most important principle that we call $C^s(\kappa)$. We also specify a weaker version of $C^s(\kappa)$ denoted by $C(\kappa)$ because in most of the applications (28, 30, 38, 41, 45, 51) we don't need the full power of $C^s(\kappa)$.

Definition 2. For $T \subset \omega^{<\omega}$ a matrix $\mathcal{A} = \langle A(\alpha, i) : \alpha \in S, i \in \omega \rangle \in \mathcal{M}(S)$ is called *T-adic* if for each $t \in T$ and $s \in (S)^{|t|}$ we have $A(s, t) \neq \emptyset$.

Definition 3. For $\kappa = \mathrm{cf}(\kappa) > \omega$ principle $C^s(\kappa)$ $(C(\kappa))$ is the following statement:
For every $T \subset \omega^{<\omega}$ and $\mathcal{A} \in \mathcal{M}(\kappa)$ we have (1) or (2) below:

(1) there is a stationary (cofinal) set $S \subset \kappa$ such that $\mathcal{A} \restriction S$ is T-adic,
(2) there are $t \in T$ and stationary (cofinal) subsets $D_0, D_1, \ldots, D_{|t|-1}$ of κ such that for every $s \in \langle D_0, \ldots D_{|t|-1} \rangle$ we have

$$A(s, t) = \emptyset.$$

Next we formulate a dual version of principles $C^s(\kappa)$ and $C(\kappa)$. Although we don't yet know any application of principles $\hat{C}^s(\kappa)$ and $\hat{C}(\kappa)$, for the sake of completeness we include their definitions here. Let us remark that we don't know whether $C^s(\kappa)$ $(C(\kappa))$ implies $\hat{C}^s(\kappa)$ $(\hat{C}(\kappa))$ or vice versa.

Definition 4. If $\kappa = \mathrm{cf}(\kappa) > \omega$, then principle $\hat{C}^s(\kappa)$ $(\hat{C}(\kappa))$ is the following statement:
For every $T \subset \omega^{<\omega}$ and $\mathcal{A} \in \mathcal{M}(\kappa)$ we have (1) or (2) below:

(1) there is a stationary (cofinal) set $S \subset \kappa$ such that for each $t \in T$ and $s \in (S)^{|t|}$

$$|A(s, t)| < \omega,$$

(2) there are $t \in T$ and stationary (cofinal) subsets $D_0, D_1, \ldots, D_{|t|-1}$ of κ such that for every $s \in \langle D_0, \ldots D_{|t|-1} \rangle$ we have

$$|A(s, t)| = \omega.$$

Let us remark that in the "plain" dual of principle $C^s(\kappa)$ we should have $|A(s,t)| = \emptyset$ in 4(1) and $|A(s,t)| \neq \emptyset$ in 4(2), but this "principle" is easily provable in ZFC.

The principles $D(\kappa)$ and $D^s(\kappa)$ that we introduce next easily follow from $C(\kappa)$ and $C^s(\kappa)$, respectively, but as their formulation is much simpler, we thought it to be worth while to have them as separate principles. We first give two auxiliary definitions.

Definition 5. If $\mathcal{A} = \langle A(\alpha, i) : \alpha < \kappa, i < \omega \rangle \in \mathcal{M}(\kappa)$, then we set

$$\hat{\mathcal{A}} = \{Y \subset \omega : |\{\alpha < \kappa : \exists i < \omega\ A(\alpha, i) \subset Y\}| = \kappa\}$$

and

$$\hat{\mathcal{A}}^2 = \{Y \subset \omega : \{\alpha < \kappa : \exists i < \omega\ A(\alpha, i) \subset Y\} \text{ is stationary in } \kappa\}.$$

Now we can formulate $D^s(\kappa)$ ($D(\kappa)$) as follows.

Definition 6. For $\kappa = \text{cf}(\kappa) > \omega$ principle $D^s(\kappa)$ ($D(\kappa)$) is the following statement:
If $\mathcal{A} \in \mathcal{M}(\kappa)$ and $\hat{\mathcal{A}}^2$ ($\hat{\mathcal{A}}$) is centered then there is a stationary (cofinal) set $S \subset \kappa$ such that $\mathcal{A} \restriction S$ is $\omega^{<\omega}$-adic.

Theorem 7. *$C^s(\kappa)$ ($C(\kappa)$) implies $D^s(\kappa)$ ($D(\kappa)$).*

Proof. We give the proof only for $D^s(\kappa)$ because the same argument works for $D(\kappa)$.

Let $\mathcal{A} \in \mathcal{M}(\kappa)$ so that $\hat{\mathcal{A}}^2$ is centered and put $T = \omega^{<\omega}$. By $C^s(\kappa)$ either 3(1) or 3(2) holds.

If $S \subset \kappa$ witnesses 3(1) for our T then $\mathcal{A} \restriction S$ is clearly $\omega^{<\omega}$-adic. So it is enough to show that 3(2) can not hold.

Assume, on the contrary, that there are $t \in T = \omega^{<\omega}$ and stationary subsets $D_0, D_1, \ldots, D_{|t|-1}$ of κ such that for each $s \in \langle D_0, \ldots, D_{|t|-1}\rangle$ we have

$$A(s,t) = \emptyset. \tag{+}$$

We can obviously assume that the sets D_i are pairwise disjoint. Let

$$X_i = \bigcup\{A(\delta, t(i)) : \delta \in D_i\}$$

for every $i < |t|$. Then clearly $X_i \in \hat{\mathcal{A}}^2$ for every $i < |t|$, while (+) implies $\bigcap_{i<k} X_i = \emptyset$, contradicting that $\hat{\mathcal{A}}^2$ is centered.

Definition 8. If $\kappa = \text{cf}(\kappa) > \omega$, then principle $F^s(\kappa)$ ($F(\kappa)$) is the following statement:
For every $T \subset \omega^{<\omega}$ and $\mathcal{A} \in \mathcal{M}(\kappa)$ (1) or (2) below holds:

(1) there is a stationary (cofinal) set $S \subset \kappa$ such that

$$|\{A(s,t) : t \in T \text{ and } s \in (S)^{|t|}\}| \leq \omega.$$

(2) there are $t \in T$ and stationary (cofinal) subsets $D_0, D_1, \ldots, D_{|t|-1}$ of κ such that if $s_0, s_1 \in (D_0, \ldots, D_{|t|-1})$ with $s_0(i) \neq s_1(i)$ for each $i < |t|$ then we have

$$A(s_0, t) \neq A(s_1, t).$$

Clearly, if t, $D_0, \ldots, D_{|t|-1}$ satisfy (2) then

$$|\{A(s,t) : s \in (D_0, \ldots, D_{|t|-1})\}| = \kappa.$$

There is a surprising connection between these principles and the dual versions $\hat{C}^s(\kappa)$ $(\hat{C}(\kappa))$ of $C^s(\kappa)$ $(C(\kappa))$, respectively.

Theorem 9. $F^s(\kappa)$ $(F(\kappa))$ *implies* $\hat{C}^s(\kappa)$ $(\hat{C}(\kappa))$.

Proof. Let $\mathcal{A} \in \mathcal{M}(\kappa)$ and $T \subset \omega^{<\omega}$ and apply $F^s(\kappa)$ to $\mathcal{A}$ and T. Assume first that there is a stationary set $S \subset \kappa$ such that the family

$$\mathcal{I} = \{A(s,t) : t \in T \text{ and } s \in (S)^{|t|}\}$$

is countable.

Now for $t \in T$, $i < |t|$ and $I \in \mathcal{I} \cap [\omega]^\omega$ set

$$D(I,t,i) = \{\alpha \in S : A(\alpha, t(i)) \supset I\}.$$

If for some $t \in T$ and $I \in \mathcal{I} \cap [\omega]^\omega$ the set $D(I,t,i)$ is stationary for each $i < |t|$ then this t and the sets $D(I,t,0), \ldots, D(I,t,|t|-1)$ witness 4(2).

So we can assume that for all $t \in T$ and $I \in \mathcal{I} \cap [\omega]^\omega$ the set

$$b(I,t) = \{i < |t| : D(I,t,i) \text{ is non-stationary in } \kappa\}$$

is not empty. Then the set

$$D = \bigcup\{D(I,t,i) : I \in \mathcal{I} \cap [\omega]^\omega, t \in T, i \in b(I,t)\}$$

is not stationary and so $S' = S \setminus D$ is stationary. We claim that S' witnesses 4(1). Assume on the contrary that $t \in T$, $s \in (S')^{|t|}$ and $I = A(s,t)$ is infinite. Then $I \in \mathcal{I} \cap [\omega]^\omega$ and $s(i) \in D(I,t,i)$ for each $i < |t|$. Since $s(i) \notin D$ it follows that $D(I,t,i)$ is stationary for each $i < |t|$, that is, $b(I,t) = \emptyset$, which is a contradiction.

Assume now that there are $t \in T$ and stationary subsets $D_0, D_1, \ldots, D_{|t|-1}$ of κ such that if $s_0, s_1 \in (D_0, \ldots, D_{|t|-1})$ with $s_0(i) \neq s_1(i)$ for each $i < |t|$ then

$$A(s_0, t) \neq A(s_1, t).$$

We show that in this case again 4(2) holds. Indeed, for each $I \in [\omega]^{<\omega}$ pick $s_I \in (D_0, \ldots, D_{|t|-1})$ such that $A(s_I, t) = I$ provided that there is such an s. Let $R = \bigcup\{s_I(i) : I \in [\omega]^{<\omega}, i < |t|\}$ and $D_i' = D_i \setminus R$ for $i < |t|$. Now if $s \in (D_0', \ldots, D_{|t|-1}')$ then for any $I \in [\omega]^{<\omega}$ we have $s_I(i) \neq s(i)$ for each $i < |t|$, hence $I = A(s_I, t) \neq A(s,t)$. As I was an arbitrary element of $[\omega]^{<\omega}$ we conclude that $|A(s,t)| = \omega$.

The following observation is almost trivial.

Proposition 10. *If $\kappa = \mathrm{cf}(\kappa) > c$ then $C^s(\kappa)$ and $F^s(\kappa)$ are valid.*

Proof. Indeed, given $\mathcal{A} \in \mathcal{M}(\kappa)$ and $T \subset \omega^{<\omega}$ there is a stationary set $S \subset \kappa$ such that for any $\alpha, \beta \in S$ and $n \in \omega$ we have $A(\alpha, n) = A(\beta, n)$. Then S witnesses 8(1) and so principle $F^s(\kappa)$ holds. If S does not witness 3(1) then for some $t = \langle n_0, \dots, n_{k-1}\rangle \in T$ we have $\bigcap_{i<k} A_{n_i} = \emptyset$. Thus t and $D_0 = D_1 = \cdots = D_{k-1} = S$ witness 3(2).

As was mentioned in section 1, our principles are of interest only for $\kappa > \omega_1$. In fact, for $\kappa = \omega_1$, they are all false!

To see that $D(\omega_1)$ (so also $C(\omega_1)$) is false we may recall that in [7] we have constructed, in ZFC, a separable, first countable $P_{<\omega}$ space X of size ω_1. (A Hausdorff space X is called $P_{<\omega}$ if the intersection of finitely many uncountable open subsets of X is always non-empty.) We can assume that the underlying set of X is ω_1 and ω is dense in X. For each $\alpha < \omega_1$ let $\{U(\alpha, n) : n < \omega\}$ be a neighbourhood base of α in X. Now consider the $\omega_1 \times \omega$-matrix

$$\mathcal{A} = \langle U(\alpha, n) \cap \omega : \alpha < \omega_1, n < \omega\rangle .$$

Then $B \in \hat{\mathcal{A}}$ if and only if there is an uncountable open set $U \subset X$ such that $U \cap \omega \subset B$. Since X is a $P_{<\omega}$ space it follows that $\hat{\mathcal{A}}$ is centered. But the space X is Hausdorff, so there is not even a two element subset S of X such that $\mathcal{A} \restriction S$ is $\omega^{<\omega}$-adic.

To show that $F(\omega_1)$ (and so also $\hat{C}(\omega_1)$) is false we need the following observation.

Theorem 11. *There is a subfamily $\mathcal{A} = \{A_\alpha : \alpha < \omega_1\}$ of $[\omega]^\omega$ such that for any $n \in \omega$ and $I_0, \dots, I_{n-1} \in [\omega_1]^{\omega_1}$ there are $\gamma_i, \delta_i \in I_i$ for $i < n$ with $\bigcap\{A_{\gamma_i} : i < n\}$ is infinite but $A_{\delta_i} \cap A_{\delta_j}$ is finite for any $i < j < n$.*

Proof. The proof is based on two lemmas which are probably well-known.

Lemma 12. *There is a function $f : [\omega_1]^2 \longrightarrow 2$ such that for any $n \in \omega$ and $I_0, \dots, I_{n-1} \in [\omega_1]^{\omega_1}$ there are $\gamma_i, \delta_i \in I_i$ for $i < n$ such that $f(\gamma_i, \gamma_j) = 0$ and $f(\delta_i, \delta_j) = 1$ for each $i \neq j < n$.*

Proof. We show that the Sierpienski coloring has this property. So let $\{r_\alpha : \alpha < \omega_1\}$ be pairwise different real numbers and for $\alpha < \beta < \omega_1$ put $f(\alpha, \beta) = 0$ iff $r_\alpha < r_\beta$. Given $n \in \omega$ and $I_0, \dots, I_{n-1} \in [\omega_1]^{\omega_1}$ let x_i be a complete accumulation point of $A_i = \{r_\alpha : \alpha \in I_i\}$ with $x_i \neq x_j$ for $i < j < n$. We may assume that $x_0 < x_1 < \cdots < x_{n-1}$. So there are $I'_i \in [I_i]^{\omega_1}$ for $i < n$ such that if $A'_i = \{r_\alpha : \alpha \in I'_i\}$ then $A'_i <_R A'_j$ whenever $i < j < n$. Now pick first $\gamma_0 \in I'_0$, then $\gamma_1 \in I'_1 \setminus (\gamma_0 + 1)$, then $\gamma_2 \in I'_2 \setminus (\gamma_1 + 1)$ and so on. Then we have $f(\gamma_i, \gamma_j) = 0$ for each $i < j < n$. Next we pick $\delta_{n-1} \in I'_{n-1}$, then $\delta_{n-2} \in I'_{n-2} \setminus \delta_{n-1}$, then $\delta_{n-3} \in I'_{n-3} \setminus \delta_{n-2}$ and so on. Then we have $f(\delta_i, \delta_j) = 1$ for $i < j < n$.

A family $\mathcal{I} \subset \mathcal{P}(\omega_1)$ is called *downwards closed* if $\mathcal{P}(I) \subset \mathcal{I}$ for each $I \in \mathcal{I}$. Given a family $\mathcal{A} = \{A_\alpha : \alpha < \omega_1\} \subset \mathcal{P}(\omega)$ and $I, J \in [\omega_1]^{<\omega}$, let

$$A[I, J] = \bigcap\{A_\alpha : \alpha \in I\} \setminus \bigcup\{A_\beta : \beta \in J\}.$$

Put $A[I] = A[I, \emptyset]$.

Clearly, if $\mathcal{A} \subset \mathcal{P}(\omega)$ then $\{I \in [\omega_1]^{<\omega} : |A[I]| = \omega\}$ is downward closed. Our next result is a converse of this statement.

Lemma 13. *If $\mathcal{I} \subset [\omega_1]^{<\omega}$ is downwards closed then there is a family $\mathcal{A} = \{A_\alpha : \alpha < \omega_1\}$ of subsets of ω such that*

$$\mathcal{I} = \{I \in [\omega_1]^{<\omega} : A[I] \text{ is infinite}\}. \qquad (\dagger)$$

Proof. For $\alpha < \omega_1$ write $\mathcal{I}_\alpha = \mathcal{I} \cap [\alpha + 1]^{<\omega}$.

We will define $A_\alpha \subset \omega$ by induction on $\alpha < \omega_1$ so as to satisfy the following inductive hypotheses $(\ddagger)_\alpha$ which is stronger than $(\dagger)$ restricted to α:

$$\mathcal{I}_\alpha = \{I \in [\alpha + 1]^{<\omega} : \forall J \in [\alpha \setminus I]^{<\omega}\ A[I, J] \text{ is infinite.}\} \qquad (\ddagger_\alpha)$$

The induction uses the following elementary fact.

Fact 14 *If $\mathcal{B}$ and $\mathcal{D}$ are countable subfamilies of $[\omega]^\omega$ such that no element of $\mathcal{B}$ is covered by the union of finitely many elements of $\mathcal{D}$ then there is a set $X \subset \omega$ such that*

(i) *$B \cap X$ is infinite for each $B \in \mathcal{B}$,*
(ii) *$B \setminus X$ is infinite for each $B \in \mathcal{B}$,*
(iii) *$D \cap X$ is finite for each $D \in \mathcal{D}$.*

Now, if A_β has been defined and $(\ddagger_\beta)$ holds for all $\beta < \alpha$ then let

$$\mathcal{B} = \{A[I, J] : I \in \mathcal{I}_\alpha, J \in [\alpha]^{<\omega} \wedge I \cap J = \emptyset\},$$

$$\mathcal{D} = \{A[I, J] : I \in [\alpha]^{<\omega} \setminus \mathcal{I}_\alpha, J \in [\alpha]^{<\omega} \wedge I \cap J = \emptyset\}$$

and apply fact 14 to get A_α. It is easy to check that $(\ddagger_\alpha)$ will be satisfied.

To get a family $\mathcal{A}$ satisfying the requirements of theorem 11 take the function f given by lemma 12 and apply lemma 13 to $\mathcal{I} = \{I \in [\omega_1]^{<\omega} : f''I \subset \{0\}\}$.

Theorem 11 yields immediately the following corollary:

Corollary 15. *There is a family $\mathcal{A} = \{A_\alpha : \alpha < \omega_1\}$ such that*

$$\hat{\mathcal{A}} = \{Y \subset \omega : |\{\alpha < \omega_1 : A_\alpha \subset Y\}| = \omega_1\}$$

is centered but for no $S \in [\omega_1]^{\omega_1}$ is $\{A_\alpha : \alpha \in S\}$ linked.

Let us remark that if $\mathcal{A}$ is almost disjoint then $\hat{\mathcal{A}}$ is centered if and only if $\mathcal{A}$ is a strong Luzin gap (i.e. there is no partition of ω into finitely many pieces such that each piece is almost disjoint to uncountably many elements of $\mathcal{A}$). If MA_{ω_1} holds, then there is no strong Luzin gap (see [7, Theorem 3.2]), so in ZFC one can not construct an almost disjoint family $\mathcal{A}$ satisfying the requirements of corollary 15.

The family $\mathcal{A}$ of 15 can be used to give counterexamples to both $D(\omega_1)$ and $C(\omega_1)$, in fact via the same matrix in $\mathcal{M}(\omega_1)$.

Corollary 16. *$C(\omega_1)$ (and so $F(\omega_1)$ too) and $D(\omega_1)$ are both false.*

Proof. Consider the family $\mathcal{A} = \{A_\alpha : \alpha < \omega_1\}$ given by 15. Put $T = \omega^{<\omega}$ and $A(\alpha, i) = A_\alpha$ for each $\alpha < \omega_1$ and $i < \omega$. Then neither 4(1) nor 4(2) can hold for $\mathcal{A} = \langle A(\alpha, i) : \alpha < \omega_1, i < \omega\rangle$ and T. Moreover, the matrix $\mathcal{A}$ clearly contradicts $D(\omega_1)$.

3 Consistency of the principles in the Cohen model.

A cardinal κ is *ω-inaccessible* if $\lambda^\omega < \kappa$ holds for each $\lambda < \kappa$. Given any infinite set I we denote by $\mathcal{C}_I$ the poset $\mathrm{Fn}(I, 2, \omega)$, i.e. the standard one adding $|I|$-many Cohen reals.

In this section we prove that if κ is a regular ω-inaccessible cardinal in some ground model V and we add λ-many Cohen reals to V, where λ is an arbitrary cardinal, then in the extension the principles $C^s(\kappa)$, $\hat{C}^s(\kappa)$ and $F^s(\kappa)$ are all satisfied. As we remarked in section 2 above the case $\kappa > \lambda$ is trivial, while the case $\kappa < \lambda$ can be reduced to the case $\kappa = \lambda$.

Since the proof of the latter is long and technical, we first sketch the main idea. So let us be given a matrix $\mathcal{A} \in \mathcal{M}(\kappa)$ and a set $T \subset \omega^{<\omega}$ in $V[G]$, where G is $\mathcal{C}_\kappa$-generic over V. In the first part of the proof we find a set $I \in [\kappa]^\omega$ and a stationary set $S \subset \kappa$ such that in $V[G \restriction I]$ the sequences $\langle A(\alpha, i) : i < \omega\rangle$ for $\alpha \in S$ have also pairwise isomorphic names with disjoint supports (contained in $\kappa \setminus I$). This reduction, carried out in lemma 22, will be the place where we use that κ is regular and ω-inaccessible in V. In the second part of the proof, using slightly different arguments for $C^s(\kappa)$ and for $F^s(\kappa)$, we show that if some $\mathcal{A} \in \mathcal{M}(S)$ has names with these properties then either S witnesses 3(1) (or 8(1), respectively) or some stationary sets $D_i \subset S$ witness 3(2) (or 8(2), respectively). In this second step we don't use that κ is ω-inaccessible or regular.

In our forcing arguments we follow the notation of Kunen [11]. Let us first recall definition [11, 5.11].

Definition 17. A $\mathcal{C}_I$-name $\dot{B}$ of a subset of some ordinal μ is called *nice* if for each $\nu < \mu$ there is an antichain $B_\nu \subset \mathcal{C}_I$ such that

$$\dot{B} = \{\langle p, \nu\rangle : \nu \in \mu \wedge p \in B_\nu\} = \bigcup\{B_\nu \times \{\hat{\nu}\} : \nu \in \mu\}.$$

We let $\mathrm{supp}(\dot{B}) = \bigcup\{\mathrm{dom}(p) : p \in \bigcup_{\nu<\mu} B_\nu\}$.

It is well-known (see e.g. lemma [11, 5.12]) that every set of ordinals in $V[G]$ has a nice name in V.

If φ is a bijection between two sets I and J then φ lifts to a natural isomorphism between $\mathcal{C}_I$ and $\mathcal{C}_J$, which will be also denoted by φ, as follows: for $p \in \mathcal{C}_I$ let $\text{dom}(\varphi(p)) = \varphi'' \text{dom}(p)$ and $\varphi(p)(\varphi(\xi)) = p(\xi)$. Moreover φ also generates a bijection between the nice $\mathcal{C}_I$-names and the nice $\mathcal{C}_J$-names (see [11, 7.12]): if $\dot{B}$ is a nice $\mathcal{C}_I$-name then let $\varphi(\dot{B}) = \{\left\langle \varphi(p), \check{\xi} \right\rangle : \left\langle p, \check{\xi} \right\rangle \in \dot{B}\}$. If I and J are sets of ordinals with the same order type then $\varphi_{I,J}$ is the natural order-preserving bijection from I onto J.

Definition 18. Assume $I, J \subset \kappa$, moreover $\dot{A}_i$ and $\dot{B}_i$ are nice $\mathcal{C}_\kappa$-names of subsets of ω for $i < \omega$, such that $\text{supp}(\dot{A}_i) \subset I$ and $\text{supp}(\dot{B}_i) \subset J$. We say that the structures of names $\left\langle I, \dot{A}_i : i < \omega \right\rangle$ and $\left\langle J, \dot{B}_i : i < \omega \right\rangle$ are *twins* if I and J have the same order type and for the order preserving bijection $\varphi_{I,J}$ we have

(1) $\varphi_{I,J}$ is the identity on $I \cap J$.
(2) $\varphi_{I,J}(\dot{A}_i) = \dot{B}_i$ for each $i < \omega$.

Definition 19. Assume that $I \subset \kappa$, G is a $\mathcal{C}_\kappa$-generic filter over V and $H = G \restriction I$. If $\dot{B}$ is a nice $\mathcal{C}_\kappa$-name of a subset of some ordinal μ we define in $V[H]$ the $\mathcal{C}_{\kappa \setminus I}$ name $\pi^H(\dot{B})$ as follows:

$$\pi^H(\dot{B}) = \{\langle p \restriction \kappa \setminus I, \check{\nu} \rangle : \langle p, \check{\nu} \rangle \in \dot{B} \wedge p \restriction I \in H\}.$$

Lemma 20. *$\pi^H(\dot{B})$ is a nice $\mathcal{C}_{\kappa \setminus I}$-name in $V[H]$ and*

$$\text{supp}(\pi^H(\dot{B})) \subset \text{supp}(\dot{B}) \setminus I,$$

moreover

$$\text{val}(\pi^H(\dot{B}), G \restriction (\kappa \setminus I)) = \text{val}(\dot{B}, G).$$

Proof. Straightforward from the construction.

Definition 21. Assume that $S \subset \kappa$. A matrix $\dot{\mathcal{B}} = \left\langle \dot{B}(\alpha, i) : \alpha \in S, i < \omega \right\rangle$ of nice $\mathcal{C}_\kappa$-names of subsets of ω is called a *nice S-matrix* if conditions (i) and (ii) below hold:

(i) putting $J_\alpha = \bigcup_{i<\omega} \text{supp}(\dot{B}(\alpha, i))$ the sets $\{J_\alpha : \alpha \in S\}$ are pairwise disjoint,
(ii) the structures of names $\{\left\langle J_\alpha, \dot{B}(\alpha, i) : i < \omega \right\rangle : \alpha \in S\}$ are pairwise twins.

We denote by $\mathcal{N}(S)$ the family of nice S-matrices.

Lemma 22. (Reduction lemma) *Assume that κ is a regular, ω-inaccessible cardinal, G is $\mathcal{C}_\kappa$-generic over V and $\mathcal{A} \in \mathcal{M}(\kappa)$ in $V[G]$. Then there are a countable set $I \subset \kappa$ and a stationary set $S \subset \kappa$ in V such that, in $V[G \restriction I]$, there is $\dot{\mathcal{B}} \in \mathcal{N}(S)$ satisfying $V[G] \models$ "$\mathcal{A}(\alpha, i) = \text{val}(\dot{B}(\alpha, i), G \restriction (\kappa \setminus I))$" for each $\alpha \in S$ and $i \in \omega$.*

Proof. Assume that

$$1_{C_\kappa} \Vdash \text{“}\dot{\mathcal{A}} = \left\langle \dot{A}(\alpha, i) : \alpha < \kappa, i < \omega \right\rangle \in \mathcal{M}(\kappa).\text{”}$$

We can assume that all the names $\dot{A}(i, \alpha)$ are nice. Let $I_\alpha = \bigcup_{i<\omega} \operatorname{supp}(\dot{A}(\alpha, i))$.

We need a strong version of Erdős-Rado Δ-system theorem saying that there is a stationary set $T \subset \kappa$ such that $\{I_\alpha : \alpha \in T\}$ forms a Δ-system with some kernel I, moreover $\sup I < \min I_\alpha \setminus I$ for each $\alpha \in T$. Although this statement is well-known we present a proof because we could not find any reference to it.

Erdős-Rado Theorem *If κ is an ω-inaccessible regular cardinal and $\mathcal{X} = \{X_\alpha : \alpha < \kappa\}$ is a family of countable sets then there is a stationary set $I \subset \kappa$ such that $\{X_\alpha : \alpha \in I\}$ forms a Δ-system.*

Proof. Since $|\cup \mathcal{A}| \le \kappa$ we can assume that $X_\alpha \subset \kappa$. Let $J = \{\alpha < \kappa : \operatorname{cf}(\alpha) = \omega_1\}$. Define the function $f : J \longrightarrow \kappa$ by the stipulation $f(\alpha) = \sup(X_\alpha \cap \alpha)$. Since X_α is countable and $\operatorname{cf}(\alpha) = \omega_1$ we have $f(\alpha) < \alpha$, i.e. the function f is regressive on the stationary set J. So by the Fodor lemma, f is constant on a stationary set $K \subset J$. Say $f''K = \{\nu\}$. For $\alpha \in K$ let $h(\alpha) = X_\alpha \cap \nu$. Since $\nu < \kappa$ it follows that the range of h is of size $|\nu|^\omega < \kappa$. But K is stationary, so there is a stationary $M \subset K$ such that h is constant on M, say $h''M = \{A\}$.

For $\alpha \in \kappa$ let $g(\alpha) = \sup X_\alpha$. Then the set

$$C = \{\beta < \kappa : g(\gamma) < \beta \text{ for each } \gamma < \beta\}$$

is club in κ. Let $I = M \cap C$. We show that $\{X_\alpha : \alpha \in I\}$ forms a Δ-system with kernel A. Let $\alpha, \beta \in I$, $\alpha < \beta$. Then $X_\alpha \cap X_\beta = (X_\alpha \cap (X_\beta \cap \beta)) \cup (X_\alpha \cap (X_\beta \setminus \beta)$.

But $(X_\beta \cap \beta) = A$ and so $(X_\alpha \cap (X_\beta \cap \beta)) = A$. Since $\beta \in C$ it follows that $g(\alpha) < \beta$, i.e. $X_\alpha \subset \beta$ and so $(X_\alpha \cap (X_\beta \setminus \beta) = \emptyset$. Putting together these two equations we obtain $X_\alpha \cap X_\beta = A$ which was to be proved.

Since $2^\omega < \kappa = \operatorname{cf}(\kappa)$ and there are only 2^ω different isomorphism types of structures of names there is a stationary set $S \subset T$ such that the structures of names $\{\left\langle I_\alpha, \dot{A}(\alpha, i) : i < \omega \right\rangle : \alpha \in S\}$ are pairwise twins.

From now on we work in $V[G \restriction I]$. Let $\dot{B}(\alpha, i) = \pi^{G \restriction I}(\dot{A}(\alpha, i))$ for $\alpha \in S$ and $i \in \omega$. Then $\operatorname{supp}(\dot{B}(\alpha, i)) \subset J_\alpha = I_\alpha \setminus I$ and the structures of names $\left\langle J_\alpha, \dot{B}(\alpha, i) : i < \omega \right\rangle$ are pairwise twins by lemma 20 above.

Thus $\mathcal{B} = \left\langle \dot{B}(\alpha, i) : \alpha \in S, i < \omega \right\rangle \in \mathcal{N}(S)$.

Definition 23. Assume that $S \subset \kappa$. A sequence $\dot{\mathcal{B}} = \left\langle \left\langle J_\alpha, \dot{B}_\alpha \right\rangle : \alpha \in S \right\rangle$ is called a *nice S-sequence* if conditions (i) and (ii) below hold:

(i) $J_\alpha \in [\kappa]^\omega$, $\dot{B}_\alpha$ is a nice C_{J_α}-name, and J_α for $\alpha \in S$ are pairwise disjoint,

(ii) the structures of names $\left\langle J_\alpha, \dot{B}_\alpha \right\rangle$ for $\alpha \in S$ are pairwise twins.

We denote by $\mathcal{S}(S)$ the family of nice S-sequences.

Lemma 24. (Homogeneity lemma)
Assume that $S \subset \kappa$ and $\dot{\mathcal{B}} = \left\langle \left\langle J_\alpha, \dot{B}_\alpha \right\rangle : \alpha \in S \right\rangle$ is a nice S-sequence. If $\varphi(x_0, x_1, \ldots, x_{n-1}, z)$ is a formula with free variables $x_0, x_1, \ldots, x_{n-1}, z$ and Z is an element of the ground model, then (1) or (2) below holds:

(1) $1_{\mathcal{C}_\kappa} \Vdash$ "$\varphi(\dot{B}_{s(0)} \dot{B}_{s(1)}, \ldots, \dot{B}_{s(n-1)}, \hat{Z})$ *for all* $s \in (S)^n$ ",
(2) *for some $r \in \mathcal{C}_\kappa$ we have*
$r \Vdash$ " *there are subsets $\dot{D}_0, \dot{D}_1, \ldots, \dot{D}_{n-1}$ of S such that*
(a) *for each $i < n$ and $A \in [S]^\omega \cap V$ we have $\dot{D}_i \cap A \neq \emptyset$,*
(b) $\neg\varphi(\dot{B}_{s(0)} \dot{B}_{s(1)}, \ldots, \dot{B}_{s(n-1)}, \hat{Z})$ *for all* $s \in (\dot{D}_0, \dot{D}_1, \ldots, \dot{D}_{k-1})$ ".

Proof. Assume that (1) fails, that is, there are $p \in \mathcal{C}_\kappa$ and $s \in (S)^k$ such that

$$p \Vdash \text{“}\neg\varphi(\dot{B}_{s(0)}, \ldots, \dot{B}_{s(n-1)}, \hat{Z})\text{”}.$$

Let $J = \bigcup_{i<k} J_{s(i)}$ and $p' = p \restriction J$ and $r = p \setminus p'$. Since the sets J_α are pairwise disjoint we can assume that $\mathrm{dom}(r) \cap J_\alpha = \emptyset$ for each $\alpha \in S$.

For $\langle \alpha, \beta \rangle \in S^2$ we denote by $\varphi_{\alpha,\beta}$ the natural order preserving bijection between J_α and J_β. For $\beta \in S$ and $i < k$ let $p(\beta, i) = \varphi_{s(i),\beta}(p \restriction J_{s(i)})$. For $i < k$ define the $\mathcal{C}_\kappa$-name $\dot{D}_i$ of a subset of S as follows: $\dot{D}_i = \{\left\langle p(\beta, i), \hat{\beta} \right\rangle : \beta \in S\}$. Then

$$V[G] \models \text{“}D_i = \mathrm{val}(\dot{D}_i, G) = \{\beta \in S : p(\beta, i) \in G\}\text{”},$$

where G is $\mathcal{C}_\kappa$-generic over V. Since the supports of $p(\beta, i)$ for $\beta \in S$ are pairwise disjoint a standard density argument gives that $D_i \cap A \neq \emptyset$ whenever $A \in [S]^\omega \cap V$, hence (a) holds.

To show (b) assume that $r \in G$ and

$$V[G] \models \text{“}u \in (D_0, \ldots, D_{k-1}).\text{”}$$

Since u is finite we have $u \in V$. Let $J^* = \bigcup_{i<k} J_{u(i)}$ and $\psi = \bigcup_{i<k} \varphi_{u(i),s(i)}$. Then ψ is a bijection between J^* and J and so it extends to isomorphisms between $\mathcal{C}_{J^*}$ and $\mathcal{C}_J$, and between the families of nice $\mathcal{C}_{J^*}$-names and of nice $\mathcal{C}_J$-names. Let Ψ be the natural extension of ψ to a permutation of κ:

$$\Psi(\nu) = \begin{cases} \psi(\nu) & \text{if } \nu \in J^*, \\ \psi^{-1}(\nu) & \text{if } \nu \in J, \\ \nu & \text{if } \nu \in \kappa \setminus (J \cup J^*). \end{cases}$$

Then Ψ extends to an automorphism of $\mathcal{C}_\kappa$, and also to an automorphism of nice $\mathcal{C}_\kappa$-names. Clearly if $q \in \mathcal{C}_{J^*}$ and $\dot{B}$ is a nice $\mathcal{C}_{J^*}$-name then $\psi(q) = \Psi(q)$ and $\psi(\dot{B}) = \Psi(\dot{B})$. Observe that $\Psi(r) = r$ and $\Psi(\hat{Z}) = \hat{Z}$.

Let $G^* = \Psi''G$. Then G^* is also a $\mathcal{C}_\kappa$-generic filter over V and since $\Psi(\dot{B}_{u(i)}) = \dot{B}_{s(i)}$ it follows that

$$\mathrm{val}(\dot{B}_{u(i)}, G) = \mathrm{val}(\dot{B}_{s(i)}, G^*). \qquad (\bullet)$$

But $p(u(i),i) \in G$, so $p \restriction J_{s(i)} = \psi(p(u(i),i)) \in G^*$. Thus $p = r \cup \bigcup_{i<k} p \restriction J_{s(i)} \in G^*$ as well. Since $p \Vdash \neg\varphi(\dot{B}_{s(0)}, \ldots, \dot{B}_{s(n-1)}, \dot{Z})$ and so $V[G^*] \models$ "$\neg\varphi(B_{s(0)}, \ldots, B_{s(n-1)}, Z)$", by (•) this implies

$$V[G] \models \text{“}\neg\varphi(B_{u(0)}, \ldots, B_{u(n-1)}, Z)\text{”}$$

which was to be proved.

Theorem 25. *If κ is a regular, ω-inaccessible cardinal then for each cardinal λ we have*

$$V^{\mathcal{C}_\lambda} \models \text{“}C^s(\kappa) \text{ and } \hat{C}^s(\kappa) \text{ hold.”}$$

Proof. We deal only with $C^s(\kappa)$ because the same argument works for $\hat{C}^s(\kappa)$. As we observed in section 2 we can assume that $\kappa \leq \lambda$. First we investigate the case $\lambda = \kappa$.

Assume that

$$1_{\mathcal{C}_\kappa} \Vdash \text{“}\dot{\mathcal{A}} = \left\langle \dot{A}(\alpha, i) : \alpha < \kappa; i < \omega \right\rangle \in \mathcal{M}(\kappa) \text{ and } T \subset \omega^{<\omega}.\text{”}$$

Applying the reduction lemma 22 and that T is countable we can find a countable set $I \subset \kappa$ and a stationary set $S \subset \kappa$ in V and a nice S-matrix $\mathcal{B}$ in $V[G \restriction I]$ such that

$$V[G] \models \text{“}\mathrm{val}(\dot{A}(\alpha, i), G) = \mathrm{val}(\dot{B}(\alpha, i), G \restriction (\kappa \setminus I))\text{”}$$

for $\alpha \in S$ and $i \in \omega$, moreover $T \in V[G \restriction I]$.

We show that for each $q \in \mathcal{C}_\kappa$ there is a condition $r \leq q$ in $\mathcal{C}_\kappa$ such that $r \Vdash$ "3(1) or 3(2) holds". Let $I' = I \cup \mathrm{dom}(q)$.

For each $t \in T$ let $\varphi_t(x_0, \ldots, x_{|t|-1})$ be the following formula:

$$\varphi(\langle B_{0,k} : k < \omega \rangle, \ldots, \langle B_{|t|-1,k} : k < \omega \rangle) \iff \bigcap_{i<|t|} B_{i,t(i)} \neq \emptyset.$$

Applying the homogeneity lemma 24 to $V[G \restriction I']$ as our ground model and to every φ_t we get that either $q \Vdash$ "3(1) holds" or $q \cup p \Vdash$ "3(2) holds". Let us remark that 24(2)(a) implies that as S is stationary, so is each D_i.

Thus we have proved the theorem in the case $\kappa = \lambda$. If $\lambda > \kappa$ and $\mathcal{A} \in (\mathcal{M}(\kappa))^{V[G]}$, where G is $\mathcal{C}_\lambda$-generic over V, then there is $J \in [\lambda]^\kappa$ such that $\mathcal{A} \in V[G \restriction J]$. The stationary sets that witness 3(1) or 3(2) in $V[G \restriction J]$ remain stationary in $V[G]$, and so we are done.

Theorem 26. *If κ is a regular, ω-inaccessible cardinal then for each cardinal λ we have*

$$V^{\mathcal{C}_\lambda} \models \text{“}F^s(\kappa) \text{ holds.”}$$

Proof. As in 25 the important case is when $\lambda = \kappa$, because the case $\lambda < \kappa$ is trivial and the case $\kappa < \lambda$ can be reduced to the case $\kappa = \lambda$.

So assume that

$$1_{\mathcal{C}_\kappa} \Vdash \text{“}\dot{A} = \left\langle \dot{A}(\alpha, i) : \alpha < \kappa, i < \omega \right\rangle \in \mathcal{M}(\kappa).\text{”}$$

Applying lemma 22 we can find a countable set $I \subset \kappa$, a stationary set $S \subset \kappa$ in V and in $V[G \restriction I]$ a nice S-matrix $\mathcal{B}$ such that $V[G] \models$ “$A(\alpha, i) = \text{val}(\dot{B}(\alpha, i), G \restriction (\kappa \setminus I))$” for each $\alpha \in S$ and $i \in \omega$, moreover $T \in V[G \restriction I]$.

We need the following lemma that is probably well-known.

Lemma 27. *If H is a $\mathcal{C}_\kappa$-generic filter over V and I, J are disjoint subsets of κ then*

$$V[H] \models \text{“}\mathcal{P}(\omega) \cap V[H \restriction I] \cap V[H \restriction J] = \mathcal{P}(\omega) \cap V.\text{”}$$

Proof. Assume that $\dot{A}$ is a nice $\mathcal{C}_I$-name, $\dot{B}$ is a nice $\mathcal{C}_J$-name, $p \in \mathcal{C}_\kappa$ and $p \Vdash$ “$\dot{A} = \dot{B}$”. We can assume that $\text{dom}(p) \subset I \cup J$. We show that for each $n \in \omega$ we have that $p \restriction I \Vdash$ “$\check{n} \in \dot{A}$” or $p \restriction I \Vdash$ “$\check{n} \notin \dot{A}$”. Indeed, if $p \restriction I \nVdash$ “$\check{n} \in \dot{A}$” then there is a condition $q \le p \restriction I$ in $\mathcal{C}_I$ such that $q \Vdash$ “$\check{n} \notin \dot{A}$” and so $p \cup q \Vdash$ “$\check{n} \notin \dot{B}$”. But $\dot{B}$ is a $\mathcal{C}_J$ name so $(p \cup q) \restriction J = p \restriction J$ forces the same statement, $p \restriction J \Vdash$ “$\check{n} \notin \dot{B}$”. But $p \Vdash$ $\dot{A} = \dot{B}$ and so $p \Vdash$ $\check{n} \notin \dot{A}$ as well. Thus p decides the elements of $\dot{A}$, in other words, $p \Vdash$ “$\dot{A} \in V$”.

To conclude the proof we show that if $q \in \mathcal{C}_\kappa$ then there is a condition $r \le q$ in $\mathcal{C}_\kappa$ such that $r \Vdash$ “8(1) or 8(2) holds”. Let $I' = I \cup \text{dom}(q)$.

For each $t \in T$ let $\varphi_t(x_0, \ldots, x_{|t|-1})$ be the following formula:

$$\varphi(\langle B_{0,k} : k < \omega \rangle, \ldots, \langle B_{|t|-1,k} : k < \omega \rangle) \iff \bigcap_{i<|t|} B_{i,t(i)} \in (\mathcal{P}(\omega))^V.$$

Applying the homogeneity lemma 24 to $V[G \restriction I']$ as our ground model we get that (A) or (B) below holds:

(A) $1_{\mathcal{C}_\kappa} \Vdash$ “ $B(s,t) \in (\mathcal{P}(\omega))^V$ for each $t \in T$ and $s \in (S)^{|t|}$,”

(B) for some $t \in T$ and $p \in \mathcal{C}_\kappa$ we have
 $p \Vdash$ “ there are subsets $\dot{D}_0, \dot{D}_1, \ldots, \dot{D}_{|t|-1}$ of S such that
 (a) for each $A \in [S]^\omega \cap V$ we have $A \cap \dot{D}_i \neq \emptyset$ for each $i < |t|$,
 (b) If $s \in (\dot{D}_0, \dot{D}_1, \ldots, \dot{D}_{|t|-1})$ we have

$$\dot{B}(s,t) \notin (\mathcal{P}(\omega))^V.\text{”}$$

Let $J_\alpha = \bigcup_{i<\omega} \text{supp}(\dot{B}(\alpha, i))$ for $\alpha \in S$ and denote by $\varphi_{\alpha,\beta}$ the natural order preserving bijection between J_α and J_β for $\langle \alpha, \beta \rangle \in S^2$.

Assume first that (A) holds. Fix $t \in T$ and $s \in (S)^{|t|}$. Write $\alpha_i = s(i)$ for $i < |t|$. Since $\mathcal{C}_\kappa$ is c.c.c , there is in V a countable set $\mathcal{I}_t \subset \mathcal{P}(\omega)$ such that $1_{\mathcal{C}_\kappa} \Vdash$ “$\bigcap_{i<k} \dot{B}(\alpha_i, n_i) \in \mathcal{I}_t$”.

Assume that $\langle \delta_0, \ldots, \delta_{|t|-1} \rangle \in (S)^k$.

Let $J^* = \bigcup_{i<k} J_{\delta_i}$, $J = \bigcup_{i<k} J_{\alpha_i}$, and $\psi = \bigcup_{i<k} \varphi_{\delta_i,\alpha_i}$. Then ψ is a bijection between J^* and J and so it lifts up to an isomorphism between $\mathcal{C}_{J^*}$ and $\mathcal{C}_J$ and between the families of nice $\mathcal{C}_{J^*}$-names and nice $\mathcal{C}_J$-names.

Let G be $\mathcal{C}_\kappa$-generic and put $G_0 = G \restriction J^*$. Since $\operatorname{supp}(\dot{B}(\delta_i, n_i)) \subset J^*$ it follows that $\operatorname{val}(\dot{B}(\delta_i, n_i), G) = \operatorname{val}(\dot{B}(\delta_i, n_i), G_0)$. Let $G_1 = \psi'' G_0$. Then G_1 is also a $\mathcal{C}_J$-generic filter and since $\psi(\dot{B}(\delta_i, n_i)) = \dot{B}(\alpha_i, n_i)$ it follows that

$$\operatorname{val}(\dot{B}(\delta_i, n_i), G_0) = \operatorname{val}(\dot{B}(\alpha_i, n_i), G_1). \tag{$\bullet$}$$

Since $1_{\mathcal{C}_\kappa} \Vdash$ "$\bigcap_{i<k} \dot{B}(\alpha_i, n_i) \in \mathcal{I}_t$", by ($\bullet$) we have $1_{\mathcal{C}_\kappa} \Vdash$ "$\bigcap_{i<k} \operatorname{val}(\dot{B}(\delta_i, n_i), G) \in \mathcal{I}_t$" as well. From this it is obvious that we have

$$1_{\mathcal{C}_\kappa} \Vdash \{\dot{B}(t, s) : t \in T \wedge s \in (S)^{|t|}\} \subset \mathcal{I} = \bigcup\{\mathcal{I}_t : t \in T\}.$$

where $\mathcal{I}$ is countable as T is.

Assume now that (A) fails and so (B) holds. Let G be $\mathcal{C}_\kappa$-generic with $p \in G$ and $\langle \gamma_0, \ldots, \gamma_{k-1} \rangle, \langle \delta_0, \ldots, \delta_{k-1} \rangle \in (D_0, \ldots, D_{k-1})$ such that

$$V[G] \models \text{"}\{\gamma_i, \delta_i\} \in [D_i]^2 \text{ for } i < k \text{ are pairs of distict ordinals"}.$$

Let $J^* = \bigcup_{i<k} J_{\gamma_i}$ and $J^\star = \bigcup_{i<k} J_{\delta_i}$. Then $J^* \cap J^\star = \emptyset$, hence by lemma 27 we have $\mathcal{P}(\omega) \cap V[G \restriction J^*] \cap \restriction V[G \restriction J^\star] = \mathcal{P}(\omega) \cap V$ and so $V[G] \models$ "$\bigcap_{i<k} B(\delta_i, n_i) \notin V$" implies that $V[G] \models$ "$\bigcap_{i<k} B(\delta_i, n_i) \neq \bigcap_{i<k} B(\gamma_i, n_i)$"

The theorem is proved.

4 Applications

We start with presenting some combinatorial applications because they are quite simple and so they nicely illustrate the use of our principles.

Kunen [12] proved that if one adds Cohen reals to a model of CH then in the generic extension there is no strictly $\subset^*$-increasing chain of subsets of ω of length ω_2. The first theorem we prove easily yields a corollary which is a generalization of Kunen's above result.

Theorem 28. *If $C(\kappa)$ holds then for each $\mathcal{A} \subset [\omega]^\omega$ of size κ either*

(a) $\exists \mathcal{B} \in [\mathcal{A}]^\kappa \; \forall B \neq B' \in \mathcal{B} \; |B \setminus B'| = \omega$

or

(b) $\exists X \in [\omega]^\omega \; |\{A \in \mathcal{A} : A \subset X\}| = |\{A \in \mathcal{A} : X \subset^* A\}| = \kappa$.

Proof. Fix a 1–1 enumeration $\{A_\xi : \xi < \kappa\}$ of $\mathcal{A}$. Let $A(\xi, 2n) = A_\xi \setminus n$ and $A(\xi, 2n+1) = (\omega \setminus A_\xi) \setminus n$. Put $T = \{\langle 2i, 2i+1\rangle : i \in \omega\}$. If $S \in [\kappa]^\kappa$ witnesses 3(1), then $\mathcal{B} = \{A_\xi : \xi \in S\}$ satisfies (a). If on the other hand $D, E \in [\kappa]^\kappa$, $D \cap E = \emptyset$, with $\langle 2i, 2i+1\rangle \in T$ show that 3(2) holds, then let

$$X = \cup\{A_\xi : \xi \in D\}.$$

Then $A_\xi \subset X$ for each $\xi \in D$ and $X \setminus i \subset A_\zeta$ for each $\zeta \in E$.

Theorem 28 yields immediately that if $C(\kappa)$ holds then there is no strictly $\subset^*$-increasing chain $\mathcal{T} \subset [\omega]^\omega$ of length κ. However, as the referee pointed out, this follows already from the weaker assumption $D(\kappa)$ and from the dual principle $\hat{C}(\kappa)$.

Theorem 29. *If $D(\kappa)$ or $\hat{C}(\kappa)$ holds, then κ is not embeddable into $P(\omega)/fin$.*

Proof. Assume that $\{A_\alpha : \alpha < \kappa\}$ is a strictly $\subset^*$-increasing chain in $[\omega]^\omega$. For $\alpha < \kappa$ and $n < \omega$ let

$$A(\alpha, n) = \begin{cases} \omega \setminus A_\alpha & \text{if } n = 0, \\ A_\alpha \setminus n & \text{otherwise.} \end{cases}$$

We show that $\mathcal{A} = \langle A(\alpha, n) : \alpha < \kappa, n < \omega\rangle$ is a counterexample to $D(\kappa)$ and $\hat{C}(\kappa)$.

To see that $\hat{\mathcal{A}}$ is centered, observe that if $Y \in \hat{\mathcal{A}}$, then there is $\alpha_Y < \kappa$ such that $A_\beta \subset A_{\alpha_Y} \subset^* Y$ for each $\alpha_Y < \beta < \kappa$. Thus if $Y_0, \ldots, Y_{n-1} \in \hat{\mathcal{A}}$ then taking $\alpha = \max\{\alpha_{Y_i} : i < n\}$ we have $A_{\alpha+1} \subset A_\alpha \subset^* \bigcap_{i<n} Y_i$. On the other hand, if $\alpha < \beta < \kappa$ then $A_\alpha \subset^* A_\beta$, thus $A_\alpha \cap (\omega \setminus A_\beta)$ is finite and so $A(\alpha, k) \cap A(\beta, 0) = \emptyset$ for some large enough k. Thus there is no S of size κ (even of size 2) such that $\mathcal{A} \restriction S$ is $\omega^{<\omega}$-adic. Thus $D(\kappa)$ fails.

Next, let $t = \langle 0, 1\rangle$ and $T = \{t\}$. Then for any $S \subseteq \kappa$ of size κ, if $s \in S^2$ is such that $s(0) < s(1)$, then $A(s, t)$ is infinite. Hence 4.(1) does not hold. Likewise if $D_0, D_1 \subseteq \kappa$ are of size κ, taking $s \in (D_0, D_1)^2$ such that $s(0) > s(1)$, $A(s, t)$ is finite. This shows that 4.(2) does not hold, i.e. $\hat{C}(\kappa)$ fails.

The next theorem can be considered as a kind of dual to 28.

Theorem 30. *If $C(\kappa)$ holds then for each $\mathcal{A} \subset [\omega]^\omega$ of size κ and for each natural number k either*

(a) *there is a family $\mathcal{B} \in [\mathcal{A}]^\kappa$ such that for each $\mathcal{B}' \in [\mathcal{B}]^k$ we have $|\bigcap \mathcal{B}'| = \omega$*

or

(b) *there are k subfamilies $\mathcal{B}_0, \ldots, \mathcal{B}_{k-1}$ of $\mathcal{B}$ of size κ such that*

$$|\bigcap_{i<k} \bigcup \mathcal{B}_i| < \omega.$$

Proof. Fix a 1–1 enumeration $\{A_\xi : \xi < \kappa\}$ of $\mathcal{A}$. Let $A(\xi, n) = A_\xi \setminus n$ and put $T = \omega^k$. If $C \in [\kappa]^\kappa$ witnesses 3(1), then $\mathcal{B} = \{A_\xi : \xi \in C\}$ satisfies (a). If $D_0, \ldots, D_{k-1} \in [\kappa]^\kappa$ and $\langle n_0, \ldots, n_{k-1}\rangle \in T$ show that 3(2) holds, then we can assume that the D_i are pairwise disjoint and if we set $\mathcal{B}_i = \{A_\xi : \xi \in D_i\}$ then we have

$$\bigcap_{i<k} \bigcup \mathcal{B}_i \subset \max_{i<k} n_i.$$

Remark. In theorem 30 we can not replace (a) with the following (slightly stronger) condition (a'):

(a') there is a family $\mathcal{B} \in [\mathcal{A}]^\kappa$ such that for each $\mathcal{B}' \in [\mathcal{B}]^{k+1}$ we have $|\bigcap \mathcal{B}'| = \omega$,

and if $k > 2$ then (b) can not be replaced by

(b') there are pairwise disjoint subsets $X_0, \ldots, X_{k-1}$ of ω such that for each $i < k$ we have $\{A \in \mathcal{A} : A \subset^* X_i\} \neq \emptyset$,

because for each $k \in \omega$ one can construct in ZFC a family $\mathcal{A} \subset [\omega]^\omega$ of size 2^ω such that $\bigcap \mathcal{A}'$ is finite for every $\mathcal{A}' \in [\mathcal{A}]^{k+1}$ but $\bigcap \mathcal{A}'$ is infinite whenever $\mathcal{A}' \in [\mathcal{A}]^k$. Indeed, let $T = 2^{<\omega}$ be the Cantor tree, and for $n < \omega$ let $C_n = 2^n$ be the nth-level of T. For each $f \in 2^\omega$ let

$$A(f) = \bigcup_{n<\omega} \{X \in [C_n]^k : f \restriction n \in X\}$$

and $\mathcal{A} = \{A(f) : f \in 2^\omega\}$. If $\mathcal{B}' \subset 2^\omega$ and $n < \omega$ then $\bigcap\{A(f) : f \in \mathcal{B}'\} \cap [C_n]^k \neq \emptyset$ iff $|\{f \restriction n : f \in \mathcal{B}'\}| \leq k$. Thus $\mathcal{A}$ satisfies our requirements. This example is due to A. Hajnal and included here with his kind permission.

Next we prove a consequence of theorem 30, but first we give a definition.

Definition 31. Let κ be a regular cardinal and $\mathcal{A} \subset [\omega]^\omega$ be an almost disjoint family. $\mathcal{A}$ is called a *κ-Luzin gap* if $|\mathcal{A}| = \kappa$ and there is no $X \in [\omega]^\omega$ such that both $|\{A \in \mathcal{A} : |A \setminus X| < \omega|\} = \kappa$ and $|\{A \in \mathcal{A} : |A \cap X| < \omega\}| = \kappa$. A *Luzin-gap* is an ω_1-Luzin gap.

An ω_1-Luzin gap can be constructed in ZFC and simple forcings give models in which there are 2^ω-Luzin gaps while 2^ω is as large as you wish. The next corollary of theorem 30 implies that one can not construct ω_2-Luzin gaps from the assumption $2^\omega \geq \omega_2$ alone .

Corollary 32. *If $C(\kappa)$ holds then there is no κ-Luzin gap.*

Proof. Assume that $\mathcal{A} \subset [\omega]^\omega$ is an almost disjoint family of size κ. Then we can not get a even a two element subfamily $\mathcal{B} \subset \mathcal{A}$ satisfying 30(a). So applying theorem 30 for this $\mathcal{A}$ and for $k = 2$ there are subfamilies $\mathcal{B} \subset \mathcal{A}$ and $\mathcal{D} \subset \mathcal{A}$ of size κ such that $(\bigcup \mathcal{B}) \cap (\bigcup \mathcal{D})$ is finite. Hence $X = \bigcup \mathcal{B}$ witnesses that $\mathcal{A}$ is not a κ-Luzin gap.

We have one more theorem of this type.

Theorem 33. *If $C(\kappa)$ holds then for each $\mathcal{A} \subset [\omega]^{\omega}$ of size κ either*

(a) *there is a centered subfamily $\mathcal{B} \subset \mathcal{A}$ of size κ,*

or

(b) *for some $k < \omega$ there are subfamilies $\mathcal{B}_0, \ldots, \mathcal{B}_{k-1}$ of $\mathcal{B}$ of size κ such that*

$$|\bigcap_{i<k} \bigcup \mathcal{B}_i| < \omega.$$

Proof. We can argue as in the proof of theorem 30 using $T = \omega^{<\omega}$ instead of $T = \omega^k$.

For $\mathcal{A} \subset \mathcal{P}(\omega)$ and $k < \omega$ let $\mathcal{A}_{(k)} = \{\cap \mathcal{A}' : \mathcal{A}' \in [\mathcal{A}]^k\}\}$.
Put $\mathcal{A}_{(<\omega)} = \bigcup_{k<\omega} \mathcal{A}_{(k)}$.

Theorem 34. *If $F(\kappa)$ holds then for each family $\mathcal{A} \subset \mathcal{P}(\omega)$ of size κ and for each natural number k either*

(a) $|\mathcal{A}_{(k)}| = \kappa$

or

(b) *there is a subfamily $\mathcal{B} \subset \mathcal{A}$ of size κ such that $|\mathcal{B}_{(k)}| \leq \omega$.*

Proof. Fix a 1–1 enumeration $\{A_\alpha : \alpha < \kappa\}$ of $\mathcal{A}$, let $T = \omega^k$ and consider the matrix $\mathcal{A}' = \langle A(\alpha, n) : \alpha < \kappa; n < \omega \rangle \in \mathcal{M}(\kappa)$ defined by the stipulation $A(\alpha, n) = A_\alpha$. Apply $F(\kappa)$. If 8(2) holds, then $|\mathcal{A}_{(k)}| = \kappa$. If $S \in [\kappa]^\kappa$ witnesses 8(1) then subfamily $\mathcal{B} = \{A_\alpha : \alpha \in S\}$ satisfies $|\mathcal{B}_{(k)}| = \omega$.

Theorem 35. *If $F(\kappa)$ holds then for each family $\mathcal{A} \subset \mathcal{P}(\omega)$ of size κ either*

(a) *there is a natural number k such that $|\mathcal{A}_{(k)}| = \kappa$*

or

(b) *there is a subfamily $\mathcal{B} \subset \mathcal{A}$ of size κ such that $|\mathcal{B}_{(<\omega)}| \leq \omega$.*

Proof. We can argue as in the proof of theorem 34 using $T = \omega^{<\omega}$ instead of $T = \omega^k$.

Now we turn to applying our principles to topology. We start with an application of the relatively weak principle $D(\kappa)$.

A. Dow [2] proved that if we add ω_2 Cohen reals to a model of GCH then in the generic extension $\beta\omega$ can be embedded into every separable, compact T_2 space of size $> c = \omega_2$. Here we show that $c = \omega_2 = 2^{\omega_1}$ together with $D(\omega_2)$ suffice to imply this statement.

First we need a lemma based on the observation that large separable spaces contain many "similar" points.

Given a topological space X and a point $x \in X$ we denote by $\mathcal{V}_X(x)$ the neighbourhood filter of x in X, that is, $\mathcal{V}_X(x) = \{U \subset X : x \in \mathrm{int}_X(U)\}$. If D is a dense subset of X let $\mathcal{V}_X(x) \upharpoonright D = \{U \cap D : D \in \mathcal{V}_X(x)\}$. We omit the subscript X if it may not cause any confusion.

In section 2 we defined the operation $\hat{\mathcal{A}}$ for $\mathcal{A} \in \mathcal{M}(\kappa)$. By an abuse of notation we define $\mathcal{A}$ for every family $\mathcal{A}$ of subsets of ω as follow:

$$\mathcal{A} = \{X \subset \omega : |\mathcal{A} \cap \mathcal{P}(X)| = |\mathcal{A}|\}.$$

Lemma 36. *Assume that X is a separable regular topological space of size $> c^{<c}$, where $c = 2^\omega$, $D \in [X]^\omega$, $\overline{D} = X$. Then there are a point $x \in X$ and a family $\mathcal{A} = \{A_\alpha, B_\alpha : \alpha < c\} \subset \mathcal{P}(D)$ such that*

(1) $\overline{A_\alpha} \cap \overline{B_\alpha} = \emptyset$ *for each* $\alpha < c$,
(2) $\hat{\mathcal{A}} \subset \mathcal{V}(x) \upharpoonright D$.

Proof. Fix an enumeration $\{D_\xi : \xi < c\}$ of $\mathcal{P}(D)$ and let $\mathcal{D}_\alpha = \{D_\xi : \xi < \alpha\}$ for $\alpha < c$. For $x \in X$ and $\alpha < c$ let $\mathcal{V}(x,\alpha) = (\mathcal{V}(x) \upharpoonright D) \cap \mathcal{D}_\alpha$. A point $x \in X$ is called *special* if there is an $\alpha < c$ such that $\mathcal{V}(x,\alpha) \neq \mathcal{V}(y,\alpha)$ for each $y \in X \setminus \{x\}$. Clearly there are at most $c^{<c}$ special points in X. Since $|X| > c^{<c}$ we can pick a point $x \in X$ which is not special. Then for each $\alpha < c$ we can find a point $x_\alpha \neq x$ in X such that $\mathcal{V}(x_\alpha, \alpha) = \mathcal{V}(x,\alpha)$. Since X is regular the points x and x_α have neighbourhoods U_α and W_α, respectively, with $\overline{U_\alpha} \cap \overline{W_\alpha} = \emptyset$. Let $A_\alpha = U_\alpha \cap D$ and $B_\alpha = W_\alpha \cap D$.

Now assume that $E \in \hat{\mathcal{A}}$ and pick $\xi < c$ with $E = D_\xi$. We can find $\alpha < c$ such that $\xi < \alpha$ and either $A_\alpha \subset E$ or $B_\alpha \subset E$. Hence $E \in \mathcal{V}(x,\alpha) \cup \mathcal{V}(x_\alpha,\alpha) = \mathcal{V}(x,\alpha)$. Therefore $E \in \mathcal{V}(x) \upharpoonright D$ which was to be proved.

Let us now recall the definition of a μ-dyadic system from [5].

Definition 37. If X is a topological space a family $\{\langle A(\alpha,0), A(\alpha,1)\rangle : \alpha \in \mu\}$ of pairs of closed subsets of X is a *μ-dyadic system* such that

1. $A(\alpha,0) \cap A(\alpha,1) = \emptyset$ for each $\alpha < \mu$,
2. for each $\epsilon \in \mathrm{Fn}(\mu, 2, \omega)$ we have $\bigcap_{\alpha \in \mathrm{dom}(\epsilon)} A(\alpha, \epsilon(\alpha)) \neq \emptyset$.

Theorem 38. *If $D(c)$ holds, X is a separable compact T_2 space of size $> c^{<c}$ then X contains a c-dyadic system, consequently X maps continuously onto $[0,1]^c$ (and so $\beta\omega$ can be embedded into X).*

Proof. Fix a countable dense subset D of X. By lemma 36 there is a family $\mathcal{A} = \{A_\alpha, B_\alpha : \alpha < c\} \subset \mathcal{P}(D)$ such that $\overline{A_\alpha} \cap \overline{B_\alpha} = \emptyset$ for $\alpha < c$ and $\hat{\mathcal{A}}$ is centered. Let $D(\alpha,0) = A_\alpha$, $D(\alpha,1) = A_\alpha$ and $D(\alpha,n) = D$ for $\alpha < \kappa$ and $n \geq 2$ and consider the $\kappa \times \omega$-matrix $\mathcal{D} = \langle D(\alpha,i) : \alpha < \kappa, i < \omega\rangle$. Since $\hat{\mathcal{A}} = \hat{\mathcal{D}}$ we can apply $D(c)$ to get a cofinal $S \subset c$ such that the family $\langle \overline{A_\alpha}, \overline{B_\alpha} : \alpha < c\rangle$ is c-dyadic. Now we can apply theorem [5, 3.18] to get the other consequences.

A topological space X is called *scattered* if every subspace of X has an isolated point. For a scattered space X denote by $X^{(\alpha)}$ the αth Cantor-Bendixon level of X. The height of X, $\text{ht}(X)$, is defined as the minimal α with $X^{(\alpha)} = \emptyset$. Following [14] we call X *thin* if all levels of X are countable.

Since the cardinality of a locally compact, scattered separable space is at most 2^ω by [13], the height of such a space is less then $(2^\omega)^+$. So under CH there is no such a space of height ω_2. I. Juhász and W. Weiss, [9, theorem 4], proved in ZFC that for every $\alpha < \omega_2$ there is a locally compact, scattered thin space X with $\text{ht}(X) = \alpha$. M. Weese asked whether the existence of such a space of height ω_2 follows from $\neg CH$. This question was answered in the negative by W. Just, who proved, [10, theorem 2.13], that if one adds Cohen reals to a model of CH then in the generic extension there are no locally compact scattered thin spaces of height ω_2. On the other hand, J. Baumgartner and S. Shelah, [1], constructed a ZFC model which contains such a space of height ω_2.

The next theorem is a generalization of the above mentioned result of Just.

Theorem 39. *If $C^s(\kappa)$ holds then there is no locally compact, thin scattered space of height κ.*

Proof. Assume on the contrary that there is such a space X. We can assume that $X^{(\alpha)} = \{\alpha\} \times \omega$ for $\alpha < \text{ht}(X)$. For each $\alpha < \text{ht}(X)$ fix compact open neighbourhoods $U(\alpha, n)$ of $\langle \alpha, n\rangle$ for $n \in \omega$ such that $U(\alpha, n) \subset \{\langle\alpha, n\rangle\} \cup \bigcup\{X^{(\beta)} : \beta < \alpha\}$ and the sets $U(\alpha, n)$ for $n < \omega$ are pairwise disjoint.

Put $A(\alpha, 2n) = U(\alpha, n) \cap X^{(0)}$ and $A(\alpha, 2n+1) = X^{(0)} \setminus \bigcup\{U(\alpha, m) : m \leq n\}$. Let

$$T = \{t \in \omega^{<\omega} : t(0) \text{ is even and } t(i) \text{ is odd for } i > 0 \}.$$

Now apply $C^s(\kappa)$ to the matrix $\langle A(\alpha, n) : \alpha < \kappa, n < \omega\rangle \in \mathcal{M}(\kappa)$ and T.

Observe that $\quad A(\beta, 2n) \cap \bigcap_{i<k} A(\alpha_i, 2n_i + 1) = \emptyset \quad$ iff

$$U(\beta, n) \cap X^{(0)} \subset \bigcup_{i<k} U(\alpha_i, n_i) \cap X^{(0)} \text{ iff}$$

$$U(\beta, n) \subset \bigcup_{i<k} U(\alpha_i, n_i).$$

Thus if $t = \langle 2n, 2n_0 + 1, \dots, 2n_{k-1} + 1\rangle \in T$ and $\langle \beta, \alpha_0, \dots, \alpha_{k-1}\rangle \in (\kappa)^{k+1}$ then $A(\beta, 2n) \cap \bigcap_{i<k} A(\alpha_i, 2n_i + 1) = \emptyset$ implies $\beta \leq \max_{i<k} \alpha_i$. This excludes 3(2).

So 3(1) holds, that is we have a stationary set $S \subset \kappa$ such that if $t = \langle 2n, 2n_0 + 1, \dots, 2n_{k-1} + 1\rangle \in T$ and $\langle \beta, \alpha_0, \dots, \alpha_{k-1}\rangle \in (S)^{k+1}$ then

$$A(\beta, n) \cap \bigcap_{i<k} A(\alpha_i, 2n_i + 1) \neq \emptyset,$$

that is

$$U(\beta, n) \setminus \bigcup_{i<k} \bigcup_{j\leq n_i} U(\alpha_i, j) \neq \emptyset.$$

But $U(\beta, n)$ is compact and each $U(\alpha, n)$ is open so it follows that for every $\beta \in S$ and $n \in \omega$ the set

$$D(\beta, n) = U(\beta, n) \setminus \bigcup\{U((\alpha, m) : \alpha \in S \setminus \{\beta\} \wedge m \in \omega\}$$

is not empty. For every such β and n let $\langle\gamma(\beta,n), m(\beta,n)\rangle \in D(\beta,n)$.

Since $X^{(\beta)}$ is dense in $X\setminus\bigcup\{X^{(\alpha)}:\alpha<\beta\}$ for every $\beta\in\kappa$ there is $k(\beta)\in\omega$ such that $\langle\beta,k(\beta)\rangle \in U(\beta^*,0)$, where $\beta^* = \min S\setminus\beta+1$. Thus $\langle\beta,k(\beta)\rangle \notin D(\beta,k(\beta))$ and so $\gamma(\beta,k(\beta)) < \beta$ for each $\beta\in S$. The set S is stationary so there are a stationary set $S'\subset S$, and ordinals $\gamma<\kappa$ and $k,m<\omega$ such that $k(\beta)=k$, $\gamma(\beta,k)=\gamma$ and $m(\beta,k)=m$ whenever $\beta\in S'$. Thus $\langle\gamma,m\rangle\in D(\beta,k)$ for each $\beta\in S'$, while $D(\beta,k)\cap D(\beta',k)=\emptyset$ for any $\{\beta,\beta'\}\in[S']^2$ by the construction. This is a contradiction, hence the theorem is proved.

Remark. It is easy to see that the proof of theorem 39 goes through if, instead of assuming that all levels of the space are countable, we only require that (i) all levels are of size $<\kappa$, (ii) there are stationary many countable levels. This observation yields that if $C^s(\kappa)$ holds, then there is no strong pcf structure.

In [10] W. Just also proved that if one adds at least ω_2 Cohen reals to a model of CH then in the generic extension there is no locally compact, scattered topological space X such that $\mathrm{ht}(X)=\omega_1+1$, $X^{(0)}$ is countable, $|X^{(\alpha)}|\le\omega_1$ for $\alpha<\omega_1$ and $|X^{(\omega_1)}|=\omega_2$. The next theorem shows how to get a generalization of this result from our principles.

Theorem 40. *If* $\mathrm{cf}(\lambda)\ge\omega_1$ *and* $F(\lambda^+)$ *holds then there is no locally compact, scattered topological space* X *such that* $\mathrm{ht}(X)=\lambda+1$, $X^{(0)}$ *is countable,* $|X^{(\alpha)}|\le\lambda$ *for all* $\alpha<\lambda$ *and* $|X^{(\lambda)}|=\lambda^+$.

Proof. Assume on the contrary that X is such a space.

We can assume that $X^{(0)}=\omega$ and that $X^{(\lambda)}=\{\lambda\}\times\lambda^+$. For each $x\in X$ choose a compact open neighbourhood $U(x)$ of X and let $B(x)=U(x)\cap\omega$. Put $\mathcal{B}=\{B(x):x\in X^{(\lambda)}\}$. Let $\mathcal{U}=\mathrm{CO}(X\setminus X^{(\lambda)})$, i.e. the family of compact open subsets of $X\setminus X^{(\lambda)}=\bigcup\{X^{(\alpha)}:\alpha<\lambda\}$. and $\mathcal{F}=\{U\cap\omega:U\in\mathcal{U}\}$. Since X is locally compact it follows that for each $\{x,y\}\in[X^{(\lambda)}]^2$ we have $U(x)\cap U(y)\in\mathcal{U}$ and so $B(x)\cap B(y)\in\mathcal{F}$.

Since $|\mathcal{F}|\le\lambda$, it follows $|\mathcal{B}_{(2)}|\le\lambda<\lambda^+$. Thus, applying theorem 34 for $k=2$ we can get a cofinal set $S\subset\lambda^+$ such that the family

$$\mathcal{I}=\{B(\langle\lambda,\alpha\rangle)\cap B(\langle\lambda,\beta\rangle):\{\alpha,\beta\}\in[S]^2\}$$

is at most countable.

Then there is $\gamma<\lambda$ such that $\mathcal{I}$ is contained in $\mathrm{CO}(\bigcup\{X^{(\gamma')}:\gamma'<\gamma\})$. Therefore $U(y)\cap U(y')\subset\bigcup\{X^{(\gamma')}:\gamma'<\gamma\}$ for each $\{y,y'\}\in[\{\lambda\}\times S]^2$ and so the sets $U(x)\cap X^{(\gamma)}$ for $x\in\{\lambda\}\times S$ are pairwise disjoint and non-empty which contradicts $|X^{(\gamma)}|\le\lambda$.

Following the terminology of [4] a Hausdorff space is called P_2 if it does not contain two uncountable disjoint open sets. Hajnal and Juhász in [4] constructed a ZFC example of a first countable, P_2 space of size ω_1 as well as consistent examples of size 2^ω with 2^ω as large as you wish . On the other hand, using a result of Z. Szentmiklóssy they proved that it is consistent with ZFC that 2^ω

is as large as you wish and there are no first countable P_2 spaces of size $\geq \omega_3$. However their method was unable to replace here ω_3 with ω_2. Our next result does just this because, as is shown in [4], every P_2 space is separable.

Theorem 41. *If $C(\kappa)$ holds then every first countable, separable T_2 topological space X of size κ contains two disjoint open sets U and V of cardinality κ.*

Proof. Let D be a countable dense subset of X. For each $x \in X$ fix a neighborhood base $\{U(x,n) : n \in \omega\}$ of x in X. Apply $C(\kappa)$ to the matrix $< U(x,n) \cap D : x \in X, n < \omega >$ and $T = \omega^2$. Since X is T_2, there is no $S \in [X]^\kappa$ satisfying 3(1). So there are $S_0, S_1 \in [X]^\kappa$ and $n, m \in \omega$ such that $U(x,n) \cap U(y,m) \cap D = \emptyset$ whenever $x \in S_0$ and $y \in S_1$. But D is dense in X, therefore $U = \bigcup\{U(x,n) : x \in S_0\}$ and $V = \bigcup\{U(y,m) : y \in S_1\}$ are disjoint open sets of size κ.

Definition 42. Let X be a topological space and $D \subset X$. We say that D is *sequentially dense in* X iff for each $x \in X$ there is a sequence S_x from D which converges to x. A space Y is said to be *sequentially separable* if it contains a countable sequentially dense subset.

Definition 43. Given a topological space $\langle X, \tau \rangle$ and a subspace $Y \subset X$ a function f is called a *neighbourhood assignment on* Y *in* X iff $f : Y \longrightarrow \tau$ and $y \in f(y)$ for each $y \in Y$.

Our next result says that under $C(\kappa)$ if a sequentially separable space X does not contain a discrete subspace of size κ, (i.e. $\hat{s}(X) \leq \kappa$ using the notation of [5]) then X does not contain left or right separated subspaces of size κ either. This can be written as $\hat{h}(X)\hat{z}(X) \leq \kappa$. Since in [6] a normal, Frechet-Urysohn, separable (hence sequentially separable) space X is forced such that $z(X) \leq \omega_1$ but $h(X) = \omega_2$, this result is not provable in ZFC. First, however, we need a lemma.

Lemma 44. *Assume that $C(\kappa)$ holds. Let X be a sequentially separable space with $Y \subset X$, $|Y| = \kappa$. If f is a neighbourhood assignment on Y in X, then either (a) or (b) below holds:*

(a) *there is $Y' \in [Y]^\kappa$ such that $f(y) \cap Y' = \{y\}$ for each $y \in Y'$ (hence Y' is discrete),*

(b) *there are Y_0, $Y_1 \in [Y]^\kappa$ such that $y \in \overline{f(x)}$ whenever $x \in Y_0$ and $y \in Y_1$.*

Proof. We can assume that $D = \omega$ is sequentially dense in X. For each $y \in Y$ choose a sequence $S_y \subset D$ converging to y. Let $A(y, 2n) = D \setminus f(y)$, $A(y, 2n+1) = S_y \setminus n$, $T = \{\langle 2n, 2m+1 \rangle : n, m \in \omega\}$ and apply $C(\kappa)$. Assume first that $Y' \in [Y]^\kappa$ witnesses 3(1) and let $x \neq y \in Y'$. Then for each $n \in \omega$ we have $(S_y \setminus f(x)) \setminus n \neq \emptyset$, i.e. $S_y \setminus f(x)$ is infinite. But S_y converges to y, so $y \notin f(x)$, and so Y' satisfies (a). Assume now that 3(2) holds. Then there are $Y_0, Y_1 \in [\omega]^\omega$ and $m \in \omega$ such that $(D \setminus f(x)) \cap (S_y \setminus m) = \emptyset$ for each $x \in Y_0$ and $y \in Y_1$. But then $S_y \setminus m \subset f(x)$ hence $y \in \overline{f(x)}$ which was to be proved.

Theorem 45. *If $C(\kappa)$ holds, X is a regular, sequentially separable space with $\hat{s}(X) \leq \kappa$ then $\hat{h}(X)\hat{z}(X) \leq \kappa$.*

Proof. Assume on the contrary that $Y \in [X]^{\kappa}$ and the neighbourhood assignment $f : Y \longrightarrow \tau_X$ witnesses that Y is left (right) separated. We can assume that $Y = \kappa$ and Y is left (right) separated under the natural ordering of κ. Since X is regular we can find a neighbourhood assignment $g : Y \longrightarrow \tau$ with $\overline{g(y)} \subset f(y)$ for each $y \in Y$. Apply lemma 44 to Y and g. Now 44(a) can not hold because $\hat{s}(X) \leq \kappa$, hence there are $Y_0, Y_1 \in [Y]^{\kappa}$ satisfying 44(b). Since both Y_0 and Y_1 are cofinal in $Y = \kappa$ under the natural ordering of the ordinals, applying left (or right) separatedness of Y we can pick $x \in Y_0$ and $y \in Y_1$ such that $y \notin f(x)$. By the choice of g this implies $y \notin \overline{g(x)}$ which contradicts 44(b).

The Sorgenfrey line L is weakly separated and is of size c with $\hat{s}(L) = \omega_1$. This shows that theorem 45 does not remain valid if you put weakly separated subspaces instead of right or left separated ones.

As an easy consequence of 45 we get the following result in which (sequential) separability is no longer assumed. We also note that under CH the assumption of X being Frechet-Urysohn is not necessary in this result.

Theorem 46. *Assume $C(\omega_2)$. If X is regular, Frechet-Urysohn space and* $s(X) = \omega$ *then* $h(X) \leq \omega_1$.

Proof. If $C(\omega_2)$ and X is separable, then by theorem 45 even $s(X) \leq \omega_1$ implies $h(X)\,z(X) \leq \omega_1$. Now, every uncountable space X which is both right and left separated contains an uncountable discrete subspace, hence every right separated subspace of X is (hereditarily) separable. So by the above if $Y \subset X$ is right separated then $|Y| \leq \omega_1$, i.e. $h(X) \leq \omega_1$.

In [8] we investigated the following question: What makes a space have weight larger than its character? To answer this question we introduced the notion of an *irreducible base of a space* and proved that any weakly separated space has such a base, moreover the weight of a space possessing an irreducible base can not be smaller than its cardinality. We asked [8, Problem 1] whether every first countable space of uncountable weight contains an uncountable subspace with an irreducible base? In theorem 50 and corollary 51 we will give a partial positive answer to this problem, using the principle $C(\kappa)$. First we recall some definitions from [8].

Definition 47. Let X be a topological space. A base $\mathcal{U}$ of X is called *irreducible* if it has an *irreducible decomposition* $\mathcal{U} = \bigcup\{\mathcal{U}_x : x \in X\}$, i.e., (*i*) and (*ii*) below hold:

(i) $\mathcal{U}_x$ is a neighbourhood base of x in X for each $x \in X$,
(ii) for each $x \in X$ the family $\mathcal{U}_x^- = \bigcup_{y \neq x} \mathcal{U}_y$ is not a base of X (hence $\mathcal{U}_x^-$ does not contain a neighbourhood base of x in X).

Definition 48. Let X be a topological space with $Y \subset X$. Similarly as above, an outer base $\mathcal{U}$ of Y in X is called *irreducible* if it has an *irreducible decomposition* $\mathcal{U} = \bigcup\{\mathcal{U}_y : y \in Y\}$, i.e., (i) and (ii) below hold:

(i) $\mathcal{U}_y$ is a neighbourhood base of y in X for each $y \in Y$,
(ii) for each $y \in Y$ the family $\mathcal{U}_y^- = \bigcup\{\mathcal{U}_z : z \in Y \setminus \{y\}\}$ does not contain a neighbourhood base of y in X.

Note that in general, a subspace Y having an irreducible outer base in X does not necessarily possess an irreducible base in itself. However, if Y is dense in an open set and the irreducible outer base of Y consists of regular open sets then clearly this is the case. Moreover, by our next result, under certain conditions we can at least find another subspace of the same size as Y that does have an irreducible base.

Lemma 49. *If X is a regular, separable space and $Y \subset X$ has an irreducible outer base in X consisting of regular open sets, then there is $Z \subset X$ with $|Z| = |Y|$ such that the subspace Z has an irreducible base.*

Proof. Let $\mathcal{B} = \bigcup\{\mathcal{B}_y : y \in Y\}$ be an irreducible outer base of Y in X consisting of regular open sets and D be a countable dense subset of X. We distinguish two cases:

Case 1 $|(\operatorname{int}\overline{Y}) \cap Y| = |Y|$.

Let $Z = (\operatorname{int}\overline{Y}) \cap Y$. Since Z is dense in the open set $\operatorname{int}\overline{Y}$, by our above remark Z has an irreducible base.

Case 2 $|(\operatorname{int}\overline{Y}) \cap Y| < |Y|$.

In this case the set $Y_1 = Y \setminus \operatorname{int}\overline{Y}$ is nowhere dense, so $D_1 = D \setminus \overline{Y_1}$ is dense in X. Let $Z = D_1 \cup Y_1$, then $|Z| = |Y_1| = |Y|$. Write $D_1 = \{d_n : n < \omega\}$ and for each $d_n \in D$ let $\mathcal{B}_{d_n}$ be a neighbourhood base of d_n in X consisting of regular open sets, that are disjoint to $Y_1 \cup \{d_m : m < n\}$. Then clearly $\bigcup\{\mathcal{B}_z : z \in Z\}$ is an irreducible outer base of Z in X consisting of regular open sets and Z is dense in X, so again we are done.

Theorem 50. *Assume $C(\kappa)$. If X is a separable, first countable, regular space with $\mathrm{w}(X) \geq \kappa$, then there is subspace $Y \subset X$ of cardinality κ that has an irreducible base.*

Proof. Let $D \subset X$ be a countable, dense subset of X. For each $x \in X$ fix a neighbourhood base $\{U(x,n) : n \in \omega\}$ consisting of regular open sets and set $V(x,n) = U(x,n) \cap D$. Since the $U(x,n)$ are regular open and D is dense, we clearly have $U(x,n) \subset U(y,m)$ iff $V(x,n) \subset V(y,m)$.

Since $\mathrm{w}(X) \geq \kappa$, by transfinite recursion on $\beta < \kappa$ we can choose points $\{x_\alpha : \alpha < \kappa\} \subset X$ such that for any $\beta < \kappa$ the family $\{U(x_\alpha, n) : \alpha < \beta, n < \omega\}$

does not contain a neighbourhood base of x_β, in other words, there is a natural number k_β such that for all $\alpha < \beta < \kappa$ and $n \in \omega$ we have

$$\neg(x_\beta \in U(x_\alpha, n) \subset U(x_\beta, k_\beta)). \qquad (*)$$

We can assume that $k_\beta = 0$ for each $\beta < \kappa$. Let $X' = \{x_\alpha : \alpha < \kappa\}$. For $x \in X'$ and $n < \omega$ put

$$A(x, 2n) = [V(x, n) \times \{0\}] \cup [(D \setminus V(x, n)) \times \{1\}]$$

and

$$A(x, 2m + 1) = [(D \setminus V(x, 0)) \times \{0\}] \cup [V(x, m) \times \{1\}].$$

Note that $A(x, 2n) \cap A(y, 2m + 1) = \emptyset$ iff $V(y, m) \subset V(x, n) \subset V(y, 0)$. Apply $C(\kappa)$ to $\langle A(x, i) : x \in X', i < \omega\rangle$ and $T = \{\langle 2n, 2m + 1\rangle : n, m < \omega\}$. By $(*)$ (and $k_\beta = 0$) there are no $D, E \in [X']^\kappa$ and $n, m \in \omega$ such that

$$V(y, m) \subset V(x, n) \subset V(y, 0) \qquad (\dagger)$$

whenever $x \in D$ and $y \in E$, because $(\dagger)$ fails if $x = x_\alpha$, $y = x_\beta$ and $\alpha < \beta$. So there is $Y \in [X']^\kappa$ such that for all $n, m \in \omega$ and $x \neq y \in Y$ the intersection of $A(x, 2n)$ and $A(y, 2m+1)$ is not empty. This means that $\neg(V(y, m) \subset V(x, n) \subset V(y, 0))$, i.e. if we set $\mathcal{B}_y = \{U(y, n) : n < \omega\}$ then it follows that $\mathcal{B} = \bigcup\{\mathcal{B}_y : y \in Y'\}$ is a an irreducible outer base of Y in X consisting of regular open sets. Now applying lemma 49 we can conclude the proof.

Unfortunately, as $C(\omega_1)$ is false, the above result is not applicable in the perhaps most interesting case when $\mathrm{w}(X) = \omega_1$. The annoying assumption of separability, however, can be circumvented as follows.

Corollary 51. *Assume $C(\kappa)$. If X is a first countable, regular space with $\mathrm{w}(X) \geq \kappa$, then there is an uncountable subspace $Y \subset X$ that has an irreducible base.*

Proof. If X is separable, then the previous theorem can be applied. If X is not separable, then X contains an uncountable left separated subspace Y and again Y has an irreducible base.

References

1. J. E. Baumgartner, S. Shelah, *Remarks on superatomic Boolean algebras*, Ann. Pure Appl. Logic, 33 (1987), no. 2, 109-129.
2. A. Dow, *Large compact separable spaces may all contain βN*, Proc. Amer. Math. Soc. 109(1990) no.n1. 275–279.
3. A. Dow , *On initially κ-compact spaces*, in: Rings of continuous functions, ed. C.E. Aull, Lecture Notes in pure and applied Mathematics, 1985., v.95, Marcel Dekker, Inc, New York and Basel, 103–108.
4. A. Hajnal, I. Juhász, *Intersection properties of open sets*, Topology and its Application, **19**(1985), 201–209.

5. I. Juhász *Cardinal functions in topology - ten years later*, Mathematical Centre Tracts 123, Amsterdam, 1980.
6. I. Juhász, L. Soukup, *How to force a countably tight, initially ω_1 compact but non-compact space?* Topology and its Applications **69**(1996), no 3, 227–250.
7. I. Juhász, Zs. Nagy, L. Soukup, Z. Szentmiklóssy, *Intersection properties of open sets, II.* to appear in the Proceedings of the Tenth Summer Conference in General Topology and Applications, TU Delft, Netherlands.
8. I. Juhász, L. Soukup, Z. Szentmiklóssy, *What makes a space have large weight?* Topology and its Applications 57(1994), no 2-3, 271–285.
9. I. Juhász, W. Weiss, *On thin-tall scattered spaces*, Colloquium Mathematicum, vol XL (1978) 63–68.
10. W. Just, *Two consistency results concerning thin-tall Boolean algebras* Algebra Universalis 20(1985) no.2, 135–142.
11. K. Kunen, *Set Theory*, North-Holland, New York, 1980.
12. K. Kunen, Inaccessibility Properties of Cardinals, Doctoral Dissertation, Stanford, 1968.
13. S. Mrówka, M. Rajagopalan, T. Soudararajan, *A characterization of compact scattered spaces through chain limits (chain compact spaces)*, Proceedings of TOPO 72, Pittsburg 1972, Lecture Notes 378, 288–297, Berlin 1974.
14. Judy Roitman, *Height and width of superatomic Boolean algebras*, Proc. Amer. Math. Soc. 94(1985), no 1, 9–14.
15. W. Weiss, *Versions of Martin's Axiom*, in Handbook of Set-Theoretic Topology, ed. K. Kunen and J.E. Vaughan, Elseview Science Publisher, B.V. 1984.

Extensions of Models of *PV*

Jan Krajíček*

Mathematical Institute and Institute of Computer Science
Academy of Sciences of the Czech Republic
Zitná 25, 115 67 Praha 1, Czech Rep
krajicek@cesnet.cz

Abstract. We prove that certain models of PV in which $NP \not\subseteq P/poly$ have a Π_1^b-elementary extension to a model of (PV and) $NP \not\subseteq coNP/poly$. If S_2 proves a particular fact about bipartite graphs then, in fact, all models of PV in which $NP \not\subseteq P/poly$ have a Π_1^b-elementary extension to a model of $NP \not\subseteq coNP/poly$.

Introduction

PV is a bounded arithmetic theory with function symbols for all polynomial time algorithms, and axiomatized by a particular set of universal formulas, cf. [3]. Models of PV are a natural environment for notions of computational complexity theory around deterministic and non-deterministic polynomial time. Major open problems in this part of complexity theory have their counterparts in bounded arithmetic and propositional logic. We are interested in proving some notorious open conjectures for a model of bounded arithmetic, and not so much in showing that some of these conjectures might be unprovable in bounded arithmetic. For a general motivation (for this author, at least) for research in this area see the preface to [4].

In a model M of the theory PV the class P of the polynomial-time sets is the class of subsets of M definable by an atomic PV-formula with parameters from M (in S_2^1 this would be provably Δ_1^b-formulas with parameters), equivalently: recognizable by a *standard* DTM with an extra input (the parameter) which may be non-standard, equivalently: recognizable by a DTM possibly with a non-standard description but whose time is bounded by a standard degree polynomial.

The class $P/poly$ is defined in the same way except that the parameters may vary with the length of the inputs, and the classes $NP, NP/poly$ and $coNP, coNP/poly$ are defined analogously using $NDTM$'s. In particular, NP-subsets of M (resp. $coNP$) are those definable by Σ_1^b-formulas (resp. by Π_1^b-formulas) with parameters, that may vary with the length in case of $NP/poly$ and $coNP/poly$.

---

* Partially supported by the *US - Czechoslovak Science and Technology Program* grant # 93025, and by grant #*A1019602* of the *AVČR*.

It is not important whether we require that the length of parameters in the non-uniform classes is polynomial in the length of the input. This is because we are concerned with definability of sets of inputs of a fixed length. In general one may restrict to those models of PV in which lengths are polynomial (with a standard degree) in one fixed length.

The problem whether PV equals to S^1_2 is closely related to the circuit complexity of NP-sets. In particular, $PV \neq S^1_2$ if $NP \not\subseteq P/poly$ (by [8]) or if there is a model of PV in which $NP \not\subseteq coNP/poly$ (by [2, 9]).[2]

Constructions of extensions of models of PV (or of S^1_2) are also closely related to length-of-proofs problems about the extended Frege systems, cf. [4, 5, 6].

In this paper we study the problem to construct a model of PV in which $NP \not\subseteq coNP/poly$. We give three versions of a construction showing that certain models of PV in which $NP \not\subseteq P/poly$ have a Π^b_1-elementary extension to a model of (PV and) $NP \not\subseteq coNP/poly$. An ultimate goal is to make the construction work under weaker assumptions on models than those in Theorem 4.

A relevant background can be found in [4]. In particular, necessary facts from all other references can be also found there.

1 Preliminaries

Given a length $n = |y|$ of $y \in M$, $SAT_n(M)$ denotes the set of satisfiable formulas in M of length n; this set is defined by a canonical Σ^b_1-formula $Sat_n(x)$ with a parameter of the same length as y. $Log(M)$ is the set of lengths of elements of M.

For a formula a and a truth assignment w the relation $w \models a$ denotes that w *satisfies* a, and is definable by a fixed open formula. We shall assume that $w \models a$ implies (in PV) that a is a formula from $SAT_n(M)$ and w is a truth-assignment to its atoms.

Let $Circuit_M$ denote the set of multi-output circuits in M and for $C \in Circuit_M$ and $a \in M$ of appropriate length, $C(a) = b$ is a function definable by a ternary PV-symbol stating that b is the output of the computation of circuit C on input a (when numbers are identified with their binary encodings).

The following lemma follows from the fact that PV can define binary search.

Lemma 1. *For any length n and any circuit $C \in Circuit_M$ there exists another circuit $C' \in Circuit_M$ such that if*

$$M \models (\forall x, |x| = n), Sat_n(x) \equiv (C(x) = 1)$$

[2] We inessentially abuse the notation here; instead of PV, which is an equational theory as defined in [3], we work with its first-order conservative extension PV_1 defined in [8, 4], and in place of S^1_2 we should use its conservative extension $S^1_2(PV)$ in the language of PV, cf. [4].

then

$$M \models (\forall x, |x| = n), Sat_n(x) \to (C'(x) \models x) \ .$$

In particular, the property that $SAT_n(M)$ is recognized in a model M of PV by a circuit is preserved to Π_1^b-elementary extensions of M.

This means that only M in which $NP \not\subseteq P/poly$ can possibly have a cofinal Π_1^b-elementary extension in which $NP \not\subseteq coNP/poly$.

Definition 2. Let M be a model of PV and assume that for some length $n \in Log(M)$ the set $SAT_n(M)$ is not recognized in M by a circuit.

A *counter-example* function (for $SAT_n(M)$ in M, tacitly) is a function ξ that assigns to any circuit $C \in Circuit_M$ with n inputs a pair

$$\xi(C) = (a, w)$$

such that

1. $w \models a$
2. $C(a) \not\models a$.

We say that ξ is in $P/poly$ of M if for every length $m \in Log(M)$ there is a circuit $D_m \in Circuit_M$ with m input bits and $2n$ output bits computing $\xi(C)$ for any C of size at most m.

Note that the statement that $SAT_n(M)$ is not recognized by a circuit of size at most m is $\Pi_1^b(\xi)$, whenever ξ is a counter-example function. Hence we have the following lemma.

Lemma 3. *Let M be a model of PV in which the set $SAT_n(M)$ is not recognized by a circuit. Let ξ be a corresponding counter-example function. Then $SAT_n(M')$ is not recognized by a circuit in any $\Pi_1^b(\xi)$-elementary, cofinal extension (M', ξ') of (M, ξ).*

In particular, if M admits a counter-example function in $P/poly$ then the set $SAT_n(M')$ is not recognized by a circuit in any Π_1^b-elementary, cofinal extension M' of M, and M' admits a counter-example function in $P/poly$.

2 An ultrapower

Theorem 4. *Let M be a countable model of PV and assume that for some length $n \in Log(M)$ the set $SAT_n(M)$ is not recognized in M by a circuit. Assume that M admits a counter-example function ξ in $P/poly$.*

Then there is a Π_1^b-elementary, cofinal extension M' of M, a model of PV, such that the set $SAT_n(M')$ is not recognized in M' by a co-non-deterministic circuit.

Proof

For $C \in Circuit_M$ let f_C be the function from M to M computed by the circuit C.

Take:

$$\mathcal{F}_M := \{f_C : SAT_n(M) \to M \mid C \in Circuit_M\}.$$

We shall construct an ultrapower of the form $\mathcal{F}_M/\mathcal{U}$, with $\mathcal{U} \subseteq exp(SAT_n(M))$ a particular ultrafilter. The following claim is obvious.

Claim 1 *For any ultrafilter $\mathcal{U}$, Loš's theorem holds for all open PV- formulas, and $\mathcal{F}_M/\mathcal{U}$ is a Π_1^b-elementary, cofinal extension of M. In particular, $\mathcal{F}_M/\mathcal{U}$ is a model of PV.*

Define a particular element of $\mathcal{F}_M$:

$$a_{\mathcal{U}} = id_{SAT_n(M)}/\mathcal{U}.$$

Claim 2 *Let $\psi(x)$ be a Π_1^b-formula with parameters from M such that:*

$$M \models \psi(a)$$

for all $a \in SAT_n(M)$. Then:

$$\mathcal{F}_M/\mathcal{U} \models \psi(a_{\mathcal{U}}).$$

Claim 2 follows by Loš's theorem for all open PV- formulas.

For a circuit $D \in Circuit_M$ with n bits of input define the set:

$$D^* := \{a \in SAT_n(M) \mid D(a) \models a\}\ .$$

Claim 3 *Assume that an ultrafilter $\mathcal{U} \subseteq exp(SAT_n(M))$ satisfies the condition:*

$$\forall D \in Circuit_M; D^* \notin \mathcal{U}$$

Then :

$$\mathcal{F}_M/\mathcal{U} \models \neg Sat_n(a_{\mathcal{U}})\ .$$

The claim follows from Loš's theorem again: an element $f_D/\mathcal{U}$ satisfies the formula $a_{\mathcal{U}}$ in $\mathcal{F}_M/\mathcal{U}$ iff $D^* \in \mathcal{U}$.

Claim 4 *$SAT_n(\mathcal{F}_M/\mathcal{U})$ is not recognized in $\mathcal{F}_M/\mathcal{U}$ by a circuit and $\mathcal{F}_M/\mathcal{U}$ admits a counter-example function in $P/poly$.*

Assume on the contrary that $SAT_n(\mathcal{F}_M/\mathcal{U})$ is recognized in $\mathcal{F}_M/\mathcal{U}$ by a circuit, hence by Lemma 1 it holds in $\mathcal{F}_M/\mathcal{U}$:

$$f_W/\mathcal{U} \models f_A/\mathcal{U} \Rightarrow f_C/\mathcal{U}(f_A/\mathcal{U}) \models f_A/\mathcal{U}$$

for some $f_C/\mathcal{U} \in Circuit_{\mathcal{F}_M/\mathcal{U}}$ and all $f_A/\mathcal{U}, f_W/\mathcal{U} \in \mathcal{F}_M$.

For an arbitrary f_C define particular f_A, f_W by:

$$(f_A(a), f_W(a)) := \xi(C(a))$$

For those $a \in SAT_n(M)$ for which $C(a)$ is a circuit with n inputs, $f_W(a) \models f_A(a)$ but $C(a)(f_A(a)) \not\models f_A(a)$ by the definition of ξ. Hence $f_C/\mathcal{U}$ cannot have the property stated earlier.

Note that by Claim 1 the circuits D_m computing ξ in M compute a counter-example function in $\mathcal{F}_M/\mathcal{U}$ as well.

Let $\mathcal{U}_0 \subseteq exp(SAT_n(M))$ consist of all sets X containing some set of the form:

$$SAT_n(M) \setminus D^*$$

for some $D \in Circuit_M$. By the hypothesis that $SAT_n(M)$ is not recognized in M by a circuit, the class $\mathcal{U}_0$ is closed under intersections and $\emptyset \notin \mathcal{U}_0$, i.e., it is a non-trivial filter. Let $\mathcal{U} \supseteq \mathcal{U}_0$ be arbitrary ultrafilter.

Define M^1 to be the countable model $\mathcal{F}_M/\mathcal{U}$. By Claims 2 and 3 no Π^b_1-formula with parameters from M defines the set $SAT_n(M^1)$ in M^1.

By Claim 4 the set $SAT_n(M^1)$ is not recognized in M^1 by a circuit. We may therefore repeat this construction countably many times to obtain a chain:

$$M \subseteq M^1 \subseteq M^2 \subseteq \dots$$

of Π^b_1-elementary, cofinal extensions (killing all potential Π^b_1-definitions of $Sat_n(x)$ with all possible parameters from all M^t) such that its union:

$$M' := \bigcup_t M^t$$

is a Π^b_1-elementary, cofinal extension of M in which $SAT_n(M')$ is not defined by any Π^b_1-formula with parameters from M', i.e., it is not recognized by a co-non-deterministic circuit.

Q.E.D.

Note that the version of the theorem with $P, NP, coNP$ in place of the non-uniform classes is a simple corollary of Herbrand's theorem.

3 A compactness argument

In this section we give another proof of Theorem 4.

Let $\pi(x)$ be a Π^b_1 - formula with parameters from M. We want to find a Π^b_1-elementary, cofinal extension of M in which $\exists x; \neg(\pi(x) \equiv Sat_n(x))$ holds. Note that we may assume w.l.o.g. that in $PV + Th_{\forall\Pi^b_1}(M)$ it holds that

$$\pi(c) \rightarrow |c| = n$$

(otherwise just replace $\pi(c)$ by $\pi(c) \wedge |c| = n$).

If already
$$M \models \exists x; \neg(\pi(x) \equiv Sat_n(x))$$
then this will be preserved in every Π_1^b -elementary extension. If
$$PV + Th_{\forall \Pi_1^b}(M) \vdash \forall x(\pi(x) \equiv Sat_n(x))$$
then by Herbrand's theorem there is a PV-symbol $f(x,y)$ and $b \in M$ such that:
$$PV + Th_{\forall \Pi_1^b}(M) \vdash \forall x(Sat_n(x) \equiv (f(x,b) \models x)) \ ,$$
so the set $SAT_n(M)$ is recognized in M by a circuit, contradicting the hypothesis of the theorem.

So the only case creating difficulties is when
$$M \models \forall x \ (\pi(x) \equiv Sat_n(x))$$
but
$$PV + Th_{\forall \Pi_1^b}(M) \not\vdash \forall x(\pi(x) \equiv Sat_n(x))$$
which implies:
$$PV + Th_{\forall \Pi_1^b}(M) \not\vdash \forall x \ (\pi(x) \rightarrow Sat_n(x))$$
(as the opposite implication is in $Th_{\forall \Pi_1^b}(M)$).

Take a new constant c and a formula
$$\pi(c) \wedge \neg Sat_n(c) \ .$$

Claim *The theory*
$$PV + Th_{\forall \Pi_1^b}(M) + \pi(c) \wedge \neg Sat_n(c)$$
does not prove that $Sat_n(x)$ is recognized by a polynomial size circuit.

Assume on the contrary that
$$PV + Th_{\forall \Pi_1^b}(M) + \pi(c) + \neg Sat_n(c) \vdash \exists D(\forall x, |x| = n); \ Sat_n(x) \rightarrow D(x) \models x$$
By the hypothesis $PV + Th_{\forall \Pi_1^b}(M) + \pi(c) + \neg Sat_n(c)$ is consistent and hence has a model N (that contains M as a submodel). Take N^* to be the unique substructure of N generated from elements of $M \cup \{c\}$ by PV-function symbols. Thus $N^* \models PV + Th_{\forall \Pi_1^b}(M) + \pi(c) + \neg Sat_n(c)$ and hence
$$N^* \models \exists D(\forall x, |x| = n); \ Sat_n(x) \rightarrow D(x) \models x$$
Moreover, N^* is a Π_1^b-elementary and cofinal (as $|c| = n$) extension of M.

However, that is a contradiction with Lemma 3, as by the hypothesis of the theorem M admits a counter-example function in *P/poly*.

By the claim we may take M^1, a Π_1^b-elementary, cofinal extension of M that is a model of $\pi(c) \wedge \neg Sat_n(c)$, and such that there is no circuit in M^1 recognizing $SAT_n(M^1)$. Then we construct a countable chain $M \subseteq M^1 \subseteq M^2 \subseteq \ldots$ killing all potential Π_1^b-definitions (with all possible parameters from all M^i) of $Sat_n(x)$. Thus $M' := \bigcup_i M^i$ is the required extension.

Q.E.D.

4 A Boolean-valued extension

Boolean-valued extensions of S_2^1 were defined in [5], see also [4, Chpt. 9.4]. For PV in place of S_2^1 the construction has a particular formulation.

Let M be a model of PV and let $(p_1, \ldots, p_n) \in M$ be a sequence of propositional atoms. Let $Circuit_M(\overline{p})$ be all circuits with one output formed from atoms p_i, and let $\mathbf{B}(\overline{p})$ be the Boolean algebra obtained by factoring $Circuit_M(\overline{p})$ by the equivalence relation $C_1 \sim C_2$ that holds for C_1, C_2 iff there is an EF-proof in M of $C_1 \equiv C_2$ (see [5] for a formalization of this notion).

Given an ultrafilter $\mathcal{G}$ on $\mathbf{B}(\overline{p})$, let $\nu_{\mathcal{G}}(C)$ be equal 1 if $(C/\sim) \in \mathcal{G}$ and equal to 0 otherwise.

Define the extension $M[\mathcal{G}]$ of M as follows. Let $Names_M(\overline{p})$ be the set of sequences $\langle C_1, \ldots, C_\ell\rangle \in M$ of elements of $Circuit_M(\overline{p})$. The elements of $M[\mathcal{G}]$ are tuples

$$\langle \nu_{\mathcal{G}}(C_1), \ldots, \nu_{\mathcal{G}}(C_\ell)\rangle$$

one for each $\langle C_1, \ldots, C_\ell\rangle \in Names_M(\overline{p})$.

For $f(x_1, \ldots, x_k)$ a PV-function and $\ell \in Log(M)$ a length, let $D^t_{f,\ell}(y_{ij})$ $(i \leq k$ and $j \leq \ell)$ be a circuit in M computing (provably in PV) the tth bit of $f(x_1, \ldots, x_k)$ for inputs x_i of length at most ℓ with bits $y_{i1}, \ldots, y_{i\ell}$. Define $f(w_1, \ldots, w_k)$ for elements w_i of $M[\mathcal{G}]$

$$w_i = \langle \nu_{\mathcal{G}}(C_{i1}), \ldots, \nu_{\mathcal{G}}(C_{i\ell})\rangle$$

to be

$$\langle \nu_{\mathcal{G}}(D^1_{f,\ell}(y_{ij}/C_{ij}), \nu_{\mathcal{G}}(D^2_{f,\ell}(y_{ij}/C_{ij}), \ldots\rangle$$

The following is a special case of [5, Thm. 5.1]. See also [7] or [5, Sec. 9.4] for another treatment of the construction.

Theorem 5. *Let M be a model of PV, $(p_1, \ldots, p_n) \in M$ propositional atoms, and let $\mathcal{G}$ be an ultrafilter on $\mathbf{B}(\overline{p})$. Assume that $\mathcal{G}$ is closed under EF-provability in M, i.e., whenever there is an EF-proof in M of D from $C_1, \ldots, C_k$ and $\nu_{\mathcal{G}}(C_i) = 1$ then $\nu_{\mathcal{G}}(D) = 1$ too.*

Then $M[\mathcal{G}]$ is a cofinal extension of M and it is a model of PV.

Moreover, if $\nu_{\mathcal{G}}(C) = 1$ whenever $C \in Circuit_M(\overline{p})$ computes the function constantly 1 in M, then $M[\mathcal{G}]$ is a Π^b_1-elementary, cofinal extension of M.

We give now another proof of Theorem 4.

Let M be a countable model of PV in which $SAT_n(M)$ is not recognized by a circuit, and that admits a counter-example function ξ in $P/poly$.

We shall denote by $y \models x$ also the circuit in M that computes on two n-bit inputs x, y whether they satisfy the relation $y \models x$. Let $\phi(x)$ be a Π^b_1-formula with parameters from M of the form $\forall z, |z| \leq |x|^k \rightarrow \phi_0(x, z)$, where ϕ_0 is open.

Let $\overline{p} = (p_1, \ldots, p_n)$ be mutually different propositional atoms in M. Consider the set T of propositional formulas of the form

$$\neg(\langle W_1, \ldots, W_n \rangle \models \overline{p})$$

and of the form

$$\phi_0(\overline{p}, \langle Z_1, \ldots, Z_m \rangle)$$

where $\overline{W} = \langle W_1, \ldots, W_n \rangle$, $\overline{Z} = \langle Z_1, \ldots, Z_m \rangle$ are all elements of $Names_M(\overline{p})$ of the length n and $m = n^k$ respectively.

Claim 1 *There is no EF-refutation of T in M.*

Assume otherwise, i.e., there is an EF-proof of

$$\bigvee_{\overline{W}} \overline{W}(\overline{p}) \models \overline{p} \quad \vee \quad \bigvee_{\overline{Z}} \neg\phi_0(\overline{p}, \overline{Z})$$

for some $\overline{W}$'s and $\overline{Z}$'s. As EF is sound in any model of PV, the $\overline{W}$'s and $\overline{Z}$'s may be combined into a circuit in M recognizing the set $SAT_n(M)$. That is a contradiction.

Claim 2 *There is an ultrafilter $\mathcal{G}$ on $\mathbf{B}(\overline{p})$ that is closed under EF-provability in M and such that*

1. $\nu_{\mathcal{G}}(C) = 1$, *for all* $C \in T$.
2. $\nu_{\mathcal{G}}(C) = 1$, *for all* $C \in Circuit_M(\overline{p})$ *computing in M constantly* 1.

Take $S \subseteq Circuit_M(\overline{p})$ the set of all circuits C' majorizing (as Boolean functions) in M some $C \in T$. By Claim 1 the subset of $\mathbf{B}(\overline{p})$ of $\sim$-classes of all $C' \in S$ is a non-trivial filter. Any ultrafilter extending this set satisfies the requirements of the claim.

Take $M^1 := M[\mathcal{G}]$ for any $\mathcal{G}$ given by Claim 2. Then, by Theorem 5, M^1 is a model of PV in which the element

$$a_{\mathcal{G}} := \langle \nu_{\mathcal{G}}(p_1), \ldots, \nu_{\mathcal{G}}(p_n) \rangle$$

is not in $SAT_n(M^1)$ but

$$M^1 \models \phi(a_{\mathcal{G}})$$

Hence $\phi(x)$ will not define $Sat_n(x)$ in any Π_1^b-elementary extension of M^1.

By the Π_1^b-elementarity and cofinality of M^1 over M and by Lemma 3, no circuit in M^1 recognizes $SAT_n(M^1)$ and M^1 admits a counter-example function in *P/poly*. We may thus repeat the construction to produce a chain $M \subseteq M^1 \subseteq M^2 \subseteq \ldots$ such that $M' := \bigcup_i M^i$ is the required model, identically as in sections 2 and 3.

Q.E.D.

5 A construction of a counter-example function

Let $E \subseteq X \times Y$ be a bipartite graph, $\lceil \log_2 |X| \rceil = n$ and $\lceil \log_2 |Y| \rceil = m$. If

$$\forall y_0, \ldots, y_n \in Y \exists x \in X; \bigwedge_j \neg(xEy_j)$$

then

$$\exists x_0, \ldots, x_m \in X \forall y \in Y; \bigvee_i \neg(x_i E y)$$

This is easily proved by a pigeon-hole argument. For the purpose of bounded arithmetic we shall relax the statement a bit, removing explicit bounds on the number of x_i's and y_j's.

Definition 6. Let $\alpha(x, y)$ be a binary predicate. $CE(u, \alpha)$ is an $\exists \Pi_1^b(\alpha)$- formula formalizing that either there is a sequence $(x_0, \ldots, x_k)$ of elements smaller than u such that

$$\forall y \leq u; \bigvee_i \neg\alpha(x_i, y)$$

or there is a sequence $(y_0, \ldots, y_\ell)$ of elements smaller than u such that

$$\forall x \leq u; \bigvee_j \alpha(x, y_j)$$

Lemma 7. *Assume that M is a model of PV in which $SAT_n(M)$ is not recognized by a circuit. Assume also that M satisfies for all open PV-formulas $\alpha(x, y)$ the statement $\forall u; CE(u, \alpha)$ with bounds $k, \ell \leq |t(u)|$, t a term.*

Then M admits a counter-example function in P/poly.

Proof

Let $\alpha(x, y)$ formalizes that y is a circuit C of size at most m with n inputs, x is a pair (a, w) of $a \in SAT_n(M)$ and $w \models a$, and $C(a) \models a$.

Take the principle $CE(u, \alpha)$ for $u := max(2^{2n}, 2^m)$. The principle provides us either with circuits $C_0, \ldots, C_\ell$ of size at most m such that for every $a \in SAT_n(M)$

$$\bigvee_j C_j(a) \models a$$

or with pairs $(a_0, w_0), \ldots, (a_k, w_k)$ of $a_i \in SAT_n(M)$ and $w_i \models a_i$ such that for every circuit C of size at most m

$$\bigvee_i C(a_i) \not\models a_i$$

The former option is, however, impossible as otherwise we could combine C_j's into one circuit recognizing $SAT_n(M)$. Hence we have the pairs (a_i, w_i) and we define the circuit D_m as follows. Given as an input a circuit C, D_m tries C on all a_i and outputs the first pair (a_i, w_i) such that $C(a_i) \not\models a_i$. Clearly D_m computes a counter-example function for circuits of size at most m.

Q.E.D.

It is open whether the combinatorial principle is provable in PV or even in S_2. A corollary of the principle, namely the tournament principle (see [4, Sec. 12.1]), is also not known to be provable in bounded arithmetic.

Theorem 8. *Assume that S_2 proves the formula*

$$\forall u; CE(u, \alpha)$$

for the Δ_1^b-formula $\alpha(x, y)$ defined at the beginning of the proof of Lemma 7. Assume also that PV has a countable model in which $NP \not\subseteq P/poly$.

Then $PV \neq S_2^1$.

Proof

Take M a countable model of PV in which $SAT_n(M)$ is not recognized by a circuit. If $M \not\models S_2^1$ then we are done. So assume that $M \models S_2^1$.

Consider the theory T formed by

$$PV + Th_{\Pi_1^b}(M)$$

together with all formulas

$$\forall y \exists x; Sat_n(x) \not\equiv \phi(x, y)$$

one for each Π_1^b-formula ϕ without parameters.

If T were consistent then any of its models is a Π_1^b-elementary extension of M in which $NP \not\subseteq coNP/poly$ and thus by [2, 9] $PV \neq S_2^1$.

On the other hand, if T is inconsistent then $PV + Th_{\Pi_1^b}(M)$ proves a disjunction of formulas of the form

$$\exists y \forall x; Sat_n(x) \equiv \phi(x, y)$$

ϕ Π_1^b-formulas without parameters. This means that in M every bounded formula is equivalent to a Σ_1^b-formula and, in particular, the $PIND$ scheme for all bounded formulas holds in M as $M \models S_2^1$. Hence $M \models S_2$ and consequently $M \models \forall u; CE(u, \alpha)$.

By Lemma 7 and Theorem 4 M has an extension M' in which $NP \not\subseteq coNP/poly$. So, by [2, 9] again, $PV \neq S_2^1$.

Q.E.D.

Acknowledgements: I discussed the idea of the first proof of the theorem with S. Buss in Toronto in Spring 1993 and I thank him for pointing out then a problem with the original version. R. Impagliazzo reminded me of a construction concerning a circuit-simulation of probabilistic computations similar to the one from Lemma 7. I thank A. A. Razborov for critical comments on the preliminary version of this paper.

References

1. Buss, S. R. (1986) Bounded Arithmetic. Naples, Bibliopolis.
2. —— (1995) Relating the bounded arithmetic and polynomial time hierarchies, *Annals of Pure and Applied Logic*, **75**: 67-77.
3. Cook, S A. (1975) Feasibly constructive proofs and the propositional calculus, in: *Proc. 7th Annual ACM Symp. on Theory of Computing*, pp. 83-97. ACM Press.
4. Krajíček, J. (1995) *Bounded arithmetic, propositional logic and complexity theory*, Encyclopedia of Mathematics and Its Applications, Vol. **60**, *Cambridge University Press*, Cambridge - New York - Melbourne, 343 p.
5. —— (1995) On Frege and Extended Frege proof systems, in: *Feasible Mathematics II*, eds. P. Clote and J. Remmel, Birkhauser, pp. 284-319.
6. —— (1995) On methods for proving lower bounds in propositional logic, in: Proceedings of the Tenth International Congress *Logic, Methodology and Philosophy of Science*, International Union of History and Philosophy of Science, Florence (August 19-25, 1995), Eds. M. L. Dalla Chiara, K. Doets, D. Mundici, J. van Benthem, to appear.
7. Krajíček, J and P. Pudlák (1990) Propositional Provability and Models of Weak Arithmetic, in: *Computer Science Logic* (Kaiserlautern, Oct. '89), E. Borger, H. Kleine Buning, M.M. Richter (eds.), LNCS 440, Springer-Verlag, pp. 193-210.
8. Krajíček, J, Pudlák, P, and Takeuti, G. (1991) Bounded arithmetic and the polynomial hierarchy, *Annals of Pure and Applied Logic*, **52**: 143–153.
9. Zambella, D. (1996) Notes on polynomially bounded arithmetic, *J. of Symbolic Logic*, **61(3)**, pp.942-966.

Convergence Laws for Random Graphs

James F. Lynch *

Department of Mathematics and Computer Science
Clarkson University
Potsdam, NY 13699-5815, USA
jlynch@sun.mcs.clarkson.edu

Abstract. A convergence law in logic pertains to a language and class of finite structures on which a probability measure has been assigned to each set of structures of a given size. It states that, for every sentence in the language, the probability that it holds for a random structure approaches a limit as the size of the structure grows. A 0-1 law states that this limit must be 0 or 1. The original convergence law was a 0-1 law for first-order logic and relational structures with a uniform probability distribution. This expository article shows how it has been extended by numerous authors to more powerful logics and the class of random structures known as random graphs. In many cases, the 0-1 law no longer holds, but a convergence law can still be proven. Full proofs are not given, but a uniform framework is provided which emphasizes ideas common to all the proofs.

1 Background

Let us begin by introducing the conventions we will use. $\mathcal{L}$ will denote a logic of some type τ. $\mathcal{C}_0, \mathcal{C}_1, \mathcal{C}_2, \ldots$ will stand for a sequence of sets of τ-structures where all structures in $\mathcal{C}_n$ have universe $\{0, 1, \ldots, n-1\}$, which we will abbreviate as n. For every $n \in \omega$, pr_n is a probability measure on $\mathcal{C}_n$. For a sentence $\sigma \in \mathcal{L}$, we put $\mathrm{pr}(\sigma, n)$ for

$$\mathrm{pr}_n(\{ \mathfrak{A} \models \sigma : \mathfrak{A} \in \mathcal{C}_n \}) \ .$$

In this article, we will examine the behavior of $\mathrm{pr}(\sigma, n)$ for growing n (given $\mathcal{L}$, $(\mathcal{C}_n)_{n\in\omega}$, and $(\mathrm{pr}_n)_{n\in\omega}$). The first theorems on this topic considered a first-order logic $\mathcal{L}$ of a purely relational type τ, where each $\mathcal{C}_n$ is the set of all τ-structures on n, and pr_n is the uniform distribution. For example, if $\mathcal{L}$ has a single relational symbol of arity r, then

$$\mathrm{pr}(\sigma, n) = \frac{|\{ \mathfrak{A} \models \sigma : \mathfrak{A} \in \mathcal{C}_n \}|}{2^{n^r}} \ .$$

Theorem 1 Fagin [11], Glebskiĭ et al. [13]. *For every* $\sigma \in \mathcal{L}$,

$$\lim_{n\to\infty} \mathrm{pr}(\sigma, n) = 0 \ \textit{or} \ 1 \ .$$

* Research supported by NSF Grant CCR-94006809

When this theorem first appeared, it did not attract much attention from the logic community. It was mostly combinatorialists and computer scientists who appreciated it, and they suggested two orthogonal directions for future research. The combinatorialists pointed out that there are many other classes of random structures besides uniformly distributed relations, and the computer scientists felt that first-order logic was not powerful enough to describe interesting computational processes.

However, when one attempts to extend the 0-1 law in either direction, it often fails. Here are some examples where it fails on classes of structures which are not purely relational (but still with a uniform distribution).

Glebskiĭ et. al. [13]: Add constants $c_1, c_2, \ldots$ to $\mathcal{L}$. Then if R is a binary relation symbol,

$$\mathrm{pr}(R(c_1, c_1), n) = \frac{1}{2} \; .$$

In fact, every dyadic rational (i.e., of the form $u/2^v$) in $[0,1]$ is $\mathrm{pr}(\sigma, n)$ for some σ in this language and sufficiently large n.

Mycielski [23]: To each $\mathfrak{A} \in \mathcal{C}$ we add the successor relation

$$\{\,(x, x+1) : x < n-1\,\}.$$

Then

$$\mathrm{pr}(\exists x \forall y (x \neq y + 1 \wedge R(x,x)), n) = \frac{1}{2} \; .$$

Again, all possible dyadic rationals are realized.

Fagin [11]: Let f be a unary function symbol. Then

$$\begin{aligned} \mathrm{pr}(\forall x (f(x) \neq x), n) &= \left(\frac{n-1}{n}\right)^n \\ &\to \frac{1}{e} \text{ as } n \to \infty \; . \end{aligned}$$

In cases where the 0-1 law fails, it is sometimes possible to prove something almost as strong. A *convergence law* states that for every $\sigma \in \mathcal{L}$, $\lim_{n\to\infty} \mathrm{pr}(\sigma, n)$ exists. The three examples above have convergence laws:

Glebskiĭ et. al. [13]: Let $\mathcal{L}_1$ be a first-order language with relational and constant symbols only. Then for every $\sigma \in \mathcal{L}_1$,

$$\lim_{n\to\infty} \mathrm{pr}(\sigma, n) = \frac{u}{2^v} \text{ for some } u, v \in \omega \; .$$

Lynch [16]: Let $\mathcal{L}_2$ be a first-order language with relational symbols only, plus one binary relational symbol which is always interpreted as the successor relation. Then for every $\sigma \in \mathcal{L}_2$,

$$\lim_{n\to\infty} \mathrm{pr}(\sigma, n) = \frac{u}{2^v} \text{ for some } u, v \in \omega \; .$$

Lynch [17]: Let $\mathcal{L}_3$ be a first-order language with unary function symbols only. Then for every $\sigma \in \mathcal{L}_3$,

$$\lim_{n\to\infty} \mathrm{pr}(\sigma, n)$$

exists, and is denoted by an expression constructed from 1, +, −, ·, /, and the Poisson operators $e^{-x}x^k/k!$.

Note that these convergence laws are still subject to the criticisms of the computer scientists and combinatorialists, since they pertain only to first-order logic and uniform distributions. In the remainder of this article, we will explore two directions of research: more powerful logics, and an important class of random structures known as random graphs. Another problem we will look at is obtaining *rate estimates*. These are approximations of how fast the probabilities of sentences converge. As we will see, rate estimates have potential applications to optimization of database queries.

2 Recursive Logics and Infinitary Logics

Recursive logics and infinitary logics were developed by Barwise [3], Moschovakis [22], and others, in order to study the model theory of infinite structures. However, they are currently the focus of much of the research in finite model theory, because of motivations from computer science. For example, various complexity classes are *captured* by recursive logics. That is, every problem in such a class can be defined by a sentence in the appropriate logic. Also, recursive logics are needed to express the queries in modern relational database languages. Before defining some of these logics, we will make a digression into database theory. There are two reasons. First, it will provide some motivation for the study of recursive and infinitary logics in finite model theory, and second, it will introduce an important equivalence relation that has been used in proofs of the convergence laws (and nonconvergence laws) and rate estimates.

A *database machine* is an abstraction of a relational database system. It consists of a control unit that interacts with a finite model. The control unit is similar to a Turing machine, but restrictions are often imposed on its power. In addition, it cannot manipulate its data (the finite model) directly. In a single step, it can only perform a first-order query. A *query* is an automorphism-invariant mapping of finite models to finite models. That is, given a collection of relations, it computes a collection of relations. A 0-ary query computes the value TRUE or FALSE. A first-order query is one that is defined by a first-order formula. Thus, a 0-ary first-order query is defined by a sentence.

The earliest relational database systems could perform only first-order queries. It was quickly realized that these were inadequate for many applications; for instance, it is not possible to test whether a relation is connected or to compute the transitive closure of a relation using only first-order queries. Consequently, relational database systems that had the power of recursion were developed. The canonical example of a database machine for testing the connectedness of a graph is the following. Let

E be the symbol for the edge relation of the graph,
R be another binary relation symbol,
$\Phi(E,R)$ be the query defined by $\{\,(x,y) : \exists z(E(x,z) \wedge R(z,y))\,\}$.

INPUT: graph $\langle V, E\rangle$.
PROGRAM: $R := \{\,(x,x) : x \in V\,\}$;
while $R \neq \Phi(E,R)$ **do** $R := \Phi(E,R)$.
OUTPUT: **if** $\forall x \forall y R(x,y)$ **then** "YES" **else** "NO."

Note that the final value of R computed by the program is the transitive closure of E. Because the ability to compute transitive closures of relations is important in many database applications, a language known as *transitive closure logic* (TCL) has been developed. It is first-order logic augmented with the transitive closure operator. Let $\sigma(\overline{x},\overline{y},\overline{z})$ be a formula of type τ whose free variables are

$$\begin{aligned}\overline{x} &= x_1,\ldots,x_i,\\ \overline{y} &= y_1,\ldots,y_j, \text{ and}\\ \overline{z} &= z_1,\ldots,z_j,\end{aligned}$$

where $i,j \geq 0$. Then $\mathrm{TC}(\sigma,\overline{x},\overline{y},\overline{z})$ is a formula with the same free variables as σ such that for any τ-structure $\mathfrak{A}$ with universe A and $\overline{a} = a_1,\ldots,a_i \in A$,

$$\{(\overline{b},\overline{c}) \in A^{2j} : \mathfrak{A} \models \mathrm{TC}(\sigma,\overline{a},\overline{b},\overline{c})\}$$

is the transitive closure of

$$\{(\overline{b},\overline{c}) \in A^{2j} : \mathfrak{A} \models \sigma(\overline{a},\overline{b},\overline{c})\}.$$

The least fixed point is another logical operator more powerful than the transitive closure. Let $\varphi(\overline{x},S)$ be a formula of type τ extended with the new k-ary relational symbol S whose free variables are $\overline{x} = x_1,\ldots,x_k$. For any τ-model $\mathfrak{A}$ with universe A, φ induces an operator $\Phi : \mathcal{P}(A^k) \to \mathcal{P}(A^k)$. Given $S \subseteq A^k$,

$$\Phi(S) = \{\,(a_1,\ldots,a_k) \in A^k : \mathfrak{A} \models \varphi(a_1,\ldots,a_k,S)\,\}.$$

Inductively, we can define the iterates of Φ:

$$\begin{aligned}\Phi^0 &= \emptyset,\\ \Phi^{i+1} &= \Phi(\Phi^i)\ .\end{aligned}$$

We say the operator Φ is monotone if $S \subseteq T$ implies $\Phi(S) \subseteq \Phi(T)$. In that case, $\Phi^i \subseteq \Phi^{i+1}$ for all $i \in \omega$. If there is some i such that $\Phi^i = \Phi^{i+1}$, then Φ^i is said to be the *least fixed point* of Φ. The formula φ is positive in S if each occurrence of S is within an even number of $\neg$ signs. It is easily seen that if φ is positive, then Φ is monotone. The Knaster-Tarski theorem (see [22]) states that if Φ is monotone then Φ has a least fixed point. Of course, if $\mathfrak{A}$ is finite, this is obvious. In fact, it must occur on or before $|A|^k$ iterations. We put $\varphi^\infty(\overline{x})$ for the least fixed point of Φ. The least fixed point logic (LFP) is first-order logic closed under the least fixed point of positive formulas.

Other operators, such as the partial fixed point, can be added to first-order logic, resulting in the logic PFP. But all the logics mentioned here are contained in the infinitary logic $L^{\omega}_{\infty\omega}$.

Definition 2. For $k \in \omega$, $L^{k}_{\infty\omega}$ is the extension of first-order logic obtained by closure under conjunction and disjunction of arbitrary sets of formulas, provided only the variables $x_1, \ldots, x_k$ occur among them. Then

$$L^{\omega}_{\infty\omega} = \bigcup_{k \in \omega} L^{k}_{\infty\omega} \ .$$

It can be shown that

$$\mathrm{TC} \subset \mathrm{LFP} \subseteq \mathrm{PFP} \subset L^{\omega}_{\infty\omega} \ .$$

The first inclusion is known to be proper. The last inclusion is easily seen to be proper because $L^{\omega}_{\infty\omega}$ can express non-recursive properties, whereas all properties expressible in the other logics are recursive. It is not known whether $LFP \subset PFP$. As shown by Abiteboul and Vianu [1], this is equivalent to showing that P is properly contained in PSPACE, so it should not be expected that it is easy to answer.

Continuing with our example, let us show how transitive closure and connectedness can be expressed in $L^{\omega}_{\infty\omega}$. Let R^i be the value of R after the ith iteration of the above program. Then

$$R^0 = \{ (x, x) : x \in V \} \text{ and}$$
$$R^i = \{ (x, y) : \exists z(E(x, z) \wedge R^{i-1}(z, y)) \} \ .$$

Inductively, we can express $R^i(x, y)$ using only three variables:

$$\exists z(E(x, z) \wedge \exists x(x = z \wedge R^{i-1}(x, y))) \ .$$

Then we can express "$\langle V, E\rangle$ is connected" using an infinite disjunction and only three variables:

$$\forall x \forall y \left(\bigvee_{i \in \omega} R^i(x, y) \right) \ .$$

A fundamental problem in the model theory of $L^{\omega}_{\infty\omega}$ is, given $k \in \omega$, to characterize those relations (i.e., queries) that can be defined by a formula in $L^{k}_{\infty\omega}$. The following equivalence relation is central to this problem.

Definition 3. Let $\mathfrak{A}$ and $\mathfrak{B}$ be models of the same type, $l \leq k$, $a_1, \ldots, a_l \in \mathfrak{A}$, and $b_1, \ldots, b_l \in \mathfrak{B}$. Then $\langle \mathfrak{A}, a_1, \ldots, a_l\rangle \equiv^k_\infty \langle \mathfrak{B}, b_1, \ldots, b_l\rangle$ if and only if for all $\sigma(x_1, \ldots, x_l) \in L^{k}_{\infty\omega}$,

$$\mathfrak{A} \models \sigma(a_1, \ldots, a_l) \Leftrightarrow \mathfrak{B} \models \sigma(b_1, \ldots, b_l) \ .$$

This equivalence relation plays a key role in many proofs of the expressibility of properties in $L^{\omega}_{\infty\omega}$, and as we will outline later, the convergence laws for $L^{\omega}_{\infty\omega}$. Since database queries are expressible in this infinitary logic, it is also useful in studying the power of database languages. For example, let M be a database machine. Given an input $\mathfrak{A}$, M computes the model $M(\mathfrak{A})$. As is often the case, we assume that $\mathfrak{A}$ and $M(\mathfrak{A})$ have the same universe. Let R be an l-ary relation in $M(\mathfrak{A})$. For inputs $\mathfrak{A}$ and $\mathfrak{B}$, $a_1, \ldots, a_l \in \mathfrak{A}$, and $b_1, \ldots, b_l \in \mathfrak{B}$, put $\langle \mathfrak{A}, a_1, \ldots, a_l \rangle \equiv_{M,R} \langle \mathfrak{B}, b_1, \ldots, b_l \rangle$ if and only if

$$M(\mathfrak{A}) \models R(a_1, \ldots, a_l) \Leftrightarrow M(\mathfrak{B}) \models R(b_1, \ldots, b_l) \ .$$

Then, for some k, $\equiv^k_\infty$ refines $\equiv_{M,R}$.

We conclude this section with a potential application of these ideas to database query optimization. When a model $\mathfrak{A}$ is the input to a database machine, it must be encoded as a string of bits. If the the universe of $\mathfrak{A}$ has n elements, then the length of this encoding is on the order of n^r for some fixed r. Let $l \leq k$. Suppose that there is a function $d\colon \omega \to \omega$ such that $d(n) \ll n^r$, and for every n, there is a collection $C_1, \ldots, C_{d(n)}$ of $\equiv^k_\infty$ classes of l-tuples on inputs $\mathfrak{A}$ of size n such that for almost all $\mathfrak{A}$, every l-tuple is in $\bigcup_{i=1}^{d(n)} C_i$. Then there is the possibility of optimizing $\sigma(x_1, \ldots, x_l) \in L^k_{\infty\omega}$. The case studied by Abiteboul, Compton, and Vianu [2] was for the partial fixed-point logic of a relational type, and the uniform probability distribution on models of that type. Using rate estimates, they proved that, for every $l \leq k \in \omega$, there is a *finite* set of $\equiv^k_\infty$ classes $C_1, \ldots, C_d$ and first-order formulas $\phi_1(x_1, \ldots, x_l), \ldots, \phi_d(x_1, \ldots, x_l)$ such that:

1. For all $\mathfrak{A}$, $a_1, \ldots, a_l \in \mathfrak{A}$, and all $i = 1, \ldots, d$,

$$\langle \mathfrak{A}, a_1, \ldots, a_l \rangle \in C_i \text{ if and only if } \mathfrak{A} \models \phi_i(a_1, \ldots, a_l) \ .$$

2. There is $c < 1$ such that

$$\mathrm{pr}\left(\exists x_1 \ldots \exists x_l \left((x_1, \ldots, x_l) \notin \bigcup_{i=1}^{d} C_i \right), n \right) < c^n \ .$$

On a parallel processor, the time complexity of a first-order query is bounded above by a constant, but the known methods for answering a fixed-point query require time n^t for some t, in the worst case. However, by exploiting facts 1. and 2. above, the average time complexity of the fixed-point query σ is

$$\begin{aligned} &\leq \mathrm{pr}\left(\forall x_1 \ldots x_l \left((x_1, \ldots, x_l) \in \textstyle\bigcup_{i=1}^{d} C_i \right), n \right) \\ &\qquad \times \max\nolimits_{1 \leq i \leq d}(\mathrm{complexity}(\langle \mathfrak{A}, x_1, \ldots, x_l \rangle \in C_i)) \\ &\qquad + \mathrm{pr}\left(\exists x_1 \ldots \exists x_l \left((x_1, \ldots, x_l) \notin \textstyle\bigcup_{i=1}^{d} C_i \right), n \right) \times \mathrm{complexity}(\sigma) \\ &\leq \max\nolimits_{1 \leq i \leq d}(\mathrm{complexity}(\phi_i)) + c^n \times n^t \leq K \end{aligned}$$

for some constant K. Thus, on average, the cost of a fixed-point query is also constant.

3 Random Graphs

Random graph theory was initiated by Erdős and Rényi [10], and has since grown into a very active branch of combinatorics. A good introductory text is Palmer [24]; for those with some familiarity with the subject, a standard reference is Bollobás [5]. We will consider only one model of random graph, but it is by far the most widely studied. Unlike our previous examples, the probability distribution on random graphs of a given size is not in general uniform.

For each $n \in \omega$, $\mathcal{C}_n$ is the collection of undirected graphs whose vertex set is n. The probability distribution pr_n on $\mathcal{C}_n$ is defined in terms of the *edge probability*, a function $p(n)$ taking values in $[0,1]$. For any $\langle n, E\rangle \in \mathcal{C}_n$,

$$\mathrm{pr}_n(\{\langle n, E\rangle\}) = p(n)^{|E|}(1-p(n))^{\binom{n}{2}-|E|} .$$

Another, perhaps more intuitive, way of regarding this probability is to construct the random graph $\langle n, E\rangle$ by choosing independently for each pair $\{i,j\} \subseteq n$, that $\{i,j\} \in E$ with probability $p(n)$. It is evident that when $p(n) = 1/2$ for all n, this is identical to the uniform distribution.

There are many possible edge probabilities, but we will restrict our attention to those of the form $p(n) = \beta n^{-\alpha}$, where $\alpha, \beta \geq 0$. This is a crude characterization of edge probabilities, but it is important because it is simple, involving only two parameters α and β, and yet by varying these parameters, a wide range of monotonically decreasing edge probabilities is covered. (Increasing edge probabilities are a symmetric case because we can replace $p(n)$ by $1-p(n)$.) Constant edge probabilities occur when $\alpha = 0$. In this case, we assume $0 < \beta < 1$. When $\alpha > 0$, the graph is said to be "sparse." As α increases, the random graph becomes sparser, i.e., its edge density decreases, until $\alpha > 2$, when almost all random graphs have no edges. Thus the interesting values of α are in the range $[0,2]$. A slight generalization of the proofs of Fagin and Glebskiĭ et. al. extends their 0-1 law to all random graphs with constant edge probabilities.

Random graph theorists are particularly interested in threshold functions. These are parameterized functions that characterize the random graph such that as the parameter passes through a certain value (the threshold), the qualitative behavior of the random graph changes. In the logics we will be considering, α is the important parameter for the edge probability. For example, the asymptotic probability that the graph has a 4-clique is 0 if $\alpha > 2/3$, 1 if $\alpha < 2/3$, and $1 - e^{-\beta^6/24}$ if $\alpha = 2/3$. In the logics we will be studying, when $\alpha > 0$, the value of β does not affect the qualitative behavior of the random graph as long as it is greater than 0, so we will assume it is 1 in this case.

Table 1. summarizes the behavior of random graphs, categorized horizontally according to α, with the sparsest graphs to the left, and vertically according to the logic: FOL (first-order logic), TCL (transitive closure logic), LFP (least fixed point logic), and $L^{\omega}_{\infty\omega}$.

	$\alpha > 1$, $\alpha = \frac{l+1}{l}$	$\alpha > 1$, $\alpha \neq \frac{l+1}{l}$	$\alpha = 1$	$\alpha \in (0,1)$, rational	$\alpha \in (0,1)$, irrational	$\alpha = 0$
$L^{\omega}_{\infty\omega}$	Con[7]	0-1[7]	Noncon	Noncon	Noncon	0-1[5]
LFP	Con	0-1	Noncon	Noncon	Noncon[9]	0-1[3]
TCL	Con	0-1	Noncon[8]	Noncon	0-1[9]	0-1
FOL	Con[6]	0-1[4]	Con[6]	Noncon[4]	0-1[4]	0-1[1,2]

Table 1: Summary of convergence laws.

Three types of behavior are distinguished: 0-1, Con, and Noncon, standing for 0-1 law, convergence law, and nonconvergence law respectively. The superscripts are cross-references to the following list of citations. They are arranged in chronological order of journal publication dates, although in many cases, they were preceded by publications in conference proceedings. In the cases that are not cross-referenced, the behavior is a simple consequence of either a 0-1 or convergence law above, or a nonconvergence result below. Perhaps the most striking feature about this table is the threshold at $\alpha = 1$. For all other values of α, the behavior of FOL and TCL is the same. For all four logics, it demarks the onset of complex behavior, i.e., nonconvergence. This is not surprising because it has been long known that the probabilities of important graph properties change when α passes through the value 1. For example, when $\alpha > 1$, the random graph is almost surely planar and not connected, but when $\alpha < 1$, it is almost surely nonplanar and connected. In fact, a much more precise analysis of this threshold has been done (see Bollobás [5]).

1. Glebskiĭ et al. [13]. Also contains the convergence law for structures with relations and constant mentioned earlier.
2. Fagin [11]. Also contains a 0-1 law for unlabelled relational structures. That is, two structures are identified if they are isomorphic.
3. Blass, Gurevich, and Kozen [4]. Actually proven for the stronger partial fixed point logic.
4. Shelah and Spencer [27]. A rate estimate when α is irrational is given in Lynch [20].
5. Kolaitis and Vardi [15].
6. Lynch [18].
7. Lynch [19].
8. Tyszkiewicz [28]. Improved to deterministic transitive closure logic in Lynch and Tyszkiewicz [21].
9. Lynch, McArthur, Tyszkiewicz, and Spencer (in preparation). They also show that the convergence law fails for LFP, which immediately imples that TCL is weaker than LFP. It was incorrectly stated in [21] that convergence holds for $L^{\omega}_{\infty\omega}$

The proofs of 0-1 or convergence law in each category $\alpha > 1$, $\alpha = 1$, $\alpha < 1$ irrational, and $\alpha = 0$ are all quite different. There is, however, a unifying principle in all the proofs. It is the use of combinatorial pebble games to characterize definable properties of structures. The original pebble game was invented by Ehrenfeucht [8] to characterize the equivalence classes of structures indistinguishable by first-order sentences of bounded quantifier rank. The quantifier rank of a first-order sentence is the depth of nesting of its quantifiers. These classes had been studied earlier by Fraïssé [12], and the game is often called the Ehrenfeucht-Fraïssé game in their honor. We will describe the game as it is played on two graphs $G_i = \langle V_i, E_i \rangle$, for $i = 0, 1$. The game consists of k rounds numbered $1, \ldots, k$, for some fixed $k \in \omega$, and is played by two players, which we will refer to as I and II. There are two sets of pebbles numbered $1, \ldots, k$. In each round r, player I places one of the pebbles numbered r on a vertex of G_0 or G_1, and then player II places the other pebble numbered r on a vertex of the other graph (G_1 or G_0). For $i = 0, 1$ and $r = 1, \ldots, k$, let $a_{i,r}$ be the vertex in V_i that is pebbled (by either player) in round r. After round k, we say that player II has won if and only if the structures $\langle \{a_{i,1}, \ldots, a_{i,k}\}, E_i \restriction \{a_{i,1}, \ldots, a_{i,k}\}, a_{i,1}, \ldots, a_{i,k} \}\rangle$, $i = 0, 1$, are isomorphic. That is, the induced subgraphs on $\{a_{i,1}, \ldots, a_{i,k}\}$ are isomorphic via $a_{0,r} \mapsto a_{1,r}$.

Definition 4. Let G_0 and G_1 be two graphs and $k \in \omega$.

1. $G_0 \sim^k G_1$ if and only if player II has a winning strategy for the k-round Ehrenfeucht-Fraïssé game on G_0 and G_1. That is, no matter how player I moves, player II can always respond so that he wins after k rounds.
2. $G_0 \equiv^k G_1$ if and only if G_0 and G_1 satisfy the same sentences of quantifier rank $\leq k$.

It is easily seen that $\sim^k$ and $\equiv^k$ are equivalence relations. The key result on the game is the following.

Theorem 5 Ehrenfeucht [8]. *Let G_0, G_1, and k be fixed. Then the following are equivalent.*

1. $G_0 \sim^k G_1$.
2. $G_0 \equiv^k G_1$.

There is a similar game for the infinitary logic $L^{\omega}_{\infty\omega}$ which does not seem to have a widely accepted name, but it is sometimes called the eternal game. Again, it is played by two players on two graphs G_0 and G_1, and there are two sets of pebbles numbered $1, \ldots, k$. The first k rounds are played exactly like the k-round Ehrenfeucht-Fraïssé game, but after that, it may continue indefinitely. In succeeding rounds, player I moves a pebble to some (possibly the same) vertex in the same graph. Then player II responds by moving the other pebble with the same number to some vertex in the other graph. The game ends and player I wins if, after any round, the mapping $a_{0,j} \mapsto a_{1,j}$ for $j = 1, \ldots, k$ is not an isomorphism.

Definition 6. Let G_0 and G_1 be two graphs and $k \in \omega$.

1. $G_0 \sim^k_\infty G_1$ if and only if player II has a winning strategy for the k-pebble eternal game on G_0 and G_1. That is, no matter how player I moves, player II can always respond so that the game continues.
2. $G_0 \equiv^k_\infty G_1$ if and only if G_0 and G_1 satisfy the same sentences in $L^k_{\infty\omega}$.

Again, $\sim^k_\infty$ and $\equiv^k_\infty$ are the same equivalence relations.

Theorem 7 Barwise [3]. *Let G_0, G_1, and k be fixed. Then the following are equivalent.*

1. $G_0 \sim^k_\infty G_1$.
2. $G_0 \equiv^k_\infty G_1$.

The original proofs of the first-order 0-1 law for $\alpha = 0$ by Fagin and Glebskiĭ et al. did not rely on pebble games. But the Ehrenfeucht-Fraïssé game provides a simpler proof by showing that, for every k, almost all graphs belong to the same $\sim^k$ class. Therefore by Theorem 5, every sentence is almost always false or almost always true. We shall not present it here because the $L^\omega_{\infty\omega}$ 0-1 law for $\alpha = 0$ by Kolaitis and Vardi follows from essentially the same proof using the eternal game instead of the Ehrenfeucht-Fraïssé game. That is, we show that almost all graphs belong to the same $\sim^k_\infty$ class. This equivalence class has a very simple axiomatization. We say a graph is k-extendible if it has at least $k-1$ vertices and, for every set of $k-1$ vertices, every possible one-vertex extension is realized. In other words, it satisfies the following finite set of first-order axioms:

$$\left\{\exists x_1 \ldots \exists x_{k-1}\Big(\bigwedge_{1\le i<j<k} x_i \neq x_j\Big)\right\} \cup$$
$$\Bigg\{\forall x_1 \ldots \forall x_{k-1}\exists y\Bigg[\Big(\bigwedge_{1\le i<j<k} x_i \neq x_j\Big) \rightarrow \tag{1}$$
$$\Big(\bigwedge_{1\le i<k} x_i \neq y \wedge \bigwedge_{1\le i\le j} E(x_j, y) \wedge \bigwedge_{j<i<k} \neg E(x_j, y)\Big)\Bigg] : j = 0, \ldots, k-1\Bigg\}$$

The proof of the $L^\omega_{\infty\omega}$ 0-1 law for $\alpha = 0$ is finished by showing:

1. $G_0 \sim^k_\infty G_1$ for any two k-extendible graphs.
2. Almost all graphs are k-extendible.

Both parts have elementary proofs. Part 1. follows because the extension axioms (1) state that player II can always match any move of player I. To prove Part 2., let θ_j, $j = 0, \ldots, k-1$, enumerate the extension axioms, where the j is as in (1). Then, letting $p(n) = \beta$ be the edge probability,

$$\mathrm{pr}\Big(\neg\Big(\bigwedge_{j=0}^{k-1} \theta_j\Big), n\Big) = \mathrm{pr}\Big(\bigvee_{j=0}^{k-1} \neg\theta_j, n\Big)$$
$$\le \sum_{j=0}^{k-1} n^{k-1} \times [\beta^j(1-\beta)^{k-j-1}]^{n-k+1} .$$

Since $0 < \beta < 1$, there is some $b < 1$ such that $\beta^j(1-\beta)^{k-j-1} \leq b$ for all $j = 0, \ldots, k-1$. Then

$$\begin{aligned} \mathrm{pr}\left(\neg\left(\bigwedge_{j=0}^{k-1} \theta_j\right), n\right) &\leq kn^{k-1}b^{n-k+1} \\ &\to 0 \text{ as } n \to \infty \ . \end{aligned}$$

Note that this also yields a rate estimate. Letting $c = \sqrt{b}$, $kn^{k-1}b^{n-k+1} < c^n$ for sufficiently large n, so

$$\begin{aligned} &\mathrm{pr}(\sigma, n) < c^n, \text{ or} \\ &\mathrm{pr}(\sigma, n) > 1 - c^n \ . \end{aligned}$$

A slight generalization of the proof shows that when $\alpha = 0$, $\mathcal{L}^{\omega}_{\infty\omega}$ almost surely collapses to first-order logic. That is, for every formula $\sigma(x_1, \ldots, x_l) \in L^k_{\infty\omega}$, there is a first-order formula $\sigma'(x_1, \ldots, x_l)$ such that

$$\lim_{n\to\infty} \mathrm{pr}(\forall x_1 \ldots \forall x_l(\sigma(x_1, \ldots, x_l) \leftrightarrow \sigma'(x_1, \ldots, x_l)), n) = 1 \ . \qquad (2)$$

To construct σ', let $\gamma_1(x_1, \ldots, x_l), \ldots, \gamma_d(x_1, \ldots, x_l)$ be an enumeration of the quantifier-free formulas that describe the possible isomorphism types of l (not necessarily distinct) vertices. For $i = 1, \ldots, d$, let

$$\phi_i(x_1, \ldots, x_l) = \gamma_i(x_1, \ldots, x_l) \wedge \bigwedge_{j=0}^{k-1} \theta_j \ .$$

Then, as in 1. above, each first-order formula ϕ_i defines an $\equiv^k_\infty$ class C_i of l-tuples, and, as in 2., for almost all graphs, every l-tuple belongs to some C_i. In fact, the probability of not belonging to some C_i is less than c^n for some $c < 1$. (Note that we have just proven the claims of Abiteboul, Compton, and Vianu mentioned earlier.) Since each C_i is an $\equiv^k_\infty$-class of l-tuples, either

1. for every graph G and vertices $a_1, \ldots, a_l$ in G, $(a_1, \ldots, a_l) \in C_i \Rightarrow G \models \sigma(a_1, \ldots, a_l)$, or
2. for every graph G and vertices $a_1, \ldots, a_l$ in G, $(a_1, \ldots, a_l) \in C_i \Rightarrow G \models \neg\sigma(a_1, \ldots, a_l)$.

Let I be the set of i such that a. holds, and $\sigma' = \bigvee_{i \in I} \phi_i$. Equation (2) now follows.

Player II's strategy in the case $\alpha = 0$ is the simplest possible. He just imitates player I's move without regard to any future moves. But this does not work for nonzero α because there are some moves that are "hard to imitate". This is true even for the first-order Ehrenfeucht-Fraïssé game. For example, take the case when $1 > \alpha > \frac{1}{2}$ and $k \geq 3$. Suppose, as illustrated below, player I places the third pebble on a_3 in one of the graphs, where a_3 is adjacent to a_1 but not to a_2. This is an easy move to imitate, because almost surely, for every two vertices in the random graph, there is a third vertex adjacent to exactly one of them. However, if player I had chosen a_3 adjacent to both a_1 and a_2, then player II will probably lose because almost surely, the random graph contains only $O(n^{3-2\alpha}) = o(n^2)$ pairs of vertices with a common neighbor.

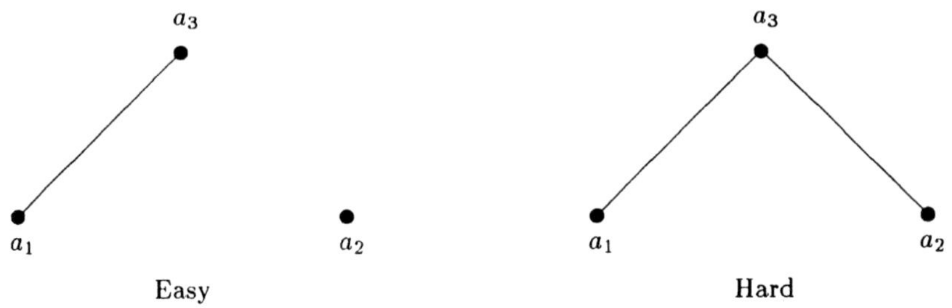

Thus player II must look ahead to possible future moves. But it is not sufficient to look ahead only one move. In the fourth round, if player I chooses a new vertex adjacent to a_1 and a_2, it is again unlikely that player II can match this move. There is, however, a probabilistic upper bound on the number of such vertices. Almost surely, there is no pair of vertices with more than $2/(2\alpha - 1)$ common neighbors. More generally, there is a probabilistic upper bound on the number of hard moves that player II must consider, and this makes a winning strategy possible.

The notions of hard moves and looking ahead were formalized by Shelah and Spencer in their proof of the first-order 0-1 law for $1 > \alpha > 0$ and irrational. The full proof is much longer than the previous proof of the 0-1 law for $L^{\omega}_{\infty\omega}$, and we will only sketch the highlights.

Definition 8. We will use the convention that if $\langle V, E\rangle$ is a graph, and $W \subseteq V$, then $\langle W, E\rangle$ is the graph $\langle W, E \restriction W\rangle$.

1. $\langle V, E, c_1, \ldots, c_i\rangle$ is a rooted graph if $\langle V, E\rangle$ is a graph, and $c_1, \ldots, c_i \in V$. We also say $\langle V, E, R\rangle$ is a rooted graph when $R \subseteq V$, and the order of the vertices in R is immaterial.
2. The weight of $\langle V, E, R\rangle$, $\zeta(\langle V, E, R\rangle)$, is $|V - R| - \alpha|E - E \restriction R|$.
3. $\langle V, E, R\rangle$ is dense if $\zeta(\langle V, E, R\rangle) < 0$.
4. $\langle V, E, R\rangle$ is rigid if $\langle V, E, S\rangle$ is dense for every set of vertices S such that $R \subseteq S \subset V$. By convention, we will say that $\langle R, E, R\rangle$ is rigid.

Note that, since α is irrational, the weight of a rooted graph cannot be 0 unless $V = R$.

Lemma 9. *If $\langle V, E, R\rangle$ and $\langle W, E, R\rangle$ are rigid, then so is $\langle V \cup W, E, R\rangle$.*

Definition 10. For a rooted graph $\langle V, E, R\rangle$, *Rig*($\langle V, E, R\rangle$) is the maximal set S such that $R \subseteq S \subseteq V$ and $\langle S, E, R\rangle$ is rigid. Note that by Lemma 9, S is unique.

The closure operator Cl in the next definition captures the notion of how far ahead player II must look when there are j moves left in the game.

Definition 11. Let $G = \langle V, E\rangle$. For $i, j \in \omega$ and $a_1, \ldots, a_i \in V$, $Cl^j(G; a_1, \ldots, a_i)$ is defined by induction on j.

$$\mathrm{Cl}^0(G; a_1, \ldots, a_i) = \langle\{a_1, \ldots, a_i\}, E, a_1, \ldots, a_i\rangle \text{ for all } i \in \omega \ .$$

Assume Cl^j has been defined for all $i \in \omega$. Fixing i and $a_1, \ldots, a_i \in V$, for all $a_{i+1} \in V$, let

$$\mathrm{Cl}^j(G; a_1, \ldots, a_{i+1}) = \langle V(a_{i+1}), E, a_1, \ldots, a_{i+1}\rangle, \text{ and}$$
$$S(a_{i+1}) = \mathrm{Rig}(\langle V(a_{i+1}), E, a_1, \ldots, a_i\rangle) \ .$$

Then

$$\mathrm{Cl}^{j+1} = \left\langle \bigcup_{a_{i+1} \in V} S(a_{i+1}), E, a_1, \ldots, a_i \right\rangle \ .$$

Lemma 12. *For every $i, j \in \omega$, there is m such that*

$$\mathrm{pr}(\forall a_1 \ldots \forall a_i \in G(|\mathrm{Cl}^j(G; a_1, \ldots, a_i)| \leq m), n) \to 1$$

as $n \to \infty$.

Because of the preceding lemma, for every k, there is some m such that with probability asymptotic to 1, for all $i \leq k$,

$$\forall a_1 \ldots \forall a_i \in G(|\mathrm{Cl}^{k-i}(G; a_1, \ldots, a_i)| \leq m) \ .$$

Therefore, the following can be expressed as a finite set of first-order axioms, where $i = 0, \ldots, k-1$:

$$\begin{aligned}
&\forall x_1 \ldots \forall x_i(\\
&\quad \text{for any graph } H = \langle W, F\rangle \text{ and } c_1, \ldots, c_{i+1} \in W \text{ such that} \\
&\qquad |\mathrm{Cl}^{k-i}(H; c_1, \ldots, c_i)|, |\mathrm{Cl}^{k-i-1}(H; c_1, \ldots, c_{i+1})| \leq m, \\
&\qquad\quad \mathrm{Cl}^{k-i}(G; x_1, \ldots, x_i) \cong \mathrm{Cl}^{k-i}(H; c_1, \ldots, c_i) \\
&\qquad \to \\
&\qquad\quad \exists y(\mathrm{Cl}^{k-i-1}(G; x_1, \ldots, x_i, y) \cong \mathrm{Cl}^{k-i-1}(H; c_1, \ldots, c_{i+1}))) \ .
\end{aligned}$$

We say that a graph is (α, k)-extendible if it satisfies all these axioms. The proof of the first-order 0-1 law for irrational α is completed by showing:

1. $G_0 \sim^k G_1$ for any two (α, k)-extendible graphs.
2. Almost all graphs are (α, k)-extendible.

The proof of the TCL 0-1 law for irrational α uses many of the concepts from the proof of the first-order 0-1 law. In essence, it shows that for almost all graphs, the transitive closure operation stops growing after a bounded number of steps. This happens in two ways: either its growth rate is so rapid that it absorbs all the tuples that it can very quickly, or its growth rate slows down until it stops absorbing new tuples. This is a generalization of the fact that, in a random graph with edge probability $n^{-\alpha}$, if $\alpha < 1$, then every pair of vertices is joined by a path of length $\lceil 1/(1-\alpha) \rceil$, but if $\alpha > 1$, then there is no path of length greater than $1/(\alpha - 1)$.

Definition 13. Let $\sigma(\overline{x}, \overline{y}, \overline{z})$ be a first-order formula in the language of graphs whose free variables are

$$\overline{x} = x_1, \ldots, x_i,$$
$$\overline{y} = y_1, \ldots, y_j, \text{ and}$$
$$\overline{z} = z_1, \ldots, z_j,$$

where $i, j \geq 0$. Let $G = \langle V, E \rangle$ be a graph and $\overline{a} = a_1, \ldots, a_i \in V$.

1. $\mathrm{TC}(G, \sigma, \overline{a})$ is the transitive closure of $\{(\overline{b}, \overline{c}) \in V^{2j} : G \models \sigma(\overline{a}, \overline{b}, \overline{c})\}$.
2. An $\overline{a}$-link is a rigid rooted graph $\langle W, F, \overline{a}, \overline{b}, \overline{c} \rangle$ where $\overline{b}, \overline{c} \in V^j$.
3. An $\overline{a}$-chain of length l is a sequence of $\overline{a}$-links $\langle \langle W_k, F_k, \overline{a}, \overline{b_k}, \overline{c_k} \rangle : 1 \leq k \leq l \rangle$ such that $\overline{b_{k+1}} = \overline{c_k}$ for $1 \leq k < l$ and for all $u, v \in W_h \cap W_k$ $(1 \leq h, k \leq l)$, $F_h(u, v)$ if and only if $F_k(u, v)$. We say the chain connects $\overline{b_1}$ to $\overline{c_l}$.

Lemma 14. *There is a finite set of $\overline{a}$-links $\mathcal{S}$ such that for almost all graphs $G = \langle V, E \rangle$, for all $\overline{a} \in V^i$ and $\overline{b}, \overline{c} \in V^j$, $G \models \sigma(\overline{a}, \overline{b}, \overline{c})$ if and only if there is $W \subseteq V$ such that $\langle W, E \restriction W, \overline{a}, \overline{b}, \overline{c} \rangle \in \mathcal{S}$.*

Lemma 15. *For almost all graphs $G = \langle V, E \rangle$, for all $\overline{a} \in V^i$ and $\overline{b}, \overline{c} \in V^j$, $(\overline{b}, \overline{c}) \in \mathrm{TC}(G, \sigma, \overline{a})$ if and only if there is an $\overline{a}$-chain connecting $\overline{b}$ to $\overline{c}$ with links in $\mathcal{S}$.*

Here is the main lemma.

Lemma 16. *There is a constant m such that for almost all graphs $G = \langle V, E \rangle$, for all $\overline{a} \in V^i$ and $\overline{b}, \overline{c} \in V^j$, if $(\overline{b}, \overline{c}) \in \mathrm{TC}(G, \sigma, \overline{a})$, then there is an $\overline{a}$-chain of length at most m connecting $\overline{b}$ to $\overline{c}$ with links in $\mathcal{S}$.*

A consequence of this lemma and the finiteness of $\mathcal{S}$ is that almost surely TCL collapses to first-order logic when $\alpha < 1$ is irrational. The 0-1 law for TCL in this case is an immediate consequence of the 0-1 law for first-order logic.

The proof of the first-order convergence law at the threshold $\alpha = 1$ will not be covered here, since it is quite involved. It has many similarities to the convergence law for unary functions [17] mentioned earlier. Some flavor of the proof can be obtained from a sketch of the $L^{\omega}_{\infty\omega}$ convergence law for $\alpha = (l+1)/l$. Thus we will finish our outline of 0-1 and convergence proofs by moving to the region $\alpha > 1$, i.e., the region of very sparse random graphs. In fact, we will outline a stronger result for all $p(n) = n^{-\alpha}$, $\alpha > 1$: for every $\sigma \in L^{\omega}_{\infty\omega}$, either

$$\text{pr}(\sigma, n) < 2^{-n^d} \text{ for some } d > 0, \text{ or} \tag{3}$$
$$\text{pr}(\sigma, n) \sim cn^{-d} \text{ for some } c > 0,\ d \geq 0 \ . \tag{4}$$

If we take $f(n) \sim n^{-\infty}$ to mean $f(n) < n^{-d}$ for all $d > 0$, then the conclusion implies $\text{pr}(\sigma, n) \sim cn^{-d}$ for some $c \in (0, \infty)$, $d \in [0, \infty]$. This kind of rate estimate is known as a *power law* in physics and engineering. When $\alpha > 1$ but not of the form $(l+1)/l$, the proof of the rate estimate shows that if $d = 0$, then $c = 1$. Thus the 0-1 law holds in this case.

When $\alpha > 1$, no version of extension axioms seems applicable. Instead, we use a theorem of Kolaitis which states that for each k, the $\equiv^k_\infty$ class of a graph is determined by counting how many components it has in each $\equiv^k_\infty$ class. In the following, $C \sqsubseteq G$ means that C is a component of G.

Theorem 17 Kolaitis [14]. *Let G_0 and G_1 be two graphs such that for every connected graph C, either*

$$|\{ C_0 \sqsubseteq G_0 : C_0 \equiv^k_\infty C \}|, |\{ C_1 \sqsubseteq G_1 : C_1 \equiv^k_\infty C \}| \geq k$$

or

$$|\{ C_0 \sqsubseteq G_0 : C_0 \equiv^k_\infty C \}| = |\{ C_1 \sqsubseteq G_1 : C_1 \equiv^k_\infty C \}| \ .$$

Then $G_0 \equiv^k_\infty G_1$.

We will actually use a weaker version of this result.

Corollary 18 *Let G_0 and G_1 be two graphs such that for every connected graph C, either*

$$|\{ C_0 \sqsubseteq G_0 : C_0 \cong C \}|, |\{ C_1 \sqsubseteq G_1 : C_1 \cong C \}| \geq k$$

or

$$|\{ C_0 \sqsubseteq G_0 : C_0 \cong C \}| = |\{ C_1 \sqsubseteq G_1 : C_1 \cong C \}| \ .$$

Then $G_0 \equiv^k_\infty G_1$.

We will use $G_0 \cong^k G_1$ to mean the condition in the Corollary is satisfied by G_0 and G_1. Thus $\cong^k$ is a refinement of $\equiv^k_\infty$.

We will categorize connected graphs in three ways:

(I) Trees with v vertices, $v < \alpha/(\alpha - 1)$.
(II) Trees with v vertices, $v = \alpha/(\alpha - 1)$.
(III) Connected graphs with v vertices and e edges, $v - \alpha e < 0$.

To see that these three classes partition the set of connected graphs, consider any connected graph that is not of type (I) or (II). If it is not a tree then $v \leq e$, which implies $v - \alpha e < 0$ since $\alpha > 1$. If it is a tree, then $v > \alpha/(\alpha - 1)$ and $e = v - 1$, again implying $v - \alpha e < 0$. Conversely, similar calculations show that any connected graph of type (III) is not of type (I) or (II).

The proof consists of three cases. Let $\sigma \in L^k_{\infty\omega}$.

Case 1. For every graph G such that $G \models \sigma$, there is some tree T of type (I) such that G has less than k components isomorphic to T.
Case 2. Not Case 1, and $\alpha/(\alpha-1)$ is not an integer.
Case 3. Not Case 1, and $\alpha/(\alpha-1)$ is an integer.

Case 1. has probability $< 2^{-n^d}$ for some $d > 0$. Combined with Corollary 18, (3) follows. In Case 2., the existence of components of type (III) has probability $\sim cn^{-d}$ for some $c > 0$, $d \geq 0$, which implies (4). Further, if $d = 0$ it can be shown that $c = 1$, i.e., the 0-1 law holds. Lastly, in Case 3., the existence of components of type (II) has a Poisson distribution, which also implies (4), but here $d = 0$ and $c \neq 1$ is possible, as we show next.

We conclude with an example of the kinds of computations used to derive asymptotic probabilities in Case 3. They rely on sieve methods that are generalizations of Ch. Jordan's formula and Bonferroni's inequalities. These formulas may be found in Bollobás [5]; their generalizations are explained in detail in Lynch [17]. As previously mentioned, similar computations are needed when $\alpha = 1$, but they are considerably more complicated. Let $\alpha = -5/4$, and consider the property that the graph has exactly 3 components isomorphic to the tree (a) below and at least 7 components isomorphic to the 5-path (b).

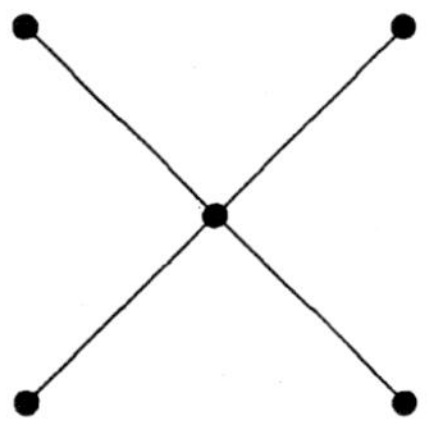

(a)

(b)

The probability of this property is

$$\begin{aligned}
&\sim \sum_{s\geq 3}\sum_{t\geq 7}(-1)^{s+t-10}\binom{s}{3}\binom{t-1}{6}\frac{n^{5s+5t)}}{s!t!} \times \left(\frac{n^{-4\alpha}}{24}\right)^{s} \times \left(\frac{n^{-4\alpha}}{2}\right)^{t} \\
&\sim \sum_{s\geq 3}(-1)^{s-3}\frac{(1/24)^s}{3!(s-3)!} \times \sum_{t\geq 7}(-1)^{t-7}\frac{(1/2)^t}{6!(t-7)!t} \\
&= \frac{(1/24)^3 e^{-1/24}}{3!} \times \left(1 - \sum_{j\leq 6}\frac{(1/2)^j e^{-1/2}}{j!}\right) .
\end{aligned}$$

4 Future Directions

As Table 1. shows, there is a complete classification of convergence laws for the four logics according to the value of α. Future research is likely to be concerned with subtler questions. For example, in our analysis, we ignored the value of β in the edge probability $\beta n^{-\alpha}$. This was because it did not affect the existence of 0-1 or convergence laws as long as it was strictly between 0 and 1. It has long been known, however, that the asymptotic probability of particular properties can be strongly affected by β. For example, in an early paper Erdős and Rényi [9] showed that for $\alpha = 1$, if $\beta < 1$, then the random graph is almost surely planar, but if $\beta > 1$, then it is almost surely nonplanar. It is not known whether planarity is expressible in $L^{\omega}_{\infty\omega}$, but the property that all components have at most one cycle is expressible in TCL, and it is almost surely false when $\beta < 1$ but almost surely true when $\beta > 1$. A metatheorem characterizing properties that have thresholds involving β would be of great interest to random graph theorists.

Other forms of edge probability should be studied. An important one is $\beta n \log_2 n$. When $\beta < 1$, the random graph is almost surely not connected, but when $\beta > 1$, it is almost surely connected. As we saw earlier, connectedness is expressible in TCL.

The original motivation of Fagin and Glebskiĭ et al. in studying 0-1 laws was to find limitations on the expressive power of logics. The general idea would be to establish a 0-1 law or convergence law for some logic $\mathcal{L}$, and then show that some property violates the law, and is therefore not expressible in $\mathcal{L}$. Potential benefits could be lower bounds in computational complexity. That is, if $\mathcal{L}$ captures some complexity class, this would show that the property is not in this class. The fond hope that this approach will eventually produce significant inexpressibility results may not be unreasonable. A recent article by Rosen, Shelah, and Weinstein [25] uses the methods described in Section 3 to prove that certain graph properties are not definable as infinitary disjunctions of first-order existential sentences that involve only a finite number of variables.

On the pessimistic side, it seems that any logic that is powerful enough to capture a complexity class is also too strong to obey a convergence law. For example, many of these logics have a symbol $\leq$, which is interpreted on n in the standard way, and the presence of a linear ordering is enough to destroy the convergence law. We have seen that random graphs have 0-1 and convergence laws for many edge probabilities of the form $\beta n^{-\alpha}$. But Compton, Henson, and Shelah [6] showed that when $\leq$ is added as a new relation, the convergence law fails for constant edge probabilities, and Dolan and Lynch [7] showed that convergence fails for any edge probability $p(n)$ as long as $p(n)n^2$ and $(1-p(n))n^2$ are bounded below by some unbounded recursive function. A phenomenon that several researchers have started to investigate is *slow variation*. This happens when $|\operatorname{pr}(\sigma, n) - \operatorname{pr}(\sigma, n+1)|$ is asymptotic to 0 for every $\sigma \in \mathcal{L}$. In at least some of the cases where the convergence law fails, including random graphs with constant edge probability and the $\leq$ relation, the slow variation law holds (Shelah [26]). Perhaps this is the kind of behavior that will serve as a discriminator of complexity classes.

References

1. S. Abiteboul and V. Vianu, Generic computation and its complexity, in: *Proc. 23rd ACM Symp. on Theory of Computing*, Association for Computing Machinery, New York, 1991, 209–219.
2. S. Abiteboul, K. J. Compton, and V. Vianu, Queries are easier than you thought, probably, in: *Proc. 11th ACM Symp. on Principles of Database Systems*, Association for Computing Machinery, New York, 1992, 32–42.
3. J. Barwise, On Moschovakis closure ordinals, *J. Symbolic Logic* **42** (1977), 292–296.
4. A. Blass, Y. Gurevich, and D. Kozen, A zero-one law for logic with a fixed point operator, *Inform. and Control* **67** (1985), 70–90.
5. B. Bollobás, *Random Graphs*, Academic Press, London, 1985.
6. K. J. Compton, C. W. Henson, and S. Shelah, Nonconvergence, undecidability, and intractability in asymptotic problems, *Ann. Pure Appl. Logic* **36** (1987), 207–224.
7. P. Dolan and J. F. Lynch, The logic of ordered random structures, *Random Structures and Algorithms* **4** (1993), 429–445.
8. A. Ehrenfeucht, An application of games to the completeness problem for formalized theories, *Fund. Math.* **49** (1961), 129–141.
9. P. Erdős and A. Rényi, On random graphs, I, *Pub. Math. (Debrecen)* **6** (1959), 290–297.
10. P. Erdős and A. Rényi, On the evolution of random graphs, *Magyar Tud. Akad. Mat. Kutató Int. Közl.* **5** (1960), 17–61.
11. R. Fagin, Probabilities on finite models, *J. Symbolic Logic* **41** (1976), 50–58.
12. R. Fraïssé, Sur quelques classifications des systéms de relations, *Pub. Sci. Univ. Alger Sér. A* **1** (1954), 35–182.
13. Y. V. Glebskiĭ, D. I. Kogan, M. I. Liogon'kiĭ, and V. A. Talanov, Range and degree of realizability of formulas in the restricted predicate calculus, *Kibernetika (Kiev)* **2** (1969), 17–28; English translation, *Cybernetics* **5** (1972), 142–154.
14. Ph. G. Kolaitis, On asymptotic probabilities of inductive queries and their decision problem, in R. Parikh, ed., *Logics of Programs '85, Lecture Notes in Computer Science* **193** (1985), Springer-Verlag, 153–166.
15. Ph. G. Kolaitis and M. Y. Vardi, Infinitary logics and 0-1 laws, *Inform. and Computation* **98** (1992), 258–294.
16. J. F. Lynch, Almost sure theories, *Ann. Math. Logic* **18** (1980), 91–135.
17. J. F. Lynch, Probabilities of first-order sentences about unary functions, *Trans. AMS* **287** (1985), 543–568.
18. J. F. Lynch, Probabilities of sentences about very sparse random graphs, *Random Structures and Algorithms* **3** (1992), 33–53.
19. J. F. Lynch, Infinitary logics and very sparse random graphs, in: *Proc. 8th IEEE Symp. on Logic in Computer Science*, IEEE Computer Society Press, Los Alamitos, 1993, 191–198. Also to appear in *J. Symbolic Logic*.
20. J. F. Lynch, An extension of 0-1 laws, *Random Structures and Algorithms* **5** (1994), 155–172.
21. J. F. Lynch and J. Tyszkiewicz, The infinitary logic of sparse random graphs, in: *Proc. 10th IEEE Symp. on Logic in Computer Science*, IEEE Computer Society Press, Los Alamitos, 1995, 54–64.
22. Y. N. Moschovakis, *Elementary Induction on Abstract Structures*, North-Holland, 1974.
23. J. Mycielski, Measure and category of some sets of models, *Notices AMS* **22** (1975), A-475.

24. E. M. Palmer, *Graphical Evolution*, Wiley, New York, 1985.
25. E. Rosen, S. Shelah, S. Weinstein, *k-Universal finite graphs*, to appear in: R. Boppana and J. F. Lynch, eds., *Logic and Random Structures, DIMACS Series in Discrete Mathematics and Theoretical Computer Science*, American Mathematical Society, Providence.
26. S. Shelah, Very weak zero one laws for random graphs with order and random binary functions, *Random Structures and Algorithms* (to appear).
27. S. Shelah and J. Spencer, Zero-one laws for sparse random graphs, *J. AMS* **1** (1988), 97–115.
28. J. Tyszkiewicz, Infinitary queries and their asymptotic probabilities. I. Properties definable in transitive closure logic, in: E. Börger et al., eds., *Proc. Computer Science Logic '91, Lecture Notes in Computer Science* **626** (1991), Springer-Verlag, 396–410.

V=L and Maximize

Penelope Maddy

Department of Philosophy
University of California
Irvine, CA 92717, USA
pjmaddy@uci.edu

The problem that interests me is easy to state: what justifies the axioms of set theory? Some observers suggest that decisions on the adoption or rejection of axiom candidates are not made on rational grounds, that they are subject only to psychological or sociological or aesthetic constraints. This may be right, but I think it is too early to concede the point; my working hypothesis is that there are sound arguments to be made on these issues. As a student of the methodology of set theory, my job is to try to isolate and elaborate these sound arguments.

In Zermelo's day, for many of the axioms in his first list, the problem of justification was somewhat less daunting than it is for the candidates in dispute today. Zermelo was faced with a fairly well-developed body of set theoretic lore; the difficulty was that paradoxes and other uncertainties lurked around its edges. His goal was to select some particularly fundamental statements to serve as starting points; this was to be done skillfully, so that the core doctrine could be deduced without the troublesome outliers. The upshot, for many of his axioms, was that certain previously-accepted claims were being promoted to axiomatic status, not that any new claims were being made. The selection of these particular statements for promotion needed justification—presumably on grounds of economy, efficiency, and likely consistency—but the problem of justifying the statements themselves was eased by their previous acceptance.[1]

The case of the axiom of choice was different. Choice was not a previously-accepted or uncontroversial claim; what needed justification was not its mere promotion, but the statement itself. With this case, the problem of what justifies an axiom arose in a more pressing and poignant form, as did the prior problem of what sorts of grounds are appropriate for such justifications. The subsequent, fascinating history is familiar, and the outcome is now stable, so I won't rehearse it here. But it is important to note that since then, with the advent of independence results and the subsequent search for new, stronger axioms, the problem of justification has become ever more acute.

Rather than talk in the abstract about the general problem of justifying axioms, I want to focus on one particularly salient case: the axiom of constructibility. The decision on V=L is the first truly momentous one after ZFC, and it is pivotal for the further development of the subject. Despite the fact that adding V=L is a safe, economical, and powerful option, settling many of the lingering

[1] This is not to say that all Zermelo's axioms, even excluding choice, were uncontroversial or that his system was immediately accepted. See Moore [1982], pp. 160-167, for details.

open questions, I think it is fair to say that most contemporary set theorists believe that it should be rejected. An instance of my working hypothesis is the assumption that there are good grounds for this consensus, and what I want to explore here is the question of what those good grounds are.

1 Naturalism

Before I take up this project directly, a few words should be said about the prior question: what sort of arguments are appropriate for the justification of an axiom candidate (or in our case, for the justification of the rejection of an axiom candidate)? The standard method of justification in mathematics is proof, but axioms cannot be proved, so where are we to turn for support? As I've mentioned, some people think we are free to choose the starting points of our proofs in any way we please, that we are unconstrained by any rational considerations (except perhaps consistency). The philosophy behind this view is what I call 'glib formalism', the stance that any relatively consistent extension of ZFC is as good as any other, that only sociology or aesthetics or personal whim prompt us to choose one over another.

Now it seems to me that this stance is not consistent with the practice of set theory; set theorists seriously committed to the project of adding new axioms to ZFC would hardly pursue the subject in the spirit they do if they believed the issue could be properly settled by a statement of personal preference or a vote at an ASL meeting (though, of course, such votes are sometimes taken). Under these circumstances, the attractions of a realistic philosophy are obvious. If set theory is the study of an objectively existing world of sets, much as botany is the study of plants and particle physics the study of the small constituents of matter, then a correct extension of ZFC is one that faithfully reflects the structure of that reality of sets; assuming V=L has the other virtues we expect from axioms— simplicity, generality, power, etc.—we should add V=L to our list of axioms if all sets are constructible; otherwise, we should reject it. The question is as substantial as any scientific question; it is a question about the properties of the objective realm of sets. On this reading, realism is called upon to support the existing practice of taking the decision on new axioms seriously.

Now it is well-known that the task of fleshing out a realistic account of mathematics in general, or of set theory in particular, is a difficult one, fraught with epistemological and ontological obstacles, but I'm not interested in those problems here.[2] In fact, I think that on closer examination the underlying strategy of justifying mathematical practice by appeal to philosophical realism should fail to satisfy either the mathematician or the philosopher of mathematics. I have various messy metaphilosophical reasons for this belief, but our subject here is V=L, so I'll just give a quick sketch of the central idea.[3]

To see why the mathematician might be wary of the strategy of justifying mathematical practice by appeal to philosophy, consider a set theorist ponder-

[2] I discuss them at length, and attempt to answer them, in [1990].

[3] For more, see [199?b], [199?c] and [199?d].

ing the addition of a new axiom. If it were somehow conclusively demonstrated that realism is not a viable account of mathematics, would he conclude that the question he is pondering is actually a trivial one, to be decided on whim? Probably not; more likely, he would continue his pondering in much the same terms as before. The lack of a suitable metaphysics with which to defend his practice might be an embarrassment, but he is unlikely to suspend the practice for that reason. In other words, his assessment of the value of his undertaking is not, after all, dependent on his belief in the correctness of realism. Realism provided a handy counter to certain challenges, but it wasn't the true justification.[4]

To see where the philosopher might also balk at the attempt to justify practice with metaphysics, consider the situation from the perspective of a philosopher who is a scientific naturalist; that is, suppose she believes that the practice of natural science is not answerable to the epistemic standards of any external, a priori standpoint, that legitimate criticisms of scientific method can only come from within scientific practice itself. Such a philosopher may well feel that the same goes for mathematics, that successful mathematical practices should not bend to criticism developed in psychology departments or philosophy seminar rooms.[5] Having come so far, our philosopher may well also conclude that considerations external to scientific or mathematical practice cannot justify those practices any more properly than they can criticize them, that justifications must also proceed from within the practice. And this means, in particular, that efforts to justify the set theorist's practice on the basis of a philosophical realism about sets is misguided and ineffective.

Both these thought experiments point toward a metaphilosophical position I call mathematical naturalism: the grounds on which to criticize and/or justify mathematical methods are to be found inside mathematical practice itself; the practice need not answer to, nor can it look for support from, any external standard. So, for example, the mathematical naturalist will come down in favor of impredicative definitions because they make possible a classical theory of real numbers, an extremely fruitful mathematical construction, but the same naturalist will dismiss as irrelevant the as-yet-unresolved metaphysical debates concerning the nature of mathematical existence that have been thought to bear on this topic. Presented with a patently successful mathematical theory, the realist might still protest, 'yes, this theory has all the mathematical virtues I can imagine, but we must also ask: is it true in the objectively existing world of sets?'. The naturalist will see this very protest as the importation of an extra-mathematical standard; if a mathematical theory has all the virtues mathematics requires, then the naturalist holds that no further question is of any methodological significance.

---

[4] Cf. Kanamori ([1994], p. 481) on the 'invariance' of mathematics under philosophical change.

[5] Of course, it isn't really a matter of where the criticisms originate, but of what methods they use: e.g., a criticism of classical analysis based on a mathematician's philosophical intuition about abstract objects is a non-starter, while a logical criticism coming from a philosopher (e.g. Bishop Berkeley) may be quite to the point.

The contrast, then, is this: where the realist will ask—is this axiom candidate true?—the naturalist will ask—would the adoption of this axiom candidate produce a fruitful theory—or better yet—would the adoption of this axiom candidate further the goals of this practice? I don't pretend that the naturalist's question is any easier than the realist's—after all, it involves such tricky notions as 'mathematical fruitfulness' and 'the goals of a mathematical practice'—but I do contend that the naturalist's question provides a more productive focus for our inquiry. While the realist is led off into philosophical worries about the nature of mathematical existence and our access to mathematical truth,[6] the naturalist concentrates instead on considerations squarely inside mathematics. Judging from historical cases, from Euler's generalization of the notion of function to the contemporary consensus on the axiom of choice, it is these practical considerations about the fruits and goals of the practice itself that are in fact decisive.[7]

Before leaving this meta-topic, let me make one last observation on the character of naturalistic arguments. Given that a naturalistic philosopher brings no special modes of argument from philosophy, every argument she gives must be based on modes of argument available to any mathematician qua mathematician; at best, she will make explicit what is already implicit. Unsatisfying as this may be in dramatic terms, a good naturalistic argument should not strike the practitioner as late- breaking news; at best, it will fall so far short of originality as to qualify as a commonplace. Given this goal, the best confirmation of success would be for the mathematician to shrug and say, 'of course, everybody knows that.' I think there is a non-trivial link between this fact and Wittgenstein's remark that 'Philosophy only states what everyone admits' ([1953], !599), but I won't drag this discussion off into the wilds of Wittgensteiniana.[8]

So, all that said, my goal here will be to sketch an argument against V=L that is naturalistic in this sense. To do this, I might try to exploit any one of various commonly heard complaints about V=L. In earlier work, I argued that V=L is worthy of suspicion as an axiom candidate, because it is closely allied with a methodological principle, Definabilism, which recommends that mathematical objects be regarded as definable or constructible in a uniform way, and Definabilism has a bad track record in the history of mathematics.[9] But grounds for suspicion are not grounds for rejection, and I am after the stronger

[6] As in my [1990].

[7] I think the 'anomalies' of [1993] are better understood from a naturalistic than from a realistic point of view. It may seem a stretch to regard the mathematical considerations that influenced opinion in those cases as evidence for the truth of the hypotheses in question, realistically understood, but it seems quite plausible to suppose that they provided good evidence for the mathematical fruitfulness of those hypotheses, good evidence for the efficacy of those hypotheses in pursuit of the goals of the relevant practices.

[8] I discuss examine this connection in [199?a].

[9] See my [1993]. There I suggested that an argument for the stronger conclusion, that V=L should be rejected, would require a different philosophical backdrop from the realism presupposed there. Naturalism is the proposed replacement.

conclusion here. The argument I will sketch rests on the common thought that V=L is bad because it is 'restrictive'; to make this go, I will have to explain why restrictiveness is bad, and why V=L is restrictive.

But first, a warning is in order: though I will do my best to present this argument as convincingly as I can, I will conclude with an examination of where it goes wrong. I do this for two reasons. First, the difficulties that arise are rather subtle and suggestive, and I hope that careful attention to their structure might lead to an improvement of the argument. And second, as a naturalistic philosopher, the best input I can get is the reactions of an audience like this one, to the general approach of the argument, to its details, to the possibilities for improvement. So I hope I will be forgiven for this rather unorthodox procedure.

2 Maximize and Unify

To begin, I want to examine what I take to be one among the many of the general goals of set theoretic practice. I will suggest that this goal, once isolated, motivates two methodological maxims, as effective means for the achievement of that goal. Both these maxims are ultimately relevant to the case against V=L. First, then, the goal.

Again, as befits a naturalistic argument, the point is quite simple and familiar. Though Cantor's set theory arose out of his work on the uniqueness of trigonometric representations, it wasn't long before he was writing that[10]

> ... pure mathematics ... according to my conception is nothing other than pure set theory.

Zermelo struck a similar theme in the paper containing his first axiomatization:

> Set theory is that branch of mathematics whose task is to investigate mathematically the fundamental notions 'number', 'order', and 'function', taking them in their pristine, simple form, and to develop thereby the logical foundations of all of arithmetic and analysis. ([1908b], p. 200)

Since then, the idea that set theory provides a 'foundation' for mathematics has become so much a part of set theoretic orthodoxy as to appear in the opening sentences of its textbooks; for example, Kunen writes:

> Set theory is the foundation of mathematics. All mathematical concepts are defined in terms of the primitive notions of set and membership ... from [the] axioms, all known mathematics can be derived. (Kunen [1980], p. xi)

[10] This quotation comes from an unpublished paper of Cantor's written in 1884 (see Grattan-Guinness [1970], p. 84). The translation is from Hallett [1984], p. 125.

Moschovakis, alert to potential metaphysical disputes,[11] puts the point more carefully to his beginning students:

> ... we will discover within the universe of sets *faithful representations* of all the mathematical objects we need, and we will study set theory on the basis of the lean axiomatic system of Zermelo **as if all mathematical objects were sets**. (Moschovakis [1994], p. 34)

The idea is that set theory provides a foundation in the following sense: every classical mathematical object can be represented by a set, and every classical mathematical theorem can be proved from the axioms of set theory.

By providing a framework broad enough to supply instances for all structures of classical mathematics, set theory brings the various subdisciplines into a common arena, so that interrelations are stressed; for example, one can hardly help being stuck by the way versions of the same assumption—the axiom of choice—turn up among the basic premises of so many distinct branches of mathematics. In some cases—e.g., when the points on a line are modelled by the set theoretic reals—set theory provides a more precise account of the structure than had previously been possible. In other cases—e.g., for the algebraic question 'are all Whitehead groups free?'—the set theoretic setting demonstrates that the question is independent of traditional assumptions, that no proof or disproof on their basis is possible. In still other cases—e.g. questions about the properties of simple sets of reals—set theory has shown how strong new hypotheses can resolve issues that baffled the analysts of the 20s. In light of the benefits, my suggestion is merely that it is one of the continuing goals of set theoretic practice to provide such a foundation.

Before drawing any consequences from this suggestion, let me pause a moment to clarify a few points often raised as objections to set theoretic foundations. The first, and oldest, harkens back to the original foundational programs of Frege and Hilbert, whose aim was to place mathematics on an unshakably secure basis. Seeing the matter in this light, Zermelo, presenting his axioms, writes with some regret:

> I have not yet even been able to prove rigorously that my axioms are consistent, though this is certainly very essential; instead I have had to confine myself to pointing out now and then that the antinomies discovered so far vanish one and all if the principles here proposed are adopted as a basis. ([1908b], p. 200-201)

Poincaré phrases the same point as a sharp criticism:

[11] E.g., in Benacerraf [1965]. Moschovakis's way of putting the point allows for the common idea that mathematics studies structures, not things; from this perspective, the job of set theory is to provide instantiations for all mathematical structures. Thus, set theoretical foundationalists can agree with their critic, MacLane, when he writes, 'a real number is not a Dedekind cut; that cut is just one possible model of a protean idea of the reals' ([1992], p. 121).

> We have put a fence around the herd to protect it from the wolves but we do not know whether some wolves were not already within the fence. (Kline [1972], p. 1186)

And the same objection appears in the work of MacLane, the most prominent contemporary critic of set theoretic foundations:

> Now in one sense a foundation is a security blanket: If you meticulously follow the rules laid down, no paradoxes or contradictions will arise. In reality there is now no guarantee of this sort of security; we have at hand no proof that the axioms ZFC for set theory will never yield a contradiction, while Gödel's second theorem tells us that such a consistency proof cannot be conducted within ZFC. ([1986], p. 406)

Since Gödel, as MacLane notes, Zermelo's hope of establishing the consistency of his axioms has been effectively dashed.

What needs emphasizing here is that the contemporary orthodoxy on set theoretic foundations does not claim to present a foundation in the sense of the early researchers, does not claim to base mathematics on unshakable premises. What's claimed is that set theory provides a unified framework in which the objects of classical mathematics can be modelled, the structures of classical mathematics instantiated, and the theorems of classical mathematics proved. Even critics like MacLane are willing to grant this much (see MacLane [1986], pp. 406, 358).

A closely related worry is expressed by Tiles. Logicians have long realized that the principles needed to found mathematics cannot all be plausibly claimed to rest on unsullied mathematical intuition. Russell, in his defense of the Axiom of Reducibility ([1910], pp. 59-60), Zermelo, in his defense of the Axiom of Choice ([1908a], pp. 186-190), Gödel, in his discussion of new set theoretic axioms ([1947], pp. 182-183)—all these thinkers have appealed to so-called extrinsic supports, that is, defenses based on the welcome consequences of an axiom candidate. Tiles claims that such defenses are inappropriate

> ... to the conception of set theory as providing a logical *foundation* for mathematics. To claim this status for set theory it is necessary to claim an independent and intrinsic justification for the assertion of set-theoretic axioms. It would be circular indeed to justify the logical foundations by appeal to their logical consequences, i.e. by appeal to the propositions for which they are going to provide the foundation. ([1989], p. 208)

But, once again, this critique is only sound if set theory is proposed as a foundation in the epistemic sense, as providing a 'secure given starting point' (Tiles [1986], p. 208). As long as set theory is only proposed—as it is here—as a unifying framework, as a shared, basic ontology, this criticism is also off the mark.

Finally, I suspect that there is another important, generally epistemological objection lingering the background. The most explicit statement of this objection

I can find comes from Mathias, a supporter of set theoretic foundations, as a conjecture on the underlying concerns of his opponents:

> Set theory is so rich a theory that it has been claimed for much of this century to be the foundation of mathematics. In ontological terms this claim is not unreasonable; but MacLane resists. I would guess that his reason is not so much that he objects to the ontology of set theory but that he finds the set- theoretic cast of mind oppressive and feels that other modes of thought are more appropriate to the mathematics he wishes to do. ([1992], p. 115)

But again, to claim that set theory provides a unified ontological setting is not to claim that only set theoretic methods should be employed in mathematics, that algebraists, analysts, number theorists, geometers should all become set theorists. To draw a comparison, to say that all entities studied by natural science are ultimately physical, subject to the laws of physics, is not to say that biologists, chemists, botanists, and geologists should all become physicists. I think Mathias puts the point quite well, so I hope I will be forgiven for quoting him at some length:

> One of the remarkable things about mathematics is that I can formulate a problem, be unable to solve it, pass it to you; you solve it; and then I can make use of your solution. There is a unity here: we benefit from each other's efforts. ... But if I pause to ask why you have succeeded where I have failed to solve a problem, I find myself faced with the baffling fact that you have thought of the problem in a very different way from me: and if I look around the whole spectrum of mathematical activity the huge variety of styles of thought becomes even more evident.
> Is it desirable to press mathematicians all to think in the same way? I say not ... Uniformity is not desirable, and an attempt to attain it, by (say) manipulating the funding agencies, will have unhealthy consequences.
> The purpose of foundational work is mathematics is to promote the unity [as opposed to the uniformity] of mathematics. ([1992], pp. 113-114)

Assuming then, that one of the goals of set theoretic practice is to provide a foundation in this sense, what methodological consequences follow? The first is immediate: if your goal is to provide a single system in which the objects and structures of mathematics can be modelled and instantiated, in which all theorems of mathematics can be proved, then you must aim for a single, fundamental theory of sets. This admonition to UNIFY is just the flip-side of another of MacLane's objections; he mentions the independence of CH, the proliferation of models achieved by forcing, the range of new axiom candidates, and concludes

> For these reasons 'set' turns out to have many meanings, so that the purported foundation of all Mathematics upon set theory totters. ([1986], pp. 358-359)

The methodological maxim UNIFY simply runs this argument in reverse: if you wish to provide a foundation, you must settle on a unique theory. If some such maxim were not in force, set theorists would not be motivated to decide the issue of V=L; it would be enough to consider alternative set theories with and without it.

The second methodological maxim that springs from the goal of set theoretic foundations is not much more distant. Contemporary pure mathematics is a vastly broad-ranging inquiry, dedicated to the notion that mathematicians should be free to study all and any structures and theories that seem to them of sufficient mathematical interest. If set theory is to found such a discipline, it should not impose any limitations of its own: the set theoretic arena in which mathematics is to be modelled should be as generous as possible; the set theoretic axioms from which mathematical theorems are to be proved should be as powerful and fruitful as possible. This desire to found mathematics without incumbering it generates the set theoretic maxim I call MAXIMIZE.

To be a bit more specific, I will focus on a particular aspect of the admonition to MAXIMIZE. Given that set theory is out to provide models for all mathematical objects and instantiations for all mathematical structures, one way in which it should MAXIMIZE is in the range of available isomorphism types. After all, perhaps the most fundamental contribution of set theory to date is its provision of a continuous structure to model the real numbers. The advice to MAXIMIZE isomorphism types is still regrettably vague; I will do my best to clarify it as I apply it.

3 Why V=L is restrictive

My aim now is to argue that V=L is restrictive, but 'restrictive' is a notoriously slippery notion. In the case of the continuum hypothesis, for example, arguments have been offered that purport to show that CH is restrictive, but arguments of similar structure purport to show that not-CH is restrictive![12] I will be trying to isolate a sense of the term that provides a bit more guidance than this. As the idea is to motivate the case against restrictive theories by appeal to MAXIMIZE, the central claim will be that restrictive theories somehow restrict isomorphism types. But to say this is not to make matters much clearer.

Let me begin, then, with a crude version of the argument that underlies the approach I'll be taking. The idea is simply this: there are things like $0^\sharp$ that are not in L. And not only is $0^\sharp$ not in L; its existence implies the existence of an isomorphism type that is not realized by anything in L.[13] These facts

[12] See my [1988], pp. 497-498, 500, for summaries.

[13] Suppose $0^\sharp$ exists, and consider the structure $(V_{\omega+1}, \epsilon)$. If $(y,S) \epsilon L$ is isomorphic to $(V_{\omega+1}, \epsilon)$, then S is well-founded and extensional on y. These notions are both absolute for L, so L thinks S is extensional and well-founded on y, and so (since L also thinks Mostowski's theorem on transitive collapse), L thinks there is a transitive A such that $(A, \epsilon) \cong (y, S)$. A given function being such an isomorphism is also absolute for L, so the two structures must actually be isomorphic. Thus, $(A, \epsilon) \cong (V_{\omega+1}, \epsilon)$, and they are both transitive, so $A = V_{\omega+1}$. But $0^\sharp \epsilon V_{\omega+1} = A \epsilon L$ and L is transitive, so $0^\sharp \epsilon L$. So L thinks $V \neq L$, which is impossible.

wouldn't carry any weight if ZFC+$\exists 0^\sharp$ were inconsistent—CONSISTENCY is an overriding maxim,[14] as inconsistent theories are useless—but accumulated evidence suggests that ZFC+$\exists 0^\sharp$ is not inconsistent.[15] So it seems that ZFC+V=L is restrictive because it rules out the extra isomorphism types available from ZFC+$\exists 0^\sharp$.

By way of contrast, consider AFA,[16] the anti- foundation axiom (see Aczel [1988]), which guarantees the existence of non-well-founded sets. It might seem that reasoning similar to what I've just rehearsed would classify ZFC as restrictive, because there are non-well-founded sets available from (ZFC-F)+AFA that are not in WF (the class of well-founded sets). But the similarity is an illusion: the new, non-well- founded sets do not realize any isomorphism types that are not realized in WF.[17]

To view the matter from another angle, consider the following style of argument in favor of V=L: V=L is to be preferred because it makes possible a deep and rich structure theory; we can say a lot about constructible sets, but very little about sets in general. Similarly, in algebra, the study of groups is richer and more productive that the study of arbitrary structures with one binary operation. The goal of mathematics, after all, is these strong structure theories, and V=L is better suited to this goal. This general line of thought is not unknown in the history of mathematics, for example, it seems to have influenced Borel's rejection of arbitrary functions and D'Alembert's earlier rejection of non-differentiable functions.

Notice, first, that this style of argument is entirely naturalistic, that is, it is based on an assessment of the goals of mathematics and the most effective means of reaching them. Let's suppose, as seems plausible to me, that a case could be made in favor of this preference for strong structure theories. Would this additional maxim then count against ZFC+$\exists 0^\sharp$? The answer, it seems to me, is no, because the new isomorphism types gained by ZFC+$\exists 0^\sharp$ are gained, as it were, for free. In moving from ZFC+V=L to ZFC+$\exists 0^\sharp$, we aren't losing anything, because we still have L itself; we can MAXIMIZE without sacrificing the strong structure theory of L. Similar responses could be given to Borel and D'Alembert.

Let me make one last remark about the virtues of this primitive argument against V=L before I attempt to tightening it up. Our two methodological

---

[14] There have been episodes in the history of mathematics when CONSISTENCY has been (temporarily) sacrificed, but given the motivating concern of axiomatic set theory with the issue of consistency, this is not a likely instance for such a sacrifice.

[15] I have in mind the relative consistency results of Jensen and Solovay [1970], the extended successful work on inner models, and the ordinary inductive evidence that no one has yet derived a contradiction from ZFC+$\exists 0^\sharp$.

[16] John Steel first recommended that I consider this case for contrast.

[17] If A is a set (possibly non-well-founded) and R is a relation on A, then in ZFC-F, it can be shown that A is equinumerous with some ordinal α; let f be a one-to-one correspondence between them. Let S be $\{< f(x), f(y) > \mid < x, y > R\}$. Then (A,R)$\cong$($\alpha$,S), which is in V. So (ZFC-F)+AFA adds no new isomorphism types. See McLarty [1993] for a related discussion.

maxims—UNIFY and MAXIMIZE—are in obvious tension. Given alternatives like ZFC+V=L and ZFC+$\exists 0^\sharp$, the easiest way to MAXIMIZE would be to allow both theories, to use whichever theory turns out to be more useful in a given situation. But UNIFY counsels against this course. The beauty of this case is that it seems possible to UNIFY—that is, to choose between ZFC+V=L and ZFC+$\exists 0^\sharp$—while still MAXIMIZING, because the choice of ZFC+$\exists 0^\sharp$ doesn't require the sacrifice of any of the content of ZFC+V=L.

So far so good. But I suspect many will have noticed a serious gap in this reasoning: we want a criterion of restrictiveness that applies to theories, like ZFC+V=L, but our discussions have centered on a model, namely L. The same switch appeared in the AFA case, when we moved from talk of ZFC to talk of WF. The pressure to make such a switch is understandable, because we are concerned with isomorphism types, and these are realized in models, not theories. To compare theories in these terms, we've actually used interpretations of one theory in another: to interpret ZFC+V=L in ZFC+$\exists 0^\sharp$, we replace a formula φ with φ^L, (that is, φ with quantifiers relativized by the requirement 'xϵL'); to interpret ZFC in (ZFC-F)+AFA, we replace φ with φ^{WF} (quantifiers relativized to 'xϵWF'). Without some further explanation, it seems this general style of argument could be turned on its head:[18] consider the theory ZFC+V=L+'there is a transitive model of "ZFC+$\exists 0^\sharp$"'; call this theory T. Obviously, T proves there is a transitive model of ZFC+$\exists 0^\sharp$; let ψ(x) say that x is in the $<_L$-least such model. Then the replacement of φ by φ^ψ is an interpretation of ZFC+$\exists 0^\sharp$ in T. But T also proves that any transitive model of ZFC+$\exists 0^\sharp$ is missing some countable ordinals,[19] and hence that there are isomorphism types not realized by anything satisfying ψ. What does MAXIMIZE now recommend?

Intuitively, what's wrong with this pro-V=L line of reasoning is that an interpretation of ZFC+$\exists 0^\sharp$ using ψ is not comparable to the interpretation of ZFC+V=L in L or the interpretation of ZFC in WF. The claim that we can UNIFY while MAXIMIZING in fact rests on the idea that interpreting ZFC+V=L in L somehow 'preserves' that theory, in a way that the proposed interpretation of ZFC+$\exists 0^\sharp$ does not seem to preserve it. To make any progress here, we need some notion of a fair interpretation of one theory in another.

To get at the idea of 'preserving' or 'fairly interpreting' a theory T, consider for a moment the point of view of the T- theorist. Obviously, his intention is to give a theory of V itself. In the cases that interest us, the interpreting theorist will not agree to this, but perhaps the two can settle on an interpretation in a substantial approximation to V. In the history of set theory, two such approximations stand out: the first is Zermelo's 'normal domains' (Zermelo [1930]), now known as 'natural models' or 'standard complete models', that is, V_κ's for κ inaccessible; the second is proper class inner models, beginning with Gödel's L. To speak crudely, the first compromises on 'tallness' and the second on 'thickness'.

[18] This sort of argument, suggested by Tony Martin, was discussed in a different connection in [199?b].

[19] 'x=$0^\sharp$' is Π^1_2, so by Shoenfield's Absoluteness Theorem, it is absolute for transitive models containing all countable ordinals.

As a first stab at an account of what it is to provide a fair interpretation of a theory extending ZFC, I propose to focus on interpretations in natural or proper class inner models.

Formally, then, suppose that we are working in the language of set theory, that α is a variable that ranges over ordinals, and that φ is a formula with one free variable.

Definition 1 *"T shows φ is an inner model" iff*
(i) for all σ in ZFC, $T \vdash \sigma^{\varphi}$, and
(ii) $T \vdash \forall\alpha\varphi(\alpha)$ or $T \vdash \exists\kappa(Innac(\kappa) \wedge \forall\alpha(\alpha < \kappa \rightarrow \varphi(\alpha)))$, and
(iii) $T \vdash \forall x\forall y((x\epsilon y \wedge \varphi(y)) \rightarrow \varphi(x))$.

This definition allows for natural models, proper class inner models, and truncations of proper class inner models at inaccessible levels (a simultaneous compromise on 'tallness' and 'thickness').

Definition 2 *φ is a fair interpretation of T in T' (where T extends ZFC) iff*
(i) T' shows φ is an inner model, and
(ii) for all $\sigma\epsilon T$, $T' \vdash \sigma^{\varphi}$.

I've required that T extend ZFC because the fair interpretations all start as models of ZFC.

On this definition, there is a fair interpretation of ZFC+V=L in ZFC+$\exists 0^{\sharp}$ (namely, xϵL), but there is no fair interpretation of ZFC+$\exists 0^{\sharp}$ in ZFC+V=L (because 'x=$0^{\sharp}$' is absolute for fair interpretations). There is also a fair interpretation of ZFC in (ZFC-F)+AFA (namely, 'xϵWF').

The next step is to add the idea of providing new isomorphism types.

Definition 3 *T' maximizes over T iff there is a φ such that*
(i) φ is a fair interpretation of T in T', and
(ii) $T' \vdash \exists x \exists R \subseteq x^2 \forall y \forall\ S \subseteq y^2((\varphi(y) \wedge \varphi(S)) \rightarrow (x, R) \not\cong (y, S))$.

If T' also extends ZFC, then clause (ii) can be replaced by the simpler

(ii)' $T' \vdash \exists x(\sim \varphi(X))$

because (in the presence of Foundation), an extra set is an extra isomorphism type. For good measure,

Definition 4 *T' properly maximizes over T iff T' maximizes over T and T does not maximize over T'.*

Using this definition, what we observed in the original, informal argument can be rephrased as: ZFC+$\exists 0^{\sharp}$ properly maximizes over ZFC+V=L, but (ZFC-F)+AFA doesn't even maximize over ZFC.[20]

[20] Our previous informal argument is not quite enough to establish this second claim, as the definition of fair interpretation now allows for (ZFC-F)+AFA to interpret ZFC in models other than WF. Given that Con(ZFC- F)$\rightarrow$Con((ZFC-F)+AFA) (see Moschovakis [1994], pp. 259- 262), (ZFC-F)+AFA cannot prove the existence of an

At this point, it's tempting to call a theory 'restrictive' when there is a (probably consistent) theory that properly maximizes over it. But this isn't quite right. For example, ZFC+'there is an inaccessible' properly maximizes over ZFC, but we wouldn't want to call ZFC restrictive on that account, because ZFC can be extended to ZFC+'there is an inaccessible'. It seems reasonable to say that ZFC doesn't go as far as it might, but not to say that it restricts. For T to be restrictive, the maximizing theory, T', should at least contradict T. So,

Definition 5 *definition: T' inconsistently maximizes over T iff T properly maximizes over T and T' is inconsistent with T.*

In fact, even this isn't quite enough. To see this, consider the supposition that there are arbitrarily large measurable cardinals:

$$T = ZFC + \forall\alpha\exists x(MC(x)\wedge x > \alpha)$$

This is surely seems a generous theory, one whose fair interpretations ought to realize lots of isomorphism types, but it can be maximized over by a theory that seems to restrict:[21]

$$T' = ZFC + \exists x(Inacc(x)\wedge x{=}\sup\{y|MC(y)\})$$

It would surely be counterintuitive to classify T as restrictive on the grounds of T', and I think the reason can be made clear in the present terms. T' provides something that T cannot prove to exist, namely, a set of measurable cardinals with an inaccessible, but not measurable supremum. In that sense, T' is richer than T. But nothing in T precludes what T' provides; that is, T can be extended to:

$$T'' = T + \exists x(\forall y(y\epsilon x\rightarrow MC(y)\wedge Inacc\ (\sup(x))\wedge \sim MC(\sup(x)))$$

Now T" maximizes over T', which suggests, once again, that though T doesn't go as far as it might, it also doesn't restrict. The term 'restrictive' ought to be reserved for theories that actually rule out a certain line of development. So,

Definition 6 *T' strongly maximizes over T iff*
(i) T' inconsistently maximizes over T, and
(ii) there is no consistent T" extending T that properly maximizes over T'.

Obviously, allowing inconsistent T" would void the definition. In most interesting cases, we won't be able to prove the consistency of a candidate for T", so the conclusion to be drawn will be that whatever evidence we have for the consistency of a candidate T" is also evidence that T' doesn't strongly maximize over T.

Finally, the proposal is:

---

inaccessible. Now suppose (ZFC-F)+AFA interprets ZFC in L rather than WF. The argument of footnote 16 then shows that (A,R) is isomorphic to a set in WF. If (ZFC-F)+AFA could prove that this set is not in L, it could prove the existence of a non-constructible, well-founded set. To see that (ZFC-F)+AFA cannot do this, modify Moschovakis's construction by beginning from L rather than WF; the result will be a model of (ZFC-F)+AFA+WF=L.

[21] This example is also due to Tony Martin.

Definition 7 *T is restrictive iff there is a consistent T' that strongly maximizes over T.*

Once again, we will often be unable to prove the consistency of a candidate T', but evidence for its consistency will be evidence for the restrictiveness of T. In our case, we know that no extension of ZFC+V=L will allow a fair interpretation of ZFC+$\exists 0^\sharp$,[22] so we can be as confident that it establishes the restrictiveness of ZFC+V=L as we are of the consistency of ZFC+$\exists 0^\sharp$. But we can do better than this by observing that we have the best possible evidence that ZFC+V$\neq$L is consistent (it is consistent if ZFC is), and thus, even better grounds on which to conclude that ZFC+V=L is restrictive.

In these terms, we can state clear grounds for claiming that ZFC+V=L is restrictive. Furthermore, MAXIMIZE will, at the very least, tell us that restrictive theories should be rejected, so we can also state clear grounds for the consensus against V=L. We are left with a wide range of theories that strongly maximize over ZFC+V=L, including ZFC+V$\neq$L, ZFC+$\exists 0^\sharp$, ZFC+MC, and for that matter, such duds as ZFC+V$\neq$L+$\sim$Con(ZFC).[23] The next step would be to examine the rational grounds for a choice between these.

4 What goes wrong

So, that's the argument that I would like to claim provides a rational grounding for the view that V=L should be rejected. The fact that it goes wrong in more complicated cases may well not surprise those long accustomed to the endless richness and variety of set theory. I suspect this very richness and variety partly accounts for the enduring fascination of set theory over the past 100 years.

In any case, what's gradually emerged is that the proposed criterion for restrictiveness produces both false negatives (theories that seem restrictive, but are not so classified) and false positives (theories that don't seem restrictive, but are so classified). What I'll be reporting is actually just the current state of play in ongoing discussions I've had with several much more knowledgeable persons, especially Tony Martin, John Steel and Sarah Resnikoff.

Let me start with the false negatives. After V=L, it is natural to ask what the criterion has to say about various theories of the form ZFC+'V=the canonical inner model with such-and-such large cardinals', the expectation being that they will be strongly maximized over by theories of the form ZFC+'the next larger large cardinal exists'. In fact, this pattern is satisfied for a considerable distance. For example, ZFC+'there are two measurable cardinals' (ZFC+2MC) provides a fair interpretation of ZFC+'V=the canonical inner model with one measurable cardinal' (ZFC+V=L[U]). Furthermore, ZFC+V=L[U] implies that there is no (uncountable) inner model of ZFC+2MC, so neither it nor any extension of

[22] In any extension of ZFC+V=L, the only candidates for fair interpretations of ZFC+$\exists 0^\sharp$ are L and models of the form L_κ for κ inaccessible. Neither of these could think $\exists 0^\sharp$, or even V$\neq$L.

[23] Thanks to John Steel for bringing this one up.

it can fairly interpret ZFC+2MC. Thus, ZFC+2MC strongly maximizes over ZFC+V=L[U], and the later is restrictive, as seems right.

But, as Steel has noted, trouble arises at the level of two Woodin cardinals, at ZFC+V=L[E], where L[E] is the canonical inner model with two Woodins. Surprisingly, this theory can be extended to provide definable proper class inner models for such theories as ZFC+'there are five Woodins', ZFC+'there is a supercompact', and so on.[24] Thus, on our definitions, none of these theories strongly maximizes over ZFC+V=L[E]. This seems wrong, perhaps for reasons along the following lines: the inner models that the extensions of ZFC+V=L[E] provide for the larger large cardinals are not 'canonical' (e.g. they are not 'iterable'), so to interpret a theory like ZFC+'there are five Woodin cardinals' in this way is not to 'preserve' it or to interpret it 'without loss'. This example suggests that defining 'fair interpretation' as interpretation in a mere inner model is too weak, a weakness that was masked in the case of V=L because the so-called 'fair' interpretation of V=L in L is, in fact, optimal.

So, it seems that any more worthy successor to the argument proposed here would need a better notion of 'fair interpretation'. I have nothing to offer at this moment, but Martin points out that the prospect of strengthening this definition raises another interesting possibility. Recall that T' maximizes over T if T' provides a fair interpretation of T and an isomorphism type not represented in that fair interpretation. The isomorphism type provision is redundant whenever T' includes Foundation and T' is inconsistent with T, that is to say, in most of the cases that interest us. The only case in which it's done any real work is that of (ZFC-F)+AFA. But if 'fair interpretation' were redefined so as to allow for a fair interpretation of (ZFC-F)+AFA in ZFC – which doesn't seem so far-fetched, given that our best understanding of models of AFA is already based on interpretations in terms of ordinary set theoretic graphs – then the extra clause would be doing no work at all and could be eliminated. The connection back to MAXIMIZE would then be simplified: instead of singling out a particular thing we want to maximize—isomorphism types—we could rest with the simple thought that if T' fairly interprets T, in the new robust sense, then whatever T provides, T' will preserve. So a better notion of fair interpretation might simplify the over- all argument considerably.

Let's turn now to the false positives. Again, the example is due to Steel. Start, this time, with ZFC+'there is a measurable cardinal' (ZFC+MC) and consider the following theory, T:

$$\text{ZFC}+\exists 0^\dagger + \forall\alpha < \omega_1\ L_\alpha[0^\dagger] \not\models ZFC\ (\text{ZFC}+\sigma).$$

$0^\dagger$ codes the canonical inner model of ZFC+MC, so T provides a truly fair interpretation of ZFC+MC. On the other hand, MC implies that if $0^\dagger$ exists,

[24] The trick, as it's been explained to me, is that the canonical model with two Woodins is Σ^1_4-correct, and inner models of the larger large cardinals can be coded by Δ^1_4 reals. The relevant extension of ZFC+V=L[E] simply adds the claim that the larger large cardinal axiom is true relativized to the formula defining the appropriate inner model.

then there is a countable α such that $L_\alpha[0^\dagger] \models ZFC$,[25] that is, ZFC+MC$\vdash\sim\sigma$. Now suppose ZFC+MC has an extension, T', with a fair interpretation of T. Then T' proves there is an inner model M$\models$T, so in particular, M$\models\sigma$. σ is Σ^1_3, and thus absolute upwards for models like M, so T'$\vdash\sigma$. But ZFC+MC$\subseteq$T', so T'$\vdash\sim\sigma$, too. Thus, T' is inconsistent. It follows that T strongly maximizes over ZFC+MC, and thus, that ZFC+MC is restrictive. I suppose most people would consider this a misclassification.

The diagnosis of what has gone wrong in this case involves a careful look at the theory T. T is a way of saying that $0^\dagger$ exists, but is not contained in any transitive set model of ZFC.[26] This isn't as strong as saying that ZFC+$\exists 0^\dagger$ is inconsistent, but it is in the same ballpark. Granted, ZFC+MC can't be extended to provide a fair interpretation of this theory, but should it? This line of thought suggests that some theories, perhaps including T, are so poorly constructed that they don't deserve to be fairly interpreted; if so, another theory's failure to do so would not count against it.

Looking back, we realize that among the theories listed as strongly maximizing over ZFC+V=L was ZFC+V$\neq$L+$\sim$Con(ZFC). This didn't cause any discomfort in the case of V=L because there were also other, more attractive theories strongly maximizing over it; at the time, I simply referred to ZFC+V$\neq$L+$\sim$Con(ZFC)as a 'dud' that appears among the range of these theories, presumably one that would not be in the running (for good reasons) when we set out to chose between them. But what if the only theories strongly maximizing over a given theory are duds? Would it still seem reasonable to classify it as restrictive?

Of course, the trick is to specify what counts as a dud and why. I don't have anything worked out to say about this, though I will come back to the problem again in a moment. However, I think it is worth remarking that we need not require that such an account be fully formalizable in set theory. Though the criterion for restrictiveness we have been considering so far is a formal one, the arguments for MAXIMIZE and UNIFY are not, and it seems to me quite unlikely that any complete theory of the various forces that influence theory choice in set theory will be fully formalizable. The point is that an argument can be sound, rational, and legitimately compelling without being formal.

In any case, the false positives I know of do seem to rest on theories that deserve to be called 'duds', for reasons that must for now remain unspecified. But before leaving the topic of unattractive theories entirely, I'd like to call attention to a more subtle worry that concerns me even in the central case in which the criterion does work, that is, in the case of V=L. Recall that another of the theories that strongly maximizes over ZFC+V=L is the simple ZFC+V$\neq$L. This

---

[25] If κ is a measurable cardinal and $0^\dagger$ exists, then V_κ is a model of ZFC to which $0^\dagger$ belongs. This model has a submodel of the form $L_\kappa[0^\dagger]$ which is also a model of ZFC. $L_\kappa[0^\dagger]$, in turn, has a countable, transitive, elementary submodel containing $0^\dagger$. This last must be of the form $L_\alpha[0^\dagger]$ for some countable α.

[26] By essentially the argument of the previous note, if $0^\dagger$ is contained in some transitive set model of ZFC, it is contained in one of the form $L_\alpha[0^\dagger]$ for some countable α.

is surely not a dud in the sense of theories like ZFC+∼Con(ZFC), but it is an unattractive theory, nevertheless. It is too weak to settle any of the outstanding open questions; though it does, strictly speaking, provide a new isomorphism type, it tells us nothing about that type, which makes it pretty much unusable. It is as if a physicist were to conclude that the known elementary particles—A, B, and C—do not exhaust the range of such particles, and then propose as a new physical theory the claim 'There are particles that are not of types A, B or C'. Not a good physical theory. Nor is ZFC+V≠L a good set theory.

This observation leads to an interesting quasi-historical question: when Scott ([1961]) proved that the existence of a measurable cardinal implies the falsity of V=L, did this increase confidence in measurable cardinals or decrease confidence in V=L? If V=L was justifiably regarded as restrictive at that time, simply due to the unattractive ZFC+V≠L, we would expect Scott's proof to increase the confidence in MC. If, on the other hand, the weakling ZFC+V≠L is not enough to establish the restrictiveness of V=L, then we would expect Scott's proof to decrease confidence in V=L, by showing that there is something of interest that is ruled out by it. On this second line of thought, for a theory to count as truly restrictive, the theory maximizing over it must have some attractions of its own: it must imply the existence of an isomorphism type about which something interesting is known, or more generally, it must have other virtues, perhaps linked to other goals of set theoretic practice.

In fact, it seems more plausible to suppose that the evidence of Scott's theorem would do both things at once. On the one hand, V=L was already under legitimate suspicion for its association with Definabilism, so implying its negation would be a plus for MC. On the other hand, MC was already known to enjoy other virtues; e.g., it implies the existence of small large cardinals, which can be regarded as efforts to extend Cantor's successful methods. On these grounds, its clash with MC would count against V=L. So the success of MC and the failure of V=L appear intertwined, and more important, considerations other than mere restrictiveness seem to be involved.

But whatever the best analysis of this episode, I think it might well be reasonable to require that the maximizing theory have independent virtues before classifying the maximized theory as restrictive. Again, I would expect these virtues to be linked to other identifiable goals of set theory, e.g., that of providing a rich theory of sets of real numbers, or of extending Cantor's original methods. If a modification of our account of 'maximizing' along these lines proved feasible and desirable, it might well solve the problem of dud theories along the way; presumably, they would not be particularly virtuous. But this is mere speculation.

In sum, then, if we hope for an account of restrictiveness along the proposed lines, we face difficulties with spelling out both aspects of the primitive underlying argument. On the one hand, we need a more robust notion of 'fair interpretation'; we have not fully captured the idea of 'preserving' the maximized theory 'without loss'. On the other hand, we also need a more discerning approach to the idea of a theory 'giving more'; to do this, a theory must be more worthy than those we've called 'duds' and perhaps it must have independent virtues of its own. I have no clear sense of the prospects for these improvements, though I seriously doubt they will be easy; but I retain the conviction that the consensus against V=L is rationally justified, and that the grounds can be clearly articulated, if perhaps not fully formalized. In any case, I hope it goes without saying that I would welcome your reflections on this or any other line of thought on these matters.[27]

References

P. Aczel [1988], *Non-Well-Founded Sets*, (Menlo Park, CA: CSLI).

P. Benacerraf [1965], 'What numbers could not be', reprinted in P. Benacerraf and H. Putnam, eds., *Philosophy of Mathematics*, (Cambridge: Cambridge University Press, 1983), pp. 272-294.

K. Gödel [1947]. 'What is Cantor's continuum problem?', reprinted in his *Collected Works*, vol. II, S. Feferman et al, eds, (Oxford: Oxford University Press, 1990), pp. 176-187.

I. Grattan-Guinness [1970], 'An unpublished paper by Georg Cantor', *Acta Mathematica* 124, pp. 65-107.

M. Hallett [1984], *Cantorian Set Theory and Limitation of Size*, (Oxford: Oxford University Press).

R. Jensen and R. Solovay [1970], 'Some applications of almost disjoint sets', in Y. Bar-Hillel, ed., *Mathematical Logic and Foundations of Set Theory*, (Amsterdam: North Holland Publishers), pp. 84-104.

A. Kanamori [1994], *The Higher Infinite*, (Berlin: Springer Verlag).

M. Kline [1972], *Mathematical Thought from Ancient to Modern Times*, (New York: Oxford University Press).

K. Kunen [1980], *Set Theory: An Introduction to Independence Proofs*, (Amsterdam: North-Holland Publishers). S. MacLane [1986], *Mathematics: Form and Function*, (New York: Springer Verlag).

—— [1992], 'Is Mathias an ontologist?', in H. Judah, W. Just, and H. Woodin, eds., *Set Theory of the Continuum*, (Berlin: Springer Verlag), pp. 119-122.

P. Maddy [1988], 'Believing the axioms', *Journal of Symbolic Logic* 53, pp. 481-511, 736-764.

—— [1990], *Realism in Mathematics*, (Oxford: Oxford University Press).

—— [1993], 'Does V equal L?', *Journal of Symbolic Logic* 58, pp. 15-41.

—— [1995], 'Naturalism and ontology', *Philosophia Mathematica* 3, pp. 248-270.

—— [199?a], 'Naturalizing mathematical methodology', to appear in M. Schirn, ed., *Philosophy of Mathematics Today*.

[27] This paper was written with the support of NSF Grant ♯SBR-9320220, which is gratefully acknowledged. My thanks also to Jeffrey Barrett, John Burgess, Matthew Foreman, Menachem Magidor, Yiannis Moschovakis, Peter Woodruff, an anonymous referee, and especially, to Tony Martin and John Steel and Sarah Resnikoff for many helpful exchanges on these topics. As usual, none of these people should be held responsible for the views expressed here.

—— [199?b], 'Set theoretic naturalism', to appear in the *Journal of Symbolic Logic*.

—— [199?c], 'How to be a naturalist about mathematics', to appear in G. Dales and G. Oliveri, eds., *Truth in Mathematics*.

A. R. D Mathias [1992], 'What is MacLane missing?', in H. Judah, W. Just, and H. Woodin, eds., *Set Theory of the Continuum*, (Berlin: Springer Verlag), pp. 113-118.

C. McLarty [1993], 'Anti-foundation and self-reference', *Journal of Philosophical Logic* 22, pp. 19-28.

G. H. Moore [1982], *Zermelo's Axiom of Choice*, (New York: Springer Verlag).

Y. Moschovakis [1994], *Notes on Set Theory*, (New York: Springer Verlag).

B. Russell and A. Whitehead [1910], *Principia Mathematica*. (Page reference is to the paperback of the second edition, (Cambridge: Cambridge University Press, 1967).) D. Scott [1961], 'Measurable cardinals and constructible sets', *Bulletin de l'Académie Polonaise des Sciences* 7, pp. 521-524.

M. Tiles [1989], *The Philosophy of Set Theory*, (Oxford: Basil Blackwell).

L. Wittgenstein [1953], *Philosophical Investigations*, (New York: MacMillan).

E. Zermelo [1908a], 'A new proof of the possibility of a well- ordering', reprinted in J. van Heijenoort, ed., *From Frege to Gödel*, (Cambridge, MA: Harvard University Press, 1967), pp. 183-198.

—— [1908b], 'Investigations in the foundations of set theory I', reprinted in J. van Heijenoort, ed., *From Frege to Gödel*, (Cambridge, MA: Harvard University Press, 1967), pp. 200-215.

—— [1930], 'Über Grenzzahlen und Mengenbereiche: neue Untersuchungen über die Grundlagen der Mengenlehre', *Fundamenta Mathematica* 16, pp. 29-47.

Towards a Categorical Foundation of Mathematics

M. Makkai*

Department of Mathematics
McGill University
Montreal QC H3A 2K6, Canada
makkai@triples.math.mcgill.ca

1. Introduction

This paper is an expanded version of a talk with the same title given at the Logic Colloquium in Haifa in August, 1995. A year later, at the time of writing these lines, one of the papers, [M2], on which the talk was based has appeared in print. The present paper is, primarily, a somewhat philosophical introduction to the research monograph [M3], the other source of the talk. For understanding the story, it is necessary to keep the contents of [M2] in view; some of [M2] will be recapitulated here. The paper is written for readers with at least a slight familiarity with Category Theory. At certain points, we make remarks referring to more advanced notions; however, the understanding of these is not essential for the main part of the paper.

Some more remarks on the background.

In the abstract for the Haifa talk, another, related development, the material of the papers [M1], was mentioned. In the talk itself, there was no time to enter the subject of sketches. Because of similar considerations of space, sketches will not be discussed in this paper either.

Higher dimensional categories played an important, albeit somewhat hypothetical, role in the talk. They were intended to form the universe of the proposed foundation of the title; on the other hand, their very definition had not yet been given in arbitrary dimensions. In the meantime, a breakthrough has taken place; John Baez and James Dolan proposed (see [BD2]), in a somewhat sketchy form, a definition for the concept of "weak n -category", for all natural numbers. I believe that the Baez/Dolan definition is indeed satisfactory. In particular, I think that their "weak n-categories" should, collectively for all n and endowed with a suitable attendant structure connecting them to each other, be adopted as forming the universe of the new foundation I am envisaging. Besides, the Baez/Dolan definition also contains very interesting conceptual innovations that make it striking and illuminating even when specialized to the otherwise well-known case of $n = 2$! On the other hand, the Baez/Dolan proposal is not the end of the work of even giving the definition of the basic concepts surrounding

* The author's research is supported by NSERC Canada and FCAR Quebec

higher dimensional categories, as the authors themselves emphasize in *loc.cit.* It turns out that the ideas surveyed in the present paper are useful in dealing with the remaining problems concerning higher dimensional categories, and in giving new insights into their structure; I refer to the work [M4] in progress for these matters.

The ideas presented here are related to the so-called structuralist position in the Foundations of Mathematics. P. Benaceraff's paper [Ben1] expresses views that have substantially influenced mine. However, as far as I am aware, the philosophical literature does not contain studies relating Category Theory to the structuralist position, despite some rather obvious points of contact.

From the philosophical point of view, it would be interesting the expand the inquiry, and probe the metaphysics of everyday objects, a multitude of which embody elements of abstraction one is used to in Mathematics. I will refrain from doing so here, but I refer to the paper [M5] (in preparation) for such a discussion.

A final general remark. In this paper, I am interested in the metatheory of an, admittedly incomplete, foundational system. For the purposes of this metatheory, I use all available tools, including ordinary Set Theory.

2. The universe and the language

I am going to call the intended new foundation the Structuralist Foundation of Abstract Mathematics, SFAM for short. Before anything else, let me emphasize that SFAM is not completed; later, I'll point to the main missing elements.

I consider SFAM as the result of thinking through, and making explicit, a foundational position behind Abstract Mathematics. I regard Category Theory to be an expression of this position. By Abstract Mathematics, I mean the mathematics of (abstract) structures, such as groups, rings, topological spaces, and in a more advanced vein, e.g., algebraic varieties in Grothendieck's sense.

Category Theory is often seen as a (mere) language for Abstract Mathematics. If we disregard the parenthetical and derogatory "mere", I find that there is much to be said for this view. Indeed, Category Theory and Mathematical Logic are concerned with the language of Mathematics in fundamentally analogous ways.

For one thing, they both have a discernible syntax. In Mathematical Logic, the imperative of providing a totally explicit syntax originates with G. Frege, who established the precise meaning of, and also the possibility of meeting, this imperative. He, and others, developed the languages of various formal axiomatic systems for mathematics. I consider First Order Logic, an all-purpose general language, to be the principal result of these efforts. On the other hand, the overall achievement of Mathematical Logic in clarifying what the qualities of the syntax of a satisfactory foundation should be, and in providing a metatheory of syntax, has, I believe, acquired a universal validity, and it cannot be ignored in any foundational inquiry. SFAM is no exception. The conclusion is that the syntax of SFAM should be approachable in an objective way, without having to adopt ontological commitments implicit in SFAM.

Historically, Category Theory has not been concerned with making its syntax explicit to a similar extent. At the basis of Category Theory, we find ***diagram manipulation*** as an informal but nevertheless distinct "language". P. Freyd's [F], written out more explicitly in [FS], is an early formalization of a combination of diagram manipulation with ordinary First Order Logic. My work in [M1] aims at pure diagram manipulation, without explicit utilization of connectives or quantifiers. E.g., "doing mathematics inside a topos" is, in my view, the same as applying the formalized diagram manipulation of loc.cit. (via (generalized) 'sketches' as basic syntactical units) to the particular setting of (elementary) toposes (*topos theory* is, of course, something much more comprehensive; it is the mathematics of toposes). The last fact should be seen as evidence of the power of pure diagram manipulation as a language; note that "doing mathematics inside a topos (with a Natural Numbers Object)" is roughly the same as doing formalized higher order number theory (analysis) (compare [LS]).

First Order Logic with Dependent Sorts (FOLDS), one of the two main themes of this paper, is the result of a search for an "all-purpose" language an applied version of which can be used for SFAM. FOLDS can be seen as a generalization and development of Freyd's [F]; see also [Bl].

Both Mathematical Logic and Category Theory intend to be global approaches to the foundations of mathematics. The Zermelo-Fraenkel formal system of set-theory (ZFC), along with its variants, is nowadays almost universally accepted as a global foundation of mathematics; every legitimate piece of mathematics is supposed to be, and as a rule, is, formalizable in ZFC. Category Theory, on the other hand, does not have a closed codification such as ZFC. The globalist character of Category Theory is revealed in its tendency to provide more and more comprehensive notions of "category". The reform through a revision of some concepts, and the extension to achieve a degree of completeness, of the existing framework constitute the other main theme of this paper.

The standard ZFC foundation possesses two prominent features: on the one hand, the intuitive appeal and clarity of its universe of discourse, the universe of (von-Neumann) pure (or regular) sets, and on the other hand, the transparency and simplicity of its language, (applied) first order logic. While Category Theory is governed by compelling concepts and intuitions, sometimes sharply at variance with the standard foundation, it does not provide clear-cut answers to the questions "what is the universe and what is the language of abstract mathematics?" The present paper tries to give partial answers to those questions in the spirit of Category Theory.

3. Abstract sets

The divergence between the standard foundation and SFAM begins with their views on sets. In SFAM's view, the world of mathematics is big and heterogeneous; a set is a relatively orderly part of the world in which an *equality predicate* reigns. The elements of a set are individuated *with respect to each other*. However, there is no global equality present for all things simultaneously. An equality predicate is an equivalence relation on the given set. In fact, the set is the underlying collection *together with* its equality predicate.

This fact is part of the explanation of why the adjective "structuralist" appears in the name of SFAM; already sets are a certain kind of structure, and other concepts of totality will be increasingly complex structures. I take it to be a tenet of structuralism that everything accessible to rational inquiry is a structure; the conceptual world consists of structures.

Another fact about sets concerns the membership predicate. In fact, in SFAM, it is not a full-fledged predicate. It is not meaningful, that is, not grammatical, to write $x \in a$ for a set a, and an arbitrary variable x. In other words, it is not meaningful to ask, for a set a and an arbitrary thing x, whether or not x is an element of a. Rather, a set a is a *type*, and a variable x may be *declared* to be of type a. "$x \in a$" will not be a predicate subject to free manipulation with the connectives and quantifiers as full-fledged predicates will be; for instance, we will never write $\neg(x \in a)$. The statement "$x \in a$" will have the role of defining *contexts* of variables, in the style of P. Martin-Löf [M-L] and Cartmell [C]. Thus, SFAM is a *type theory*; the difference to ordinary (unramified, simple, Russell-Ramsay) type-theory is that types are *variable*; more precisely, they (may) *depend on* variables.

Thus in SFAM, both equality and membership are denied the free reign they enjoyed in the standard foundation.

In SFAM, we will adhere to a very strict idea of "type". An entity belonging to a type cannot be discussed without reference to the type; the type logically precedes the entity, and the type is inseparable from the entity.

As we said, a set is a type; actually, for a set a, we have the type $\mathrm{E}(a)$ of its elements. The type $\mathrm{E}(a)$ a *dependent type*, has the invariable element E , and the variable element a in it. When a variable x is declared to be of type $\mathrm{E}(a)$, in symbols x: $\mathrm{E}(a)$, which is synonymous with $x \in a$, the relationship established between x and a is completely rigid, or invariable. For instance, it will not be possible for a variable to be of two types that are different syntactically. (This is in contrast to [M-L] and [C]; see also below). What we just said is the syntactical counterpart of the metaphysical intuition that an element of a set may enter into relationships with other things only as an element of the set, only through the set itself.

The above example E(a) of a dependent type is only one of an infinite number of such types in SFAM.

Dependent types are well-known from P. Martin-Löf's work [M-L]. See also J. Cartmell [C] that pertains, similarly to SFAM, to an essentially first-order framework, albeit without quantifiers. However, both the philosophy and the actual syntax of dependent types I am proposing to use for SFAM differ from those in the cited sources. I have not been able to find in the literature views on types that accord in all the essentials with the views I am advocating here.

Note that the character of a set as "type" in the way described above is, at first sight, at variance with the idea behind the comprehension principle of Set Theory. In that principle, we have, in the first place, certain entities, of a certain arbitrary type, say, and in the second place, we form a set whose elements are the given entities. This would mean, in our context, attributing a new type, namely E(a) for the set a thus "comprehended", to entities of an already established type. This, directly at least, is not allowed in SFAM. Instead, we will have a somewhat elaborate "relational structure" connecting the entities of type E(a) with the original entities. This structure will also reflect our specific way of dealing with equality.

As a result, comprehension, although possible, is much less "cheap" in SFAM than in Set Theory. This is not such a bad thing, considering the treacherous nature of (unrestricted) comprehension, trivial-looking instances of which lead to contradictions (Russell's paradox). While axiomatic Set Theory is based on one selection of instances of (unrestricted) comprehension, ones that are "Cantorian", or fit the "limitation of size" doctrine (see [H]), SFAM will be based on another selection of instances, including ones that (or rather, whose direct counterparts) are not admissible in Set Theory. SFAM is envisaged to be a "non-Cantorian" theory of totalities. The actual laying down of a reasonably complete set of axioms of SFAM awaits future work.

In Abstract Mathematics, we find the intuitive idea of *abstract set*, one whose elements are characterless, nevertheless distinct, points. An "abstract" structure is one whose underlying set is an abstract set. An "abstract" group is given by an abstract set A, and by a binary operation on A, assumed to satisfy certain conditions, the group axioms. The elements of the group, which are mere points to begin with, enter into meaningful relations via the equality predicate of the underlying set and the group-operation, which is a function of the form $A \times A \to A$.

The sets of SFAM are abstract sets in the above intuitive sense. But then, one might object that in Mathematics, even in Abstract Mathematics, we also need "non-abstract" structures, ones whose underlying sets are not abstract sets. For instance, we want to consider the group of automorphisms of a given structure; the elements of this group are certainly not characterless points.

The answer to this is that in the group of automorphisms, the elements are characterless surrogates of the automorphisms, rather than the automorphisms themselves. There is an additional structure ensuring the presence of a relation which establishes a bijection between the automorphisms and the characterless points. It is as if we had abstracted away from the concrete nature of the automorphisms to get the elements of the group.

The reader who is familiar with mathematics done in a topos will recognize that, in fact, in a topos all objects behave like abstract sets. For instance, the categorical product of two objects, which corresponds to the idea of the Cartesian product of two sets, is just another object, equipped with the two projection maps; the complex of the three objects and two arrows involved have a so-called universal property characteristic of products (see [ML]). In the "topos of abstract sets", the product-set, instead of having the ordered pairs (a, b) with a and b elements of the two sets we began with, has elements that are characterless points; however, the projection maps together with the universal property characteristic of products ensure that those elements are in fact in a "structural", built-in bijection with the said ordered pairs. The way the topos structure deals with specifying the powerset of an abstract set is similar; the powerset is just another abstract set, but there are "structure maps" around ensuring that its elements are in a bijection with the subsets of the given set. Note that the universal properties alluded to are statements formulated by the use of quantifiers in the first-order language of the category; they express a relation of the arrows involved in the product or in the powerset to the totality of all arrows in the category. Thus the "internal" structure of the Cartesian product given by the fact that its elements are suitable ordered pairs is replaced by an "external" relation of the product set and the projections to other sets and functions; a similar thing can be said concerning the powerset.

In a topos, it is possible to construct the group of automorphisms of a structure (e.g., field) in the topos. This will be an abstract group; however, the construction will provide an accompanying structure of various maps giving a faithful representation of the fact that the elements are in a bijective relation with the automorphisms of the structure, and that under this relation, the group-operation corresponds to composition of automorphisms.

On the other hand, the regular (well-founded) sets of ZFC set-theory are *concrete sets*. A regular set has elements that are themselves are regular sets. A regular set is individuated *by* its elements, rather than the other way around as with abstract sets. Each element x of a regular set A is individuated 'completely' by the membership relation, and by the equality relation deduced from it, *without reference to* A. The well-founded nature of the membership relation ensures that the determination of x is indeed independent of A. Each regular set x can be uniquely picked out from the totality of all regular sets by structural properties alone, indeed by referring only to elements of x, to elements of the elements of x, and so on, through the transitive closure of x. In particular, the structured totality of regular sets has no non-trivial automorphism. In a way, regular sets are at an opposite end from being "abstract"; they are "concrete".

Here is an illustration how Abstract Mathematics gets into conflict with "concrete sets". Consider a hypothetical book on group theory, for instance, and suppose we had something like the following in it. "Let G and H be arbitrary groups, and consider the intersection X of their underlying sets", after which a construction involving the data mentioned, in particular the set X, is described. I am sure everyone used to group-theory should be taken aback by such a passage. One would say that what the common elements of the groups are is purely incidental to the essence of the groups, and thus the datum of the intersection X can have no significant role in a mathematically meaningful construction based on the given groups G and H. Now, the intersection of two sets is a typical operation on sets that is meaningful only under the free reign of an (untyped) membership relation, but not for abstract sets. In a more advanced manner, we would say that the proposed construction is very likely not going to be invariant under isomorphism of groups; we take up this point later.

When one is told, as in Mathematics explicitly based on Set Theory one is, that "the underlying sets of structures are regular sets", one may acquiesce in that view on the basis that it is necessary (one is told) for having a consistent foundation. But one will feel that the original intuitions of Abstract Mathematics have been altered.

4. Functions, categories and isomorphisms.

The usual kind of equality of sets is not meaningful in SFAM. That equality, defined through extensionality, relies on the untyped (free-reigning) membership relation, or alternatively, on the untyped equality relation of all things, neither of which is available in SFAM. (We may put the "definition" of equality of sets in two ways:

$$X = Y \Leftrightarrow \forall x(x \in X \leftrightarrow x \in Y) \ ,$$

$$X = Y \Leftrightarrow \forall x \in X \cdot \exists y \in Y \cdot y = x \ \ \& \ \ \forall y \in Y \cdot \exists x \in X \cdot x = y \ .$$

(In the first form, the untyped $\in$, in the second, the untyped $=$ appear.)

The upshot is that we do not have available a *set* of sets; sets do not form a set (not, in this instance, for size considerations, but rather, because of the lack of an available equality of sets). Rather, sets will form a new type of totality: a *category*.

What connects sets is *functions*. Functions are a new kind of entity; for any sets X and Y, we have the idea of a function from X to Y. We write $f : X \to Y$ for "f is a function from X to Y", i.e., "f is a function with domain equal to X, and range included in Y". "Function" is a primitive concept in SFAM, unlike "function" in the standard foundation. A further primitive is *composition* of (composable) functions, making sets and functions into a *category* in the usual sense. Below, in section 8, we will see the exact type-theoretic treatment of the concept of category that is adopted in SFAM.

(I do not see any a *priori* justification why we must have, e.g., functions as primitives. For instance, we could take instead relations, a generalization of function. However, when doing so we end up with essentially the same theory. The Category of Sets as sets-and-functions and the "Category" of Sets as sets-and-relations are interdefinable. On the other hand, there are *a posteriori* justifications for preferring functions over relations; the theory is nicer with functions than with relations. We have to consider it a genuine discovery that the "right" structure of sets is the category of sets. This discovery is due to F. W. Lawvere; it culminates in the concept of the Lawvere-Tierney concept of (elementary) topos [J], [MM].)

Set, the category of sets (and functions) gives rise to many further categories, such as the category of groups, that of topological spaces, etc. In the General Algebra of structures (("universal") algebras, relational structures, topological structures, etc.), the categories of those structures are nowadays recognized to play a basic, albeit possibly "trivial" role, the role of a "mere language". In [M6], I tried to show that the exposition of an algebraic version of propositional logic, involving as it does Boolean algebras, Heyting algebras, etc., is considerably enlightened by the systematic use of the categories of the various kinds of structure involved. In general, I subscribe to the view that General Algebra requires Category Theory.

In any category, we have the notion of isomorphism. An *isomorphism* of objects X and Y is a pair of arrows $X \underset{g}{\overset{f}{\rightleftarrows}} Y$ such that $gf = 1_X$ and $fg = 1_Y$ (Note that we have *equality* of arrows here. Let us note that equality will be allowed only for arrows with the same domain and the same codomain. For more details on the SFAM notion of "category", see section 8).

A single arrow $X \xrightarrow{f} Y$ is called an *isomorphism* (or if necessary, an *isomorphism arrow*) if there is a (necessarily unique) g making the pair (f, g) an isomorphism. We say that X and Y are *isomorphic* if there is an isomorphism between them. Note that the isomorphism arrows in a given category form another category, called the *underlying groupoid* of the given category (in a *groupoid*, every arrow is an isomorphism). The usual concepts of isomorphism for various mathematical structures ("homeomorphism" for topological spaces, for instance) are the same as isomorphism in the respective categories.

One of the first intuitions of Abstract Mathematics is that two isomorphic structures are, essentially, *the same*. SFAM adopts isomorphism of objects in a category (in particular, isomorphism, that is, equipollence, of sets) to play the role of *equality* for those objects, in the sense that

(**Principle of Isomorphism**) *all grammatically correct properties of objects of a fixed category are to be invariant under isomorphism.*

Here, a property $P(Y)$ of the variable object Y in a given category is ***invariant under isomorphism*** if $P(Y)$ and $Z \cong Y$ imply $P(Z)$.

The just-stated principle is the analog, formulated for isomorphism in place of ordinary equality, of Leibniz's indiscernibility of identicals, the usual axiom of equality that allows to substitute any thing by any other thing equal to it. At this point, when I am not presenting a closed codification of SFAM, the Principle of Isomorphism appears as a requirement on the language of SFAM. Of course, the principle restricts the possible expressions allowed in SFAM. In particular, "ordinary" equality of objects (in particular, "ordinary" equality of sets) is immediately disallowed in SFAM, since for a fixed object X, the predicate $Y = X$ of the variable Y is, obviously, (usually) not invariant under isomorphism.

The last-stated consequence of the Principle of Isomorphism is an expression of a wide-spread understanding in Category Theory that excludes equality of objects in a category from meaningful concept-formation. But more positively, the Principle of Isomorphism itself appears to be a generally accepted idea in Abstract Mathematics. We say things like these:

"two groups that are isomorphic share all structural properties; they are structurally indistinguishable";

"(abstract) group theory studies only structural properties of groups, ones that are invariant under isomorphism".

The basic character of the Principle of Isomorphism is that of a *constraint* on the language of Abstract Mathematics; a welcome one, since it provides for the separation of sense from nonsense. But the fact that isomorphism is the real equality in Abstract Mathematics is also an *experience*. E.g., the structure theorems of the subject essentially always assert the *isomorphism*, often by "canonically given" isomorphism mappings, of a "source" structure to a "target" structure, the latter possessing a specific description; one almost never finds the *equality* in the sense of underlying set theory of those two structures asserted.

The aim of SFAM is to have a self-contained language in which only structural properties of, say, groups can be formulated.

Let us note that in the practice of Category Theory, we usually expect that all meaningful properties of an object in a fixed category are invariant under isomorphism.

Some more remarks on isomorphism as equality. Unlike ordinary equality, isomorphism as a relation between objects is not a *primitive* notion; it is derived, by logical means, from the primitives of the concept of category. Indeed, we may say that the individuation of objects does not happen through the *relation* of isomorphism, but rather, by the total structure of isomorphism arrows, the underlying groupoid of the category. Because of this, the Principle of Isomorphism as stated above is the tip of an iceberg; it does not describe fully what is going on.

As it happens, in category-theoretic concept-formation (such as "monoidal category", for instance), it is never the relation of isomorphism (of the objects $A \otimes (B \otimes C)$ and $(A \otimes B) \otimes C$, for instance) that appears directly; rather, we pick out "canonical" isomorphism arrows (for instance, the associativity isomorphisms $\alpha_{A,B,C} : A \otimes (B \otimes C) \xrightarrow{\cong} (A \otimes B) \otimes C)$ that we incorporate into the concept as part of the data.

As a further example, consider the fact that the construction of the direct product $G \times H$ of groups is invariant under isomorphism. This, in actual fact, says more than what the above formulation of the Principle of Isomorphism states; it is more than just saying that if $G \cong G'$ and $H \cong H'$, then $G \times H \cong G' \times H'$. It also contains the fact that any pair of automorphisms $g : G \xrightarrow{\cong} G$, $h : H \xrightarrow{\cong} H$ induces an automorphism $g \times h : G \times H \xrightarrow{\cong} G \times H$.

The *structure of individuation* of objects in a category (in particular, that of sets) is something more elaborate than a mere primitive relation.

Probably this is a good place to remark that Abstract Mathematics is not so much a restricted part of Mathematics, the one dealing with abstract structures, as a specific way of looking at all that is mathematical. The natural numbers are indispensable in any mathematics, abstract or concrete. Now, the view of SFAM of the natural numbers is that they are the elements of *any* structure of the form $(N,\ 0 \in N,\ S : N \to N)$ that is an initial object in the category of all structures of the said form (this is what is called the (Lawvere) Natural Numbers Object; it is the same, under a certain amount of set theory, as a model of the (categorical) second order Peano Axioms). Any object isomorphic to an initial object is again one; the concept of Natural Numbers Object is invariant under isomorphism. In SFAM, certainly there is no uniquely determined totality of all natural numbers; and the names "one", "two", ... do not name absolutely fixed entities. P. Benaceraff of [Ben1] should approve of SFAM's dealing with the natural numbers, although Benaceraff of [Ben2] may not.

5. A revision of category theory

We are facing the question of the totality of all categories. What kind of totality is this? It cannot be a category itself. The argument for this is similar to the one we saw above concerning why sets do not form a set. More specifically, the concept of an *isomorphism* between two categories is clearly seen to involve the idea of equality of objects; how would one otherwise formulate the 1-1 character of the isomorphism? Therefore, categories cannot be *isomorphic*; as a consequence, they cannot be objects of a category.

As we expect from the progression from the sets as totalities with equality to categories, the kind of totality categories form will be a more complicated version of category; in fact, it will be a "2-dimensional category" of a certain kind. But before we get to the general notion of "2-dimensional category", we need the particular ingredients of the "2-dimensional category" of categories. And now we find a surprise.

The classical concept of morphism of categories is *functor*. Categories, functors, together with natural transformations, endowed with several composition operations, form the classical 2-category *Cat* of categories (at the moment, we are ignoring size considerations); see [ML]. However, upon inspection, the concept of "functor" turns out to be deficient from the point of view of SFAM.

A functor, as a function assigning (among others) a definite value-object to an argument-object, is involved with *equality* of objects, via the requirement that the value-object be *uniquely* determined. In a more positive manner, we should have, according to the Principle of Isomorphism stated above, that

(*) for a "functor" $F : \mathbf{X} \to \mathbf{A}$, and an object $X \in \mathbf{X}$, if A is a value of F at X, and B is another object of $\mathbf{A}$ isomorphic to A, $B \cong A$, then B too is a value of F at X.

Of course, functors will not satisfy this requirement. The notions of "anafunctor", and "saturated anafunctor", are introduced to deal with the problem.

Here are some definitions, made in ordinary Set Theory, for the sake of definiteness. Let $\mathbf{X}$, $\mathbf{A}$ be categories. An *anafunctor* $F : \mathbf{X} \xrightarrow{a} \mathbf{A}$ is given by data as in (i) and (ii):

(i) a set $|F|(X, A)$ for any $X \in Ob(\mathbf{X})$, $A \in Ob(\mathbf{A})$

(the elements of $|F|(X, A)$ are called *specifications*; $s \in |F|(X, A)$ *specifies* A as a possible value of F at X);

(ii) for $(f : X \to Y) \in Arr(\mathbf{X})$, A, $B \in Ob(\mathbf{A})$, $s \in |F|(X, A)$, $t \in |F|(Y, B)$, an arrow $F_{s,t}(f) : A \to B$.

The data are required to satisfy conditions (iii) to (v):

(iii) for any $X \in Ob(\mathbf{X})$, there is $A \in Ob(\mathbf{A})$ such that $|F|(X, A)$ is inhabited (non-empty);

(iv) $s \in |F|(X, A) \Rightarrow F_{s,s}(1_X) = 1_A$;

(v)

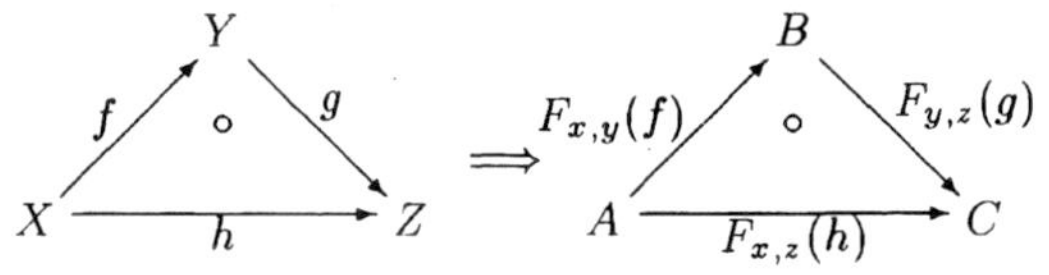

$$(x \in |F|(X, A), \quad y \in |F|(Y, B), \quad z \in |F|(Z, C)).$$

Although at the time of working on [M2] I did not know about the fact, G. M. Kelly [K] had already defined the notion that I call anafunctor, in a brief passage, as a side remark. He did not name it, but he characterized the concept as a description of a common situation in which one, seemingly possessing all the data needed, wants to define a functor, but in which, for the specification of the functor, one still needs to make arbitrary choices that usually require the axiom of choice. What I do in [M2], and Kelly did not do, is develop the theory of anafunctors, and thereby demonstrate that one does not have to make those arbitrary choices, and that one can live with anafunctors in place of functors. The result is a category theory which is more "canonical" than the the usual one. In technical terms, when using anafunctors, to a large extent one can avoid the use, ubiquitous in the usual theory, of the Axiom of Choice.

Let me emphasize that [M2] is developed in ordinary Gödel-Bernays set-theory, mostly without the Axiom of Choice, but at any rate, with a careful keeping track of the uses of choice, and even of the principle of the excluded middle (which is regarded, after R. Diaconescu [D], a particular case of Choice). In SFAM proper, the problem with functors is bound to be graver, and the need for something like anafunctors greater, since the kind of choice we need, Global Choice, *cannot even be meaningfully stated in* SFAM. The reason is precisely the lack of equality on the entities, objects in a category, among which we want to make *unique* choices. In contrast, the local, (set-)version of choice remains perfectly meaningful in SFAM.

Whereas the anafunctors are more general than functors, *saturated* anafunctors, the ones we really want because of the requirement (*), are a special class within anafunctors, almost disjoint from that of functors; I am not aware in the literature of a mention of the property of saturation, under any name.

To introduce this concept, let $F : \mathbf{X} \xrightarrow{a} \mathbf{A}$, $X \in Ob(\mathbf{X})$, $A \in Ob(\mathbf{A})$ and $s \in |F|(X, A)$, $t \in |F|(X, B)$. Then we have $f = F_{s,t}(1_X) : A \to B$, and $g = F_{t,s}(1_X) : B \to A$, and the laws easily imply that f and g are inverses of each other; (f, g) is an isomorphism. Thus, we see that any two possible values of F at an object X, A and B here, are isomorphic. Now, forget t but keep the other data; and take an arbitrary isomorphism $f : A \xrightarrow{\cong} B$. F is *saturated* if *there is a unique* $t \in |F|(X, B)$ such that $f = F_{s,t}(1_X)$, for all choices of the parameters X, A, B and s involved.

I regard this condition as the correct rendering of the Principle of Isomorphism, for the notion of (ana)functor. The principle expresses that if we have one possible value of $F(X)$, and we change it by a specific isomorphism to another object, then the new object is also a possible value of $F(X)$, *by a uniquely determined specification.* We see again that the Principle of Isomorphism in its general form by itself is only the tip of an iceberg. Simply requiring of an anafunctor that for a fixed argument-object, being a possible value-object be invariant under isomorphism, although logically possible, would clearly not give a good notion.

Let me mention that, as a consequence of the definition, we can see that in a saturated anafunctor one can also change the argument-object along an isomorphism, with keeping the value-object fixed.

To illustrate the effect of anafunctors, consider a category C with binary products. In category theory, we want to consider the product functor

$$P : \mathbf{C} \times \mathbf{C} \to \mathbf{C}$$

$$(X, Y) \mapsto X \times Y$$

Of course, we do not have, in general, a well-defined product $X \times Y$; we only have the notion of product diagram, given by a universal (extremal) property: certain diagrams of the shape

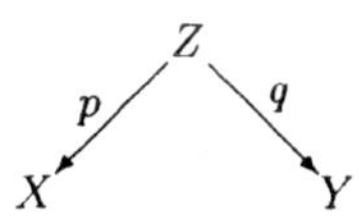

(1)

qualify as products, and there is at least one such for any X and Y. This is the point where one uses "global choice", and one picks a particular product-object Z, and projections p and q, and call them $X \times Y$, $\pi_{X,Y}$ and $\pi'_{X,Y}$, respectively. Having made these choices, the definition of P is natural. It is also true that it does not matter how we make the choices, since the resulting functors are all *isomorphic* (by natural transformations). The fact remains, however, that we need Choice to define P. On the other hand, we have the anafunctor, also called

P, defined as follows: $P : \mathbf{C} \times \mathbf{C} \xrightarrow{a} \mathbf{C}$:

$$|P|((X, Y), Z) = \{(p, q) : (1) \text{ is a product}\}$$

for $(p, q) \in |P|((X, Y), Z)$, $(r, s) \in |P|((A, B), C)$, $f : X \to A$, $g : Y \to B$:

$P_{(p,q),(r,s)}((f, g)) =$ the unique h for which P is in fact a saturated anafunctor.

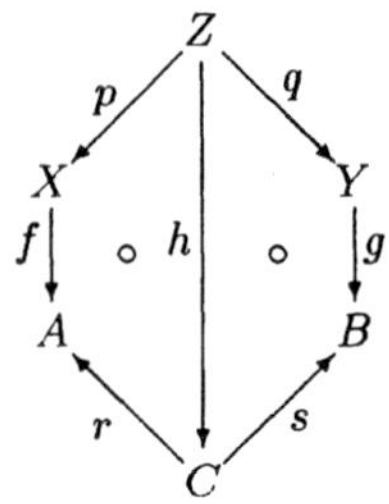

One defines *natural transformations* of anafunctors. For fixed categories $\mathbf{X}$ and $\mathbf{A}$, anafunctors $\mathbf{X} \xrightarrow{a} \mathbf{A}$ and their natural transformations form a category $Ana(\mathbf{X}, \mathbf{A})$. Every $F : \mathbf{X} \xrightarrow{a} \mathbf{A}$ is canonically isomorphic in $Ana(\mathbf{X}, \mathbf{A})$ to a saturated anafunctor $F^{\#}$; saturated anafunctors are enough. Every ordinary functor is an anafunctor in an obvious way; and by using Choice, but not without it, every anafunctor is isomorphic to a functor. (This is a way of putting what Kelly was after; note however that for stating this we need the notion of natural isomorphism of anafunctors, which Kelly did not consider.) In section 10, we will define $F^{\#}$ for an ordinary functor F.

In [M2], "basic category theory" is redone with using (saturated) anafunctors in place of functors. For instance, one obtains a bicategory (see the next section) *SanaCat* of categories, saturated anafunctors, and their natural transformations; under the Global Axiom of Choice, *SanaCat* is biequivalent to *Cat*, the classical 2-category of categories.

6. 2-dimensional categories

Abstracting from categories, functors and natural transformations, together with the basic operations on them, leads to the concept of 2-category; see [ML]. In a 2-category $\mathcal{A}$, we have three kinds of entities: objects (0-cells) $\mathbf{A}, \mathbf{B}, \cdots$ arrows (1-cells)

$$F : \mathbf{A} \to \mathbf{B}$$

(thus, each arrow has a source (domain), and a target (codomain)), and 2-cells as in

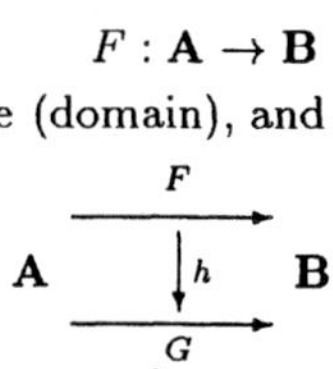

(2)

(thus, a 2-cell has *parallel* source and target 1-cells). For a fixed pair of 0-cells $\mathbf{A}, \mathbf{B}$, the 1-cells (1), and the 2-cells (2) form a category $\mathcal{A}(\mathbf{A}, \mathbf{B})$; this involves the so-called vertical composition of 2-cells:

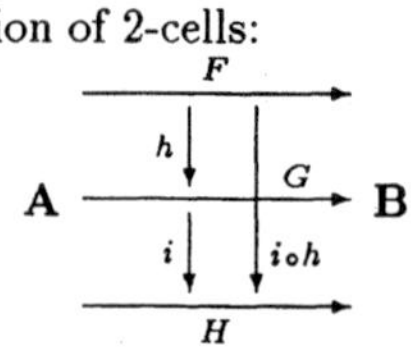

Furthermore, for any 0-cells $\mathbf{A}, \mathbf{B}, \mathbf{C}$, we have a *composition functor*

$$\circ_{\mathbf{A},\mathbf{B},\mathbf{C}} : \mathcal{A}(\mathbf{A},\mathbf{B}) \times \mathcal{A}(\mathbf{B},\mathbf{C}) \to \mathcal{A}(\mathbf{A},\mathbf{C})$$

$$(\mathbf{A} \xrightarrow{F} \mathbf{B}\ ,\ \mathbf{B} \xrightarrow{G} \mathbf{C}) \mapsto \mathbf{A} \xrightarrow{GF} \mathbf{C}$$

involving the so-called *horizontal composition* of 2-cells:

$$\mathbf{A} \underset{H}{\overset{F}{\rightrightarrows}} \!\Downarrow h\ \mathbf{B} \underset{I}{\overset{G}{\rightrightarrows}} \!\Downarrow i\ \mathbf{C} \quad \mapsto \quad \mathbf{A} \underset{IH}{\overset{GF}{\rightrightarrows}} \!\Downarrow ih\ \mathbf{C}$$

We require that composition be associative; in particular, in the situation

$$\mathbf{A} \xrightarrow{F} \mathbf{B} \xrightarrow{G} \mathbf{C} \xrightarrow{H} \mathbf{D},$$

we have

$$H(GF) = (HG)F.$$

(To complete the definition of "2-category", certain stipulations on identity arrows need to be added only.)

In [Be], the concept of *bicategory* is introduced. As Benabou demonstrated, bicategories occur naturally at many points of Category Theory; in particular, the concept of (not necessarily strict) monoidal (or tensor) category (see [ML]) is the same as a one-object bicategory. The concept of bicategory can be seen as *the result of replacing equalities of objects in a category by canonical (structural) isomorphisms*, a move in the direction of SFAM. In particular, one way a bicategory differs from a 2-category is that, in place of the required equality (3), in which $H(GF)$ and $(HG)F$ are objects of the category $\mathcal{A}(\mathbf{A},\mathbf{D})$, we take up, as part of the structure, an isomorphism 2-cell

$$\alpha_{F,G,H} : H(GF) \xrightarrow{\cong} (HG)F,$$

an isomorphism in the category $\mathcal{A}(\mathbf{A},\mathbf{D})$. The complete definition of "bicategory" requires further canonical isomorphisms, and also, the so-called coherence conditions that connect, e.g., various instances of $\alpha_{-,-,-}$.

As it was mentioned at the end of the last section, when, working in ordinary Set Theory, we replace functors by (saturated) anafunctors, and make the necessary other adjustments, we obtain from *Cat* a bicategory *AnaCat* (*SanaCat*). In particular, now we *do not* get a 2-category.

Just like *Set* spawns a large class of other categories of set-based structures, so does *Cat* give rise to 2-categories, and bicategories, of structured categories. In [S], we find the view that, similarly to the need for the theory of categories in General Algebra that treats of set-based structures, the algebra of structured categories should be based on the theory of bicategories. In the work of the

Australian School (G. M. Kelly, R. Street, and others), we find a realization of this idea.

In any bicategory, in particular, in any 2-category, there is a notion of *equivalence* (of 0-cells), a higher-dimensional analog of isomorphism in a category. For 0-cells $\mathbf{A}, \mathbf{B}$, an *equivalence of* $\mathbf{A}$ *and* $\mathbf{B}$ is a quadruple (F, G, i, j) as in

$$\mathbf{A} \underset{G}{\overset{F}{\rightleftarrows}} \mathbf{B} \qquad i : GF \xrightarrow{\cong} 1_{\mathbf{A}}, \quad j : FG \xrightarrow{\cong} 1_{\mathbf{B}}$$

involving the *identity* 1-cells $1_{\mathbf{A}} : \mathbf{A} \to \mathbf{A}$, $1_{\mathbf{B}} : \mathbf{B} \to \mathbf{B}$. We write $(F, G, i, j) : \mathbf{A} \simeq \mathbf{B}$. A 1-cell F is called an *equivalence* (arrow) if it can be extended to an equivalence (F, G, i, j). Similarly for the pair (F, G) being an equivalence. $\mathbf{A}$ and $\mathbf{B}$ are *equivalent*, in symbols $\mathbf{A} \simeq \mathbf{B}$, if there is an equivalence of them.

The usual concept of *equivalence of categories* is that of equivalence in *Cat*. It is generally recognized, in exact analogy to sets and set-based structures in relation to the notion of isomorphism, that the "right notion" of "equality" for categories, resp. category-based structures is equivalence of categories, resp. equivalence in the corresponding bicategory. This principle acts, again, in two different ways. First, as the *constraint* on properties of objects in a bicategory, which we may call the Principle of Equivalence, asserting that

any (meaningful) property of an object in a bicategory is invariant under equivalence.

Secondly, as the *experience* that usually, especially in "serious" representation theorems, one gets that a given category can be represented in a certain desired way *up to equivalence*, but not up to isomorphism.

It is obvious that for the purposes of SFAM, the concept of bicategory requires a (further) modification: the composition functors (3) have to be replaced by (saturated) anafunctors, and a similar, more minor change, is needed with respect to the identity 1-cells. The resulting (saturated) *anabicategories* are worked out in [M2].

By making certain "arbitrary" choices, any anabicategory can be reduced to a bicategory; and vice versa, there is a canonical process of saturating any anabicategory, in particular, any bicategory. Thus, in the presence of the Global Axiom of Choice, saturated anabicategories are essentially "equivalent" (in fact, also in a technical sense, (bi)equivalent) to bicategories.

It is instructive to see the definition of the monoidal category, that is, one-object bicategory, of R-modules, for R any commutative ring, and compare it with its saturated anaversion. The latter can be defined directly, by appropriately modifying the usual definition. In fact, the modifications are seen to consist in eliminating all arbitrary choices the usual definition has to make (see [ML]).

7. Higher dimensional categories

The foregoing strongly suggests that there should be an infinite hierarchy of "n-dimensional categories", for $n = 0, 1, 2, \cdots$, such that n-dimensional categories "form" (are the objects of) an $n+1$-dimensional category. We have discussed this hierarchy for $n = 0$ ('set'), 1 ('category') and 2, in which case several concepts have been advanced, with the final suggestion that from the point of view of SFAM, 'saturated anabicategory' should be the right one.

Let me emphasize that from the point of view of SFAM, the said hierarchy is one of ever increasing generality. For example, every set is, or gives rise to, a category, namely the discrete category with objects the elements of the set, and only identity arrows. On the other hand, a category does not give rise to a set; the totality of objects is not a set; it is not provided, by the data of the category, with an equality predicate. More strongly, from the point of view of SFAM, a category is not an ordinary, "set-based" structure. This can be seen from the fact that the structural identity relation for categories is not isomorphism, but something else, namely equivalence of categories.

The notion of n-category, the direct analog of 2-category for arbitrary n, is rather clear. In fact, we have the general theory of enriched categories [K2]; and, recursively, an $n+1$-category is a category enriched in the (ordinary) category of n-categories.

The case of the hierarchy continuing the concept of bicategory, which, following [BD1], we will call *weak n-category*, is quite a different story; in fact, no definition of "weak n-category" has gained general acceptance yet (however, see below for some good news in this regard). The general idea of what a weak n-category should be is reasonably clear: one should "weaken" the definition of n-category, by replacing all equalities by "canonical" arrows: arrows that are part of the structure. The difficulty is in giving explicitly the so-called "coherence conditions" the canonical arrows should satisfy. In [BD1] we read: "It is clear that new ideas are needed to do so without a combinatorial explosion, since already the explicit definition of a tricategory [in [GPS]] takes six pages, and that of a triequivalence 13 pages!".

Let us note that the present high interest in weak n-categories, and in higher dimensional category theory in general, is driven mainly by actual and potential applications in topology, theoretical physics, and in particular, quantum groups; it has relatively little to do with the foundational concerns of the present paper. See [BD1].

Upon contemplation, one sees, already for $n = 2$ that the "real" definition of bicategory is not the official one, but rather one that stipulates "all coherence conditions" at once, i.e. (roughly) that "all diagrams" built, in the "free" or "generic" case, from instances of the associativity isomorphisms and other canonical isomorphisms should commute. It is an important *coherence theorem* (due to S. Mac Lane, formulated for the one-object case, that is, for monoidal categories; see [ML]) that this stronger condition is a consequence of the official definition, which is thus nothing but a finite (equational) axiomatization of the totality of all coherence conditions.

This suggests that, possibly, the right approach to the definition of weak n-category is to aim at formulating all coherence conditions at once, regardless the fact that this might give a very "theoretical" definition. It would then be a separate, and still very important, project to find a (hopefully) finite and concise set of coherence conditions that would be enough to imply all coherence conditions.

J. Baez and J. Dolan have produced a very interesting definition of "weak n-category" [BD2]. The similarity type of the Baez/Dolan n-category is radically larger than the usual one. In the usual n-categories, the domain and codomain of a $k+1$-cell are both k- cells; in a Baez/Dolan n-category, there are $k+1$-cells whose domains are arbitrary "pasting diagrams" of k-cells, not the composites of the latter. In addition, the coherence conditions are replaced, and the combinatorial complexity eliminated, by the requirement of the existence of cells satisfying certain universal properties (this phenomenon is well-known from Grothendieck's concept of fibration, relative to the concept of pseudo-functor; see [SGA1]).

The Baez/Dolan definition for $n = 2$ can be shown to be closely related to the concept of "saturated anabicategory" mentioned above. However, Baez and Dolan do not use the ana-versions of morphisms, in particular, of "functor".

As I mentioned in the introduction, the Baez/Dolan definition is only a beginning, albeit a very promising one. Among others, the definition of the Baez/Dolan $n+1$-category of all n-categories is still missing. I believe that thinking in the spirit of anafunctors will help give the latter definition.

8. Dependent types

Dependent types are familiar from [M-L] and [C]; there is a further extensive literature of them. Their use for first-order logic seems to be new, however. First Order Logic with Dependent Sorts (FOLDS; I use "sort" in place of "type") is a variant of ordinary First Order Logic (FOL); the basic metatheory of FOLDS is an extension (generalization) of that of FOL.

A similarity type, or vocabulary, for FOLDS is a *one-way category*; a small category $\mathbf{L}$ is one-way if it has the following two properties:

(i) $\mathbf{L}$ *has finite fan-out*: for any $K \in \mathbf{L}$, there are only finitely many arrows with domain equal to K.

(ii) **L** *is reverse well-founded*: there is no infinite sequence $<K_n \xrightarrow{f_n} K_{n+1}>_{n\in N}$ of composable proper (non-identity) arrows ($f_n \neq 1_{K_n}$) .

Immediate consequences of the definition are the following:

(iii) **L** is skeletal: any two isomorphic objects are identical;

(iv) the only arrow from an object to itself is the identity;

and in fact, as a consequence of (iii) and (iv),

(v) **L** has no circuit of positive finite length consisting of proper arrows; there is no $<K_n \xrightarrow{f_n} K_{n+1}>_{n<N}$ with $N \in N$, $K_N = K_0$, and $f_n \neq 1_{K_n}(n < N)$.

It is easy to see that a small category is one-way iff it satisfies (i) and (v); and if the category is finite, then it is one-way iff it satisfies (iii) and (iv).

If **L** is a one-way category, the set $Ob(\mathbf{L})$ of objects is partitioned as in

$$Ob(\mathbf{L}) = \bigcup_{i<\ell} \mathbf{L}_i$$

into non-empty *levels* $\mathbf{L}_i$, for $i < \ell$, ℓ the *height* of **L**, $\ell \leq \omega$, such that $\mathbf{L}_0$ consist of the objects A for which there is no proper arrow with domain A, and such that, for $i > 0$, $\mathbf{L}_i$ consists of those objects A for which all proper arrows $A \to B$ have $B \in \mathbf{L}_{<i} = \bigcup_{j<i} \mathbf{L}_j$, and there is at least one proper arrow $A \to B$ with $B \in \mathbf{L}_{i-1}$. All proper arrows go from a level to a lower level. Of course, the height of a finite one-way category is finite.

For **L** a one-way category, an **L**-*structure* is a functor $\mathbf{L} \to Set$.

There is a way of introducing FOLDS that follows Martin-Löf's idea of dependent types closely, without any preconceptions about similarity types; this is then seen to lead precisely to the notion of similarity type we just introduced without prior motivation (see [M3]).

The recognition of role of finite one-way categories in syntax is due to F. W. Lawvere, who pointed out their role in connection with the sketch-syntax of [M1]. Their role in FOLDS is related to their role in the sketch-syntax. I should note that the use of infinite one-way categories, in the sense used here, is essential (see below). Here is an example for a FOLDS similarity type, called $\mathbf{L}_{cat}$:

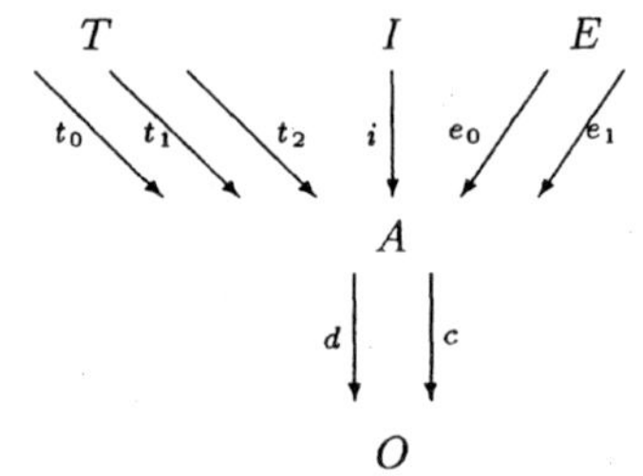

$$dt_0 = dt_2,\ ct_0 = dt_1,\ ct_1 = ct_2$$
$$di = ci$$
$$de_0 = de_1,\ ce_0 = ce_1$$

The proper arrows of $\mathbf{L}_{cat}$ are the arrows shown, and their composites; the equalities shown identify some of these composites.

Note that any small category $\mathbf{C}$ gives rise to an $\mathbf{L}_{cat}$-structure

$$M[\mathbf{C}] = M : \mathbf{L}_{cat} \to Set$$

in a natural way. $M(O)$ is the set of objects of $\mathbf{C}$; $M(A)$ is the set of arrows; $M(d)$ and $M(c)$ map an arrow to its domain and codomain, respectively. $M(T)$ is the set of commutative triangles, i.e., tuples

$$(X, Y, Z, f : X \to Y, g : Y \to Z, h : X \to Z)$$

such that $h = gf$. In other words, the elements of $M(T)$ are commutative diagrams of the form

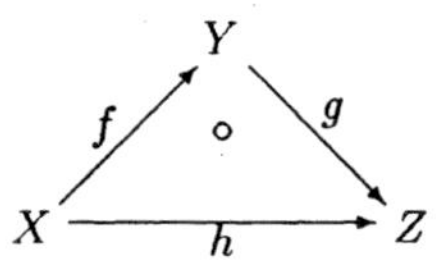

(1)

$M(I)$ is the set of identity arrows, $M(E)$ is the set of pairs (f, f) of equal arrows. With τ standing for (1), $M(t_1)(\tau) = f, M(t_2)(\tau) = g, M(t_3)(\tau) = f$. The rest of the definition of M should be clear. Note that M is indeed a functor.

Of course, not every $\mathbf{L}_{cat}$-structure is a category. On the other hand, for two categories construed as $\mathbf{L}_{cat}$-structures, a natural transformation from one to the other is precisely the same as a functor from one category to the other.

Let $\mathbf{L}$ be a one-way category; we fix $\mathbf{L}$ for a while. The objects of $\mathbf{L}$ are called *kinds*. Let us write K_p for $dom(p)(p \in Arr(\mathbf{L}))$. We use the notation $K|\mathbf{L}$ for the set of all proper arrows $p : K \to K_p$ with domain K. The set $K|\mathbf{L}$ will figure as the *arity* of the symbol K. In particular, the ones with empty arity are exactly the level-0 kinds.

We are going to define what *sorts* are, and what *variables of* a given sort are; we will write $x : X$ to denote that the variable x is of sort X. Every sort will be of the form $K(<x_p>_{p \in K|\mathbf{L}})$, with K a kind (the displayed sort is then said to be *of* the kind K), with variables x_p indexed by the elements p of the arity $K|\mathbf{L}$ of K; additional conditions will have to be satisfied. Let $k \in N$, and suppose we have defined sorts of kinds on levels less than k, and variables of such sorts. Then, for a kind K on level $k, K(<x_p>_{p\in K|\mathbf{L}})$ is a sort (of the kind K) iff for each $p \in K|\mathbf{L}$, we have $x_p : K_p(<y_{p,q}>_{q\in K_p|\mathbf{L}})$ with $p : K \to K_p$ (note that K_p is on a lower level than K), and for every $q \in K_p|\mathbf{L}$, $y_{p,q} = x_{qp}$:

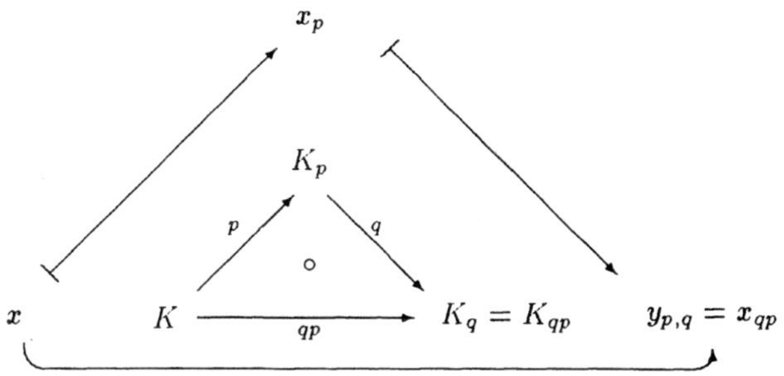

For every sort $X = K(<x_p>_{p\in K|\mathbf{L}})$ thus specified, we declare certain symbols as variables of sort X; the only important things about this declaration are that (i) variables of sort X have to be new, so that every variable uniquely determines its own sort, and that (ii) there are enough (infinitely many) variables of each sort.

Note that every variable "carries" its own sort with it. This is in contrast with the practice of most of the relevant literature (see e.g. [C]), where variables are "locally" declared to be of certain definite sorts, but by themselves, they do not carry sort information. For a sort

$$X = K(<x_p>_{p\in K|\mathbf{K}}), \; Var(X) \underset{def}{=} \{x_p : p \in K|\mathbf{K}\};$$

and if $x : X$, $Dep(x) \underset{def}{=} Var(X)$; x *depends on* the variables in $Dep(x)$.

Let $\mathbf{L} = \mathbf{L}_{cat}$. Since $O|\mathbf{L}$ is empty, $O(\emptyset)$, with $\emptyset$ the empty sequence of variables, is a sort; we write simply O. We have $A|\mathbf{L} = \{d, c\}$; the sorts of the kind A are of the form $A(X, Y)$, with $X, Y : O$; here, X is indexed by d, Y by c.

$T|\mathbf{L} = \{f_1, f_2, f_3, t_1, t_2, t_3\}$, with $f_1 = dt_1 = dt_3$, $f_2 = ct_1 = dt_2$, $f_3 = ct_2 = ct_3$. A sort of the kind T will have the form $T(X, Y, Z, f, g, h)$; applying the condition in the definition with $p = t_1$, we get that $f : A(X, Y)$ must be the case. In summary, the variables in $T(X, Y, Z, f, g, h)$ have to line up as in

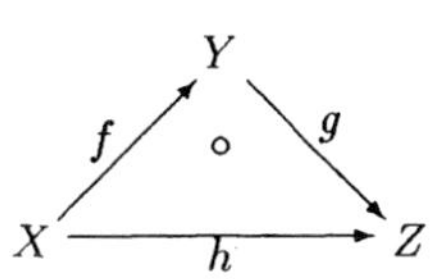

We define *formulas* φ and the set $Var(\varphi)$ of the *free variables of* φ by a simultaneous induction.

The symbols $\mathbf{t}$ ("true"), $\mathbf{f}$ ("false") are formulas; $Var(\mathbf{t}) = Var(\mathbf{f}) = \emptyset$.

The sentential connectives $\wedge, \vee, \rightarrow, \neg, \leftrightarrow$ can be applied in an unlimited manner; $Var(\;)$ for the compound formulas formed using connectives is defined in the expected way; e.g., $Var(\varphi \wedge \psi) = Var(\varphi) \cup Var(\psi)$.

Suppose φ is a formula, x is a variable *such that there is no* $y \in Var(\varphi)$ *with* $x \in Dep(y)$. Suppose $x : X$. Then $\forall x : X.\varphi$, $\exists x : X.\varphi$ are formulas;

$$Var(\forall x : X.\varphi) \underset{def}{=} Var(\exists x : X.\varphi) \underset{def}{=} (Var(\varphi) - \{x\}) \cup Dep(x).$$

All formulas are obtained as described.

Here is an example of a sentence (formula without free variables) over $\mathbf{L}_{cat}$:

$$\forall X : O.\forall Y : O.\forall Z : O.\forall f : A(X,Y).\forall g : A(Y,Z).\exists h : A(X,Z).\exists t : T(X,Y,Z,f,g,h).\mathbf{t}$$

The sentence, referred to below as (2), expresses the existence of the composite $h : X \rightarrow Z$ of composable arrows $f : X \rightarrow Y$, $g, Y \rightarrow Z$.

The sorts are interpreted in structures as certain sets. Let K be a kind, M an $\mathbf{L}$-structure. By $M[K]$ we mean the set of all tuples $<a_p>_{p \in K|\mathbf{L}}$ with $a_p \in M(K_p)$ (where $p : K \rightarrow K_p$), and such that for every $p \in K|\mathbf{L}$ and $q \in K_p|\mathbf{L}$, $M(q)(a_p) = a_{qp}$:

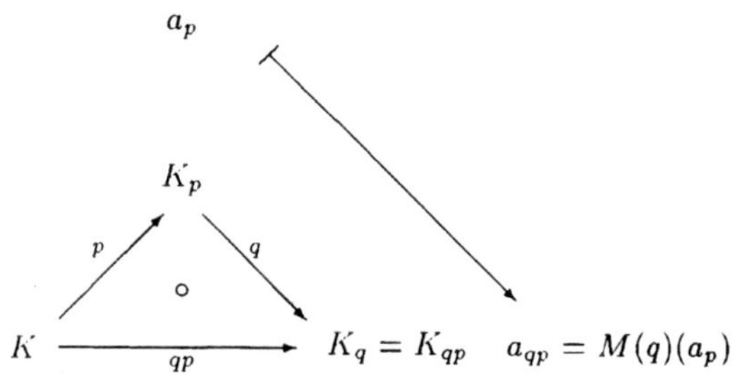

The elements of the set $M[K]$ are called *contexts* for K in M.

For instance, for $\mathbf{L} = \mathbf{L}_{cat}$, when $K = A$ and M is $M[\mathbf{C}]$ for a category $\mathbf{C}$, then $M[A]$ is the set of pairs (X,Y) of objects; when $K = T$, $M[T]$ is the set of all not necessarily commutative triangles in $\mathbf{C}$.

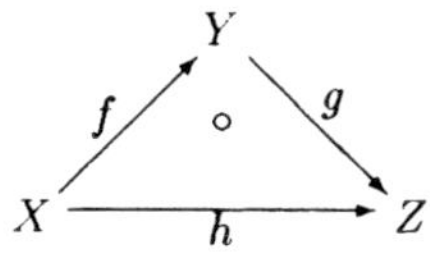

The set $M(K)$ is "fibered over" $M[K]$; $M(K)$ is the disjoint union of sets $MK(\mathbf{a})$, one for each $\mathbf{a} = <a_p>_{p \in K|\mathbf{L}} \in M[K]$;

$$MK(\mathbf{a}) \underset{def}{=} \{a \in M(K) : (Mp)(a) = a_p \text{ for all } p \in K|\mathbf{L}\},$$

the *fiber over* $\mathbf{a}$ (it is clear that for any $a \in M(K)$,

$$\mathbf{a} \underset{def}{=} <(Mp)(a)>_{p \in K|\mathbf{L}} \in M[K]).$$

Let $X = K(<x_p>_{p \in K|\mathbf{L}})$ be a sort. An interpretation of X in M is given by a context $\mathbf{a} = <a_p>_{p \in K|\mathbf{L}} \in M[K]$, with a_p *assigned to* x_p, which means the additional condition $a_p = a_{p'}$, every time $x_p = x_{p'}$, $(p, p' \in K|\mathbf{L})$; the interpretation itself is the fiber $MK(\mathbf{a})$ over $\mathbf{a}$.

Returning to the example started above, the interpretation of $A(X,Y)$ is $\hom_{\mathbf{C}}$ (A, B) when A, B are objects assigned to the variables X, Y respectively. The interpretation of the sort $T(X,Y,Z,f,g,h)$ must be given by a (not necessarily commutative) diagram of the shape

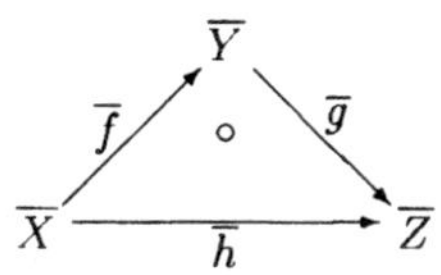

in $\mathbf{C}$, and of course, if, say, X is the same variable as Y, then the object $\overline{X}$ must be the same as $\overline{Y}$. The interpretation itself is a singleton set if the displayed diagram commutes, and the empty set otherwise.

For the full definition of the semantics of FOLDS, we need the concept of context (of variables). A *context* is a finite set $\mathcal{Y}$ of variables such that, for all $y \in \mathcal{Y}$ we have that $Dep(y) \subset \mathcal{Y}$.

For instance, in the case $\mathbf{L} = \mathbf{L}_{cat}$, the set $\{X, Y, Z, f, g, h, t\}$, under the conditions

$$X : O, Y : O, Z : O, f : A(X,Y), g : A(Y,Z), z : A(X,Z), t : T(X,Y,Z,f,g,h),$$

is a context.

Note that for any formula φ, $Var(\varphi)$ is a context.

Let $\mathcal{Y}$ be a context, and M an **L**-structure. We define the set $M[\mathcal{Y}]$, the set of (legitimate) valuations of $\mathcal{Y}$ in M.

For a variable $y \in \mathcal{Y}$, let us display the sort of y in the notation

$$y : K_y(<x_{y,p}>_{p \in K_y|\mathbf{L}}).$$

We define $M[\mathcal{Y}] \underset{def}{=}$

$$\left\{ <a_y>_{y \in \mathcal{Y}} \in \prod_{y \in \mathcal{Y}} M(K_y) : (Mp)(a_y) = a_z \text{ whenever } y \in \mathcal{Y},\ p \in K_y|\mathbf{L} \text{ and } z = x_{y,p} \right\}.$$

The definition says that the elements of $M[\mathcal{Y}]$ are compatible valuations of the variables in $\mathcal{Y}$, where "compatibility" refers to the fact that if two variables y and z from $\mathcal{Y}$ are in a relation of dependence, $z \in Dep(y)$, in a particular way given by the "place" p of the kind K_y (that is, $z = x_{y,p}$), then the corresponding elements of the family have to be related by $M(p)$.

By recursion on the complexity of the formula φ, we define $M[\mathcal{Y} : \varphi]$, *the interpretation of* φ in M *in the context* $\mathcal{Y}$, whenever $\mathcal{Y}$ is a context such that $Var(\varphi) \subset \mathcal{Y}$; we will have that $M[\mathcal{Y} : \varphi] \subset M[\mathcal{Y}]$.

$$M[\mathcal{Y} : \mathbf{t}] \underset{def}{=} M[\mathcal{Y}] ;$$

$$M[\mathcal{Y} : \mathbf{f}] \underset{def}{=} \emptyset .$$

For the propositional connectives, the clauses are the expected ones; *e.g.*,

$$<a_y>_{y \in \mathcal{Y}} \in M[\mathcal{Y} : \psi \wedge \theta] \underset{def}{\Longleftrightarrow}$$

$$<a_y>_{y \in \mathcal{Y}} \in M[\mathcal{Y} : \psi] \text{ and } <a_y>_{y \in \mathcal{Y}} \in M[\mathcal{Y} : \theta].$$

The interpretation of formulas $\forall x : X.\psi$, $\exists x : X.\psi$ will be according to the readings "for all x in X, ψ", and "there is x in X such that ψ". Thus, quantification in FOLDS is a relativized quantification. Here, the sort X is interpreted as a set according to what was said above. For the precise clause, we need a bit more notation.

Let $M, \mathcal{Y}$ and $\mathbf{a} = <a_y>_{y \in \mathcal{Y}} \in M[\mathcal{Y}]$ be as above; assume

$$Var(\forall x : X.\psi) = Var(\exists x : X.\psi) = (Var(\varphi) - \{x\}) \cup Dep(x) \subset \mathcal{Y}.$$

Part of $\mathcal{Y}$ may be discarded. Let $\mathcal{Y}' = \mathcal{Y} - \{x\} - \{y \in \mathcal{Y} : x \in Dep(y)\}$. Then, still, $\mathcal{Y}'$ is a context, and $Var(\forall x \psi) \subset \mathcal{Y}'$ (the reason is that, since $\forall x : X.\psi$ is well-formed, if $y \in Var(\psi)$, then $x \notin Dep(y)$). Also, $\mathcal{Y}' \cup \{x\}$ is a context, and $Var(\psi) \subset \mathcal{Y}' \cup \{x\}$.

Let $X = K(<x_p>_{p \in K|\mathbf{L}})$. Each $x_p \in \mathcal{Y}'$, so $\tilde{a} = <a_{x_p}>_{p \in K|\mathbf{L}}$ is defined, and as easily seen, $\tilde{a} \in M[K]$. Therefore, we have the fiber $MK(\tilde{a})$ of $M(K)$.

Let $\mathbf{a} = <a_y>_{y \in \mathcal{Y}} \in M[\mathcal{Y}]$. Define $\mathbf{a}' = <a_y>_{y \in \mathcal{Y}'}$. For any $a \in MK(\tilde{a})$, let $\mathbf{a}'[a/x]$ be $<b_y>_{y \in \mathcal{Y}' \cup \{x\}}$ for which $b_y = a_y$ when $y \in \mathcal{Y}'$, and $b_x = a$. It is immediately seen that $\mathbf{a}'[a/x] \in M[\mathcal{Y}' \cup \{x\}]$. We define

$$\mathbf{a} \in M[\mathcal{Y} : \forall x : X.\psi] \Leftrightarrow \mathbf{a}'[a/x] \in M[\mathcal{Y}' \cup \{x\} : \psi] \text{ for all } a \in MK(\tilde{a}) ,$$

and

$$\mathbf{a} \in M[\mathcal{Y} : \exists x : X.\psi] \Leftrightarrow \mathbf{a}'[a/x] \in M[\mathcal{Y}' \cup \{x\} : \psi] \text{ for some } a \in MK(\tilde{a}) .$$

This completes the definition of the standard, *Set*-valued semantics of FOLDS.

As usual, we also write $M \models \varphi[\mathbf{a}]$ for $\mathbf{a} \in M[\mathcal{Y} : \varphi]$.

Returning to the example of $\mathbf{L}_{cat}$, the reader will easily write down all axioms for "category" in the form of sentences over $\mathbf{L}_{cat}$; the sentence (2) above is an example. Let us call the resulting finite set of sentences Σ_{cat}. Since equality is not treated as "logical", Σ_{cat} includes axioms concerning equality, the kind E. It will *almost*, but not quite, be the case that an $\mathbf{L}_{cat}$-structure M satisfies Σ_{cat} iff it is of the form $M = M[\mathbf{C}]$ for a category $\mathbf{C}$; for explanation, see below.

9. Formal systems

FOLDS may be regarded as a restricted form of ordinary first order logic. Given a one-way category $\mathbf{L}$, we may consider the multi-sorted first order language with sorts the objects of $\mathbf{L}$, with unary sorted operation symbols the arrows of $\mathbf{L}$, and with equality predicates, one for each sort; let us refer to this language as *First Order Logic* (FOL) *over* $\mathbf{L}$. The formulas of FOLDS over $\mathbf{L}$ can obviously be translated into FOL over $\mathbf{L}$. This consideration will immediately imply the fact that the compactness theorem, a kind of abstract completeness, holds for FOLDS.

We have completeness theorems of appropriate formal systems for FOLDS. Completeness is stronger than compactness; it is also essential for the purpose of adopting FOLDS as a language for an axiomatic system.

We have the classical and the intuitionistic versions of FOLDS. The third version, coherent logic, is a proper part of both classical and intuitionistic logic. Categorical logic, mainly in the pioneering work of Andre Joyal, has shown the fundamental theoretical role of coherent logic in relation to both classical and intuitionistic logic, for example, in connection with completeness theorems (see [MR1]). The coherent fragment of FOLDS continues to play a basic role.

In [M3], I worked out the completeness theorems for systems in a format following [LS]. This format is the one that fits best a categorical framework for the metatheory of FOLDS.

The Lambek-style system employ *entailments.* An entailment is an entity of the form $\theta \underset{\mathcal{X}}{\Rightarrow} \sigma$, with θ and σ formulas, and $\mathcal{X}$ a context such that $Var(\theta) \cup Var(\sigma) \subset \mathcal{X}$. The intended meaning of the entailment $\theta \underset{\mathcal{X}}{\Rightarrow} \sigma$ is that of $\forall\mathcal{X}(\theta \to \sigma)$, where $\forall\mathcal{X}$ is the string of universal quantifiers $\forall x$, with x ranging over all $x \in \mathcal{X}$ in an appropriate order. The formal system consists of *rules of inference*, in any instance of which zero, one or two entailments, the *premisses*, yield another, the *conclusion.* (The inference rules with zero premisses are *axiom schemes.*) We have systems for each of coherent, intuitionistic and classical logic. We are not going to describe the systems in full; all the less since they present no surprises in comparison to ordinary first order logic; see [M3] for the details. The most characteristic rules are the quantifier rules; the ones for the universal quantifier are as follows:

$$\frac{\theta \underset{\mathcal{X}\cup\{x\}}{\Longrightarrow} \sigma}{\theta \underset{\mathcal{X}}{\Longrightarrow} \forall x : X.\sigma} \qquad (x \notin \mathcal{X}).$$

The display contains two rules: one is the upper entailment implying the lower one, the other vice versa. The upper entailment is assumed to be well-formed; in particular, $\mathcal{X}$ and $\mathcal{X}\dot{\cup}\{x\}$ are both contexts, and $Var(\sigma)$ is a subset of $\mathcal{X}\dot{\cup}\{x\}$; also $x \notin \mathcal{X}$. It is interesting to note that these facts imply that $\forall x : X.\sigma$ is well-formed.

A *theory* T in FOLDS is a pair $T = (\mathbf{L}, \Sigma)$, where $\mathbf{L}$ is a one-way category, and Σ is a set of "axiom" entailments over $\mathbf{L}$. For an entailment α over $\mathbf{L}$, we write $T \vdash \alpha$, and say that α is *deducible from* T, if α can be obtained from the axioms in Σ by repeated use of the rules of inference. On the other hand, $T \models \alpha$ if every (ordinary, set-valued) model of T is a model of α according to the intended meaning of entailments.

The (Gödel) completeness theorems for the coherent fragment, and for classical version, of FOLDS say that $T \vdash \alpha$ and $T \models \alpha$ are equivalent, for any T and α within the respective logic. Completeness for the classical system is a consequence of that for the coherent fragment, just like for ordinary first-order logic (see [MR1]).

Besides the classical, intuitionistic and coherent versions of FOLDS, there is a fourth that is important: this is the *regular* fragment. In this, the only formulas (*regular* formulas) are the ones built of t, $\wedge$ and $\exists$. *Regular* entailments are $\theta \underset{\mathcal{X}}{\Rightarrow} \sigma$ for regular formulas θ, σ.A *regular theory* is one whose axioms are regular entailments. The regular fragment of FOLDS is the one which comes closest to the essentially algebraic fragment [Co] of first order logic, which has $\exists$ restricted to the unique existential quantifier. In FOLDS, the uniqueness condition for $\exists$ is inappropriate in most situations, because of the lack of equality. All the significant types of categorical structure are axiomatized in the regular fragment, provided we employ the appropriate similarity type. For instance, the concept of "category with binary product" can be axiomatized in FOLDS over $\mathbf{L}_{cat}$, although not in regular FOLDS. But if we add a new kind to $\mathbf{L}_{cat}$, one for product diagrams, with appropriate arrows, the concept becomes axiomatizable in regular FOLDS. Regular FOLDS is closely related to sketch-syntax of [M1].

Completeness theorems for intuitionistic logic employ "non-standard" semantics. For any category $\mathbf{S}$, the concept of $\mathbf{S}$-valued $\mathbf{L}$-structure is automatically available, in the form of functors $\mathbf{L} \to \mathbf{S}$. In order for this concept to be of value, $\mathbf{S}$ needs to have a certain amount of structure; the minimum is $\mathbf{S}$ being a regular category (see [B]). In applications, $\mathbf{S}$ is usually much more than regular. For the modeling of intuitionistic logic, $\mathbf{S}$ is taken to be a Grothendieck topos. In particular, for Kripke-style modeling, $\mathbf{S}$ is a category of the form $Set^{\mathbf{C}}$, for some small category (e.g., a poset). For modeling in H-valued sets, for H a complete Heyting algebra, $\mathbf{S}$ is the topos of H-valued sets, or equivalently, the topos of sheaves over H.

$\mathbf{S}$-valued semantics uses the interpretation of formulas in FOLDS in the category $\mathbf{S}$. This is based on the available structure of $\mathbf{S}$; if $\mathbf{S}$ is a topos, all of FOLDS can be interpreted in $\mathbf{S}$. This interpretation will be fairly clear to readers familiar with categorical semantics; see e.g. [MR1].

For the model theory of intuitionistic FOLDS, and indeed, for the metatheory of FOLDS in general, including the proof of the Gödel completeness theorems, it is good to put FOLDS into a framework of fibrations (see [Gr], [M7]). A theory in FOLDS becomes a *quantificational fibration* with a base category having finite limits (a terminal object, and pullbacks), with a class of distinguished arrows in the base category stable under pullback, with an appropriate structure in the fibers (corresponding to the propositional connectives), and with quantification *along the distinguished arrows only* in the form of adjoints to substitution-functors between the fibers. Upon the said identification of theories with certain categorical structures (fibrations), an interpretation of one theory in another, and models of theories in a category $\mathbf{S}$, become morphisms of the appropriate categorical structures (we associate with the category $\mathbf{S}$ a fibration in a standard way; see [M7]).

The said fibrational framework represents a generalization of FOLDS, with the result that the general theory can be applied to other things, notably the concept of bisimulation in theoretical computer science.

10. Equivalence in FOLDS

The main pont about FOLDS is that it comes with an appropriate notion of *structural equality*. For each similarity type $\mathbf{L}$ for FOLDS, we have the concept of **L**-*equivalence* of **L**-structures. Properties expressible in FOLDS over **L** are invariant under **L**-equivalence. There is a strong converse to the last statement, which shows that the syntax of FOLDS and **L**-equivalence are precisely matched. The issues of invariance are addressed in the next section; in this section, the meaning of **L**-equivalence is explored.

Let M, P be **L**-structures, $h : P \to M$ a natural transformation. h induces the "componentwise" map $h_{[K]} : P[K] \to M[K]$; and for any $\mathbf{x} \in P[K]$, and for $\mathbf{a} = h_{[K]}(\mathbf{x})$, the map

$$h_{\mathbf{x}} : PK(\mathbf{x}) \longrightarrow MK(\mathbf{a})$$

between the corresponding fibers. h is called *fiberwise surjective* if $h_{\mathbf{x}}$ is surjective for all $\mathbf{x} \in P[K]$.

Let M, N be **L**-structures. An **L**-*equivalence* of M and N is a triple (P, h, k) where

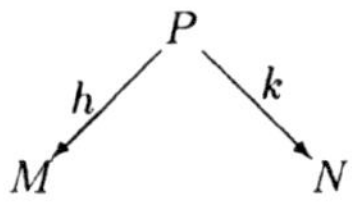

(1)

and both h and k are fiberwise surjective maps; for this, we write

$$(P, h, k) : M \simeq_{\mathbf{L}} N.$$

M and N are **L**-*equivalent*, $M \simeq_{\mathbf{L}} N$, if there exists an **L**-equivalence between them.

It is important, e.g., for FOLDS in an "intuitionistic (constructive) world", that the concept of **L**-equivalence has a natural extension to **S**-valued structures, for any regular category **S**. When $M : \mathbf{L} \to \mathbf{S}$, and $K \in \mathbf{L}$, there is a natural definition of $M[K]$ the object of K-contexts in M, as a certain finite limit in **L**. Then $h : P \to M$ a natural transformation of **S**-valued **L**-structures, is said to be fiberwise surjective if in the induced commutative diagram

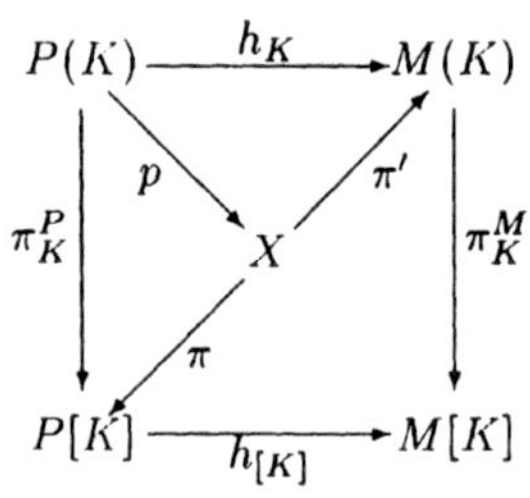

in which X is a pullback with projections π and π', the arrow p is *surjective*, that is, a regular epimorphism (one says for this that the outside rectangle is a *quasi-pullback*).

L-equivalence is an equivalence relation. Reflexivity and symmetry are clear. For transitivity, we use composition of **L**-equivalences as spans; that is, given the equivalences $(Q, h, k) : M \simeq_{\mathbf{L}} N$, $(R, \ell, m) : N \simeq_{\mathbf{L}} P$, we construct the pullback S with projections as in

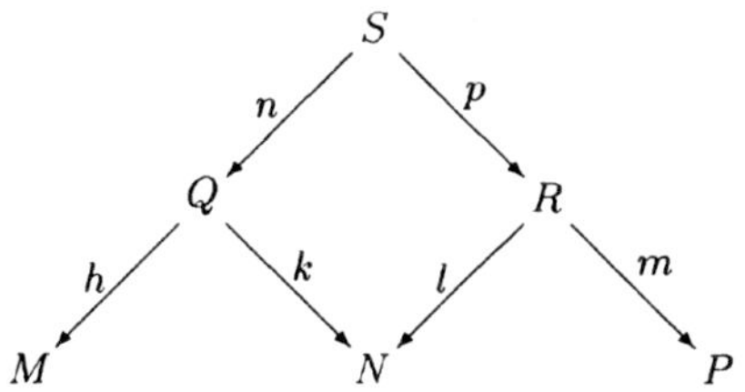

the desired composite equivalence is $(S, hn, mp) : M \simeq_{\mathbf{L}} P$.

To understand **L**-equivalence, we should note that, for *Set*-valued models at least, a simplified concept, that of a "relational **L**-equivalence", suffices. The **L**-equivalence $(P, h, k) : M \simeq_{\mathbf{L}} N$ is *relational* if the arrows h and k are jointly monic. Equivalently, a relational equivalence is given by a family $\rho = <\rho_K>_{K \in \mathbf{L}}$ of relations $\rho_K \subset MK \times NK$ such that, with

$$\mathbf{a}\rho_{[K]}\mathbf{b} \underset{\text{def}}{\Longleftrightarrow} \forall p \in K|\mathbf{L} \cdot a_p \rho_{K_p} b_p,$$

for $\mathbf{a} = <a_p>_{p \in K|\mathbf{L}} \in M[K]$, $\mathbf{b} = <b_p>_{p \in K|\mathbf{L}} \in N[K]$, the following hold:

(i) For any $p : K \to K_p$, $a \in MK$, $b \in NK$

$$a\rho_K b \Rightarrow (Mp)(a)\rho_{K_p}(Np)(b).$$

(ii) For any $K \in \mathbf{K}$, $\mathbf{a} \in M[K]$, $\mathbf{b} \in N[K]$,

$$\mathbf{a}\rho_{[K]}\mathbf{b} \;\&\; a \in MK(\mathbf{a}) \Rightarrow \exists b \in NK(\mathbf{b}).a\rho_K b.$$

$$\mathbf{a}\rho_{[K]}\mathbf{b} \;\&\; b \in NK(\mathbf{b}) \Rightarrow \exists a \in MK(\mathbf{a}).a\rho_K b.$$

Let us take the example of the similarity type L of one-sorted relational structures (essentially, the only one that standard model theory deals with). L can be construed as a one-way category **L** with two levels, on the bottom level a single object, the "sort", and on level-1 the relation symbols, with a symbol for the equality relation added. From each relation symbol of arity k, there are k arrows to the sort. Clearly, each L-structure can be construed as an **L**-structure, although the correspondence is not bijective; the L-structures correspond to **L**-structures in which the level-1 fibers are either empty or singletons, and in which the equality is real equality. Now, the point is that, for **L**-structures coming from L-structures, relational **L**-equivalences are the same as isomorphisms in the usual sense.

In the clause (ii), there is a formal resemblance to the "back-and-forth" systems for equivalence in the logic $L_{\infty\omega}$; see, e.g. [Ba]. As the previous paragraph shows, when we look at the simplest case, the notion of relational equivalence is not at all $L_{\infty\omega}$-equivalence; it is ordinary isomorphism. Nevertheless, the resemblance does lead to a unified view of back and forth properties and FOLDS equivalence, and to an interesting group of facts about infinitary first order logic with dependent sorts; see [M3].

Another concept one is reminded of by the above is the notion of bisimulation in theoretical computer science; see [MPW]. This connection is genuine. The more general version of the basic theory, with the quantificational fibrations at the basis, yields as special cases of FOLDS equivalence some of the concepts of bisimulation treated in the literature. In this context, the language of FOLDS is seen to generalize the Henessy-Milner language associated to bisimulation.

If two *Set*-valued **L**-structures are **L**-equivalent, then there is a relational equivalence between them. The proof of this uses Choice, which indicates that the concept of relational equivalence is not optimal. Indeed, it does not work when we pass to **S**-structures for more general **S**. In fact, it does not work even in the *Set*-valued case when we want to keep track of individual elements in the structures under the equivalence.

Let us consider the case $\mathbf{L} = \mathbf{L}_{cat}$. At the end of section 8, we mentioned the theory $T_{cat} = (\mathbf{L}_{cat}, \Sigma_{cat})$ whose models are *almost* the same as categories. The fact is that every model M of T_{cat} is **L**-equivalent to $M[\mathbf{C}]$ for a category **C**. The construction of **C** involves cutting down every non-empty level-2 fiber in M to a singleton, and taking appropriate equivalence classes of equivalence relations induced by E. Indeed, the underlying facts here apply to general FOLDS similarity types. Since **L**- equivalence is treated as the structural equality for **L**-structures, it is fair to say that T_{cat} axiomatizes the notion of category.

From now on, we will write **C** for $M[\mathbf{C}]$ as well; the context will clarify any potential ambiguity.

We have that, for categories **C** and **D**, **L**-equivalence coincides with ordinary equivalence of categories:

$$\mathbf{C} \simeq \mathbf{D} \Leftrightarrow \mathbf{C} \simeq_{\mathbf{L}_{cat}} \mathbf{D}. \tag{2}$$

The construction from an ordinary equivalence to an **L**-equivalence is a canonical one; it uses the anafunctor-saturation (see section 5) of the functors involved. In fact, it suffices to have a single equivalence functor $F : \mathbf{C} \to \mathbf{D}$, with the properties that it is fully faithful (induces bijections $\mathbf{C}(C, C') \to \mathbf{D}(FC, FC')$), and surjective on objects up to isomorphism; the saturation of F will yield an **L**-equivalence. Conversely, an **L**-equivalence gives rise to a categorical equivalence by the use of the axiom of choice.

The case of more complicated categorical concepts is more interesting; a passage to the ana-concepts is involved.

The simplest such structures are the "free-living" functors; two categories connected by a functor. The most well-known type of such a structure is fibration. In [MR2], G.E. Reyes and the author have invesitgated "free-living" functors with properties derived from universal properties as a basis for categorical formulations of versions of intuitionistic and modal logics; these "modal categories" (actually, functors) are not fibrations in general.

There is a concept of equivalence of functors as follow:
$F : \mathbf{X} \to \mathbf{A}$ and $G : \mathbf{Y} \to \mathbf{B}$ are *equivalent*, in notation $F \simeq G$, if there are equivalence-functors E_0, E_1 and an isomorphism e as in the diagram

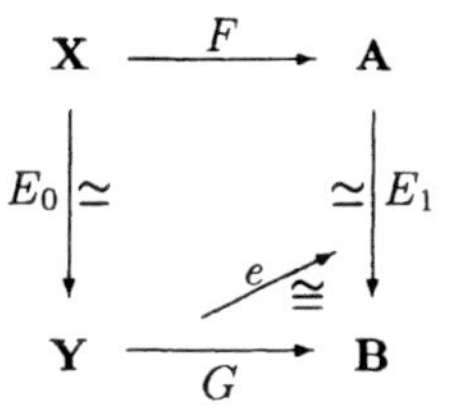

$$e \colon GE_0 \xrightarrow{\cong} E_1F$$

This notion is the combination of (the least equivalence relation containing) isomorphism of parallel functors, and equivalence of categories. In several places in the literature, it is recognized that this is the "right" notion of equivalence for functors. An early example of this is Kelly's paper [K] quoted before. When one has a theorem to the effect that a free-living functor of a certain kind can be presented, or constructed, in a certain way, then one expects that the result will be only *equivalent* to the original. This is the case when the theorem needs non-canonical choices in its proof. An example for this is the Uniform Sketchability Theorem of [MP].

The properties appearing in [MR2] of functors are all invariant under equivalence. The concept of fibration is not invariant under equivalence; however, restricted to fibrations, the additional properties of fibrations usually considered are.

The FOLDS similarity type to be considered in connection with functors is the following one-way category $\mathbf{L}_{fun}$:

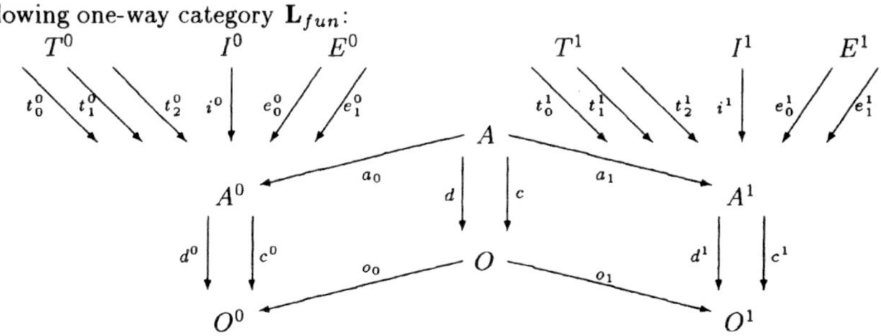

Here, we have two copies of $\mathbf{L}_{cat}$; the rest is subject to the following equalities:

$$d^0 a_0 = o_0 d \ , \ c^0 a_0 = o_0 c \ , \ d^1 a_1 = o_1 d \ , \ c^1 a_1 = o_1 c \ .$$

Every free-living functor is, in a natural way, a functor $\mathbf{L}_{fun} \to Set$. For instance, O and A are interpreted as the graphs of the object-function and the arrow function of the given functor.

Now, we do not have the fact that the equivalences in the usual categorical sense and the FOLDS sense are the same as in the case of categories. However, every functor $F : \mathbf{X} \to \mathbf{A}$ gives rise to the saturated anafunctor $F^{\#} : \mathbf{X} \xrightarrow{a} \mathbf{A}$ mentioned at the end of section 5. $F^{\#}$ is also an $\mathbf{L}_{fun}$-structure. $F^{\#}$ interprets the two copies of $\mathbf{L}_{cat}$ in $\mathbf{L}_{fun}$ just as F does.

$$F^{\#}(A) \underset{\text{def}}{=} \{(X, A, \mu) : X \in Ob(\mathbf{X}),\ A \in Ob(\mathbf{A})\ , \mu : F(X) \xrightarrow{\cong} A\} \ ,$$

with $(X, A, \mu) \overset{F^{\#}o_0}{\mapsto} X$, $(X, A, \mu) \overset{F^{\#}o_1}{\mapsto} A$; and

$$F^{\#}(A) \underset{\text{def}}{=} \left\{ ((X, A, \mu),\ (Y, B, \nu),\ f, g) :\ f : X \to Y,\ g : A \to B,\ \begin{array}{ccc} FX & \xrightarrow{Ff} & FY \\ \mu \downarrow & o & \downarrow \nu \\ A & \underset{g}{\longrightarrow} & B \end{array} \right\}$$

In other words, $|F^{\#}|(X, A) = Iso(FX, A) = \{\mu \in \hom(FX, A) : \mu \text{ is iso}\}$, and, with the notation used for $F^{\#}(A)$, $F_{\mu,\nu}(f) = g$ determined by the commutative square. The displayed item in $F^{\#}(A)$ is mapped to f by $F^{\#}(a_0)$, and to g by $F^{\#}(a_1)$. One can prove (using the Axiom of Choice) that the saturated anafunctors are, up to isomorphism as $\mathbf{L}$-structures, the same as the $F^{\#}$ for functors F.

Now, the main fact is that with $\mathbf{L} = \mathbf{L}_{fun}$, for any functors F and G,

$$F \simeq G \Leftrightarrow F^{\#} \simeq_{\mathbf{L}_{fun}} G^{\#}. \tag{3}$$

We may look at this fact as saying that the passage to the saturation "straightens out" the somewhat *ad hoc* notion of equivalence for functors into something that follows a general logical pattern.

There is a similar situation in the case of the notion of bicategory. We have the notion of *biequivalence* of bicategories, the recognized "correct" notion of structural equality for bicategories; we write $\mathcal{A} \simeq \mathcal{B}$ for the biequivalence of bicategories $\mathcal{A}, \mathcal{B}$. We have particular finite one-way category $\mathbf{L}_{bicat}$ (having four levels, rather than three as in the previous two examples) so that a bicategory may be naturally considered as a $\mathbf{L}_{bicat}$-structure. For every bicategory $\mathcal{A}$, we have a particular anabicategory $\mathcal{A}^{\#}$, the *saturation* of $\mathcal{A} \cdot \mathcal{A}^{\#}$ is obtained by saturating the composition functors, and the identities (as functors from 1, the terminal category), and by other necessary adjustments. $\mathcal{A}^{\#}$ is also an $\mathbf{L}_{bicat}$-structure. The basic fact it:

$$\mathcal{A} \simeq \mathcal{B} \Leftrightarrow \mathcal{A}^{\#} \simeq_{\mathbf{L}_{bicat}} \mathcal{B}^{\#}. \tag{4}$$

The verification of the facts (2),(3),(4), though by no means difficult, consist in increasingly complex calculations. I believe that they are all special cases of a (future) general result on categorical structures of arbitrary dimension. In this, each structure in question is presented in the form of an **L**-structure, for a specific one-way category **L**. The categorical consideration provides one notion of equivalence, based on the morphisms, natural transformations, etc., in arbitrary dimensions connected to the original concept. The other notion of equivalence is that of **L**- equivalence. The would-be theorem is that the two coincide. Here, of course we mean that the two *relations* of equivalence coincide; the fuller notions, the two concepts of equivalence-carriers will have a more complex relation to each other.

I now have, for any fixed $n \in N$, a presentation of the Baez/Dolan n-categories as $\mathbf{L}_n$-structures for a particular (infinite) one-way category $\mathbf{L}_n$. Baez and Dolan themselves provided a notion of equivalence of n-categories. I am confident that it will turn out that the Baez/Dolan n-equivalence, and $\mathbf{L}_n$-equivalence coincide, although I still have to do the calculations for this.

Using **L**-equivalence for $\mathbf{L} = \mathbf{L}_n$ and also for another, unspecified, FOLDS signature **L**, I can give a precise formulation to the conjecture that the Baez-Dolan notion of n-category is "essentially finite", in the sense that by one can change the signature $\mathbf{L}_n$ to a finite FOLDS signature, with the concept of n- category remaining the same "up to equivalence".

11. Invariance under L-equivalence

Let **L** be a one-way category, M, N, **L**-structures, $\mathcal{X}$ a context of variables and $\mathbf{a} \in M[\mathcal{X}]$, $\mathbf{b} \in N[\mathcal{X}]$ valuations. We write $(M, \mathbf{a}) \simeq_{\mathbf{L}} (N, \mathbf{b})$ if there exist $(P, h, k) : M \simeq_{\mathbf{L}} N$ and $\mathbf{x} \in P[\mathcal{X}]$ such that $\mathbf{x} \overset{h}{\mapsto} \mathbf{a}$, $\mathbf{x} \overset{k}{\mapsto} \mathbf{b}$. Each FOLDS-formula θ is preserved under **L** equivalence;

$$\mathbf{a} \in M[\mathcal{X} : \theta] \ , \quad (M, \mathbf{a}) \simeq_{\mathbf{L}} (N, \mathbf{b}) \Rightarrow \mathbf{b} \in N[\mathcal{X} : \theta] \ .$$

($Var(\theta) \subset \mathcal{X}$). The proof of this is a straightforward verification, by induction on the complexity of θ.

The generalization of this fact for intuitionistic logic, and for interpretations in a category suitable for intuitionistic logic ("Heyting category": see [MR2]) is, somewhat surprisingly, also true (the clause for $\exists$ is the natural one to show directly; in classical logic, the one for $\forall$ is a consequence since $\forall$ is expressible in terms of $\exists$; in the intuitionistic case, $\forall$ is not expressible in terms of $\exists$).

We have a converse of this fact. Let φ be a formula of FOL over $\mathbf{L}$ (see beginning of section 9); assume that the variables of φ are among the ones in $\mathcal{X}$, a context of FOLDS over $\mathbf{L}$ (a variable of sort $X = K(\cdots)$ in FOLDS is understood, when in FOL over $\mathbf{L}$, to be a variable of sort K). Let S be any theory in FOL over $\mathbf{L}$. Then, if φ is invariant under $\mathbf{L}$-equivalence for models of S:

$$M, N \models S, \quad M \models \varphi[\mathbf{a}/\mathcal{X}], \ (M, \mathbf{a}) \simeq_{\mathbf{L}} (N, \mathbf{b}) \Rightarrow N \models \varphi[\mathbf{b}/\mathcal{X}],$$

then there is a FOLDS formula θ, $Var(\theta) \subset \mathcal{X}$, such that φ and θ are equivalent under S:

$$M \models S, \ \ \mathbf{a} \in MI[\mathcal{X}] \Rightarrow M \models \varphi[\mathbf{a}/\mathcal{X}] \leftrightarrow M \models \theta[\mathbf{a}/\mathcal{X}].$$

This result is a special case of the Invariance Theorem stated below. Applied to $\mathbf{L}_{cat}$, it together with the relation 10.(2), implies P.Freyd's theorem [F] characterizing first order formulas of category theory that are invariant under equivalence of categories.

To obtain similar results for more complex categorical structures, a stronger form of the Invariance Theorem is needed. To state it, we need the concept of an *interpretation* of a theory $T = (\mathbf{L}, \Sigma)$ in FOLDS in an ordinary first order theory S.

For the sake of definiteness, we restrict ourselves to classical FOLDS (the case of intuitionistic logic is also relevant). There is a "conceptual category" $[S]$ associated with S, made up of formulas of S. See [MR1]. An interpretation of T in S is a functor $\mathbf{L} \to [S]$, that is, on $[S]$-valued structure which is a model of T, one that makes all axioms in Σ true.

There are two things lacking for this description to be an explicit definition. One is the $\mathbf{S}$-valued semantics, already mentioned at the end of section 8; the other is the definition of the particular $\mathbf{S} = [S]$.

Another way of giving the definition would proceed by associating an appropriate quantificational fibration (mentioned in section 9) with T, and one with S too, and defining interpretation as the direct notion of structure-preserving morphism from one to the other.

However, the concept of an interpretation $I : T \to S$ of T in S is, we claim, intuitively clear. It is given by imitating the definition of "Set-valued model of T", and replacing in that definition sets and functions by S-definable sets and S-definable functions between the latter. Each object $K \in \mathbf{L}$ is mapped by I to an abstract definable set $[\mathbf{x} : \varphi(\mathbf{x})]$ (read "the set of $\mathbf{x}$ such that φ"), where φ is a formula of S, and $\mathbf{x}$ is a finite tuple of variables containing all the free variables of φ. If K is mapped to $[\mathbf{x} : \varphi(\mathbf{x})]$, K' to $[\mathbf{y} : \psi(\mathbf{y})]$, $\mathbf{x}$ and $\mathbf{y}$ are disjoint tuples, an arrow $f : K \to K'$ is mapped to an abstract definable function $<\mathbf{x} \mapsto \mathbf{y} : \theta(\mathbf{x}, \mathbf{y})>$ (read " the function that maps $\mathbf{x}$ to $\mathbf{y}$ determined by the condition $\theta(\mathbf{x}, \mathbf{y})$"); of course, it has to be provable in the theory S that the said prescription is indeed a function defined on $[\mathbf{x} : \varphi(\mathbf{x})]$, with range contained in $[\mathbf{y} : \psi(\mathbf{y})]$. Further, the composition law of the category $\mathbf{L}$ has to be respected: every time $gf = h$ in $\mathbf{L}$, it has to be provable in S that the "composite" of the definable functions corresponding to f and g is equal, in the sense of provability in S, to that corresponding to h. Finally, under the mapping described, formulas over $\mathbf{L}$ acquire meanings that are formulas over L; it is required that, in the sense of provability, the axioms of T become true in S.

Given any interpretation $I : T \to S$ as described, there is associated, with any model M of S, a model MI of T. With I construed as a functor $I : \mathbf{L} \to [S]$, and M as a functor $[S] \to Set$, MI is the composite $M \circ I : \mathbf{L} \to Set$.

The relevant example for the above concepts is the case of the FOLDS theory $T = T_{sanafun} = (\mathbf{L}_{fun}, \Sigma_{sanafun})$ of (free-living) saturated anafunctors, and the ordinary theory $S = S_{fun}$, over the same $L = \mathbf{L}_{fun}$, of all (free-living) functors. The construction $F \mapsto F^{\#}$ of the saturation of F, described in section 10, is the same as the passage $F \mapsto FI$ for a suitable, and obvious, interpretation

$$I : T_{sanafun} \to S_{fun}.$$

For simplicity, we are going to talk about invariance of sentences (formulas without free variables) only. Note that an interpretation $I : T \to S$ maps any FOLDS-sentence θ over $\mathbf{L}$ to a sentence $I(\theta)$ such that, for any $M \models S$, $M \models I(\theta)$ iff $MI \models \theta$. Thus, for instance, any sentence θ of FOLDS over $\mathbf{L}_{cat}$ is translated by (1) into a sentence $\theta^* = I(\theta)$ such that, for any functo F,

$$F \models \theta^* \Leftrightarrow F^{\#} \models \theta.$$

Let us say that an S-sentence φ is *invariant* under $\mathbf{L}$-equivalence (relative to I) if the implication

$$M, N \models S,\ M \models \varphi,\ MI \simeq_{\mathbf{L}} NI \Rightarrow N \models \varphi$$

holds universally.

Invariance Theorem Given an interpretation $T = (\mathbf{L}, \Sigma) \to S$ of a FOLDS theory in a FOL theory, an S-sentence φ is invariant under $\mathbf{L}$- equivalence if and only if it is expressible in FOLDS over $\mathbf{L}$: there is a FOLDS sentence θ over $\mathbf{L}$ such that $S \models \varphi \leftrightarrow I(\theta)$.

The proof of the Invariance Theorem uses standard tools of model theory [CK].

As an application, we can, e.g., characterize the first-order sentences that may be asserted of a structure of the form $\mathbf{X} \xrightarrow{F} \mathbf{A}$, consisting of two categories and a functor. Such a sentence is invariant under the equivalence explained in the last section iff it is equivalent, for structures of the given kind, to θ^*, for some FOLDS-sentence θ over $\mathbf{L}_{fun}$, where θ^* is the translate of θ for which $F \models \theta^* \Leftrightarrow F^{\#}\theta$.

Of course, there is a similar characterization theorem for sentences invariant under biequivalence of bicategories.

The full form of the Invariance Theorem refers to formulas with free variables rather than sentences.

The Invariance Theorem is, essentially, a preservation theorem in the sense of [CK], and as such, it can be construed as a purely syntactic result. Regarded purely syntactically, it then has a natural version for intuitionistic logic. The latter is easily translated into an equivalent semantical version, with reference to Kripke's semantics for intuitionistic logic. [M3] establishes the intuitionistic form the Invariance Theorem.

References

[BD1] J.C. Baez and J. Dolan, Higher dimensional algebra and topological quantum field theory. *Journal of Mathematical Physics* **36**(1995), no. 11, 6073-6105.

[BD2] J.C. Baez and J. Dolan, n-Categories - Sketch of a Definition. Letter to Ross Street, Nov. 29, 1995; corrected version as of Dec. 3, 1995.

[B] M. Barr, Respresentations of categories. *J. Pure and Applied Algebra* **41** (1986), 113-137.

[Ba] J. Barwise, Back and forth thru infinitary logic. In: *Studies in Model Theory*, ed. M.D. Morley. MAA Studies in Mathematics, Vol. 8. The Mathematical Association of America, 1973; pp. 5-34.

[Be] J. Benabou, Introduction to bicategories. In: Lecture Notes in Math., no. **47**, Springer Verlag, 1967; pp. 1-77.

[Ben1] P. Benaceraff, What numbers could not be. *Philosophical Review* **74** (1965), 47-73.

[Ben2] P. Benaceraff, Mathematical truth. *The Journal of Philosophy* **LXX**, no. 19, 1973, 661-675.

[Bl] G. Blanc, Equivalence naturelle et formules logiques en theorie des categories. *Archiv. math. Logik* **19** (1978), 131-137.

[C] J. Cartmell, Generalised algebraic theories and contextual categories. *Annals of Pure and Applied Logic* **32** (1986), 209-243.

[CK] C.C. Chang and H.J. Keisler, *Model Theory*. Third edition, North-Holland, 1990.

[Co] M. Coste, Localisation, spectra and sheaf representation. In: *Applications of sheaves*, Lecture Notes in Math., no. **753**, Springer Verlag, 1979; pp. 212-238.

[D] R. Diaconescu, Axiom of choice and complementation. *Proc. Amer. Math. Soc.* **51** (1975), 176-178.

[F] P. Freyd, Properties invariant within equivalence types of categories. In: *Algebra, Topology and Category Theories*, ed. A. Heller and M. Tierney, Academic Press, New York, 1976; pp. 55-61.

[FS] P.J. Freyd and A. Scedrov, *Categories, Allegories*. North-Holland, 1990.

GPS] R. Gordon, A.J. Power and R. Street, Coherence for tricategories. *Memoirs Amer. Math. Soc.* **117**(1995), no. 558.

[Gr] A. Grothendieck, Expose VI, *Revetements Etales et Groupe Fondamental*; SGA1. Lecture Notes in Mathematics 224, Springer-Verlag, 1971; pp. 145-194.

[H] M. Hallett, *Cantorian Set Theory and Limitation of Size*. Oxford University Press, 1984.

[J] P.T. Johnstone, *Topos Theory*. Academic Press, 1977.

[K1] G.M. Kelly, Complete functors in homology I. Chain maps and endomorphisms. *Proc. Cambridge Phil. Soc.* **60** (1964), 721-735.

[K2] G.M. Kelly, *Basic Concepts of Enriched Category Theory*. London Math. Soc. Lecture Notes Series **64**, Cambridge University Press, 1982.

[LS] J. Lambek and P. Scott, *Introduction to Higher Order Categorical Logic*. Cambridge University Press, 1986.

[ML] S. Mac Lane, *Categories for the Working Mathematician*. Springer-Verlag, 1971.

[MM] S. Mac Lane and I. Moerdijk, *Sheaves in Geometry and Logic. A First Introduction to Topos Theory*. 2nd corrected printing. Springer-Verlag, 1994.

[M1] M. Makkai, Generalized sketches as a framework for completeness theorems. To appear in three parts in *J. Pure and Applied Algebra*.

[M2] M. Makkai, Avoiding the axiom of choice in general category theory. *J. Pure and Applied Algebra* 108(1996), 109-173.

[M3] M. Makkai, *First-Order Logic With Dependent Sorts with Applications to Category Theory.* Research monograph, to appear in the series **Lecture Notes in Logic** of Springer-Verlag.

[M4] M. Makkai, On the Baez/Dolan concept of n-category. In preparation.

[M5] M. Makkai, What are abstract objects? In preparation.

[M6] M. Makkai, *An Algebraic Look at Propositional Logic.* AILA Preprint no. **18** -iuglio/dicembre 1994. Associazione Italiana de Logica e sue Applicazioni.

[M7] M. Makkai, The fibrational formulation of intuitionistic predicate logic I: Completeness according to Godel, Kripke and Lauchli. *Notre Dame Journal of Formal Logic* **34** (1993), 334-377 (Part 1) and 471-498 (Part 2).

[MR1] M. Makkai and G.E. Reyes, *First Order Categorical Logic.* Lecture Notes in Mathematics, no **611**, Springer-Verlag, 1977.

[MR2] M. Makkai and G.E. Reyes, Completeness theorems for intuitionistic and modal logic in a categorical setting. *Annals of Pure and Applied Logic* **72** (1995), 25-101.

[MP] M. Makkai and Robert Pare, *Accessible Categories.* Contemporary Mathematics, V. **104**. American Mathematical Society, 1989.

[M-L] P. Martin-Löf, An Intuitionistic Theory of Types. In: *Proc. Bristol Logic Colloquium*, North-Holland, 1973.

[MPW] R. Milner, J. Parrow and D. Walker. A calculus of mobile process II. Information and Computation 100 (1) 41-77.

[S] R. Street, Fibrations in bicategories. *Cah. de Top. et Geom. Diff.* **21**(1980), 111-160.

Strongly Minimal Sets and Geometry

David Marker *

Department of Mathematics
University of Illinois at Chicago
851 South Morgan Street, Chicago, IL 60607-7045, USA
marker@math.uic.edu

These lectures will provide a brief introduction to the model theory of strongly minimal sets. The first two sections will develop the basic combinatorial geometry of strongly minimal sets. In sections three and four we will show how the pregeometry of the strongly minimal set detects the presence of ambient algebraic structure. Finally we will show how these ideas come together in Hrushovski's proof ([H1]) of the Mordell-Lang conjecture for function fields.

Strongly minimal sets are just the beginning of the story in *geometric model theory*. My hope is that by concentrating on strongly minimal sets I can give a reasonably self-contained introduction to this important and beautiful subject. We assume only that the reader is familiar with the treatment of ω-stable theories given in [CK] or [S]. A full development of geometric model theory is given in [P1].

My own appreciation of geometric model theory was slow in coming. I owe a great deal to Anand Pillay, Elisabeth Bouscaren and John Baldwin for the numerous conversations it took to enlighten me.

1 Strongly minimal sets and pregeometries

Let $\mathcal{L}$ be a first order language and let M be an $\mathcal{L}$-structure. Recall that a formula $\phi(\bar{v})$ with parameters from M is said to be *strongly minimal* if for any elementary extension N of M and any formula $\psi(\bar{v})$ with parameters from N exactly one of $\{\bar{a} \in N : N \models \phi(\bar{a}) \wedge \psi(\bar{a})\}$ and $\{\bar{a} \in N : N \models \phi(\bar{a}) \wedge \neg\psi(\bar{a})\}$ is infinite. We say that a subset D of M^n is strongly minimal if it is defined by a strongly minimal formula. We will often consider D as a structure in its own right by taking all of the structure induced from definable subsets in M^n (by "definable" I will always mean "definable with parameters" unless I specify otherwise).

If $A \subset M$ and $b \in M$ we say that b is *algebraic* over A if there is an $\mathcal{L}$-formula $\phi(v)$ with parameters from A such that $M \models \phi(b)$ and $\{a \in M : M \models \phi(a)\}$ is finite. The set of all elements of A algebraic over A is called the *algebraic closure* of A and denoted $\mathrm{acl}(A)$. The algebraic closure relation on a strongly minimal set determines a pregeometry.

* Partially supported by NSF grant DMS-9306159 and an American Mathematics Society Centennial Fellowship.

Definition Let X be a set and let $\mathrm{cl} : \mathcal{P}(X) \to \mathcal{P}(X)$ be an operator on the power set of X. We say that (X, cl) is a *pregeometry* if the following conditions are satisfied.

i) (monotonicity) If $A \subseteq X$, then $A \subseteq \mathrm{cl}(A)$.

ii) (transitivity) If $A \subseteq X$, then $\mathrm{cl}(\mathrm{cl}(A)) = \mathrm{cl}(A)$.

iii) (exchange) If $A \subseteq X$, $a, b \in X$ and $a \in \mathrm{cl}(A \cup \{b\})$, then $a \in \mathrm{cl}(A)$ or $b \in \mathrm{cl}(A \cup \{a\})$.

iv) (finite nature of closure) If $A \subseteq X$ and $a \in \mathrm{cl}(A)$, then there is a finite $A_0 \subseteq A$ such that $a \in \mathrm{cl}(A_0)$.

In any structure M conditions i), ii) and iv) are true of algebraic closure. If D is a strongly minimal subset of M^n we consider the operator $X \mapsto \mathrm{acl}(X) \cap D$. We abuse notation by also calling this operator acl. The important observation of Baldwin and Lachlan [BL] is that exchange holds in any strongly minimal set.

Lemma 1 *If D be a strongly minimal set, then (D, acl) is a pregeometry.*

Proof We need only verify that exchange holds. Suppose $a, b \in D$, $A \subseteq D$, $a \in \mathrm{acl}(A, b)$. For notational simplicity we assume that $A = \emptyset$.

Suppose $\phi(a, b)$ and $|\{x : \phi(x, b)\}| = n$. Let $\psi(w)$ be the formula "$|\{x : \phi(x, w)\}| = n$". If $\psi(w)$ defines a finite set, then $b \in \mathrm{acl}(\emptyset)$ and $a \in \mathrm{acl}(\emptyset)$. Thus we may assume that $\psi(w)$ defines a cofinite subset of D.

If $\{y : \phi(a, y) \wedge \psi(y)\}$ is finite we are done (as $b \in \mathrm{acl}(a)$). Thus we assume that $|D - \{y : \phi(a, y) \wedge \psi(y)\}| = l$. Let $\chi(x)$ be the formula expressing

$$|D - \{y : \phi(x, y) \wedge \psi(y)\}| = l.$$

If $\chi(x)$ defines a finite set, then $a \in \mathrm{acl}(\emptyset)$ as desired. Thus we assume $\chi(x)$ defines a cofinite set.

Choose $a_1, \ldots, a_{n+1}$ such that $\chi(a_i)$. The set $B_i = \{w \in D : \psi(w) \wedge \phi(a_i, w)\}$ is cofinite for $i = 1, \ldots, n+1$. Choose $\hat{b} \in \bigcap B_i$. Then $\phi(a_i, \hat{b})$, for each i. So $|\{x : \phi(x, \hat{b})\}| \geq n + 1$, contradicting the fact that $\psi(\hat{b})$.

If (X, cl) is a pregeometry we say that A is *independent* if $a \notin \mathrm{cl}(A \setminus \{a\})$ for all $a \in A$ and B is a *basis* for Y if $B \subseteq Y$ is independent and $Y \subseteq \mathrm{acl}(B)$. In any pregeometry any two bases for Y have the same cardinality which is called the *dimension*. If $A \subseteq X$ we also consider the *localization* $\mathrm{cl}_A(B) = \mathrm{cl}(A \cup B)$ and notions of independence over A and dimension over A. We let dim (B) and dim (B/A) denote the dimension of B and dimension of B over A respectively.

We say that a pregeometry (X, cl) is a *geometry* if $\mathrm{cl}(\emptyset) = \emptyset$ and $\mathrm{cl}(\{x\}) = \{x\}$ for any $x \in X$. If (X, cl) is a pregeometry, then there is a natural associated geometry. Let $X_0 = X \setminus \mathrm{cl}(\emptyset)$. Consider the equivalence relation $\sim$ on X given by $a \sim b$ if and only if $\mathrm{cl}(\{a\}) = \mathrm{cl}(\{b\})$. By exchange, $\sim$ is an equivalence relation. Let $\hat{X}$ be $X_0/\sim$. Define $\hat{\mathrm{cl}}$ on $\hat{X}$, by $\hat{\mathrm{cl}}(A/\sim) = \{b/\sim : b \in \mathrm{cl}(A)\}$. It is easy to check that $(\hat{X}, \hat{\mathrm{cl}})$ is a geometry.

We distinguish some properties of pregeometries that will play an important role.

Definition Let (X, cl) be a pregeometry.

i) We say that (X, cl) is *trivial* if $\mathrm{cl}(A) = \bigcup_{a \in X} \mathrm{cl}(\{a\})$ for any $A \subseteq X$.

ii) We say that (X, cl) is *modular* if for any finite dimensional closed $A, B \subseteq X$

$$\dim\,(\mathrm{A} \cup \mathrm{B}) = \dim\,\mathrm{A} + \dim\,\mathrm{B} - \dim\,(\mathrm{A} \cap \mathrm{B}).$$

iii) We say that (X, cl) is *locally modular* if (X, cl_a) is modular for some $a \in X$.

We give several examples for strongly minimal sets.

1) Let D be a set with no structure. Then for all $a \in D$, $\mathrm{acl}(a) = \{a\}$ and $\mathrm{acl}(\emptyset) = \emptyset$. Thus (D, acl) is a trivial geometry.

2) Let $D \models Th(\mathbf{Z}, s)$, where $s(x) = x + 1$. Then $\mathrm{acl}(\emptyset) = \emptyset$, $\mathrm{acl}(a) = \{s^n(a) : n \in \mathbf{Z}\}$ and $\mathrm{acl}(A) = \{s^n(a) : a \in A, n \in \mathbf{Z}\}$. Thus (D, acl) is trivial pregeometry that is not a geometry.

3) (projective geometry) Let F be a division ring and let V be an infinite vector space over F. We view V as a structure in the language $\mathcal{L} = \{+, 0, \lambda_a : a \in F\}$ where $\lambda_a(x) = ax$. Then V is a strongly minimal set and for any set $A \subseteq V$ the algebraic closure of A is equal to smallest F-subspace spanned by A. The usual dimension theorem for intersections of linear subspaces shows that this pregeometry is modular. This is not a geometry since $\mathrm{cl}(\emptyset) = \{0\}$ and for any $a \in V \setminus \{0\}$, $\mathrm{cl}(a)$ is the line through a and 0. To form the associated geometry we take as points the lines through 0. The closure of a set of lines is the set of all lines in their linear span. Thus the associated geometry is just the projective space associated to V. If dim V = n, then the projective space has dimension $n - 1$.

4) (affine geometry) Let V and F be as above. We define a second geometry on V where the closure of a set A is the smallest affine space containing it and $\mathrm{cl}(\emptyset) = \emptyset$. (An affine space is a translate of a linear space). Here $\mathrm{cl}(\{a\}) = \{a\}$. So this is a geometry. Let $a, b, c \in V$ be non-colinear. Then dim (a, b, c, c+b−a) = 3, while dim (a, b) = dim (c, c+b−a) = 2 and $\mathrm{cl}(a, b) \cap \mathrm{cl}(c, c + b - a) = \emptyset$ as these are parallel lines. Thus the geometry is not modular. If we localize at zero, then the pregeometry is exactly example 3) so this is locally modular.

For $F = \mathbf{Q}$ we can view this as the algebraic closure geometry of a strongly minimal set by viewing V as a structure in the language $\{\tau\}$ where

$$\tau(x, y, z) = x + y - z.$$

(For arbitrary F, add function symbols for $ax + y - az$ for each $a \in F$.)

5) Let K be an algebraically closed field of infinite transcendence degree. We claim that (K, acl) is not modular. Let k be an algebraically closed subfield of

transcendence degree n. We will show that even localizing at k the geometry is not modular. Let a, b, x be algebraically independent over k. Let $y = ax + b$. Then dim $(\mathrm{k(x,y,a,b)/k}) = 3 + \mathrm{n}$ while dim $(\mathrm{k(x,y)/k}) =$ dim $(\mathrm{k(a,b)/k}) = 2$. But $\mathrm{acl}(k(x,y)) \cap \mathrm{acl}(k(a,b)) = k$ contradicting modularity. To see this suppose $d \in \mathrm{acl}(k(a,b))$ and y is algebraic over $k(d,x)$. Let $k_1 = \mathrm{acl}(k(d))$. Then there is $p(X,Y) \in k_1[X,Y]$ an irreducible polynomial such that $p(x,y) = 0$. By model completeness $p(X,Y)$ is still irreducible over $\mathrm{acl}(k(a,b))$. Thus $p(X,Y)$ is $\alpha(Y - aX - b)$ for some $\alpha \in \mathrm{acl}(k(a,b))$ which is impossible as then $\alpha \in k_1$ and $a, b \in k_1$.

Algebraically closed fields are the only known natural examples of non-locally modular strongly minimal sets. Zilber conjectured that every non-locally modular strongly minimal set is essentially an algebraically closed field. By Macintyre ([Mac]) we know that every infinite ω-stable field is algebraically closed This was refuted when Hrushovski ([H2]) showed that there are non-locally modular strongly minimal sets which do not even interpret infinite groups. Hrushovski ([H3]) later showed that there are strongly minimal sets which are proper expansions of an algebraically closed field. For example he showed that there are strongly minimal structures $(D, +, \cdot, \oplus, \otimes)$ where $(D, +, \cdot)$ is algebraically closed field of characteristic $p \geq 0$ and $(D, \oplus, \otimes)$ is an algebraically closed field of characteristic $q \neq p$.

Although Zilber's conjecture is false it provided the motivation for a great deal of important work. One related problem is still open.

Cherlin-Zilber Conjecture Suppose G is an infinite simple group of finite Morley rank. Then G interprets an algebraically closed field F and G is definably isomorphic to an algebraic group over F.

We next give a useful characterization of modularity. Let (X, cl) be a pregeometry.

Lemma 2 *The following are equivalent.*

i) (X, cl) *is modular.*

ii) *If* $A \subseteq X$ *is closed,* $b \in X$ *and* $x \in \mathrm{cl}(A, b)$, *then there is* $a \in A$ *such that* $x \in \mathrm{cl}(a, b)$.

iii) *If* $A, B \subseteq X$ *are closed and* $x \in \mathrm{cl}(A, B)$, *then there are* $a \in A$ *and* $b \in B$ *such that* $x \in \mathrm{cl}(a, b)$.

Proof

i)$\Rightarrow$ ii) By the finite nature of closure we may assume that dim A is finite. If $x \in \mathrm{cl}(b)$ we are done, so we may assume $x \notin \mathrm{cl}(b)$. By modularity

$$\dim\ (\mathrm{A, b, x}) = \dim\ \mathrm{A} + \dim\ (\mathrm{b, x}) - \dim\ (\mathrm{A} \cap \mathrm{cl(b, x)})$$

and

$$\dim\ (\mathrm{A, b, x}) = \dim\ (\mathrm{A, b}) = \dim\ \mathrm{A} + \dim\ \mathrm{b} - \dim\ (\mathrm{A} \cap \mathrm{cl(b)}).$$

Since dim (b, x) = dim (b) + 1, there is $a \in A$ such that $a \in \mathrm{cl}(b, x) \setminus \mathrm{cl}(b)$. By exchange, $x \in \mathrm{cl}(b, a)$.

ii)$\Rightarrow$ iii) We may suppose A and B are finite dimensional. We proceed by induction on dim A. If dim A is zero then iii) holds. Suppose that $A = \mathrm{cl}(A_0, a)$ where dim $\mathrm{A_0}$ = dim A − 1. Then $x \in \mathrm{cl}(A_0, B, a)$. By ii) there is $c \in \mathrm{cl}(A_0, B)$ such that $x \in \mathrm{cl}(c, a)$. By induction there is $a_0 \in A_0$ and $b \in B$ such that $c \in \mathrm{cl}(a_0, b)$. Again by ii), there is $a^* \in \mathrm{cl}(a_0, a) \subseteq A$, such that $x \in \mathrm{cl}(a^*, b)$.

iii)$\Rightarrow$ i) Suppose $A, B \subset X$ are finite dimensional and closed. We prove i) by induction on dim A. If dim A = 0, then we are done. Suppose $A = \mathrm{cl}(A_0, a)$ where dim $\mathrm{A_0}$ = dim A − 1, and we assume, by induction, that

$$\dim\,(\mathrm{A_0}, \mathrm{B}) = \dim\,\mathrm{A_0} + \dim\,\mathrm{B} - \dim\,(\mathrm{A_0} \cap \mathrm{B}).$$

First, assume that $a \in \mathrm{cl}(A_0, B)$. Then dim $(\mathrm{A_0}, \mathrm{B})$ = dim (A, B) and, since $a \notin A_0$, dim A = dim $\mathrm{A_0}$ + 1. Since $a \in \mathrm{cl}(A_0, B)$, by iii) there is $a_0 \in A$ and $b \in B$ such that $a \in \mathrm{cl}(a_0, b)$. Since $a \notin \mathrm{cl}(a_0)$, by exchange, $b \in \mathrm{cl}(a, a_0)$. Thus $b \in A$. But $b \notin A_0$, since otherwise $a \in A_0$. Therefore

$$\dim\,(\mathrm{A} \cap \mathrm{B}) = \dim\,(\mathrm{A_0} \cap \mathrm{B}) + 1$$

as desired.

Next, suppose that $a \notin \mathrm{cl}(A_0, B)$. In this case we need to show that $A \cap B = A_0 \cap B$. Suppose $b \in B$ and $b \in \mathrm{cl}(A_0, a) \setminus \mathrm{cl}(A_0)$. Then by exchange, $a \in \mathrm{cl}(A_0, b)$, a contradiction.

Remark: If V is a vector space, then it is easy to see that if $a_1, \ldots, a_n$ is a basis for A and $b_1, \ldots, b_m$ is a basis for B. Then any x in the span of $A \cup B$ is the sum of a linear combination of the a_i and a linear combination of the b_j. Thus condition iii) above holds.

Exercise: If K is an algebraically closed field, $x, a_0, \ldots, a_n$ are algebraically independent and

$$y = \sum_{i=0}^{n} a_i x^i$$

then y is not algebraic over $k(x)$ for k a subfield of $\mathrm{acl}(a_0, \ldots, a_n)$ of dimension less than $n + 1$.

We conclude this section by illustrating the close connection between Morley rank and dimension for strongly minimal sets.

Let D be a strongly minimal set and let $A \subseteq D$. There are two kinds of types in $S_1(A)$. We say that $p \in S_1(A)$ is an *algebraic* type if there is a formula $\phi(v, \bar{w})$ and $\bar{a} \in A$ such that $\phi(v, \bar{a})$ defines a finite set and $\phi(v, \bar{a}) \in p$. There will be a minimal n such that p contains a formula $\phi(v, \bar{b})$ which defines a set of size n. It is easy to see that this formula isolates p.

Let $p_1 = \{\neg\phi(v,\bar{a}) : \bar{a} \in A$, and $\phi(v,\bar{a})$ defines a finite set$\}$. By strong minimality p_1 is complete and consistent. Clearly p_1 is the unique non-algebraic one type. Clearly $\mathrm{RM}(p_1) = 1$, while $\mathrm{RM}(p) = 0$ if p is algebraic.

We inductively define the type of n-independent elements $p_n \in S_n(A)$ as follows:

$\phi(v_1,\ldots,v_{n-1},\bar{a})$ is in p_{n+1} if and only if the formula

$$|\{w : \neg\phi(v_1,\ldots,v_{n-1},w,\bar{a})\}| \leq m$$

is in p_{n-1} for some m. Since p_{n-1} is a complete type, strong minimality implies that

- either "$|\{w : \phi(v_1,\ldots,v_{n-1},w,\bar{a}\}| \leq m$"
- or "$|\{w : \neg\phi(v_1,\ldots,v_{n-1},w,\bar{a}\}| \leq m$"

is in p_{n-1} for some m.

Let $q \in S_n(A)$. We let dim q be the maximum m such that there are $i_1,\ldots,i_m$ such that q restricted to $v_{i_1},\ldots,v_{i_m}$ is p_m. We reorder the variables so that $i_j = j$. If $\bar{b}$ is a realization of q then $b_1,\ldots,b_m$ are independent over A and $b_{m+1},\ldots,b_n \in \mathrm{acl}(A,b_1,\ldots,b_m)$.

We argue by induction that $\mathrm{RM(p_n)} = \mathrm{n}$. We may assume that D is saturated and A is infinite. Suppose $q \in S_n(A)$ and $q \neq p_n$. Suppose $\bar{b} = (b_1,\ldots,b_n)$ is a realization of q, (reordering the indicies if necessary) $b_1,\ldots,b_m$ are algebraically independent over A and $b_{m+1},\ldots,b_n \in \mathrm{acl}(A,b_1,\ldots,b_m)$. Then $b_1,\ldots,b_m$ realizes p_m and, by induction, $\mathrm{RM(p_m)} = \mathrm{m}$. Thus $\mathrm{RM(q)} = \mathrm{m}$. It follows that $\mathrm{RM(p_n)} \leq \mathrm{n}$.

If $\bar{b} = (b_1,\ldots,b_n)$ is a realization of p_n, then $b_1,\ldots,b_{n-1}$ are independent over A, so $\mathrm{RM(p_n)} \geq \mathrm{n} - 1$. Suppose $\mathrm{RM(p_n)} = \mathrm{n} - 1$ and $\phi(v_1,\ldots,v_n)$ is a formula (with parameters from A) that isolates p_n among types of Morley rank $n-1$. Since $(b_1,\ldots,b_{n-1})$ realizes p_{n-1} and b_n is not algebraic over $b_1,\ldots,b_{n-1}$, the formula

$$\exists^{>N} x\ \phi(v_1,\ldots,v_{n-1},x)$$

is in p_n for every $N > 0$. By strong minimality, there is an N such that p_n contains a formula asserting

$$|\{x : \neg\phi(v_1,\ldots,v_{n-1},x)\}| = N.$$

Since A is infinite, we can find $a \in A$, such that $\phi(b_1,\ldots,b_{n-1},a)$. But then $\mathrm{tp}(b_1,\ldots,b_{n-1},a/A)$ has rank $n-1$, is not equal to p_{n-1} and contains $\phi(v_1,\ldots,v_n)$, a contradiction.

It is easy to easy to see that $\mathrm{RM(E^n)} \geq \mathrm{RM(E)n}$ for any definable set E. Thus $\mathrm{RM(D^n)} \geq \mathrm{n}$. If $q \neq p_n$, then $\mathrm{RM(q)} < \mathrm{n}$. Since there must be some type of rank at least n, that type must be p_n. Since p is isolated once we discard types of rank less than n, $\mathrm{RM(p_n)} = \mathrm{n}$. Thus $\mathrm{RM(D^n)} = \mathrm{n}$.

Thus for any $p \in S_n(A)$, $\mathrm{RM(p)} = \dim \mathrm{p}$. Also $\mathrm{RM(\bar{a}/A)} = \dim (\bar{a}/A)$. From this, and the fact that the corresponding equation is clearly true for dimension, we see that in strongly minimal sets we have the following version of the Lascar equality

$$\mathrm{RM}(\bar{a}, \bar{b}/A) = \mathrm{RM}(\bar{a}/A) + \mathrm{RM}(\bar{b}/A, \bar{a}).$$

The equivalence of Morley rank and dimension will allow us to conclude that in strongly minimal sets Morley rank is definable.

Lemma 3 *Let D be strongly minimal. Suppose $C \subseteq D^{m+n}$ is definable. Let $C_a = \{x \in D^n : (a, x) \in C\}$ for $a \in D^m$. The set $Y_{n,k} = \{a \in D^m : \mathrm{RM}\ \mathrm{C_a} \geq \mathrm{k}\}$ is definable for each $k \leq n$.*

Proof We prove this by induction on n.

Suppose $n = 1$. We first note that there is a number N such that $|C_a| < N$ or $|D \setminus C_a| < N$, for all $a \in D^m$; since otherwise the type

$$\Gamma(\bar{w}) = \{\exists v_1, \ldots, v_{2s} \bigwedge_{i \neq j} v_i \neq v_j \wedge \bigwedge_{i=1}^{s} \phi(\bar{w}, v_i) \wedge \bigwedge_{i=s+1}^{2s} \neg\phi(\bar{w}, v_i) : s = 1, 2, \ldots\}$$

is consistent and a realization violates strong minimality.

Thus $\mathrm{RM}(\mathrm{C_a}) \geq 1$ if and only if $|C_a| > N$. So $Y_{1,1}$ is definable. Clearly $Y_{1,0} = D^m$.

Suppose $n = s + 1$. We work by induction on k. Clearly $Y_{n,0}$ is definable. For $a \in D^m$, let $B_a = \{\bar{b} \in D^s : \exists y\ (\bar{b}, y) \in C_a\}$. Clearly if $\mathrm{RM}(\mathrm{B_a}) \geq \mathrm{k}$, then $\mathrm{RM}(\mathrm{C_a}) \geq \mathrm{k}$. Suppose $\mathrm{RM}(\mathrm{B_a}) < \mathrm{k}$. If $\bar{b} \in B_a$ and $(\bar{b}, c) \in C_a$, then $\dim(\bar{\mathrm{b}}, \mathrm{c}) = \dim \mathrm{b} + \dim(\mathrm{c}/\bar{\mathrm{b}})$. Let $A_a = \{\bar{b} \in D^s : \{y : (\bar{b}, y) \in C_a\} \text{ is infinite}\}$. As above, there is an N (independent of a) such that

$$\bar{b} \in A_a \leftrightarrow |\{y : (\bar{b}, y) \in C_a\}| > N.$$

Thus A_a is definable and $\mathrm{RM}(\mathrm{C_a}) \geq \mathrm{k}$ if and only if $\mathrm{RM}(\mathrm{A_a}) \geq \mathrm{k} - 1$.

Thus $\mathrm{RM}(\mathrm{C_a}) \geq \mathrm{k}$ if and only if $\mathrm{RM}(\mathrm{B_a}) \geq \mathrm{k}$ or $\mathrm{RM}(\mathrm{A_a}) > \mathrm{k} - 1$. So, by induction, $Y_{n,k}$ is definable.

Finally we make one very useful definition.

Definition Let $v = v_1, \ldots, v_n$, $b = b_1, \ldots, b_m$ and suppose $\phi(v, b)$ has Morley rank k and Morley degree 1. If $b \in B$ and $p \in S_n(B)$ is the unique type of rank k containing $\phi(v, b)$, then we call p the *generic* type for $\phi(v, b)$ over B.

2 Families of plane curves

Let D be a strongly minimal set. In this section we will consider families of strongly minimal subsets of D^2. We first consider two illustrative examples. Suppose V is a $\mathbf{Q}$ vector space. Let $E = \{(x, y, z) \in V^3 : y = mx + z\}$ where $m \in \mathbf{Q}$. For $a \in V$, let $E_a = \{(x, y) : (x, y, a) \in E\}$. We think of E as describing the *family of plane curves* $\{E_a : a \in V\}$. We call V the *parameter space* for the family E. Note that in this case the parameter space is rank one. Indeed, if E is a family of plane curves in V^2 of higher rank, then for many distinct a and b E_a and E_b agree except perhaps on a finite set.

On the other hand suppose K is an algebraically closed field. Fix $n \in N$ and consider

$$E = \{(x, y, z_0, \ldots, z_{n-1}) : y = x^n + z_{n-1}x^{n-1} + \cdots + z_1 x + z_0\}.$$

If

$$E_a = \{(x, y) : y = x^n + \sum_{i=0}^{n-1} a_i x^i\}$$

for $a = (a_0, \ldots, a_{n-1}) \in K^n$, then $\{E_a : a \in K^n\}$ is an n-dimensional family of strongly minimal sets.

In this section we will examine families of plane curves and use them to give an alternative characterization of local modularity. To make these notions precise we must first digress and discuss M^{eq} and canonical bases.

Let M be any $\mathcal{L}$-structure. We associate to M a new structure M^{eq} in many sorted language $\mathcal{L}^{\mathrm{eq}} \supseteq \mathcal{L}$. For each $\emptyset$-definable equivalence relation E on M^n, we add a new sort S_E and a new n-ary function symbol f_E. We interpret S_E as M^n/E and f_E as the quotient map

$$x \mapsto x/E.$$

We can think of M as being the sort $S_=$. Clearly if $M \prec N$, then $M^{\mathrm{eq}} \prec N^{\mathrm{eq}}$. Moreover if $\mathcal{A}$ is an $\mathcal{L}^{\mathrm{eq}}$-structure elementarily equivalent to M^{eq}, then $\mathcal{A}$ is N^{eq} for some $N \equiv M$. Also note that if $\sigma \in Aut(M)$, then there is a unique extension of σ to $\widehat{\sigma} \in Aut(M^{\mathrm{eq}})$. Clearly every automorphism of M^{eq} restricts to an automorphism of M.

One function of M^{eq} is that any structure interpretable in M is isomorphic to a structure definable in M^{eq}. For our purposes the most important property of M^{eq} is the existence of canonical bases.

Definition Suppose M is saturated. Let $X \subseteq M^n$ be definable. We say that A is a *canonical base* for X if σ fixes X setwise (ie. $X = \{\sigma(x) : x \in X\}$) if and only if σ fixes A pointwise (ie. $\sigma(a) = a$ for $a \in A$) for all $\sigma \in Aut(M)$.

If $p \in S_n(M)$ then A is a canonical base for p if $\sigma(p) = p$ if and only if σ fixes A pointwise for all $\sigma \in Aut(M)$.

Note that if A is a canonical base for X and $\mathrm{dcl}^{\mathrm{eq}}(A) = \mathrm{dcl}^{\mathrm{eq}}(B)$, then B is also a canonical base for X.

Suppose X is defined by the formula $\phi(\bar{x}, \bar{a})$. Let E be the equivalence relation

$$\bar{a}\ E\ \bar{b} \Leftrightarrow (\phi(\bar{x}, \bar{a}) \leftrightarrow \phi(\bar{x}, \bar{b})).$$

Let $\alpha = \bar{a}/E \in M^{\mathrm{eq}}$. Then α is a canonical base for X.

We next show that if D is strongly minimal, $A \subseteq D$ and $p \in S_n(A)$. Then p has a canonical base $\alpha \in M^{\mathrm{eq}}$.

Suppose p has rank r and Morley degree 1. There is a formula $\phi(v, b) \in p$ such that ϕ has Morley rank r and Morley degree one. By definability of rank there is a formula $\Psi(w)$ such that $\Psi(b)$ and

$$\Psi(w) \Leftrightarrow \mathrm{RM}(\{x : \phi(x, w)\}) = r.$$

Let $\theta(w)$ be the formula

$$\Psi(w) \wedge \forall z \neg\big[\Psi(z) \wedge \mathrm{RM}(\{x : \phi(x, b) \wedge \phi(x, z)\}) = r \wedge \mathrm{RM}(\{x : \phi(x, b) \wedge \neg\phi(x, z)\}) = r\big].$$

Clearly $\theta(b)$ holds.

We can define an equivalence relation on $Y = \{y \in D^m : \theta(y)\}$ by

$$y \sim z \Leftrightarrow \mathrm{RM}\{x : \phi(x, y) \wedge \phi(x, z)\}) = r.$$

Let α be the equivalence class $b/\sim$.

Then $\sigma(\alpha) = \alpha$ iff $\sigma(b) \sim b$. But $\sigma(p) = p$ iff $\sigma(p)$ is the generic of $\phi(v, b)$. Since $\sigma(p)$ is the generic of $\phi(v, \sigma(b))$, $\sigma(p) = p$ if and only if $\sigma(b) \sim b$. Thus α is a canonical base for p. Clearly $\alpha \in \mathrm{dcl}^{\mathrm{eq}}(b)$.

If p has rank r and Morley degree $d > 1$ and $\phi(v, b)$ is a rank r degree d formula in p, then a similar argument shows that $\alpha \in \mathrm{acl}^{\mathrm{eq}}(b)$.

We summarize these arguments in the following theorem.

Theorem 1 *Suppose $A \subseteq D$ and $p \in S_n(A)$, then there is a canonical base for p in $\mathrm{acl}^{\mathrm{eq}}(A)$. If p has Morley degree 1, then the canonical base for p is in $\mathrm{dcl}^{\mathrm{eq}}(A)$.*

We next need to consider the computation of ranks in D^{eq}. The key is the following unpublished lemma of Lascar and Pillay.

Lemma 4 *Let D be a strongly minimal set and let $D_0 \subseteq D$ be infinite. Suppose E is an $\emptyset$-definable equivalence relation on D^m. Let $a \in D^m$ and $\alpha = a/E$. There is a finite $C \subset D^k$ (for some k) such that an automorphism of D fixing D_0 fixes α if and only if it fixes C setwise.*

Proof By adding D_0 to our language we assume that $\mathrm{acl}(\emptyset)$ is infinite.

Our first claim is that there is $b = (b_1, \ldots b_m) \in D^m$ algebraic over α such $a\ E\ b$. Choose b such that $b\ E\ a$, and $j = |\{i \leq m : b_i$ is algebraic over $\alpha\}|$ is maximal. We must show that $j = m$. Suppose not. By reordering the variables we may assume that $b_1, \ldots, b_j$ are algebraic over α and b_i is not algebraic over α for $i > j$. Let

$$Y = \{x \in D : \exists y_{j+2} \ldots \exists y_n\ (b_1, \ldots, b_j, x, y_{j+2}, \ldots, y_n) \in E\ \bar{a}\}.$$

Clearly $b_{j+1} \in Y$. If Y is finite, then any element of Y is algebraic over $b_1, \ldots, b_j, \alpha$, and hence algebraic over α. Thus by choice of $\bar{b}$, Y is infinite.

If Y is infinite, then since D is strongly minimal, Y is cofinite. In particular there is $d \in \mathrm{acl}(\emptyset) \cap Y$. But then we can find $d_{j+2}, \ldots, d_m$ such that $(b_1, \ldots, b_j, d, d_{j+2}, \ldots, d_m)/E = \alpha$ and $b_1, \ldots, b_j, d$ are algebraic over α contradicting the maximality of j.

Let C be the set of all conjugates of $\bar{b}$ under automorphisms fixing α. Then C is fixed setwise by any automorphism which fixes α and D_0. If $\bar{c} \in C$, then $\bar{c}/E = \alpha$. Thus α is fixed under all automorphisms which permute C. In particular an automorphism fixes $\bar{a}/E$ setwise if and only if it fixes C setwise.

Lemma 2.2 gives an easy description of Morley rank in D^{eq}. If p is a $\mathcal{L}_{\mathrm{eq}}$-type over D. Let α realize p. By lemma 2.2 there is a finite set C such that the set C is interdefinable with α over D. Then $\mathrm{RM}(p)$ is equal to the dimension of C.

Definition We say that a strongly minimal set D is *linear* if for all $p \in S_2(D)$, the canonical base for p has rank at most one.

Suppose $\phi(v_1, v_2, b)$ is strongly minimal. Let $\mathcal{F}$ be the family of sets C_β where β and b realize the same type. There is a natural equivalence relation on $\mathcal{F}$, $C_\alpha \sim C_\beta$ if and only if $C_\alpha \triangle C_\beta$ is finite. If p is the generic type of ϕ, then the Morley rank of the canonical base of p intuitively corresponds to the dimension of $\mathcal{F}/\sim$. Thus D is linear if and only if there is no family of plane curves of dimension greater than one. Algebraically closed fields are nonlinear since we have the family of curves $C_{(a,b)} = \{(x, y) : y = ax + b\}$, while vector spaces are linear.

We next show that the linear strongly minimal sets are exactly the locally modular ones.

Theorem 2 *Let D be a strongly minimal set. The following are equivalent.*

i) *For some small $B \subset D$, the pregeometry D_B is modular.*
ii) *D is linear.*
iii) *D is locally modular (ie. there is $b \in D$ such that D_b is modular).*

Proof

i)$\Rightarrow$ ii) Adding the parameters B to the language we assume that $B = \emptyset$ and D is modular. Let $p \in S_2(A)$ be strongly minimal. Let $\phi(v_1, v_2, \bar{a})$ be a strongly minimal formula in p. Let b_1, b_2 realize p.
Let $X = \mathrm{acl}(\bar{a}) \cap \mathrm{acl}(b_1, b_2)$. By modularity

$$\dim X = \dim\,(\bar{a}) + \dim\,(b_1, b_2) - \dim\,(\bar{a}, b_1, b_2).$$

Since $\dim\,(\bar{a}, b_1, b_2) = \dim\,(\bar{a}) + 1$ and $1 \le \dim\,(b_1, b_2) \le 2$, $\dim X \le 1$. Thus

$$\begin{aligned} \dim\,(b_1, b_2/\bar{a}) &= \dim\,(b_1, b_2, \bar{a}) - \dim\,(\bar{a}) \\ &= \dim\,(b_1, b_2) - \dim X \\ &= \dim\,(b_1, b_2/X). \end{aligned}$$

Thus dim $(b_1, b_2/X) = 1$. Let c be a canonical base for p, then, by 2.1, $c \in \mathrm{acl}^{\mathrm{eq}}(X)$, so $\mathrm{RM}(c) \leq 1$.

ii)$\Rightarrow$ iii) Let $b \in D - \mathrm{acl}(\emptyset)$. We will show that D localized at b is modular. We will use the fact that 1.2ii) is equivalent to modular. Suppose B is a finite dimensional closed set. Suppose $a_1 \in \mathrm{acl}(a_2, B, b)$. We must find $d \in \mathrm{acl}(B, b)$ such that $a_1 \in \mathrm{acl}(a_2, d, b)$. Clearly we may assume that $a_1 \notin \mathrm{acl}(B, b)$, $a_2 \notin \mathrm{acl}(B, b)$ and $a_1 \notin \mathrm{acl}(a_2, b)$ (else we are done). Thus dim $(a_1, a_2/b) = 2$ and dim $(a_1, a_2/Bb) = 1$.

Let c be a canonical base for the type of a_1, a_2 over $\mathrm{acl}^{\mathrm{eq}}(B, b)$.
By 2.1 $c \in \mathrm{acl}^{\mathrm{eq}}(B, b)$.

claim $c \in \mathrm{acl}^{\mathrm{eq}}(a_1, a_2)$.
Since c is the canonical base for p, $\mathrm{RM}(a_1, a_2/c) = 1$.
Since $\mathrm{RM}(a_1, a_2/c) < \mathrm{RM}(a_1, a_2)$, $\mathrm{RM}(c/a_1, a_2) < \mathrm{RM}(c)$.
By ii) $\mathrm{RM}(c) = 1$, thus $c \in \mathrm{acl}^{\mathrm{eq}}(a_1, a_2)$.

Since $\mathrm{RM}(a_1, a_2/b) = 2$ and $\mathrm{RM}(a_1, a_2/c) = 1$, $c \notin \mathrm{acl}(b)$.
Thus, since $b \notin \mathrm{acl}(\emptyset)$, $b \notin \mathrm{acl}(c)$. Thus a_1 and b realize the same type over c and, by saturation, there is $d \in D$ such that $tp(a_1, a_2/c) = tp(b, d/c)$. Then

$$d \in \mathrm{acl}(b, c) \subseteq \Big(\mathrm{acl}(a_1, a_2, b) \cap \mathrm{acl}(Bb)\Big)$$

and $d \notin \mathrm{acl}(b)$.

We claim that $d \notin \mathrm{acl}(a_2, b)$. If $d \in \mathrm{acl}(a_2, b)$, then, since $d \notin \mathrm{acl}(b)$, $a_2 \in \mathrm{acl}(d, b) \subseteq \mathrm{acl}(B, b)$, a contradiction.

Thus since $d \in \mathrm{acl}(a_1, a_2, b) \setminus \mathrm{acl}(a_2, b)$, $a_1 \in \mathrm{acl}(a_2, b, d)$ as desired.

We next give an "incidence geometry" interpretation of local modularity. This material will not be used in the subsequent sections.

Definition Suppose P and L are sets and $I \subseteq P \times L$. We think of P as a set of *points*, L as a set of *lines*, and I as the incidence relation (ie. $I(p, l)$ if and only if p is on l). We say that (P, L, I) is a *quasi-design* if

i) for any p, $\{l : I(p, l)\}$ is infinite

ii) for any l, $\{p : I(p, l)\}$ is infinite, and

iii) for any $l_1 \neq l_2$, $\{p : I(p, l_1) \text{ and } I(p, l_2)\}$ is finite.

If in addition

iv) for any $p_1 \neq p_2$, $\{l : I(p_1, l) \text{ and } I(p_2, l)\}$ is finite

then we call (P, L, I) a *pseudo-plane.*

We say that (P, L, I) is *complete-type definable* if there is a complete type $r(x, y)$ such that $P = \{x : \exists y\ r(x, y)\}$, $L = \{y : \exists x\ r(x, y)\}$ and I is the set of realizations of r.

Lemma 5 *Let M be any saturated structure. If there is a complete type definable quasi-design in M^{eq}, then there is a complete type definable pseudo-plane.*

Proof Let $c \in L$ and let $B = \{b_1, b_2, \ldots\} \subset P$, such that $r(b_i, c)$ for all i. Let σ be any automorphism of M fixing B pointwise. Since L is the set of realizations of a type, $\sigma(c) \in L$ and $r(b_i, \sigma(c))$ for all i. Thus by iii) $c = \sigma(c)$. Hence $\sigma(c) \in \mathrm{dcl}^{\mathrm{eq}}(B)$.

Let n be maximal such that there are distinct $b_1, \ldots, b_n$ such that $r(b_i, c)$ and $c \notin \mathrm{acl}^{\mathrm{eq}}(b_1, \ldots, b_n)$. Let $d \in M^{\mathrm{eq}}$ be the unordered set $\{b_1, \ldots, b_n\}$. Let r^* be the type of (d, c).

Since $b_i \in \mathrm{acl}(d) \setminus \mathrm{acl}(c)$, $d \notin \mathrm{acl}(c)$. Thus i) holds. By choice of n, $c \notin \mathrm{acl}(d)$. If $r^*(d_1, c_1)$ and $r^*(d_1, c_2)$ then $d_1 = \{b_1, \ldots, b_n\}$ where $r(b_i, c_j)$ for $i = 1, \ldots, n$ and $j = 1, 2$. Clearly since $\{b : r(b, c_1) \wedge r(b, c_2)\}$ is finite, $\{d : r^*(d, c_1) \wedge r^*(d, c_2)\}$ is finite. By choice of n if $r(d_1, c), r(d_2, c)$ and $d_1 \neq d_2$, then $c \in \mathrm{acl}(d_1, d_2)$. By saturation, there are only finitely many c such that $r(c, d_1)$ and $r(c, d_2)$. Thus r^* determines a pseudoplane.

Proposition 1 *Let D be strongly minimal. If D is non-locally modular then there is a complete-type definable pseudoplane.*

Proof By 2.4 it suffices to show that there is a complete-type definable quasi-design. Suppose D is not locally modular. Then by 2.3 there is a strongly minimal $p \in S_2(D)$ such that if c is the canonical base for p, then RM(c)=2. Let $a = (x, y)$ realize p and let r be the type of (a, c). Since $\mathrm{RM}(p) = 1$, $a \notin \mathrm{acl}(c)$. Since $\mathrm{RM}(a, c) = \mathrm{RM}(c) + 1 \geq 3$, $c \notin \mathrm{acl}(a)$. Suppose $c \neq c_1$ and both (a, c) and (a, c_1) realize r. Let $\phi(\bar{v}, \bar{b})$ be a strongly minimal formula in the type with canonical base c and let $\phi(\bar{v}, \bar{b}_1)$ be a formula in the type with canonical base c_1. Since $c_1 \neq c$, $\{\bar{v} : \phi(\bar{v}, \bar{b}) \wedge \phi(\bar{v}, \bar{b}_1)\}$ is finite. Thus $\{x : r(x, c) \text{ and } r(x, c_1)\}$ is finite.

3 Algebraic structure

A remarkable insight of Hrushovski is that in many situations algebraic structure can be detected from the geometry of forking. In this section we will give one version of this result and apply it to show that a locally modular strongly minimal set interprets a group. We will also state a theorem of Hrushovski and Pillay describing groups interpretable in locally modular strongly minimal sets.

Throughout this section D will be a saturated strongly minimal set. Our main tool is the *group configuration*. Suppose we have elements a, b, c, x, y, z in D^{eq} such that:

i) $\mathrm{RM}(a) = \mathrm{RM}(b) = \mathrm{RM}(c) = \mathrm{RM}(x) = \mathrm{RM}(y) = \mathrm{RM}(z) = 1$;
ii) any pair of elements has rank 2;
iii) $\mathrm{RM}(a, b, c) = \mathrm{RM}(c, x, y) = \mathrm{RM}(a, y, z) = \mathrm{RM}(b, x, z) = 2$;
iv) any other triple has rank 3.

Look at Diagram 1. Conditions iii) and iv) assert that each line has rank two while any three non-colinear points have rank three.

There is one easy way that a group configuration arises. Suppose G is a strongly minimal abelian group (indeed strongly minimal groups must be abelian). Let a, b, x be independent elements of G. Let $c = ba$, $y = cx$ and

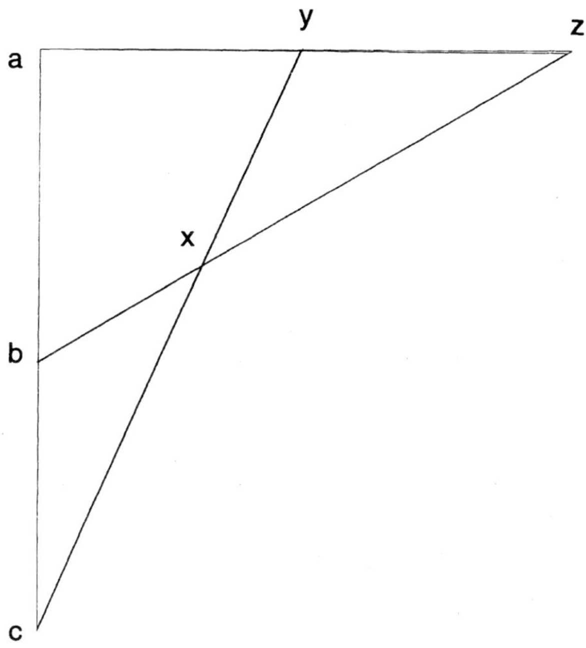

Diagram 1
The group configuration

$z = bx$, then $y = az$ and it is easy to check that conditions i)-iv) hold. Remarkably this observation has a converse.

Theorem 3 (Hrushovski) (see [Bo1] or [P1] chapter 5) *Suppose there is a group configuration in D^{eq}. Then D interprets a strongly minimal abelian group.*

Let D be a saturated non-trivial locally modular strongly minimal set. By adding a parameter to our language we assume that D is modular. We will show how to use theorem 3.1 to find a group in D^{eq}.

Since D is non-trivial we can find a finite $A \subset D$ and $b, c \in D \setminus \mathrm{acl}(A)$ such that $c \in \mathrm{acl}(A, b) \setminus (\mathrm{acl}(A) \cup \mathrm{acl}(b))$. By modularity

$$\dim (b, c, A) = \dim (b, c) + \dim (A) - \dim C$$

where $C = \mathrm{acl}(A) \cap \mathrm{acl}(b, c)$. Since $\dim (b, c, A) = \dim A + 1$ and $\dim (b, c) = 2$, $\dim C = 1$. Thus there is $a \in C$ with $\dim a = 1$. Note that $\dim (a, c) = \dim (a, b) = 2$, but $\dim (b, c/a) = 1$.

Choose $y, z \in D$ such that (b, c) and (y, z) realize the same type over $\mathrm{acl}^{\mathrm{eq}}(a)$ and (y, z) are independent from (b, c) over a

$$(\text{ie. } \dim (y, z/a, b, c) = \dim (y, z/a) = \dim (b, c/a) = 1).$$

Since $a \in \text{acl}(b,c)$, $z \in \text{acl}(b,c,y)$. Thus dim $(b,c,y) = 3$ and dim $(c,y) = 2$. Similarly dim $(b,z) = 2$. By modularity

$$\dim (c,y,b,z) = \dim (c,y) + \dim (b,z) - \dim X$$

where $X = \text{acl}(c,y) \cap \dim \text{acl}(b,z)$. Thus dim $X = 1$ and there is $x \in \text{acl}(c,y) \cap \text{acl}(b,z)$ with dim $x = 1$.

It is easy to see that a, b, c, x, y, z is a group configuration. Thus D interprets a rank 1 abelian group.

The next theorem summarizes a much finer version of this result (see [P1] chapter 5)]). It says that not only is there an interpretable group, but the group determines most of the structure of the the strongly minimal set.

Theorem 4 (Hrushovski) *Let D be a non-trivial locally modular strongly minimal set. Then there is a rank 1 abelian group G definable in D^{eq} such that G acts definably as a group of automorphisms of the generic type of D.*

In §5 we will use a structure theorem for groups interpretable in a locally modular strongly minimal set. This result is true in a much more general setting. We say that a stable theory T is *1-based* if for any a and B the canonical base for the type of a over $\text{acl}^{\text{eq}}(B)$ is contained in $\text{acl}^{\text{eq}}(a)$.

Lemma 6 *If M is interpretable in a locally-modular strongly minimal set then $Th(M)$ is 1-based.*

Indeed a finite Morley rank theory T is 1-based if and only if every strongly minimal set is locally modular (see [P1] chapter 2).

Theorem 5 (Hrushovski-Pillay) (see [HP] or [P1] chapter 4) *Suppose G is a 1-based group. Then every definable subset of G^n is a finite boolean combination of cosets of definable subgroups.*

We conclude by stating a generalization of the group configuration which detects the presence of a field. A *field configuration* is a collection of elements $a, b, c, x, y, z \in D^{\text{eq}}$ such that

i) $\text{RM}(a) = \text{RM}(b) = \text{RM}(c) = 2$ and $\text{RM}(x) = \text{RM}(y) = \text{RM}(z) = 1$;

ii) $\text{RM}(a,b) = \text{RM}(a,c) = \text{RM}(b,c) = 4$ and $\text{RM}(a,b,c) \leq 5$;

iii) $\text{RM}(c,x,y) = \text{RM}(b,x,z) = \text{RM}(a,y,z) = 3$;

iv) any three non-colinear points (see diagram 1) are independent;

v) If we replace a, b, c by a^*, b^*, c^* with $\text{RM}(a^*) = \text{RM}(b^*) = \text{RM}(c^*) = 1$, then a^*, b^*, c^*, x, y, z is not a group configuration.

Suppose K is an algebraically closed field and $\alpha_1, \alpha_2, \beta_1, \beta_2, x$ are algebraically independent. Let a, b, c be elements of the rank 2 group of affine transformations where a is the transformation $z \mapsto \alpha_1 z + \alpha_2$ and b is $z \mapsto \beta_1 z + \beta_2$. Let c be the composition. Let $y = cx$ and $z = bx$. Then $az = y$. Note that $\text{RM}(a/z,y) = \text{RM}(b/x,z) = \text{RM}(c/x,y) = 1$. Thus we have a field configuration. Remarkably this is the only way this can happen.

Theorem 6 (Hrushovski) (see [Bol]) *If there is a field configuration in* D^{eq}, *then* D *interprets an infinite field.*

By a result of Macintyre ([Mac], see also [Po]), any infinite field interpretable in an ω-stable structure is algebraically closed.

4 Zariski geometries

Zariski geometries were introduced by Hrushovski and Zilber in [HZ1], [HZ2] and [Z]. In addition to providing an important class of strongly minimal sets where Zilber's conjecture is true, they answer the metamathematical question: Can one characterize the topological spaces that arise from the Zariski topology on an algebraic curve?

We say that a topological space is *Noetherian* if there are no infinite descending chains of closed sets. If K is a field, then the *Zariski topology* on K^n is given by taking the solutions to systems of polynomial equations as the basic closed sets. Since the polynomial ring $K[X_1, \ldots, X_n]$ has no infinite ascending chains of ideals, the Zariski topology is Noetherian.

A closed set X is *irreducible* if there are no proper closed subsets X_0 and X_1 such that $X = X_0 \cup X_1$. A simple König's lemma argument shows that in any Noetherian topological space, every closed set X is a union of finitely many irreducible closed sets, called the irreducible components of X.

In Noetherian topological spaces we can inductively assign an ordinal

$$\dim X = \sup\{\dim Y + 1 : Y \text{ is a nonempty, closed, irreducible proper subset of } X\}.$$

The dimension of a reducible closed set is the maximum dimension of an irreducible component.

Definition A *Zariski geometry* is an infinite set D and a sequence of Noetherian topologies on $D, D^2, D^3, \ldots$ such that the following axioms hold.

(Z0) [Coherence]: i) If $f : D^n \to D^m$ is defined by $f(x) = (f_1(x), \ldots, f_m(x))$ where each $f_i : D^n \to D$ is either constant or a coordinate projection, then f is continuous.

ii) Each diagonal $\Delta^n_{i,j} = \{x \in D^n : x_i = x_j\}$ is closed.

(Z1) [Weak QE]: If $C \subseteq D^n$ is closed and irreducible, and $\pi : D^n \to D^m$ is a projection, then there is a closed $F \subset \overline{\pi(C)}$ such that $\pi(C) \supseteq \overline{\pi(C)} \setminus F$.

(Z2) [Uniform one-dimensionality]: i) D is irreducible.

ii) Let $C \subseteq D^n \times D$ be closed and irreducible. For $a \in D^n$, let $C(a) = \{x \in D : (a, x) \in C\}$. There is a number N such that for all $a \in D^n$, either $|C(a)| \leq N$ or $C(a) = D$. In particular any proper closed subset of D is finite.

(Z3) [Dimension theorem]: Let $C \subseteq D^n$ be closed and irreducible. Let W be a non-empty irreducible component of $C \cap \Delta^n_{i,j}$. Then $\dim C \leq \dim W + 1$.

The basic example of a Zariski geometry is a smooth algebraic curve C over an algebraically closed field where C^n is equipped with the Zariski topology. In this case Z0 is clear, Z1 follows from quantifier elimination and Z2 follows from the fact that C is strongly minimal. The verification of Z3 uses the smoothness of C (though a weaker condition suffices). The remarkable result of Hrushovski and Zilber is that with some natural additional assumptions the converse holds.

First one must see how model theory enters the picture. Given a Zariski geometry D, let $\mathcal{L}_D$ be language with an n-ary relation symbol for each closed subset of D^n. Let $\mathcal{D}$ be D viewed in the natural way as an $\mathcal{L}_D$-structure.

Lemma 7 *The theory of $\mathcal{L}_D$ admits quantifier elimination and D is a strongly minimal set. Moreover, Morley rank in $\mathcal{D}$ is exactly dimension.*

Zariski geometries may be locally modular. If X is an infinite set we can topologize X^n by taking the positive quantifier free definable sets in the language of equality (allowing parameters) as the closed sets. This determines a trivial Zariski geometry on X. If K is a field we could topologize K^n by taking the affine subsets (ie. cosets of subspaces). This is a non-trivial locally modular Zariski geometry. If a Zariski geometry is non-locally modular, then Zilber's conjecture holds

Theorem 7 *Suppose D is a non-locally modular Zariski geometry, then $\mathcal{D}$ interprets an infinite algebraically closed field K. If $X \subseteq K^n$ is definable in $\mathcal{D}$, then X is definable using only the field structure of K (we say that K is a pure field).*

Hrushovski and Zilber give a more refined version of 4.2 with additional geometric information. By a *family of plane curves in D* we mean closed sets $X \subseteq D^m$ and $C \subset D^2 \times X$ such that for all $a \in X$, if we let $C(a)$ denote $\{(x, y) \in D^2 : (x, y, a) \in C\}$, then $C(a)$ is a one dimensional irreducible closed subset of D^2. We say that a family of plane curves is *ample* whenever for p and q are independent generic points of D^2, there is a plane curve $C(a)$ with $p, q \in C(a)$. An ample family is called *very ample* if for p, q (not necessarily independent) generic points in D^2 there is a curve $C(a)$ with $p \in C(a)$ and $q \notin C(a)$. In this case, we say that the family *separates points.*

We say that a Zariski geometry D is (very) ample if there is a (very) ample family of curves on D. If D is ample, then from 2.3 we see that D is non-locally modular. The converse is also true (see for example [M1]). Very ample Zariski geometries are intimately related to algebraic curves.

Theorem 8 *If D is a very ample Zariski geometry, then there is an interpretable field K, C a smooth quasi-projective curve defined over K and a definable bijection $f : D \to C$ such that the induced maps $f^n : D^n \to C^n$ are homeomorphisms for all n.*

The proof of theorem 4.2 is a quite delicate application of both the group configuration and the field configuration. Suppose we have a rich family of plane curves through a point (p,p). If X and Y are two curves through (p,p) we define the composition $X \circ Y$ by $\{(x,z) : \exists y\ (x,y) \in Y, (y,z) \in X\}$. The key idea is to define an approximate notion of "tangency" and consider the "operation" $(X,Y) \mapsto Z$ where Z is tangent to $X \circ Y$ at (p,p). This gives rise to the group configuration. A more subtle construction, using the group we just found, allows us to find the field configuration. A similar type of construction in a much simpler setting is given in [MP].

We refer the reader to [M1] for a more detailed survey of Zariski geometries.

We conclude this section by giving one further setting where this work applies. Let K be a differentially closed field (see [S] or [M2]). There is a natural topology on K^n given by taking solutions to algebraic differential equations as the basic closed sets. The Ritt basis theorem implies that this topology is Noetherian. For any $D \subseteq K^n$ we topologize D^n by taking the induced topology.

Theorem 9 (Hrushovski-Sokolovic) *Let D be a strongly minimal subset of K^n. Then there is a finite $X \subset D$ such that $D \setminus X$ is a Zariski geometry.*

Using quantifier elimination in differentially closed fields it is clear that Z0, Z1 and Z2 hold. Hrushovski and Sokolovic show how to reduce dimension calculations to calculations in classical algebraic geometry and deduce Z3.

If D is a non-locally modular strongly minimal subset of K^n, then by 4.2, D interprets an algebraically closed field F. The field F must be of finite Morley rank (while K has Morley rank ω). There is one natural finite Morley rank field interpretable in K, namely the field of constants $C_K = \{x \in K : x' = 0\}$. Using quantifier elimination and stability (see [M2]) one can see that any subset of C_K^n definable in K is already definable in C_K using only the field structure. Sokolovic, building on work of Cassidy (see [P2]), showed this is the only interpretable field of finite rank.

Theorem 10 (Sokolovic) *If F is an infinite field of finite Morley rank interpretable in a differentially closed field K, then F is definably isomorphic to C_K.*

Corollary 1 *If K is differentially closed and $D \subset K^n$ is a non-locally modular strongly minimal set, then D interprets a field definably (in K) isomorphic to C_K.*

5 The Mordell-Lang conjecture for function fields

In 1993 Hrushovski [H1] found an application of geometric model theory to diophantine geometry in his proof of the Mordell-Lang conjecture for function fields. What, to me, is truly remarkable about Hrushovski's proof, is that it seems to flow naturally from the stream of ideas in modern model theory. In this section I will give some of the background of the Mordell-Lang conjecture and outline Hrushovski's proof in one basic case. I refer the reader to [L] for a more detailed history of the problem.

We will work inside a large algebraically closed field $\mathbf{K}$. An abelian variety A is a projective variety equipped with a rational map $\mu : A \times A \to A$ making A into a group. We usually write the group additively. The simplest example of an abelian variety is an elliptic curve. Let E be the projective plane curve given by the equation $Y^2Z = X(X-Z)(X+\lambda Z)$. We can think of E as the plane curve $y^2 = x(x-1)(x-\lambda)$ together with one point O at infinity (this is the point with homogeneous coordinates (0,1,0)). We must define the group law on E. The zero of the group will the point O. If P and Q are distinct points on E, consider the line l through P and Q (see diagram 2). Since E is a cubic curve the line l will intersect E in exactly three points (counting multiplicity). If R is the third point of E on l, then we say $P + Q + R = O$. Note that if P has affine coordinates (x, y), then $-P$ will have coordinates (x, y). To add a point S to itself we take l to be the tangent line to E at S. Usually l will intersect E at S with multiplicity 2. In this case there is a second point T on E such that R is also on l and $S+S+T = 0$. Otherwise, l will intersect E at S with multiplicity 3 and $S+S+S = 0$. Though it is non-trivial to verify associativity, this defines a group law on E and addition is a rational map (see [Si1] or [SiT]).

One dimensional abelian varieties are isomorphic to elliptic curves. It is not easy to give algebraic descriptions of higher dimensional abelian varieties. However, if we look at complex abelian varieties there is a very easy topological description. If A is a complex abelian variety of dimension d, there are $\alpha_1, \ldots, \alpha_{2d} \in \mathbf{C}^d$ linearly independent over $\mathbf{R}$ such that A is analytically isomorphic to $\mathbf{C}^d/\Lambda$ where Λ is the lattice $\mathbf{Z}\alpha_1 + \cdots + \mathbf{Z}\alpha_{2d}$ (see [R]). The next lemma summarizes some properties of abelian varieties which we will need. The first principles are easy to verify for complex analytic varieties using the topological description.

Lemma 8 *Let A be an abelian variety of dimension d.*

i) A is a divisible abelian group. If $\mathbf{K}$ has characteristic zero or n is prime to the characteristic of $\mathbf{K}$, then A has n^{2d} n-torsion points.

ii) (rigidity (see [Mi1]) *If A is defined over an algebraically closed field k, then every abelian subvariety of A is also defined over k.*

Definition Suppose Γ is a subgroup of A. We say that Γ is *finite rank* if there is a finitely generated subgroup Γ_0 such that

$$\Gamma \subseteq \{g \in A : ng \in \Gamma_0 \text{ for some } n = 1, 2, \ldots\}.$$

For example, taking $\Gamma_0 = \{0\}$, the torsion subgroup of A is of finite rank.

We can now state the full Mordell-Lang conjecture.

Mordell-Lang Conjecture (characteristic zero) Suppose $\mathbf{K}$ has characteristic zero, A is an abelian variety, Γ is a finite rank subgroup of A and X is a proper subvariety of A. Then $X \cap \Gamma$ is a finite union of cosets of subgroups of A.

This conjecture implies the Mordell Conjecture. Suppose C is a curve of genus $g > 1$ defined over a number field k. The Mordell Conjecture asserts that C has only finitely many k-rational points (i.e. points with coordinates in k).

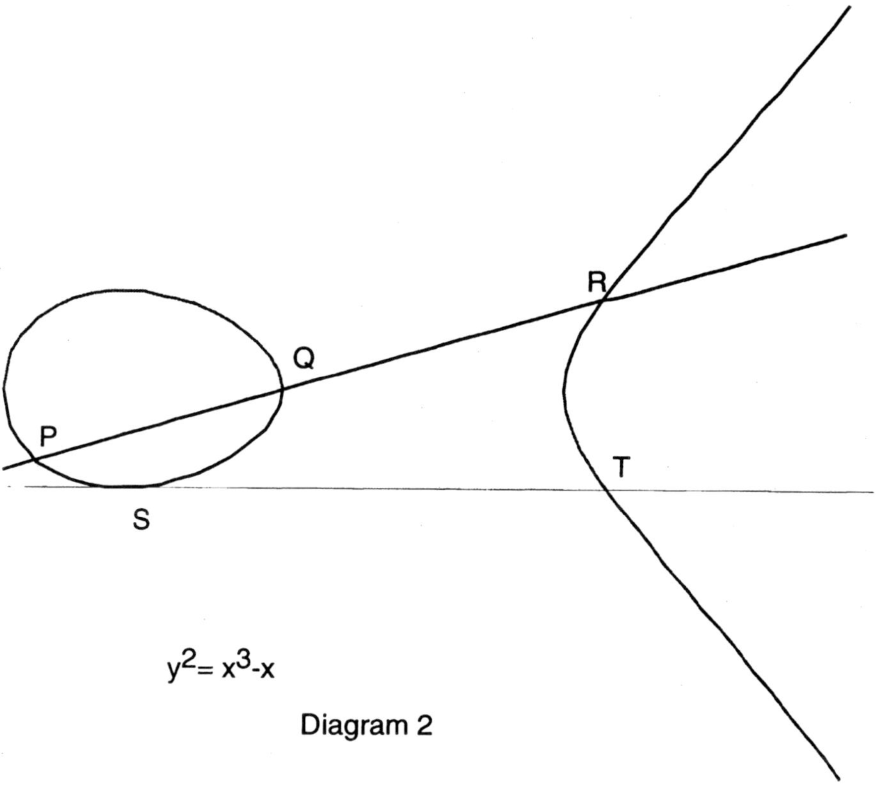

Diagram 2

To any curve C of genus $g \geq 1$ we can associate a g-dimensional abelian variety $J(C)$ defined over k called the Jacobian of C (see [Mi2]). The curve C is a subvariety of $J(C)$ and $J(C)$ is the smallest abelian variety in which C embeds. If C has genus 1, then C is an elliptic curve and $J(C) = C$.

Let C have genus $g > 1$. Let Γ be the k-rational points of $J(C)$. The Mordell-Weil theorem (see [L]) asserts that Γ is a finitely generated group. Thus $\Gamma \cap C$ is a finite union of cosets of subgroups of Γ. If any of these subgroups is infinite, then the Zariski closure of that coset is also a coset and the Zariski closure must be the entire curve C. But then there would be a group structure defined on C and C would be an abelian variety contradicting the fact that $J(C) \supset C$ and $J(C)$ is the smallest abelian variety in which C embeds. Thus $C \cap \Gamma$ is finite and C contains only finitely many k-rational points.

The Mordell conjecture was proved by Faltings who also proved the Mordell-

Lang conjecture in case Γ is finitely generated. The full characteristic zero Mordell-Lang conjecture (even in the case where A is a semi-abelian variety) was proved by McQuillen ([Mc]) building on Faltings' work as well of work of Raynaud, Hindry and Vojta.

In number theory, when studying question about number fields, it is often insightful to ask the same question about finitely generated extensions of algebraically closed fields (we call these *function fields*). Long before Faltings, Manin [Ma] proved the function field case of the Mordell conjecture.

Theorem 11 (Manin) *Let k be algebraically closed of characteristic zero and let $K \supset k$ be finitely generated over k. Let C be a curve of genus $g > 1$ defined over K. Then either C has only finitely many K-rational points or C is isomorphic to a curve defined over k.*

So far we have only considered characteristic zero. What about characteristic $p > 0$? The obvious generalization of the Mordell-Lang conjecture to characteristic p is false. Suppose C is a curve of genus $g > 1$ defined over $\mathbf{F}_p$ and let $J(C)$ be its Jacobian. Let α be a generic point of C. Let Γ be the group of $\mathbf{F}_p(\alpha)$-rational points of $J(C)$. By the Lang-Neron theorem (the function field version of the Mordell-Weil theorem (see [Si2])) Γ is finitely generated. If σ is the Frobenius automorphism $x \mapsto x^p$, then $\sigma^n(\alpha) \in C$ for all n. Thus $C \cap \Gamma$ is infinite, but C is not a coset of a subgroup. In this case our curve C is defined over the prime field. This leaves open the possibility that one could prove a function field version of the Mordell-Lang conjecture in all characteristics. Prior to Faltings, Buium ([Bu]) proved a characteristic zero function field version. Abramovich and Voloch [AV] had made progress in characteristic p. Finally Hrushovski proved the complete function field version (in fact Hrushovski's proof works in the more general case of "semi-abelian" varieties).

Theorem 12 (Mordell-Lang Conjecture for function fields) *Let k be an algebraically closed field with $K \supset k$. Let A be an abelian variety defined over K , X a subvariety of A and Γ a finite rank subgroup of A. Suppose $X \cap \Gamma$ is Zariski dense in X. Then there is a subabelian variety $A_1 \subseteq A$, an abelian variety B defined over k, a surjective homomorphism $g : A_1 \to B$ and a subvariety X_0 of B defined over k, such that $g^{-1}(X_0)$ is a translate of X.*

For the remainder of this section we will outline the proof of a special case of Hrushovski's theorem which illustrates the use of model theoretic ideas.

Let k be algebraically closed of characteristic zero and let $K \supset k$. Let A be an abelian variety of dimension $d > 1$ defined over K with no non-trivial subabelian varieties (such an abelian variety is called *simple*). Let Γ be the torsion points of A and let $C \subset A$ be a curve on A. We will show that either $C \cap \Gamma$ is finite or A is isomorphic to an abelian variety defined over k.

We are free to replace K by a larger field. Following Manin and Buium we see that it is useful to replace K by a rich differential field. We can find a derivation $\delta : K \to K$ such that $k = \{a \in K : \delta(a) = 0\}$, the field of constants of δ. We next replace K by the differential closure of (K, δ). Since k is already algebraically closed, passing to the differential closure of K does not add any new constants. Thus we may assume that K is differentially closed and k is the field of constants of K.

The next idea is to replace Γ by a "small" group $\Sigma \supseteq \Gamma$ definable in the differentially closed field. We then have a chance of using our model theoretic tools. What do I mean by "small"? Recall that K has Morley rank ω and if V is a d-dimensional variety then V has Morley rank ωd. The next result is a compilation of work of Manin and Buium refined by Hrushovski and Sokolovic in the model theoretic setting.

Theorem 13 *Let K be a differentially closed field and let A be an abelian variety defined over K. There is a definable group homomorphism $\mu : A \to K^n$ for some n such that $\Sigma = \ker(\mu)$ has finite Morley rank. Moreover, if A is simple we can choose μ so that $\ker(\mu)$ has no proper infinite definable subgroups.*

Since K^n has no torsion, $\Gamma \subseteq \Sigma = \ker(\mu)$. We will argue that either $\Sigma \cap C$ is finite or A is isomorphic to an abelian variety defined over k. Replacing Γ by Σ allows us to use all of the tools from finite Morley rank group theory.

Let $X \subseteq \Sigma$ be a strongly minimal set containing 0. By Zilber's indecomposability theorem (see [Po2] 2.b) the subgroup of Σ generated by X is definable and for some fixed n every element of that group is a sum of n elements of X. By 5.4, since A is simple, X generates Σ. Thus Σ is in the definable closure of X and, if we consider X with all of its induced structure, Σ is interpretable in X.

We now break into cases depending on whether or not X is locally modular.

Case 1 X is locally modular.

In this case Σ is a one-based group. By the theorem 3.4, every definable subset of Σ is a Boolean combination of cosets of definable subgroups. Since Σ has no infinite definable subgroups, X must differ from Σ by a finite set. Thus Σ itself is strongly minimal.

Consider $\Sigma \cap C$. If $\Sigma \cap C$ is infinite it differs from Σ by a finite set. Thus the Zariski closure of Σ is C plus a finite set. But the torsion points of A are Zariski dense in A, thus the Zariski closure of $\Sigma = A$. Since $C \subset A$ we have a contradiction. Thus $\Sigma \cap C$ is finite.

Case 2 X is non-locally modular.

As we pointed out in 4.6, in a differentially closed field any non-locally modular strongly minimal set interprets the field of constants. So we have a strongly minimal set X which interprets the field of constants k and the group Σ. Using Hrushovski's analysis of non-orthogonality in groups (see [Po2] 2.e), it follows that there is a group G interpretable in k, and a definable homomorphism h from Σ onto G. Since Σ is minimal h has finite kernel.

Using quantifier elimination and stability one sees that k is a pure algebraically closed field. That is: any subset of k^n defined in the differentially closed field K is already definable using the field structure of k. Thus the group G is interpreted in the pure algebraically closed field k. But then there is an algebraic group G_0 defined over k such that G is isomorphic to $G_0(k)$, the k-rational points of G_0 ([Po2] 4.e).

Using the fact that Σ has only finitely many $|\ker(h)|$-torsion point, one constructs a dual homomorphism $g : G_0(k) \to \Sigma$ with finite kernel. Modding out by the kernel we may assume that g is an isomorphism. Since the derivation is trivial on k, g must be a rational map. Since $g : G_0(k) \to A$ and $G_0(k)$ is Zariski dense in $G_0(K)$, $g : G_0(K) \to A$. Since Γ is Zariski dense in A, g is surjective.

Let $N \subset G_0$ be a maximal linear algebraic subgroup. By a theorem of Chevalley ([Sh]) G_0/N is an abelian variety. Since there are no nontrivial homomorphisms from a linear group into an abelian variety, $N \subseteq \ker g$. Thus we can replace G_0 by the abelian variety G_0/N. Since N is defined over k, we may assume that G_0 is an abelian variety defined over K. The rigidity of abelian varieties implies that every abelian subvariety of G_0 is definable over k. In particular $\ker(g)$ is defined over k. Thus A is isomorphic to the abelian variety $G_0/\ker(g)$ which is defined over k.

We conclude with a few words on what is needed to prove the full theorem. The proof of Hrushvoski's theorem when A is not simple or Γ is a more general finite rank group requires a slightly more subtle analysis of the finite Morley rank groups that arise. We refer the reader to [Bo2] for a detailed account of the proof in this case. In characteristic p we work with separably closed fields rather than differentially closed fields. Separably closed fields are stable but, unlike differentially closed fields, are not ω-stable. This leads to a number of complications because we must now consider types rather than definable sets. We must also use more general geometric model theory (see [Pi1]). On the other hand rather than the complicated group $\ker(\mu)$ from lemma 4.4, one uses $\bigcap_{m=1}^{\infty} mA$. Messmer's ([Me]) analysis of fields interpretable in separably closed fields replaces corollary 4.6 work in this case.

There are a number of very good discussions of Hrushovski's proof. I suggest that the interested reader consult [Bo2], [P3], [Po2] or [H4] in addition to the original article [H1].

References

[AV] D. Abramovich and J. Voloch, Towards a proof of the Mordell-Lang conjecture in positive characteristic, Duke Math. J. 5 (1992).

[BL] J. Baldwin and A. Lachlan, On strongly minimal sets, J. Symbolic Logic 36 (1971).

[Bo1] E. Bouscaren, The group configuration-after E. Hrushovski, *The Model Theory of Groups*, A. Nesin and A. Pillay ed., Notre Dame Press, 1989.

[Bo2] E. Bouscaren, The proof of the Mordell-Lang conjecture, in [BoL], to appear.

[BoL] E. Bouscaren and D. Lascar, *Stability Theory and Algebraic Geometry: an introduction*, Springer Lecture Notes, to appear.

[Bu] A. Buium, Intersections in jet spaces and a conjecture of S. Lang, Annals of Math. 136 (1992).

[CK] C.C. Chang and H. J. Keisler, *Model Theory*, North Holland, 1977.

[CS] G. Cornell and J. Silverman, *Arithmetic Geometry*, Springer, 1986.

[H1] E. Hrushovski, The Mordell-Lang conjecture, J. AMS 9 (1996).

[H2] E. Hrushovski, A new strongly minimal set, Ann. Pure and Applied Logic 62 (1993).

[H3] E. Hrushovski, Strongly minimal expansions of algebraically closed fields, Israel J. Math. 79 (1992).

[H4] E. Hrushovski, Stability and its uses, Current Developments in Mathematics, R. Bott et. al. ed., International Press Inc., 1996.

[HP] E. Hrushovski and A. Pillay, Weakly normal groups, *Logic Colloquium '85*, North Holland, 1987.

[HS] E. Hrushovski and Z. Sokolovic, Minimal sets in differentially closed set, Trans. AMS, to appear.

[HZ1] E. Hrushovski and B. Zilber, Zariski geometries, J. AMS 9 (1996).

[HZ2] E. Hrushovski and B. Zilber, Zariski geometries, Bull. AMS 28 (1993).

[L] S. Lang, *Number Theory III*, Springer Verlag, 1991.

[Mac] A. Macintyre, On ω_1-categorical theories of fields, Fund. Math. 71 (1971).

[Ma] J. Manin, Rational points of algebraic curves over function fields, AMS Translations ser. II 59 (1966).

[M1] D. Marker, Zariski Geometries, to appear in [BoL].

[M2] D. Marker, Model Theory of Differential Fields, in [MMP].

[MMP] D. Marker, M. Messmer and A. Pillay, *Model Theory of Fields*, Lecture Notes in Logic 5, Springer Verlag, 1996.

[MP] D. Marker and A. Pillay, Reducts of $(\mathbf{C}, +, \cdot)$ which contain $+$, J. Symbolic Logic 55 (1990).

[Mc] M. McQuillen, Division points on semi-abelian varieties, Inventiones Math., 120 (1995), 143-159.

[Me] M. Messmer, Groups and fields interpretable in separably closed fields, Trans. AMS 344 (1994).

[Mi1] J. Milne, Abelian varieties, in [CS].

[Mi2] J . Milne, Jacobian varieties, in [CS]

[P1] A. Pillay, *Geometric Stability Theory*, Oxford University Press, 1996.

[P2] A. Pillay, Differential algebraic groups and the number of countable differentially closed fields, in [MMP].

[P3] A. Pillay, Model theory and diophantine geometry, preprint.

[Po1] B. Poizat, *Groupes Stables*, Nur Al-Maniq wal-Ma'rifah, 1987.

[Po2] B. Poizat, H. L. M. (Hrushovski-Lang-Mordell), *Séminaire Bourbaki*, to appear.

[R] M. Rosen, Abelian varieties over $\mathbf{C}$, in [CS]

[S] G. Sacks, *Saturated Model Theory*, Benjamin, 1972.

[Sh] I. Shafarevich, *Basic Algebraic Geometry*, Springer, 1977.

[Si1] J. Silverman, *The arithmetic of elliptic curves*, Springer 1986.

[Si2] J. Silverman, *Advanced topics in the arithmetic of elliptic curves*, Springer 1994.

[SiT] J. Silverman and J. Tate, *Rational Points on Elliptic Curves*, Springer, 1992.

[Z] B. Zilber, Zariski type structures, unpublished lecture notes, University of Illinois Chicago, 1991.

Computationally-Sound Proofs

Silvio Micali

Department of Computer Science
Massachusetts Institute of Technology
Cambridge, MA 02139, USA
silvio@theory.lcs.mit.edu

Abstract. This paper puts forward a new notion of a proof based on computational complexity, and explores its implications to computation at large.

Computationally-sound proofs provide, in a novel and meaningful framework, answers to old and new questions in complexity theory. In particular, given a random oracle or a new complexity assumption, they allow us to prove that verifying is easier than deciding; to provide a quite effective way to prove membership in computationally hard languages (such as $\mathcal{C}o$-$\mathcal{NP}$-complete ones); and to show that every computation possesses a short certificate vouching its correctness.

1 Introduction

A new notion. Proofs are fundamental to our lives, and as for all things fundamental we should expect that answering the question of what a proof is will always be an on-going process. Indeed, we wish to put forward the new notion of a *computationally-sound proof* (*CS proof* for brevity) which achieves new and important goals, not attained or even addressed by previous notions.

Informally, a CS proof of a statement S consists of a short string σ, very easy to verify and as easy to find as possible, offering a strong computational guarantee about the verity of S. By "very easy to verify" we mean that the time necessary to inspect a CS Proof of a statement S is poly-logarithmically shorter than that required to decide S. By "as easy to find as possible" we mean that a CS proof of a *true* statement (i.e., for the purposes of this paper, *derivable* in a given axiomatic theory) can be computed in a time essentially comparable to that needed to decide the statement. Finally, by saying that the guarantee offered by a CS proof is "computational" we mean that false statements either do not have any CS proofs, or such "proofs" are practically impossible to find.

Implementations of CS proofs. The value of a new notion, of course, crucially depends on whether it can be sufficiently exemplified. We provide two main implementations of our notion. The first consists of an explicit construction, based on a random oracle, that provably yields a CS proof system without any complexity assumption. The second implementation, which is the subject of a forthcoming paper, is cryptography-based, and yields a CS proof system under a new complexity assumption.

Applications of CS proofs. CS proofs provide, in a new and meaningful framework, very natural answers to some of our oldest questions in complexity theory. In particular, they imply that verifying is poly-logarithmically easier than deciding. More importantly, they actually imply that this is *almost always* the case, rather than a phenomenon possibly occurring for some special theorems. CS proofs also provide a quite effective way for proving membership in computationally hard languages (e.g., $\mathcal{C}o\text{-}\mathcal{NP}$ complete ones).

In addition, CS proofs have novel and important implications for computational correctness. In particular, they imply that every computation possesses a short certificate vouching its correctness. Cryptography-based CS proofs also imply that any heuristic or program for a $\mathcal{NP}$-complete problem is cryptographically checkable. This application at the same time extends and demonstrates the wide applicability of Blum's [8] original framework for checking program correctness.

We wish to emphasize that the above mentioned applications of CS proofs to computation at large have been obtained by means of surprisingly simple arguments. Indeed, after setting up the stage for the new notion, the results about computational correctness fall with the ripeness of a Newtonian apple. This simplicity, in our opinion, eloquently vouches for the power of the new perspective.

Roots of CS proofs. In conceiving and constructing CS proofs, we have benefited from the research effort in interactive and zero-knowledge proofs. In particular, the notion of a probabilistically-checkable proof [3] [13] and and that of a zero-knowledge argument [11] have been the closest sources of inspiration in conceiving the new notion itself. In exemplifying the new notion, most relevant is a construction of Kilian's [22]; also relevant have been the works of [14] and [6].

2 Prior Notions and New Goals

The problem of defining the intuitive notion of an efficient proof has definitively attracted a lot of attention in the last three decades; indeed, the notions of $\mathcal{NP}$, $\mathcal{IP}$, and $\mathcal{PCP}$ are notable fruits of this long research effort. Yet, our developing CS proofs has been motivated by our perceiving and wishing to express some new aspects of an efficient proving process. These additional aspects are informally presented in this section, and our way to capture them formally in the next one.

Prior to sketching out our new goals, we recall some prior notions of an efficient proof, both because they have been inspiring for our venture, and because they will, by contrast, help us express what our new desiderata are. We actually find it useful going as far back as recalling the "classic notion of a proof," but in a way that instrumentally paves the road to our new notion.

2.1 Prior Notions of a Proof

Proceeding directly to give the definition of a CS Proof is attractive in that it would minimize the number of possible objections. (After all, one can define anything.) We feel, however, that CS Proofs are better understood "in context", that is, by comparing them to the notions that preceded them.

To be sure, no such comparison can be objective. Classical proofs, for instance, and CS Proofs are so *different* that any comparison presupposes reducing them first to a common "denominator". In so doing (apples and oranges as they are), one must introduce some distortion or oversimplification (e.g., overemphasizing some of their aspects at the expense of others), and rely heavily on the fact that "words" may remain the same, while their "meanings" do not. But the alternative would be not to compare at all.

Accordingly, below we provide a summary of prior notions for pure comparison purposes (no survey intended), with the understanding that its value is that of a *subjective* summary. In a sense, it is the author's account of how he came to think of CS Proofs.

The Classical Notion of a Proof

What is true? What does it mean to prove that something is true? These are questions that go hand in hand. Indeed, as formalized in the first half of this century by a brilliant series of works, the classical notion of a proof [1] is inseparable from that of a true statement. Given any finite set of axioms and inference rules,

[1] Thinking that the intuitive notion of a proof has remained unchanged from the times of classic Greece (i.e., thinking that people like Peano, Zermelo, Frankel, Church, Turing, and Gödel have only contributed its rigorous formalization and the discovery of its inherent limitations) is certainly appealing, but unrealistic. Personally, we believe that no notion so fundamental and so human can remain, not even intuitively, the same across so different spiritual experiences and historical contexts. No doubt, our yearning for permanence (dictated by our intrinsically transient nature) predisposes us to perceive more continuity in our endevours than may actually exist.

Having said this, the author firmly believes that the above mentioned formalizations of the notion of a classical proof are the only meaningful ones!)

the corresponding true statements form a *semi-recursive* set. (Again, throughout this work *true* is considered equivalent to *derivable.*) In the expressive and elegant approach of Turing, such sets possess two equivalent characterizations particularly important for our enterprise, one in terms of deciding algorithms and one in terms of (proof-)verifying algorithms:

1. A language (set of binary strings) L is semi-recursive if and only if there exists a *(deciding)* Turing machine D such that

$$L = \{x \ : \ D(x) = YES\}.$$

2. A language L is semi-recursive if and only if there exists a total *(verifying)* Turing machine V such that

$$L = \{x \ : \ \exists \sigma \in \{0,1\}^* \text{such that } V(x,\sigma) = YES\}.$$

Deciding the verity of a statement is thus a purely algorithmic process, and classical proofs —the σ's of the second definition— are just strings.

Stating that classical proofs are "just string" might appear quite diminishing. In an ordinary sense, proofs possess additional characteristics (insightfulness, elegance, usefulness, etc.) and may even be emotionally charged, but, formally speaking, a classical proof is a string satisfying special syntactic constraints. Call it "Turing machine" or "procedure", any proof system that is formal enough must specify in sufficient detail a *verification algorithm.* Proofs then are those strings that (relative to a given statement) are "accepted" by such an algorithm.[2]

We should also clarify that, though we all experience abbreviations and "reusability" of proofs to be essential to the mathematical enterprise (we rarely, if ever, deal with proofs from "scratch"), in this paper proofs are thought of as self-contained and abbreviation-free.[3]

Finally, let us recall that, for the purposes of this paper a "true" theorem simply is one derivable in a given theory. Thus, for variety of discourse, we may call a member of a semirecursive language a *theorem* or a *true statement*, and a classical proof a *derivation.*

[2] The "non-logician" should realize that this is also the case for formalizations other than Turing machines. For instance, roughly, a classical proof of a statement can be viewed as a sequence of "lines", each being an axiom (out of a given set) or obtained from previous lines according to one inference rule out of a finite set (e.g., if one line is "A" and another is "A implies B", then it is fine to write down a new line consisting of "B") whose last line is the statement in question. Here too, therefore, proofs are strings satisfying some proper syntactic constraints.

[3] Modularity is however so important that we actually wonder whether it is possible to develop a meaningful and formal model to measure the "centrality" and "usefulness" of lemmas and theorems in a given theory (e.g., one where weights are algorithmically associated to each derivable statement).

Proofs Versus Efficient Proofs

The two classical definitions of semi-recursiveness recalled above are equivalent in the sense that they identify the same sets.[4] Of course, they are syntactically different, but they are indistinguishable from the point of view of computational convenience. Indeed, classical proofs are more a way of expressing what is *in principle true* rather than a way of capturing what is *efficiently provable.* In fact, providing a derivation is certainly a way to convince someone that a given theorem is true, but is not necessarily at all efficient: a classical proof may be arbitrarily long, or its relative verifying algorithm may take arbitrarily many computational steps to verify it.

The above criticism, of course, unfairly presupposes the more modern comple xity-theoretic point of view. Indeed, going back to a recent past, one might legitimately object that what is true has nothing to do with how long it takes to determine it. It is thus better to say that the advent of complexity theory has allowed us to perceive a *new concept* by distinguishing two notions, "*truth*" and "*proof*," in what used to be —at least formally if not intuitively— an undifferentiated one.

To be sure, proving and deciding may have always been psychologically different notions. (Personally, we perceive the first as the solitary process of determining what is true, and the second as the social process of conveying to others the results of this determination.) But from a computational point of view, *if one discards efficiency*, proving would just be a synonym of deciding, whose only *raison d'être* would be variation of discourse. Indeed, if the time needed for "critically receiving" a proof were not substantially less than that of deciding, one might as well run a deciding algorithm rather than receiving a proof from someone else.

But realizing that proving should be efficient (for being a meaningful and distinct notion), and finding the "right" notion of an efficient proof are two very different things; particularly because what we perceive as "right" changes over time.

2.2 Complexity, Polynomial Time and Classical Proofs

Intuitively, we expect that the complexity of —say— deciding primality grow as a function of integer size. Indeed, underlying any notion of complexity lies

[4] Indeed, it is immediately seen that if a language L possesses a verifying Turing machine V, then it also possesses a deciding algorithm D, namely, the one that, on input a string x, looks for a classical proof that $x \in L$ by examining, in lexicographic order, all strings, stopping only when one is found. Conversely, the history of the accepting computation of a deciding algorithm D on input x is itself a string that is "easy to verify."

the question of how inputs are represented. Complexity assumes an infinity of inputs and that these inputs are coded in binary.[5]

A bit more formally, let $f(n)$ be a non decreasing function from Z^+ to Z^+, and let L be an infinite language. We say that L's complexity is $O(f(n))$ if there is a (deciding) Turing machine M such that, for (any n and) any n-bit input belonging to L, halts outputting YES within $f(n)$ steps, and rejects otherwise. (By a step of a Turing machine we mean any of its elementary operations —such as a state transition, reading or writing a symbol, etc.) By saying that a language is polynomial-time (decidable) we mean that it has complexity $O(n^c)$ for some constant c.

Sometimes, a different definition of input length is used. For instance, it is traditional -as well as natural– to assume that a graph of n nodes has input length n. "Correctly", however, this length should be $O(nlogn)$: such a graph has at most n^2 edges, and $\log m$ bits suffice to give distinct names to m objects. It should be noticed, however, that this "slight" difference in input length does not affect which graph languages are recognizable in polynomial time. (Indeed, a large part of the success of the notion of polynomial time is due to robustness properties such this.)

Given the nature of the most familiar examples, one may wonder whether the complexity point of view applies to "all proofs" and "all theorems" or to just a subset of them of a more-or-less combinatorial flavor. For instance: What is the complexity of Fermat's last theorem? What is the complexity of establishing whether a given multivariate polynomial of degree 4, $P(x_1, ..., x_n)$, has real roots? What is the length of such inputs?

To address such questions, one should realize that complexity is a very *language-dependent* notion: it all depends on how one groups together *infinitely many* strings, rather than on their combinatorial nature. (In particular, the complexities of a language and a sub-language can be vastly different.)

In fact, though addressing a combinatorial problem, . speaking of the complexity of establishing whether a *single* specific graph is Hamiltonian is not meaningful in a complexity framework: a trillion steps performed on a given graph G is a "constant number" of operations. (It is like speaking of the asymptotic behaviour of a function f whose only known value is that at point 3.) By contrast, speaking about the complexity of deciding Hamiltonicity is meaningful: one considers the infinite language consisting of all Hamiltonian graphs and then see how the computational effort of deciding membership in it grows with input size.

Similarly, the question of the complexity of Fermat's last theorem (or that

[5] Decision algorithms would appear "much faster" if inputs were coded in unary; while adopting any base greater than 2 would result in a constant savings in input length (which, for instance, would not affect the notion of polynomial time).

of a given polynomial P having real roots) is not well defined. The problem is not that Fermat's last theorem is not of a combinatorial nature. Nor does the problem consist of how one can measure the input size of Fermat's last theorem. (Its statement, in fact, though making assertions about infinitely many integers, has a finite description. Thus, choosing a suitable encoding scheme yields a clear input size: the length of the string encoding the statement.) The problem is that, in order to discuss complexity, one needs to be dealing with an infinite language. But this does not mean that the question of complexity does not apply to a classical setting! It only means that we must *re-phrase* the question in a proper manner.

While in complexity theory one is usually conscious of selecting "arbitrarily" which languages to investigate, in a classical setting one "ordinarily" assumes to be dealing with a single set of axioms (and inference rules) —the "natural" ones— and thus with a single language: that of all statements derivable in such a *fixed* system.

One can thus both formally and meaningfully consider "all theorems" as a language. All that is needed is an encoding for statements and proofs, so that a (verifying) Turing machine can be specified. This language is obviously undecidable, but complexity applies quite naturally if one considers proper sublanguages of it by bounding proof length: for instance, the language of all statements whose proof is at most n^2 bit-long (where n is the statement length), which we may conjecture to be not polynomial-time decidable.

As we shall see, for any possible language L, CS Proofs succeed in proving membership in L by means of strings that are very short, very easy to verify, and as easy to produce as possible. In particular, therefore, CS Proofs apply to the language of "all theorems," by replacing any classical proof by a much shorter string.

NP, IP, PCP, and Zero-Knowledge Arguments

A survey paper discussing in greater detail these prior complexity-theoretic approaches to the notion of a proof would be very valuable. But, as already said, ours is not such a paper: all we want is to put forward and discuss a new notion of a proof. We thus recall these prior notions only to the extent necessary to illustrate and provide context for our new one.

Non-deterministic Polynomial time. The first attempt to capture what is efficiently provable is Cook's [12], and independently Levin's [25], beautiful notion of $\mathcal{NP}$. Informally, a language L belongs to the class $\mathcal{NP}$ if every string x in it possesses a *short* (i.e., polynomially-long in the length of x) derivation, and its verifying algorithm is efficient (i.e., runs in polynomial time). Thus, $\mathcal{NP}$-proofs are a special type of classical proofs.

Interactive proofs. A non-classical notion of a proof, that of an *interactive proof*, was subsequently put forward by Goldwasser, Micali, and Rackoff [18] and independently by Babai and Moran [4]. According to their point of view, proofs are *processes* rather than syntactic objects. Because the notion of a CS proof "mirrors" to a large extent that of an interactive proof, it will be useful to recall the latter a bit. Informally, an interactive proof is a *game* between two algorithms: a Prover P, who can perform an arbitrary amount of computation, and a Verifier V, who is bound to computing in polynomial time. Given the same statement as an input, the two algorithms compute *interactively* (i.e., they send each other messages during the proving process), and *probabilistically* (i.e., they can toss coins as part of their strategies).

A Prover-Verifier pair computing as above, (P, V), is said to be an *interactive proof-system* for a language L if, on input a string x, the following two conditions hold:

1. (Completeness): If $x \in L$, then P always convinces V (i.e., V outputs YES at the end of any such computation); and,
2. (Soundness): If $x \notin L$, then, for any false prover P', the probability of P' convincing V (solely computed over the coin tosses made by the two algorithms on input x) is negligible.

It is easily seen that the class of languages having an interactive proof-system, $\mathcal{IP}$, contains $\mathcal{NP}$, but it may contain much more. (Indeed, thanks to the works of Fortnow, Karloff, Lund, and Nisan [26] and Shamir [37] we now know that $\mathcal{IP} = PSPACE$.)

Probabilistically checkable proofs. We shall talk about $\mathcal{PCP}$ [3] [13] in some detail when constructing CS proofs. For now, let us just recall the notion of a *probabilistically checkable proof* at a very high level. Quite succinctly, Babai Fortnow, Levin and Szegedy [3] and Feige, Goldwasser, Lovasz, Safra and Szegedy [13] have, independently, provided explicit algorithms transforming an $\mathcal{NP}$-witness, σ, into a new proof (i.e., string), τ, which is polynomially longer, but whose correctness can be detected in probabilistic poly-logarithmic time, by properly sampling it in a few locations rather than by reading it in its entirety. Probabilistically checkable proofs can thus be viewed as a special type of syntactic proofs.

Zero-knowledge arguments. As introduced by Brassard, Chaum, and Crépeau's [11], *zero-knowledge argument* are special proof systems. ' These systems do not make a larger class of theorems efficiently provable. Rather, in the spirit of Goldwasser, Micali, and Rackoff's prior notion of a *zero-knowledge proof* [18], they aim at minimizing the "amount of knowledge" conveyed in a "proof" of membership in an $\mathcal{NP}$ language.

Informally, a zero-knowledge argument is an interactive protocol enabling a Prover to convince a Verifier that an input belongs to an $\mathcal{NP}$-language, but without conveying any more knowledge than the mere fact that a given witness of such a membership exists. (In particular, no Verifier may learn such a witness in its entirety, nor whether w has more 0's than 1's, etc.)

In a zero-knowledge argument, both Prover and Verifier are polynomial-time machines. Further, when they interact on input a string x belonging to an $\mathcal{NP}$-language L, it is assumed that the Prover also has (on a special tape inaccessible to the Verifier) an $\mathcal{NP}$-witness, w, of $x \in L$.

At an intuitive level, the Prover sends the Verifier what may be conceptualized as "special encodings" of w,[6] and then answers some special questions of the Verifier about such encodings. The properties enjoyed by such an interaction are the following: (1) the Verifier, no matter how long he might compute, cannot learn anything about the specific witness w possessed by the Prover, and (2) if the Prover properly answers the Verifier's questions, the Verifier is guaranteed that, unless the Prover succeeds in breaking the special encryption scheme, the Prover must possess a witness of $x \in L$.

In sum, therefore, zero-knowledge arguments have two distinct aspects, correctness and zero-knowledgeness, and privilege the second rather than the first. In fact, a zero-knowledge argument offers an *unconditional* guarantee of zero-knowledgeness but only a *conditional* guarantee of correctness. (I.e., if the Prover could break the special scheme, he could cheat the Verifier into believing false statements about $\mathcal{NP}$-language membership. But, knowing that the Prover is bound to polynomial-time computation, the Verifier also knows that he will not succeed in such a breakage, but with negligible probability.)

It should be noticed that these guarantees are the exact reverse of those offered by the type of zero-knowledge proofs used earlier by Goldreich, Micali, and Wigderson to prove that any $\mathcal{NP}$-language has a zero-knowledge proof-system

[6] Each special encoding scheme, E, is actually sent by the Verifier to the Prover, is probabilistic, and enjoys the following two properties: (a) each ciphertext is equally likely to encode any cleartext (i.e., for any ciphertext C and any cleartext x, there is the same number of random strings r —of a given length— such that $C = E(x, r)$); however, (b) without succeeding in breaking the scheme, he who has generated a ciphertext C by encoding a cleartext x with a random string r, cannot decode C in any other way (i.e., he cannot find a different cleartext x' and a random string r' such that $E(x', r') = C$).

[20]. In fact, their zero-knowledge proofs offer an *unconditional* guarantee of correctness (i.e., every cheating prover, no matter how much computational power it may have, has but a negligible probability of convincing its verifier that a false statement is true), but only a *conditional* guarantee of zero-knowledgeness, that is, one holding for any Verifier that is bound to polynomial-time computation.[7]

2.3 New Goals For Efficient Proofs

A General Objective. A common objective of all prior notions of an efficient proof is guaranteeing efficient verification. We too share this objective, but provide a different interpretation of what "efficient verification" should mean. In addition, believing that a proof system should specify both the process of *proving* and that of *verifying*, we demand that proving too be efficient.

At the highest and informal level, our objective is

finding the right relationship between deciding, efficiently proving, *and* efficiently verifying *that a statement is true.*

As it is often the case, realizing and stating the objective carries in itself the means to reach it. It implicitly states that the right notion of proving should arise from that of deciding. As we have already said, we view deciding as the (solitary) process of *convincing ourselves* of what is true, and proving as the (social) process of *convincing others* of what is true, and we believe that one should be able to convince himself for being able to convince others.

Accordingly, the overall usefulness (i.e., the efficiency) of the proof of a theorem arises only when contrasted with the effort necessary for deciding it. In other

[7] That is, given much longer time to compute, the Verifier could reconstruct any piece of hidden knowledge, and thus a proof of the given statement in its entirety. Zero-knowledge proofs of this type, as introduced by [18], are called *computational zero-knowledge proofs*. In the same paper [18], also other types of zero-knowledge proofs (called *perfect* and *statistical* zero-knowledge proofs) have been introduced and explicitly constructed. These latter zero-knowledge proofs offer an unconditional guarantee of *both* correctness and zero-knowledgeness. Though this would appear to be a preferable notion, only a few notable languages are known to possess zero-knowledge proofs of this type (e.g., graph isomorphism and graph non-isomorphism [20]). Fortnow [15] has actually shown that if a zero-knowledge proof of this type were found for a single $\mathcal{NP}$-complete problem, then the polynomial-time hierarchy would collapse at the second level. Some researchers have interpreted this result as saying that this type of zero-knowledge proof is unlikely to be powerful enough so as to encompass all $\mathcal{NP}$ languages (i.e., that, most likely, one has to live with some form of conditional guarantee if one wishes to have a general enough notion of zero-knowledge proofs). By contrast, based on a very weak cryptographic assumption, [20] have shown that any language in $\mathcal{NP}$ possesses a computational zero-knowledge proof system.

words, our main and informal goal consists of arguing that the right notions of *efficiently proving* and *efficiently verifying* are not definable in "absolute terms" (as in some sense it has been done so far to express "efficient verifiability"), but must be expressed "relative to" that of *deciding.*

Three Main Goals. We articulate the above vast and vague objective into three slightly smaller and more concrete goals for an efficient proof-system.

1. **Efficient Verifiability.** We should construct proof-systems so that, for *all* theorems, the complexity of verification is poly-logarithmic in the complexity of deciding.
2. **Efficient Provability.** We should construct proof-systems so that the prover's computational complexity is polynomially close to that of deciding.
3. **Universality:** We should construct proof-systems capable of efficiently proving membership in *any* semi-recursive language.

As we shall point out in the sequel, our CS proofs also achieve additional goals, but we do not consider them essential to the notion of an efficient proof. Let us now explain the novelty of our goals by contrasting them with what was achievable by some prior proof systems.

Efficient Verifiability

Relative Efficiency. As in previous proof-systems, we too require that verification be "efficient." Differently from those previous notions, however, we do not express this efficiency requirement in absolute terms. Rather, in accordance to our premise, we regard a proof-system as making verification of a given statement efficient if it enables the computational effort of the verifier to be much easier than that he would have to invest, without any help from a prover, in order to establish the verity of the same statement. (I.e., rather than requiring that verifying be *easy* —e.g., polynomial-time, like in $\mathcal{NP}$ or $\mathcal{IP}$,— we require that it be *easier* than deciding.)

Having settled on a relative measure, we now face the problem of quantifying what computational savings are deemed to make verification efficient. We choose to consider verification efficient if its complexity is polylogarithmic in the complexity of deciding. Though somewhat arbitrary, our choice stems from two simple reasons: "logarithmic" because we wish that the advantage of verification over decision be *substantial* (whenever the decision time is substantial!), and "poly" because we wish such an advantage to be reasonably *independent* from any specific computational model.

Ubiquitous Efficiency. There is an additional, novel, and crucial aspect to our goal of efficient verifiability; namely, that such convenience should arise for *all* theorems, and not for just some rare ones. Let us explain.

According to prior notions, one might consider expressing that verification is easier than decision by saying that $\mathcal{P} \neq \mathcal{NP}$. The latter statement, however, might only entail the existence of a super-polynomial gap between decision and verification for some *rare* inputs, such as certain instances of satisfiability. (Indeed, satisfiability appears to be "easy on most inputs.")

According to our present point of view, instead, a proof-system does not make verification sufficiently efficient unless it makes it polylogarithmically easier than deciding for *all* theorems. (Of course, however, for certain trivial statements, "polylogarithmically easier" may be no savings at all in absolute terms! Indeed, for some polynomials Q and some sufficiently small values v, $Q(\log v)$ may be greater than v.)

Efficient Provability

Implicitly or explicitly, proofs involve *two* agents, a Prover and a Verifier. We thus believe that the right notion of a proof should require efficiency for *both* agents and that the efficiency of a prover should not be measured in absolute terms, but relatively to the complexity of the decision problem at hand.

To us, measuring prover efficiency relative to decision complexity is a quite natural choice. Indeed, based on our intuition that the complexity of convincing someone else cannot be lesser than that of convincing ourselves (and based on our view that deciding is the process of convincing ourselves), we believe that the complexity of proving cannot be (at least much) lower than that of deciding; while it could be much greater. We thus regard as necessary properties of an efficient proof-system not only that (1) the Prover succeeds in convincing the Verifier whenever the theorem at hand is true (the older *completeness* condition of an interactive proof-system), but also that (2) the amount of computation needed by the Prover to convince the Verifier is reasonably close to that needed to decide that the given theorem is true. We refer to the simultaneous holding of these two properties as *feasible completeness*.

Feasible completeness a novel *requirement* for proof-systems, but we may wonder whether some prior proof-systems "happened" to enjoy it anyway. The answer is possibly no. Consider, for instance a $\mathcal{NP}$-language L (preferably not $\mathcal{NP}$-complete[8]) decidable by an algorithm D in, say, $n^{\log n}$ time. Then, in the

[8] Above, we assume that L is not $\mathcal{NP}$-complete to avoid raising two issues at once. Indeed, due to our current complexity measures, $\mathcal{NP}$-proving membership in a $\mathcal{NP}$-complete language *appears* feasible. In fact, because of the self-reducibility, if L is $\mathcal{NP}$-complete and decidable in $n^{\log n}$ time, then a $\mathcal{NP}$-witness of $x \in L$ is findable

$\mathcal{NP}$ mechanism, proving that a given string x belongs to L entails finding a polynomially-long and polynomial-time inspectable witness w_x. But the complexity necessary to find such an *insightful* string may vastly exceed that of running algorithm D on input x for $n^{\log n}$ steps! Indeed, it may be as high as $O(2^n)$. In other words, while a few months of hard work may suffice for proving to ourselves (i.e., for deciding) that a given mathematical statement is true, it is conceivable that a life time may not be enough for finding an explanation followable by a verifier with a limited attention span (e.g., a child).

Efficient provability might also not hold for the $\mathcal{IP}$ proof mechanism. Indeed, often the best way to prove membership in an $\mathcal{IP}$-language consists of invoking the mentioned and general $\mathcal{IP} = PSPACE$ protocol, which is extremely wasteful of prover resources.

Realizing the importance of feasible completeness in a proof-system allows us to raise a variety of intriguing questions about $\mathcal{NP}$ and $\mathcal{IP}$.[9] But our point is not determining which proof-systems *happen* to enjoy feasible completeness. Our point is that feasible completeness is a *must* for every "sufficiently right" proof-system.

Universality

The proof-systems discussed in Subsection 2.1.3 have only a limited "range of action." For instance, an interactive proof-system (P, V) is defined only with respect to proving membership in a *specific* language L. Different languages have therefore had different interactive proof-systems, or none. Indeed, even considering the classes of all languages having a proof-system of a *given type*, among those discussed in that subsection, one obtains sets of languages (e.g., $\mathcal{NP}$, $\mathcal{IP}$, etc.) quite small with respect to the set of all recursive languages.

In contrast with those proof-systems, we consider *universality* (i.e., the capability of acting on the entire range of recursive languages) to be a necessary property of every "sufficiently right" proof-system. By this we do not just mean

in $poly(n) \cdot n^{\log n}$ time. However, as we shall see in subsection 6.5, $\mathcal{NP}$ may not enjoy feasible completeness even when one focuses solely on $\mathcal{NP}$-complete languages.

[9] For instance, in an intuitive language,

Q1: *What is the computational complexity required from any $\mathcal{IP}$-prover of Unsatisfiability?*

Q2: *What is the complexity required from any $\mathcal{NP}$-prover for, say, Graph Isomorphism?*

Q3: *Are there $\mathcal{NP}$-languages L, such that proving membership in L may require much less computation from an $\mathcal{IP}$-prover than from an $\mathcal{NP}$-prover?*

(I.e., can giving a Prover "more freedom" save him much work?) In particular,

Q3′: *What is the computational complexity required from any $\mathcal{IP}$-prover of Satisfiability?*

that every recursive language should admit a proof-system of the "right" type. We actually mean that a "right" proof-system should be able to prove membership in *any* recursive language. That is, for any language L and any member x of L, on input x and a suitable description of L, a right proof-system should be able to prove, *efficiently*, that x belongs to L.

As we shall see, we consider a deciding algorithm for L to be a suitable description of L (and one that immediately allows us to assess the efficient provability and efficient verifiability of a proof system).

3 Computationally-Sound Proofs

In this section we put forward the notion of a CS proof to approximate in a reasonable manner the goals set forth in the previous section. We actually present three main types of CS proof-systems, according to the type of randomness source they use. All types of CS proofs, however, share the same paradigm.

A new paradigm. Similarly to $\mathcal{IP}$, a CS proof-system consists of a pair of algorithms: a Prover and a Verifier. Differently from $\mathcal{IP}$, however, in a CS proof-system Prover and Verifier do not interact with one another. Rather, on input the statement of a theorem, the Prover outputs a string, called a CS proof of the theorem. The Verifier computes on inputs the statement of the theorem and an alleged CS proof of it, and either accepts or rejects.

So far, therefore, CS proofs are *syntactic objects*, much in the spirit of classical proofs. There is, however, a major difference between CS proofs and classical ones. Classical proofs ensure that all *provable* statements are *true*. CS proofs break with this tradition: they allow the existence of false proofs, but they ensure that these are computationally impossible to find. That is,

False CS proofs may exist, but they will "never" be found.

Indeed, each CS proof specifies a security parameter, controlling the amount of computing resources necessary to "cheat" in the proof, so that these resources can be made arbitrarily high. Accordingly, CS proofs are meaningful only if we believe that the Provers that have produced them, though more powerful than their corresponding Verifiers, are themselves computationally bounded.[10]

From a practical point of view, this shift of paradigm is *adequate*. Indeed, as long we restrict our attention to physically implementable Provers, no physical

[10] Those familiar with the relevant literature may appreciate that the transition from an interactive proof-system to a CS proof-system is analogous to the transition from perfect zero-knowledge proof-system to a computational zero-knowledge proof-system[18], which has proved to be a more flexible and powerful notion [20].

process in our Universe can perform $2^{1,000}$ steps of computation, at least as long the human race will exist. Thus, "practically speaking," all Provers are computationally bounded.

From a theoretical point of view, the new paradigm (though contrary to a long established tradition) is *natural*, in the sense that the new notion of a proof allows us to answer in an appealingly simple way, many of the oldest questions in complexity theory, and some new and fundamental ones as well. After all, "the right notion is the one that allows us to prove the right theorem in the right way!"

We also wish to clarify that the possibility of "proving a false statement" does not arise during the proving process (as in the case of a zero-knowledge argument[11]), Indeed, there is no interaction in a CS proof-system between Provers and Verifiers. Thus, if one wished to look for a false CS proof, he could do so "off-line", alone, and having all the elements that might be needed, or deemed useful, for such a search. And given that false CS proofs are strings, if such strings exist (and they do!), they can be surely found, *but only if more than a prescribed (and arbitrarily high) amount of computation is performed.*

If this may provide adequate protection against a cheating prover that tries to compute a good-looking CS proof, σ, of a false statement, what about such a string σ that "already exists out there"? For instance, what if, while walking on a beach we find a sand pattern that looks like the sequence of bits of such a σ?

Our answer is that though CS proofs are strings, and strings can be written in a variety of ways (on the sand, in a DNA sequence, or a plain piece of paper), we must consider the *process* that has generated them. For instance, in the above hypothetical example, the grains of sand have been arranged in such a σ-shape by natural elements such as wind, waves, sun, etc. We can thus view planet Earth as a computer, and Earth's age as its computing time, so that, in a final analysis, such a σ has been found in a few billion years: an unlikely event if we have chosen our parameters so that the age of the Universe is negligible with respect to the time necessary to find a good-looking CS proof of a false statement.

In sum,

CS proof-systems are deliberately inconsistent,
but indistinguishable, from a practical point of view, from consistent systems.

Three types of CS proof-systems. A CS proof-system may be probabilistic, because its Prover and Verifier share a *common source of randomness.* We actually distinguish three types of CS proof-systems, according to the type of

[11] Where the Prover may break an encoding scheme provided to him by the Verifier only at the beginning of an individual proving process.

randomness source Prover P and Verifier V may share (roughly said, a random oracle, a random string, and no randomness at all):

1. *CS proof-systems with a random oracle.* In this type of CS proofs, when having a given statement as an input, both P and V have access to the same random oracle. A bit more precisely (given that a random oracle can be viewed as an infinite string of random bits), P and V have oracle access to a random function $f : \Sigma^k \rightarrow \Sigma^k$, where k is a security parameter. Equivalently said, therefore, P and V can access in a single step any bit of a long random string, that is, one having length exponential in the security parameter k.
2. *CS proof-systems with a random string.* Here P and V have access to a short random string, that is, one having length polynomial in a security parameter k.
3. *Deterministic CS proof-systems.* In such systems P and V do not share any source of randomness.

The order of presentation of these three types of CS proof-systems is determined by the strength of the assumptions needed for their explicit construction. Indeed, we show how to construct explicitly CS proof-systems of the first type without relying on any additional assumption, CS proof-systems of the second type based on a new (but to us reasonable) complexity assumption, and CS proof-systems of the third type based a *very strong* complexity assumption.

All three types of CS proof-systems are *non-interactive*, in the sense that, to convince the verifier or the verity of a given statement, the Prover just sends him a single message (the CS proof), without any need to exchange messages back and forth (not even once).

As the reader may notice, however, it is possible to define *interactive CS proof-systems* which can actually be constructed more simply (in particular, their construction can be immediately obtained from that of subsection 3.1.5.) and under more traditional complexity assumptions (in particular, the existence of collision-free hash functions). We refrain from defining such proof-systems because we wish to "contain" this already long paper and because non-interactive CS proof-systems are more interesting and powerful. For instance, CS proofs with a random string and deterministic CS proofs yield some results on *checking* that are not possible to obtain with interactive CS proofs (or CS proofs with a random oracle for that matter).

3.1 CS Proofs With A Random Oracle

Significance of the model. As announced above, in a CS proof-system with a random oracle Prover and Verifier are given oracle access to a random function from $\{0,1\}^k$ to $\{0,1\}^k$, where k is an integer representing a security parameter (controlling the "trustworthiness" of the system).

Though somewhat "impractical", CS proofs of this type present a main advantage:

> *Even if $\mathcal{NP} = \mathcal{P}$, they guarantee that, given a sufficient amount of randomness in the proper form, fundamental intuitions like verification being poly-logarithmically easier than decision are indeed true.*

Let us explain. He who is concerned about truth, but not about time, does not need proofs and provers: he may be equally happy to run a decision algorithm for establishing whether a given statement holds. Proofs are in fact mechanisms that aim at quickly and "critically" transfer truths that have been hard to obtain. Thus,

> *Proofs cannot properly exist as a separate notion (i.e., separate from that of decision) unless they succeed in making verification of truth MUCH easier than deciding truth.*

The author's inclination to believe that $\mathcal{P} \neq \mathcal{NP}$ is only based on (1) his *a priori certainty* that proofs are a meaningful and separate notion, and (2) his *inclination* to believe that $\mathcal{NP}$ is a reasonable approximation to the notion of a proof. But if it turned out that $\mathcal{P} = \mathcal{NP}$, to us this would only mean that $\mathcal{NP}$ did not provide an adequate approximation. Indeed, fundamental intuitions such as proofs being an independently meaningful notion could not be shaken by a formal result such as $\mathcal{P} = \mathcal{NP}$. It is thus important to establish meaningful models in which we can *show* that *proofs do exist as an independent notion.*

CS proofs with a random oracle provide us with such a model. Better yet, they actually achieve, in their setting, *all the goals* set forward in subsection 2.2. It is thus legitimate to ask:

Where does their power come from?

That is, how can a random function help in making proofs an independently meaningful notion? The intuitive answer is that a random function

$$f : \Sigma^k \to \Sigma^k$$

essentially consists of an exponentially-long (in k) string. Thus, having f available as an *oracle* rather than as a *string* allows one to have *polynomial-time* (actually, constant-time) rather than *exponential-time* access to its bits. And it is precisely this exponential speed-up that will be converted, thanks to our specific construction, to the much more interesting exponential gap between decision and verification.

Preliminary Definitions

Oracles and oracle-calling algorithms. By an *oracle*, we mean a function $f : \Sigma^a \to \Sigma^b$, for some positive integers a and b. When the domain Σ^a and the codomain Σ^b are understood, by a *random oracle* we mean a function randomly selected among those mapping Σ^a into Σ^b.

Let A be an algorithm capable of making oracle calls. If A makes calls to a single oracle, we emphasize this fact by writing $A_{(\cdot)}$; similarly, if A makes calls to a pair of oracles, we may emphasize this fact by writing $A_{(\cdot,\cdot)}$; etc. If f is an oracle, we write A_f to denote the algorithm obtained by having algorithm $A_{(\cdot)}$ make its calls to oracle f; similarly, we write A_{f_1,f_2}, if (f_1, f_2) is the pair of oracles to which $A_{(\cdot,\cdot)}$ makes its calls; and so on.

For complexity purposes, in a computation of $A_{(\cdot,\ldots,\cdot)}$, the process of writing down a query σ to one of its oracles, f, and receiving $f(\sigma)$ in response is counted as a single step. (No result of this paper would change in an essential way if this call would "cost" $poly(a, b)$ steps whenever $f : \{0,1\}^a \to \{0,1\}^b$.)

An algorithm that, in any possible execution, makes exactly N calls to each of its oracles will be refereed to as a *N-call algorithm*.

A special language. Let $\mathcal{L}$ denote the language consisting of the triplets $\mathbf{q} = (\bar{M}, x, t)$, where $\bar{M}$ is the description (in some standard encoding) of a Turing machine M, x is a binary input to M such that $M(x) = YES$, and t is the *binary* representation of an upperbound to the number of steps M makes on input x.

Definition of a CS proof-system with a random oracle. Let (P, V) be a pair of Turing machines capable of oracle calls, the second of which runs in polynomial-time. At the start of an execution of (P, V), both machines have two common inputs, a binary *input* $\mathbf{q}$ (an alleged member of $\mathcal{L}$) and a unary *security parameter* k, and access to a common oracle f. The execution consists of running P on $\mathbf{q}$ and k with access to f, so as to produce a binary output $\mathcal{C}$, and then running V on $\mathbf{q}, k$, and $\mathcal{C}$ with access to f.

We say that (P, V) is a *CS proof-system with a random oracle* if there exist a sequence of 6 positive constants, $c_1, \ldots, c_6$ (refereed to as the *fundamental constants* of the system), such that the following two properties are satisfied:

1′. *Feasible Completeness.* $\forall$ integers $n > 1$, $\forall$ n-bit input $\mathbf{q} = (\bar{M}, x, t) \in \mathcal{L}$, $\forall$ unary integers k, and $\forall$ oracle $f : \Sigma^{k^{c_1}} \to \Sigma^{k^{c_1}}$,
(*i*) P_f halts within $(nkt)^{c_2}$ computational steps, and outputs a binary string $\mathcal{C} = P_f(\mathbf{q}, k)$ whose length is $\leq (nk \log t)^{c_3}$, and
(*ii*) $V_f(\mathbf{q}, k, \mathcal{C}) = YES$.

2'. *Computational Soundness.* $\forall\, k > n^{c_4}$, $\forall \mathbf{q} \notin \mathcal{L}$, and $\forall$ deterministic (cheating) algorithm P' making $\leq 2^{c_5 k}$ oracle calls, for a random oracle $\mathcal{R} : \Sigma^{k^{c_1}} \to \Sigma^{k^{c_1}}$,

$$Prob_{\mathcal{R}}(V_{\mathcal{R}}(\mathbf{q}, k, P'_{\mathcal{R}}(\mathbf{q}, k)) = YES) < 2^{-c_6 k}.$$

If (P, V) is a CS proof-system with a random oracle, the process of running P on an input $(\bar{M}, x, t)$ with a security parameter k and a random oracle $\mathcal{R}$ will be refereed to as a *random-oracle CS proof*, and the output of P will be refereed to as a *random-oracle CS witness* or a *random-oracle CS certificate* (of security k) of $M(x) = YES$. If it is clear from the context that we are dealing with CS proofs with a random oracle, we may actually simplify our language by dropping the qualifications "with a random oracle" and "random-oracle."

Discussion

Deterministic Cheating. In the above definition of a CS proof-system with a random oracle, we have considered cheating provers P' to be deterministic. This choice simplifies the notational burden of later sections a bit, but does not cause any loss of generality. Indeed, since we are not concerned about the size of the description of P', nor about its running time (except for the number of oracle calls it may make), we could easily "wire-in" any "lucky" sequence of coin tosses if probabilism might help cheating.

Verified Cheating. Without loss of generality, the cheating provers of computational soundness always halt: either outputting nothing, or outputting *verified* certificates. That is, for all oracles f, if $\mathcal{P}'_f(q', k) = \mathcal{C}' \neq \varepsilon$, then, prior to outputting $\mathcal{C}'$, $\mathcal{P}'$ has appropriately called oracle f so as to verify that $\mathcal{V}_f(q', k, \mathcal{C}') = YES$.

Choosing the security parameter. Informally, the security parameter k controls the probability of something going wrong, and CS proofs become meaningful only when k is chosen large enough. Now, what should our considerations be when choosing k?

In the physical world as we know it, no one can perform 2^{300} computational steps, and any probability less than 2^{-300} is undistinguishable from 0. This suggests choosing our security parameter so that, by making less that 2^{300} oracle calls, one cannot cheat with probability greater than 2^{-300}.

The above choice of k may be particularly adequate when dealing with a single theorem statement, but if one considers all the false statements which are 300-bit long, then the probability that a cheating prover, in a relatively few oracle calls, may construct a good-looking CS certificate of correctness for at least one of them may be constant, because there may be up to 2^{300} of such false statements. It may thus be preferable to ensure that, with probability $> 1-2^{-300}$, no false statement, no matter how long, has a feasibly-found CS certificate of correctness. This is easily accomplished by choosing k as a (fixed) moderately-growing function of n, the input length. Indeed, we recommend in our definition that k should grow polynomially in the input length: $k > n^{c_4}$.

Finally, letting k grow with n maximizes the meaningfulness of a CS proof-system (P, V) in a theoretical sense. Loosely speaking, when "CS-proving" that $(\bar{M}, x, t) \in \mathcal{L}$, we should choose k so that the effort necessary for cheating, even with a miniscule probability of success, is much larger than the decision time, t. In fact, a Verifier who is willing to check a CS witness for "$M(x) = YES$ within t steps," is implicitly acknowledging that in the world there is already someone —namely, the honest P himself— who can perform $poly(k, t)$ steps of computation.

In other words, after having argued that "feasible proving" and "efficient verifying" should not be expressed in absolute terms, but relative to the complexity of deciding, we now argue that the "hardness of cheating" should not be expressed in absolute terms either. Rather, we should require that cheating is "much harder than deciding." But, what values may be reasonably considered to be much greater than t (and actually achieved by an explicit construction)?

Our answer is $t^{O(\log t)}$. In fact, such difficulty of cheating is easily enforceable by choosing, as we do, k to grow polynomially in n. For instance, assume that one chooses $c_4 \geq 2$, thereby lowerbunding k by n^2. Such a lowerbound guarantees that the number of oracle calls necessary to a cheating prover to cheat with probability of success $2^{c_6 k}$ must exceed $2^{c_5 n^2}$. Now, the fact that $n > \log t$ (because the binary value t is part of the n-bit input $\mathfrak{q}$) implies $t^{c_5 \log t}$.

In addition to the above reasons, choosing k to grow polynomially in n is also essential to our proof of Theorem 1, where we show that an explicit pair $(\mathcal{P}, \mathcal{V})$ is a CS proof-system with a random oracle.

Achieving our goals. Let us now see how our goals for the "right" notion of a proof have been "hard-wired" in the definition of CS proofs.

1. *Efficient Verifiability.* Our first goal required that, for every theorem, verifying should be polylogarithmically easier than deciding. This goal is approximated by a CS proof-system (P, V) in the following sense.
   Let L be any recursive language, x be any member of L, and D any deciding algorithm for L. Then, the theorem $x \in L$ can be verified by running D on input x and verifying that $D(x) = YES$. Assume now that the latter computation takes t steps. Then, by choosing a proper security parameter k and running P on inputs $(\bar{D}, x, t, k)$ and access to a random oracle $\mathcal{R}$, one obtains a CS proof, σ, of $x \in L$ such that (a) σ's length is bounded by a fixed

polynomial in $\bar{D}$, k, and $\log t$, and (*b*) σ is accepted by $V_{\mathcal{R}}$ running on input $(\bar{D}, x, k)$. Therefore, being V polynomial-time, there is a fixed polynomial Q such that V verifies σ within $Q(|\bar{M}|, |x|, \log t, k)$ steps.
In sum, no matter what algorithm one specifies for deciding whether $x \in L$, there is a corresponding CS proof of $x \in L$ which is verifiable in a time that is polylogarithmic in the number of steps taken by that algorithm on input x (and polynomial in all other parameters).

2. *Efficient Provability.* The second goal called for the complexity of proving being polynomially closer to that of deciding. This property is immediately guaranteed by the feasible-completeness of a CS proof-system. Feasible completeness in fact states that there exists a fixed polynomial $Q(\cdot, \cdot, \cdot)$ such that, if an algorithm D decides in t steps that a string x belongs to a recursive language L, then a CS prover can, on input $\mathbf{q} = (\bar{D}, x, t)$ and security parameter k, demonstrate that $x \in L$ within $Q(|\mathbf{q}|, k, t)$ steps.
3. *Universality.* Our third goal called for proof-systems capable of proving membership in all possible semi-recursive languages. In apparent contrast with this requirement, a CS proof-system is defined to prove membership only in the very special language $\mathcal{L}$. But $\mathcal{L}$ is designed so as to encode membership questions relative to any possible semi-recursive language. In fact, to each semi-recursive language L corresponds a (deciding) Turing machine M_L so that $x \in L$ if and only if M_L, on input x, outputs YES in some number of steps t. Thus, $x \in L$ if $(M_L, x, t) \in \mathcal{L}$, thus achieving the third goal.

The explicitness of t. There are two reasons for explicitly including in $\mathcal{L}$ the number of steps, t, within which M_L accepts x. First, our explicitly constructed CS proof-systems while certifying that $M_L(x) = YES$ also certify within how many steps M_L outputs YES. Second, as we have already argued in this subsection, the value t —or, at least, an upperbound for it— must be known in order to choose meaningfully the security parameter k.

An additional property. CS proofs also satisfy additional properties. In particular, like in a classical proof but unlike in an interactive proof, a Verifier who has inspected a CS proof can easily convince someone else of the verity of its corresponding statement. Indeed, being CS proofs strings checkable with a publically available oracle, all the Verifier has to do is to send someone else the same string he has successfully verified. (We do not wish, however, to make this property a requirement for any "sufficiently right" notion of a proof. Indeed, there are important contexts —in particular, cryptographic ones— in which it may be preferable or advantageous that a convinced verifier cannot convince anyone else. Indeed, this is one of the main advantages of a zero-knowledge proof.)

Back to zero-knowledge arguments. We can now better contrast CS proofs with Brassard, Chaum and Crépeau's zero-knowledge arguments. Taking aside zero-knowledge considerations, there are the following differences:

- Zero-knowledge arguments are interactive.
  (Incidentally, interactive versions of CS proofs can be also defined and constructed in an easy manner, though we omit doing so not to burden further this paper.)
- Zero-knowledge arguments may not enjoy Efficient Verifiability.
  (In fact, $\mathcal{NP}$ may not satisfy Ubiquitous Efficiency, an important sub-requirement of Efficient Verifiability.)
- Zero-knowledge arguments may not enjoy Efficient Provability.
  (Indeed, on input a member x of a $\mathcal{NP}$-language L, it is assumed that their Provers possess, "for free", a $\mathcal{NP}$-witness of $x \in L$. But, as we have already pointed out in Subsection 2.2.2, the time necessary for a Prover to find such a witness may vastly exceed that necessary to decide that $x \in L$.)
- Zero-knowledge arguments do not enjoy Universality.
  (Indeed, they are only designed to work for $\mathcal{NP}$-languages.)

(We wish to point out, however, that it might be possible to change the definition and the design of zero-knowledge arguments so as to enjoy most, if not all, of the properties they miss so far. Presumably, this could be accomplished by using Kilian's construction, the ideas of our paper, and the non-interactive zero-knowledge proofs of [7] and [6]. Investigating further this possibility is beyond the scope of the present paper.)

A paradox. CS proofs are paradoxical in that a computationally-bounded prover appears able to "prove more theorems" than an unbounded one. Indeed, if we choose k's value as a suitable function of the input length, then a properly-bounded CS prover can demonstrate membership in any $EXPTIME$ language to a Verifier whose running time is upperbounded by a fixed polynomial of the input length alone. By contrast, the unbounded prover of an interactive proof-system can only prove membership in $PSPACE$ languages to a polynomial-time verifier, and it is widely believed that $PSPACE$ is a proper subset of $EXPTIME$.

A few years ago, this seemed puzzling to us.

Deconstructing our construction

Warning. The reader hoping to see a catchy, simple example of a CS Proof system is about to be disappointed. Our construction of a CS proof-system is hardly practical, and we do not know of any simpler ones. In my "defense" I can only say that not everything needs to be practical (I trust that most abstract mathematicians would agree), and that things important often become simpler with time.

Because of the construction's complexity, we shall use the present subsection to provide an informal introduction to it, and to highlight its ideas. This subsection, to be sure, is itself quite long (though hopefully less tedious than the construction itself and its proof). This is purposely so. We in fact wish to give the reader a chance to bypass subsections 3.1.5 and 3.1.7 altogether. We encourage him, however, to glance at 3.1.6 (because it is possible to believe that the construction works for the wrong reasons!).

Our construction is based on an earlier one of Kilian's [22], which is itself based on Merkle's trees and probabilistically-checkable proofs. Let us thus start by recalling these other notions.

Probabilistically-checkable and samplable proofs. Very recently, Babai Fortnow, Levin and Szegedy [3] and Feige, Goldwasser, Lovasz, Safra and Szegedy [13] have put forward, independently and with different aims,[12] some related and important ideas sharing a common technique: *proof-samplability*. A bit more precisely, they present an explicit algorithm transforming a $\mathcal{NP}$-witness, σ, into a new proof (i.e., string), τ, which is polynomially longer, but whose correctness can be detected by properly sampling it in a few locations rather than by reading it in its entirety.

Though the original constructions have been greatly improved (in particular, by Arora and Safra [2]), the simpler technique of [13] suffices for the goals of this paper. Actually, for simplicity purposes, we shall rely on the following "stripped-down" version of their technique, which we call *samplable proofs*.[13]

[12] The authors of [3] focus on proofs of membership in $\mathcal{NP}$-languages, and show that it is possible to construct Verifiers that work in time poly-logarithmic in the length of the input. (Since in such a short time the Verifier could not even read the whole input —and thus check that the proof he is going to sample actually relates to the "right" theorem,— these authors have devised a special error-correcting format for the input, and assume that it is presented in that format. An input that does not come in that format can be put into it in polynomial-time.)

The authors of [13] use proof-samplability to establish the difficulty of finding approximate solutions to important $\mathcal{NP}$-complete problems. (With this goal in mind, these other authors do not mind Verifiers working in time polynomial in the length of the input, and do not use or need that inputs appear in any special format.)

[13] Indeed, a probabilistically-checkable proof is a richer and more flexible notion, allowing one to specify a wide range for the number of queries the Verifier may make, and for the number of coins it may toss. Moreover, the careful reader may realize that, even restricting one's attention to $PCP(\log(n)^{O(1)}, log(n)^{O(1)})$, the PCP techniques yield a reacher structure than samplable proofs; for instance, samplable proofs do

Definition. A *sampling proof-system* consists of an elementary interactive proof-system, (SP, SV), where both the sampling prover SP and the sampling verifier SV run in probabilistic polynomial time. On input x (a n-bit string belonging to a given $\mathcal{NP}$-language L) and w_x (a $\mathcal{NP}$-witness that indeed $x \in L$), SP computes a (slightly longer) string w'_x, a samplable proof that $x \in L$. On input strings x (candidate member of L) and random access to σ (candidate samplable proof for $x \in L$), Verifier SV, after accessing only poly-log(n) bit-locations of σ, outputs either ACCEPT or REJECT with the following constraints. If $x \in L$ and $\sigma = SP(x, w_x)$ for some correct $\mathcal{NP}$-witness w_x, then SV always outputs ACCEPT. But if $x \notin L$, then, $\forall \sigma$, SV outputs REJECT with probability $\geq 1/2$.

The reader familiar with the recent advancements on probabilistically-checkable proofs —in particular the beautiful papers of Arora and Safra [2], Arora, Lund, Motwani, Sudan, and Szegedy [1], Sudan [39], Polishuck and Spielman [30],— will realize that, for the sake of simplicity, we are sacrificing quite a bit of efficiency.[14]

Merkle's trees. Merkle trees [27] are based on binary trees and collision-free hash functions.

- A *binary tree* is a tree in which every node has at most two children (in which case, each one of them is said to be the sibling of the other).
- A *collision-free hash function* is, informally speaking, a polynomial-time computable function H mapping binary strings of arbitrary length into reasonably short ones, so that it is computationally infeasible to find two different strings x and y for which $H(x) = H(y)$. Such pair of strings x and y is called a *collision (for H)*. (Popular candidate collision-free hash function is the standardized *secure hash function* [36] and Rivest's MD4 [35].)
- A *Merkle tree*, informally speaking, is a binary tree whose nodes store (i.e., are associated to) values, some of which computed by means of a one-way hash function H in a special manner. A leaf node can store any value, but each internal node should store a value that is the one-way hash of the concatenation of the values in its children. (I.e., if an internal node has a 0-child storing the value U and a 1-child storing a value V, then it stores

not guarantee that any particular bit of the samplable proof is accessible by the Verifier with positive probability —only the final (and proper) ACCEPT and REJECT probabilities are guaranteed.

[14] Indeed, Polishchuk and Spielman show that a $\mathcal{NP}$-witness with length n possesses a probabilistically-checkable version that is only $n^{1+\epsilon}$-bit long for any constant $\epsilon > 0$. Nonetheless, we shall ignore polynomial improvements in the running time of sampling provers and poly-log improvements in the running time of sampling verifiers. Unlike for approximation theory, in fact, improvements in the efficiency of proof-samplability will only affect the efficiency of CS proofs.

the value $H(UV)$. If a child of an internal node does not exist, we assume by convention that it stores a special value, denoted by EMPTY.)
If the one-way hash function produces k-bit outputs, then each internal node of a Merkle tree, including the root, stores a k-bit value. Except for the root value, each value stored in a node of a Merkle tree is said to be a 0-value, if it is stored in a node that is the 0-child of its parent, a 1-value otherwise.

The crucial property of a Merkle tree is that, unless one succeeds in finding a collision for H, *it is impossible to change any value in the tree without also changing the root value.* In particular, one cannot change the original values stored in the leaves without changing also the root value.

This property allows a party A to "commit" to n values, $V_1, \ldots, V_n$ (for simplicity assume $n = 2^a$ for some integer a), by means of a single k-bit value. That is, A stores value V_i in the ith leaf of a full binary tree of depth d, and uses a collision-free hash function H to build a Merkle tree, thereby obtaining a k-bit value RV stored in the root. This value RV "implicitly defines" what the n original values were. Assume that A gives RV, but not the original values, to another party, B, at some point in time. Then, whenever, at a later point in time, A wants to "prove" to B what the value of, say, V_i was, he may just reveal all n original values to B, so that B can recompute the Merkle tree and the verify that the newly computed root-value indeed equals RV.

More interestingly, A may "prove" what V_i was by revealing just $d+1$ (i.e., $\log n + 1$) values: V_i together with its *authentication path*, that is, the values stored in the siblings of the nodes along the path from leaf i (included) to the root (excluded), $Y_1, \ldots, Y_d$. B verifies the received alleged leaf-value V_i and the received alleged authentication path $Y_1, \ldots, Y_d$ as follows. Letting $i_1, \ldots, i_d$ be the binary expansion of i, B sets $V = X_1$ and computes the values $X_2, \ldots, X_d$ as follows: if $i_j = 0$, she sets $X_{j+1} = H(Y_j X_j)$; else, she sets $X_{j+1} = H(X_j Y_j)$. Finally, B checks whether the computed k-bit value X_d equals RV.

Kilian's construction. In [22], Kilian presents a zero-knowledge argument for $\mathcal{NP}$, (P, V), exhibiting a polylogarithmic amount of communication, where prover P uses Merkle trees in order to provide to V "virtual access" to a samplable proof.

In essence, disregarding zero-knowledge considerations, the polynomial-time P, as in any zero-knowledge argument, possesses a polynomially-long witness, w, proving that a given input x belongs to a given $\mathcal{NP}$-language L. P thus transforms w in a longer, but still polynomially-long in the length of x, samplable proof w'. In order to obtain a logarithmic amount of computation, P cannot send V witness w, nor can he send him the longer samplable proof w'. Rather, P uses a Merkle tree with a collision-free hash function H, producing k-bits outputs, as above to compute a k-bit string, RV, that commits him to w', and then sends RV

to the verifier V. (For instance, disregarding further efficiency considerations, if w' is n-bit long and , for simplicity, n is a power of 2, the ith bit of w' is made to be the original value V_i in the above described construction, the Merkle tree is a full binary tree of depth $\log n$, and RV is the k-bit value stored in its root.)

Verifier V runs as a subroutine a sampling verifier SV. When SV wishes to consult the jth bit of w', V asks P for it, and P responds by providing the original value b_j together with its authentication path. V then checks whether b_j's authentication path is correct relative to RV, and, if so, he is assured that b_j is the original value because he trusts that P, being polynomial-time, cannot find a collision for H. V then feeds b_j to SV. The computation proceeds this way until V finds that an authentication path is incorrect, in which case it halts and REJECTS, or until SV halts, in which case V REJECTS if SV does, and ACCEPTS otherwise. Because SV "virtually" accesses a logarithmic (in n) number of bits of w', and because each such a virtual access is answered by $k \log n$ bits of authentication path, the overall amount of communication is logarithmic in n and thus in the length of x.

Notice that the above construction only shows how a Verifier can be given virtual access to w'. Let us reiterate that, in order to obtain a communication efficient zero-knowledge argument, Kilian's construction is actually more complicated, but the additional zero-knowledge constraint is irrelevant for our goals.

Our first variant. Essentially this same construction (minus its zero-knowledge addition) was independently found by the author [29], but our version included a small, essentially conceptual, variant which is crucial to the present enterprise. Indeed, while Kilian aimed at providing more efficient zero-knowledge arguments for $\mathcal{NP}$, we aimed (as we do now) at providing a more general and powerful notion of a proof.

Our variant follows from the following two, simple but important, observations: (1) realizing that probabilistically-checkable proofs are more general than traditionally claimed, and (2) considering proofs of membership in a given language as individual problems. Let us explain.

1. The salient features of samplable proof-systems, although traditionally claimed for $\mathcal{NP}$ languages, actually hold in a more general context. Indeed, the results of [13] and [3] (sometimes referred to as "poly-PCP") are mostly seen as a "reduction of PCP to NP." To us this is a *big understatement* of their results. (For instance, it would not allow us to prove Theorem 1 at all.) In fact, a more careful reading of those papers yields the following

   Theorem 0 ([13], [3]): There exists a deterministic algorithm $SP(\cdot,\cdot)$, a probabilistic algorithm $SV(\cdot,\cdot)$ running in time polynomial in the first input and polylogarithmic in the length of the second input, and two polynomials

$\ell(\cdot,\cdot)$ and $\lambda(\cdot,\cdot)$ such that, for any polynomial-time relation R over $\Sigma^* \times \Sigma^*$ (i.e., $R(x,y) = 1$ may not imply any a priori bound for the length of y relative to that of x),

(a) for any strings x and y such that $R(x,y)$ holds, algorithm $SP(x,y)$ halts within $\ell(|x|,|y|)$ steps outputting a string y'; and algorithm SV, on inputs x and $|y|$ and random access to y', flips at most $\lambda(|x|,|y|)$ coins and accepts; and

(b) For any string x such that $\forall y\ R(x,y) = 0$, and for any string σ, the probability (computed over SV's coin tosses) that algorithm SV, on inputs x and $|\sigma|$ and random access to σ, flips at most $\lambda(|x|,|y|)$ coins and accepts is at most $1/2$.

2. If the first observation consists of realizing that samplable proofs apply "above $\mathcal{NP}$, the second observation consists of realizing that one can meaningfully use cryptography, in particular the infeasibility of finding collisions in a collision-free hash function, in order to keep honest a Prover capable of proving membership in languages "above $\mathcal{NP}$".
Erroneously adopting a complexity-classes perspective, using cryptography "against" such a Prover was not considered meaningful. Though re-evoking in a convincing way past conceptual barriers is impossible, one could have raised the following "objection":

> because a collision-free hash function H is polynomial-time computable, finding two strings x and y such that $H(x) = H(y)$ is "within $\mathcal{NP}$". Thus, how can a Prover with more than $\mathcal{NP}$-power be kept honest by the difficulty of finding collisions?[15]

This objection vanishes if one views a Prover, working on an input string x and a given language L, not as an universal mechanism for proving membership in L or in a given complexity class, but as an *individual process/device*, endowed with an *individual amount of computational resources*, sufficient for the *individual input* at hand.
Often, this amount of computational resources can be upperbounded in a natural and absolute manner, no matter what the complexity of the language L may be. For example, if x is n-bit long and L is $PSPACE$-complete, then there is a positive constant c such that 2^{n^c} steps certainly suffice to determine whether $x \in L$.
Thus, no matter how high n may be, it is meaningful, for a Verifier that has interacted with a Prover so as to become convinced that a specific n-bit input x belongs to L, to believe that he has interacted with some agent capable 2^{n^c} steps of computation: indeed, $\Omega(2^{n^c})$ could be L's complexity, and the agent has convinced the Verifier to be capable of proving membership in L!

[15] Notice that cryptography was used against the Prover of a zero-knowledge argument, but that such a Prover was defined to be polynomial-time!

But it is also meaningful, for the *same* Verifier, after the *same* interaction, to believe the *same* agent incapable of —say— $2^{(n^c)^2}$ steps of computation. Thus, if, for some constant d, the complexity of collision-finding for a k-bit-output hash function H is $\Omega(2^{k^d})$, and if k is greater than $(n^c)^{2/d}$, it is totally meaningful to believe that the same agent cannot find a k-bit collision for H.

The above two remarks yield what may be called an *interactive* CS proof-system, $(\mathcal{P}, \mathcal{V})$ (a notion that, as already said, we refrain from formalizing for more quickly getting to the notion we really care about). Namely, by adopting a proper representation of "the input" and by using collision-free hash functions whose output is suitably larger than the length of the input at hand, we can use Kilian's construction to prove membership into any language, of any complexity class, while satisfying both Efficient Verifiability and Efficient Provability.

In essence, on input $\mathbf{q} = (\bar{M}, x, t) \in \mathcal{L}$, $(\mathcal{P}, \mathcal{V})$ works as follows. First, $\mathcal{P}$ runs machine M on input x so as to generate, in t steps, the history (i.e., sequence of instantaneous configurations), σ, of an accepting computation of M. Such an history σ is then thought of as a proof that $M(x) = YES$. This proof will not be insightful, and, because no restriction is put on M, can be arbitrarily long relative to x.

Consider now the following relation R: $R(x, \sigma) = 1$ if and only if σ is the (t-step) history of an accepting computation of M on input x. Then, R is $poly(|x|, |\sigma|)$ computable. Thus, due to Theorem 0, σ can be put in a probabilistically-checkable form τ by algorithm SP (the prover of a sampling proof-system (SP, SV)) within $poly(|x|, t)$ steps. The so obtained τ is efficiently checkable by algorithm SV with $poly(|x|, \log t)$ coin tosses.

But, while proving that "$M(x) = YES$ in t steps," τ is again too long. Thus, assuming that finding a collision in an k-bit-output collision-free hash function requires 2^{k^d} for some constant d, a collision-free hash function H whose output has size $(\log t)^{2/d}$ is chosen and, instead of handing τ to verifier $\mathcal{V}$, $\mathcal{P}$ gives him virtual access to τ using, as above, a Merkle tree with a hash function H. This increases moderately (i.e., by a factor polynomial in $\log t$) the amount of computation of $\mathcal{P}$, but makes cheating practically impossible for a malicious Prover (conceived as an individual process/device endowed with an individual amount of computational resources).

In sum, without any contradiction, we may keep honest a Prover working on a given individual problem by rescaling and choosing a much harder individual problem.

Our second variant. The CS proof-system with a random oracle presented later on differs from the above outlined $(\mathcal{P}, \mathcal{V})$ in another way.

On the minor side, rather than an k-bit-output collision-free hash function, the new construction uses a random oracle-function mapping $2k$-bit strings to k-bit strings. Indeed, finding collisions for such a "function" is provably hard (and can be quite precisely computed).

More importantly, we use our random oracle for obtaining a non-interactive CS proof system. Notice, in fact, that the proof system $(\mathcal{P}, \mathcal{V})$ informally described so far is interactive. Indeed, during the proving process, the sampling verifier SV keeps on computing which bits in the samplable proof it wishes to see; verifier $\mathcal{V}$ sends then these requests to prover $\mathcal{P}$; and prover $\mathcal{P}$ responds with the requested bits and their authentication paths.

Throughout the entire process, subroutine SV flips coins; more precisely, it uses the same random tape RT fed to it by verifier $\mathcal{V}$ at the start of SV's computation. Because RT is genuinely random, when the input statement "$M(x)$ =YES in t steps" is false and verifier $\mathcal{V}$ interacts with a malicious prover $\mathcal{P}'$ that does not succeed in finding a collision for the random oracle, SV and thus $\mathcal{V}$ accept with probability at most 1/2.

This probability can be made at most 2^{-k} by repeating the whole process k times, each time using an independently-selected random tape RT.

Now it can be seen that if k (and thus the number of bit positions virtually accessed by the sampling verifier SV throughout its k runs) is sufficiently small with respect to the length of the samplable proof (which we would like to be our case in order to satisfy Efficient Verifiability), and if a malicious Prover knew in advance which k random tapes the Verifier is going to feed SV, then he could cheat with a much higher probability, even with probability 1. However, roughly said, if the k tapes are selected *after* the malicious Prover provides the root value RV, then, roughly said, unless he succeeds in finding at least one collision for the random oracle, his probability of cheating is still 2^{-k}. This continues to be true even if he receives the so selected k tapes in their entirety, rather than just the bit-requests that the sampling verifier computes from them.

This suggests replacing interaction in the above proof-system as follows. On input the statement "$M(x)$ =YES in t steps", Prover $\mathcal{P}$, as before, (a) computes a classical proof of it by running M on input x and listing the sequence of instantaneous configurations of the computation, (b) puts this classical proof in samplable form, and (c) stores this samplable proof in the leaves of a suitable binary tree and constructs a corresponding Merkle tree, using the random oracle in lieu of a collision-free hash function, so as to compute a root value RV. At this point, $\mathcal{P}$ uses the random oracle (actually, an independent random oracle "extracted" from the given one) on input RV so as to compute k suitably-long random tapes, $RT_1, \ldots, RT_k$. He then runs ("in his head") verifier $\mathcal{V}$ and its subroutine SV as in the whole process described above, for k times, using RT_i as SV's random tape in the ith iteration. Therefore, he computes (in his head) all the bit-locations of the samplable proof that SV requests to access, and outputs, as a CS proof with a random oracle for "$M(x)$ =YES in t steps", the value RV and the requested bits, each with its own authentication path relative to RV. Such proof can be verified, in the obvious way, by using verifier $\mathcal{V}$ (with subroutine SV) and the same random oracle.

Consider now a malicious Prover trying to "CS-prove with a random oracle" a false statement. Of course, he can choose a root value RV' of his liking and consult the oracle so as to see whether he can produce a good-looking CS proof relative to RV'. However, roughly said, because for each RV' (as long as he does not succeed in finding a collision for the random oracle), his chance of finding a good-looking proof is at most 2^{-k}, we expect that he tries 2^k times before he succeeds. Thus, if k is large enough, and the running time of the malicious prover is properly and meaningfully upperbounded, his chance of finding a CS proof of a false statement is negligible.

Our variant is reminiscent of a step used by Fiat and Shamir [14]. Indeed, they construct their digital signature scheme by starting with an interactive two-party protocol, in which the first party sends a first message to the second party and the second party responds with a random string, and then replacing the random message of the second party by evaluating a one-way function on the first party's message. (Prior to that, Manuel Blum actually suggested to the author using a random oracle in order to replace interaction in a zero-knowledge proof-system, though this suggestion never appeared in print.[16])

A conceptual contribution. We believe the main contribution of this paper to be a *conceptual* one. Namely (in addition to new but natural requirements such as feasible completeness) our paper provides a complexity-based notion of a proof which is *universal* (i.e., applicable to all recursive languages) and deliberately *inconsistent*, but in a *controllable* (and meaningfully advantageous) way.

It has been surprising to many (and to us in particular) why such a notion had not been put forward much longer ago. Indeed, the main algorithmic ideas of our construction (1) samplable proofs [13] [3], (2) Merkle's trees [27] and (3) the Fiat-Shamir signature scheme [14], have been individually known for quite a while. In addition, (1) and (2) had already been used *together* in [22]. (To these ideas we just added the two simple remarks mentioned above[17] and the use of

[16] His suggestion was in fact the starting point of the non-interactive zero-knowledge proofs of [7] and [6]. The suggestion did not enter the final version of these papers because, for their specific task of zero-knowledge proofs of membership in $\mathcal{NP}$ languages, we succeeded in replacing random oracles by short random strings, under quite-standard complexity assumptions.

[17] I.e., realizing that probabilistically-checkable proofs are more general than tradition-

random oracles.) Yet, their techniques have been successfully applied to many contexts (approximability of $\mathcal{NP}$-complete problems, zero-knowledge arguments for $\mathcal{NP}$, etc.), but not to distilling a new notion of a proof.

Perhaps, we have been focusing on consistency and efficient verifiability for so long that we became unable (the author first) to see any other issues and desiderata in the notion of a proof.

Having presented all the ideas entering in our construction at an intuitive level, let us now proceed more formally.

Our construction

The readers finding this (semi-) algorithmic rendition of the above ideas somewhat superfluous has my full solidarity: I hate "programming" in all its approximations. However, failure to provide an explicit algorithmic form would have upset plenty of other readers. (One simply cannot please everyone!)

From one oracle to two oracles. According to our definition, in a CS proof-system with a random oracle Prover and Verifier have oracle-access to a single function f, where feasible completeness holds for any f, and computational soundness for a random f.

It will be easier, however, to exemplify a CS proof-system with a random oracle $(\mathcal{P}, \mathcal{V})$, where $\mathcal{P}$ and $\mathcal{V}$ have oracle-access to two distinct functions: f_1 and f_2, where feasible completeness holds for any possible choice of f_1 and f_2, while computational soundness when f_1 and f_2 are random and independent.

Oracle-access to these two functions can be simulated by accessing a single, properly selected, function f: to ensure that f_1 and f_2 are randomly and independently selected when f is random, we arrange that whenever $(i, x) \neq (j, y)$ no query made to f in order to compute $f_i(x)$ coincides with a query made to f in order to compute $f_j(x)$. For instance, if, for $i = 1, 2$, $f_i : \{0,1\}^{a_i} \rightarrow \{0,1\}^{b_i}$ (for some positive integer values a_i and b_i, $i = 1, 2$), letting f map $\{0,1\}^{a_1+a_2}$ into $\{0,1\}^{b_1+b_2}$, allows us to achieve our goal quite straightforwardly.

Length bounds. From Theorem 0 we distill the following

Definition: Let SP, SV, ℓ and λ be as in Theorem 0. Then, we shall refer to (SP, SV) as a *sampling proof system*, and to ℓ and λ as its *length bounds* (respectively, for the samplable proof produced by SP and the random tape used by SV).

---

ally claimed, and considering proofs of membership in a given language as individual problems.

Note that in our setting we deal with membership in the special language $\mathcal{L}$, consisting of triplets $\mathbf{q} = (\bar{M}, x, t)$ such that $M(x)$ accepts within t steps, and we are interested in putting in samplable form a proof y consisting of the history of the computation of $M(x)$, so that y's length is bounded by a fixed polynomial in t. In fact, y is describable by a Turing-machine tableau of side t. Thus, if we denote by n the binary length of $\mathbf{q}$, then $|x| + |\bar{M}| \leq n$, $t < 2^n$, and y's length is bounded by a fixed polynomial in n. Accordingly, in our setting, we can view length bounds ℓ and λ as polynomials in the single variable n.

Notation. We denote the empty word by ε, the set $\{0,1\}$ by Σ, the set of all natural numbers by $\mathcal{N}$, the set of all positive integers by Z^+, the concatenation of two strings x and y by $x|y$ (or more simply by xy), and the complement of a bit b by $\bar{b}$.

We denote the length of a binary string α by $|\alpha|$. If α is a n-bit string whose ith bit is α_i, then for any integer j between 1 and n we let $\alpha[1 \cdot\cdot j]$ denote the j-bit string $\alpha_1 \cdots \alpha_{j-1}\alpha_j$. We let $\alpha[1 \cdot\cdot \bar{j}]$ the string obtained by complementing the last bit of $\alpha[1 \cdot\cdot j]$, that is, $\alpha[1 \cdot\cdot \bar{j}] = \alpha_1 \cdots \alpha_{j-1}\bar{\alpha_j}$. Finally, if α is any binary string and i a natural number, we let $\alpha[i+1 \cdot\cdot]$ denote α after deleting its first i bits (i.e., if α is a n-bit string, $\alpha[i+1 \cdot\cdot] = \alpha[i+1 \cdot\cdot n]$.)

If N is a power of two, we let $\mathcal{T}_N$ denote the complete binary tree with N leaves, labeled so that v_ε is the root, v_0 and v_1 are, respectively, the left and right children of v_ε, and, $\forall \alpha \in \{0,1\}^i$ and $\forall i < \log N$, $v_{\alpha 0}$ and $v_{\alpha 1}$ are, respectively, the left and right children of node v_α. Consequently, $v_{\alpha[1 \cdot\cdot j]}$ and $v_{\alpha[1 \cdot\cdot \bar{j}]}$ are siblings whenever $0 < |\alpha| \leq \log N$ and $0 < j \leq |\alpha|$. The leaves of $\mathcal{T}_N$ are thought to be ordered "from left to right." Within the context of a tree $\mathcal{T}_N$, we denote by $[j]$ the $\log N$-bit binary representation of integer j (thus, $[j]$ consists of the binary representation of j with the right number of leading zeros). Accordingly, the jth leaf of $\mathcal{T}_N$ is node $v_{[j]}$.

In the following protocol we associate values to the nodes of $\mathcal{T}_N$ so as to obtain a Merkle tree (but constructed with a random-oracle function rather than a collision-free hash function). In so doing, we shall consistently denote the value of v_α as R_α. Thinking of each node of $\mathcal{T}_N$ as having its own memory location, we may also say that "R_α has been stored in node v_α."

Algorithms $\mathcal{P}$ and $\mathcal{V}$.

Common inputs: $\mathbf{q} = (\bar{M}, x, t)$, and k

respectively, a n-bit triplet in $\mathcal{L}$ and a unary security parameter.

Common subroutines: SP and SV,

respectively prover and verifier of a given samplable proof-system (SP, SV) with length bounds ℓ and λ.

Common oracles: $f_1 : \Sigma^{2k} \to \Sigma^k$ and $f_2 : \Sigma^{k+n} \to \Sigma^{k\lambda(n)}$.

(*Comment:* When randomly selected, oracle f_1 is used as a collision-free hash function in Merkle tree $\mathcal{T}_n$, and thus to enable $\mathcal{P}$ to commit to a samplable proof of $\mathbf{q} \in \mathcal{L}$. Oracle f_2 is used to generate k random tapes for SV.)

$\mathcal{P}$'s output: $\mathcal{C}$, a CS certificate that $\mathbf{q} \in \mathcal{L}$.

$\mathcal{V}$'s additional input: $\mathcal{C}$.

Algorithm $\mathcal{P}$

$P1$. (Commit to a samplable proof of $\mathbf{q} \in \mathcal{L}$.)

$P1.1$ (Find a proof σ of $x \in \mathcal{L}$.)
Run machine M on input x so as to generate the history, σ, of the $\leq t$-step accepting computation of $\mathcal{M}$.
(*Comment:* σ can be considered a proof that $x \in \mathcal{L}$)

$P1.2$ (Put σ in a samplable form τ.)
Run algorithm SP on input $\mathbf{q}$ and σ so as to obtain a samplable proof, τ.
(*Comment:* Because $|\mathbf{q}| = |(\bar{M}, x, t)| = n$, $|\tau| \leq \ell(n)$.)

$P1.3$ (Commit to τ by means of a k-bit value R_ε.)
Assume, for simplicity only, that $|\tau|/k = N$, where N, an integral power of 2. Then, sub-divide τ into the concatenation of N k-bit strings, $\tau = \tau_1 \cdots \tau_N$, and compute a value R_ε by associating to the vertices of tree $\mathcal{T}_N$ the following values. For $0 \leq j < N$, assign to the jth leaf, $v_{[j]}$, the k-bit value

$$R_{[j]} = \tau_j. \tag{1}$$

Then, in a bottom-up fashion, assign to each interior node v_α of $\mathcal{T}_N$ the k-bit value

$$R_\alpha = f_1(R_{\alpha 0}|R_{\alpha 1}). \tag{2}$$

(*Comment:* R_ε is the k-bit value assigned to the root of $\mathcal{T}_N$.)

$P2$. (Build a CS certificate $\mathcal{C}$ of $\mathbf{q} \in \mathcal{L}$.)

$P2.1$ (Start certificate.)
$\mathcal{C} \leftarrow R_\varepsilon$.

$P2.2$ (Choose k random tapes for SV.)
Call $TAPE$ the random $k\lambda(n)$-bit string obtained by computing $f_2(\mathbf{q}\,|R_\epsilon)$; divide this string in k disjoint segments, each $\lambda(n)$-bit long, and call $TAPE_j$ the jth such segment.

$P2.3$ (Run SV k times with virtual access to τ.)
For $j = 1, \ldots, k$, run SV with $TAPE_j$ as the random tape, inputs $\mathbf{q}$ and $|\tau|$, and virtual access to τ. Whenever SV wishes to access bit-location i of τ, perform the following instructions:

$P2.3.1$ (Find the index, I, of the substring of τ containing b_i.)
Let I be the smallest positive integer p such that $i \leq pk$;

$P2.3.2$ (Add leaf I to the CS certificate.)
$\mathcal{C} \leftarrow \mathcal{C}|R_{[I]}$; and

$P2.3.3$ (Add to the certificate the siblings of the path between leaf I and the root of $\mathcal{T}_N$.)
Set $\alpha = [I]$; set $v_j = R_{\alpha[1 \cdot\cdot \bar{j}]}$ for $j = 1, \ldots, \log N$;
set $SIBLINGPATH_I = (v_1, \ldots, v_{\log N})$; $\mathcal{C} \leftarrow \mathcal{C}|SIBLINGPATH_I$.
(*Example:*
if $I = 3$ and $N = 8$, then $SIBLINGPATH_I = (R_{010}, R_{00}, R_1)$.)

$P3.$ (Output the certificate.)
Output $\mathcal{C}$, a CS certificate of $\mathbf{q} \in \mathcal{L}$ *relative to root-value* R_ϵ.

Algorithm $\mathcal{V}$

$V1.$ (Read and delete the root of $\mathcal{T}_N$ from the certificate, and compute k random tapes for SV.)
Set $ALLEGED - ROOT = \mathcal{C}[1 \cdot\cdot k]$ and reset $\mathcal{C} \leftarrow \mathcal{C}[k+1 \cdot\cdot]$. Then compute the $k\lambda(n)$-bit string $TAPE = f_2(\mathbf{q}|R_\epsilon)$; divide $TAPE$ into k non-overlapping segments, each $\lambda(n)$-bit long, and call $TAPE_j$ the jth such segment.

$V2.$ (Run SV k times with virtual access to samplable proof τ.)
For $j = 1, \ldots, k$ run SV with $TAPE_j$ as the random tape, input $\mathbf{q}$, and virtual access to samplable proof τ. Whenever SV wishes to access bit-location i of τ, do

$V2.1$ (Find the index, I, of the segment of τ containing b_i, and read the value of leaf I from the certificate.)
Set $I = \lceil i/k \rceil$, $\alpha = [I]$, and $R_\alpha = \mathcal{C}[1 \cdot\cdot k]$.

$V2.2$ (Delete the value of leaf I from the certificate.)
$\mathcal{C} \leftarrow \mathcal{C}[k+1 \cdot\cdot]$.

$V2.3$ (Check whether the certificate contains the siblings of the path between leaf I and the root of $\mathcal{T}_N$, and remove them.)

For $m = 1$ to $\log N$, set $R_{\alpha[1 \cdot\cdot \bar{m}]} = \mathcal{C}[1 \cdot \cdot k]$ and reset $\mathcal{C} \leftarrow \mathcal{C}[k+1 \cdot \cdot]$. Then, for $m = \log N, \ldots, 1$, compute $R_{\alpha[1 \cdot\cdot m-1]}$ as follows:

$$R_{\alpha[1 \cdot\cdot m-1]} = \begin{cases} f_1(R_{\alpha[1 \cdot\cdot j]} | R_{\alpha[1 \cdot\cdot \bar{j}]}) & \text{if } \alpha_m = 1 \\ f_1(R_{\alpha[1 \cdot\cdot \bar{m}]} | R_{\alpha[1 \cdot\cdot m]}) & \text{if } \alpha_m = 0 \end{cases}$$

and check whether the computed value R_ε equals the read value $ALLEGED-ROOT$.

(*Example:* If $N = 8$ and $I = 3$, then the verifier computes

$$\begin{aligned} R_{011} &= \tau_3, \\ R_{01} &= f_1(R_{010}|R_{011}), \\ R_0 &= f_1(R_{00}|R_{01}), \text{ and} \\ R_\varepsilon &= f_1(R_0|R_1). \end{aligned}$$

In this example, the values of τ_3, R_{010}, R_{00}, and R_1 are part of the certificate provided by the prover, and the values of R_{011}, R_{01}, R_0, and R_ε are computed by the verifier.

$V3$. (Accept if and only if the sibling path have always been correct and if SV has always accepted.)
If each check performed in Verification Step 2.3 has been passed, and if SV has output YES in each of its k runs, then output YES. Else, output NO.

A flawed proof that the construction works. We prove that $(\mathcal{P}, \mathcal{V})$ is a CS proof-system with random oracle in the next subsection. The proof of feasible completeness is quite straightforward, but our proof of computational soundness is not particularly compact nor intuitive, due to our insistence in proving even intuitive properties. Though we might have exaggerated in details, our desire to be meticulous derives from the fact that more intuitive and simple "proofs" of computational correctness are actually flawed in *fundamental* ways. Indeed, their being intuitive derives from the traditional use of Merkle trees, which is quite different from ours.[18] Based on previous use of Merkle trees, we may be tempted of quickly "deriving" computational soundness from the following

Irrelevant (and informal) Fact: the value R_ε, computed in Proving Step 1.3 and included in the certificate, in practice commits the Prover to at most one possible string τ because, without making $2^{O(k)}$ oracle calls, the chances of finding two different strings τ and τ' that "Merkle-hash" to R_ε is negligible.

[18] Traditionally, one uses a Merkle tree to prevent someone else from cheating. (E.g., in the original application [27], a signer uses a Merkle tree to authenticate n values of his choice, so as to make it infeasible for an impostor to authenticate different values.) In our case, instead, the Prover uses a Merkle tree to prevent *himself* from cheating, which complicates the proof a great deal.

The above statement is certainly true, and, once properly formalized, not hard to prove. It is not, however, directly relevant to the scenario at hand. It applies, instead, to settings where one party wishes to secretly commit to a string τ that will be later revealed in its entirety.[19]

Unfortunately, the setting of Theorem 1 is quite different: prover $\mathcal{P}$ never de-commits R_ε by revealing the entire samplable proof τ. Indeed, because τ may be too long, and because feasible completeness must be satisfied, $\mathcal{P}$ only reveals relatively few bits of τ: those requested by the samplable verifier in its k simulated and virtual runs. It is thus irrelevant that a cheating prover cannot feasibly find two strings τ and τ' that tree-hash to the same k-bit value.

Equally irrelevant is the fact that successful cheating is highly improbable if a malicious prover *first* selects a suitable value R_ε, and *then* tries to answer satisfactorily all bit-requests that are induced by R_ε (via the k random tapes it specifies).

Rather, we must prove that it is infeasible for a cheating prover, on input $q' \notin \mathcal{L}$, to compute a k-bit value R'_ε (possibly obtained without Merkle-hashing any string) *together with*, somehow, (a) bit-values (corresponding to requests made by SV based on the tapes specified by R_ε) that lead SV to accept at every run, and (b) authentication paths corresponding to said bit-values that correctly merge with R_ε. This is what we are going to do next.

The construction works

Theorem 1: $(\mathcal{P}, \mathcal{V})$ is a CS proof-system with a random oracle.

Proof of Feasible Completeness.

It is immediately seen that subproperty (ii) (completeness) holds. That is, for all $\mathbf{q} = (\bar{M}, x, t) \in \mathcal{L}$, for all security parameter k, and for all oracles f_1 and f_2, the certificate output by $\mathcal{P}$ convinces $\mathcal{V}$. Subproperty (i), that is the fact that $\mathcal{P}$ performs only polynomially many (in n, k, and t) for producing a certificate, follows as easily. Indeed, $\mathcal{P}$ invests t steps of computation for running M on input x, a number of steps polynomial in $\mathbf{q}$'s length and t for obtaining the samplable proof τ, less than $t \log t$ queries to the random oracle for generating the Merkle tree, and much fewer steps for running $\mathcal{V}$ "in his head" and answering its queries.

(Also recall that each call to our f_1 and f_2 can be simulated very efficiently even if one were given access only to a single oracle $f : \Sigma^{3k+n} \rightarrow \Sigma^{k(\lambda(n)+1)}$). □

[19] For instance, a party A secretly chooses a string τ, Merkle-hashes it into a k-bit string R_ε, gives R_ε to party B, and permits him to look at a few bits of τ proving their values via authentication paths relative to R_ε. After that, B is asked to bet on whether or not A's secret string possesses a given property. After B's bet, A reveals τ, so that B can verify whether it Merkle-hashes to R_ε.

Proof of Computational Soundness.

Initial notation. To simplify the proof of computational soundness, we assume, without loss of generality, any of our oracle-calling algorithms does not query any of its oracles twice about the same string σ.

In our proof we shall also consider oracle-calling algorithms that are probabilistic. The sequence of coin tosses of a probabilistic oracle-calling algorithm A shall be always denoted by $C(A)$. Notice that the notion of a N-call algorithm include probabilistic ones.[20]

Probabilities of events occuring in the executions of an oracle-calling algorithm A are computed over the random choice of a specified subset of its oracles and, if A is probabilistic, over the choice of $C(A)$. To indicate the probability of an event E taken over "the possible choices of $X, Y, \ldots$" we write

$$Prob_{X,Y,\ldots}(E).$$

Let $\mathcal{A}_{(\cdot)}$ be a N-call algorithm querying its oracle about strings in a finite set D, let R be another finite set, and let $f : D \to R$ be a function. Then, an execution of $\mathcal{A}_f$ identifies an element of the Cartesian product R^N; that is, the sequence $T = r_1, \ldots, r_N$, where r_i is f's answer to $\mathcal{A}$'s ith query.

Viceversa, because $\mathcal{A}$ does not, within the same execution, query twice its oracle about the same string, given an R-valued sequence $T = r_1, \ldots, r_N$, we can envisage executing $\mathcal{A}$ so that the ith query to the oracle is answered by r_i. The ensemble of executions so generated will be denoted by $\mathcal{A}_T$.

It should be noticed that the set of ensembles $\mathcal{A}_T$ and the set of ensembles $\mathcal{A}_f$ may be different. (If $\mathcal{A}$ is deterministic, then $\mathcal{A}_T$ consists of a single execution, in which case it is immediate to see that there exist functions f such that $\mathcal{A}_T = \mathcal{A}_f$. However, if $\mathcal{A}$ is probabilistic, it may happen that in one execution of $\mathcal{A}_T$, the ith query, q, is answered by a value v, while in a different execution the same string q, asked as the jth query, where $j \neq i$, is answered by a value other than v. In this case for no function f is $\mathcal{A}_f$ equal to $\mathcal{A}_T$.) Nonetheless, it is simple to notice that, whether or not $\mathcal{A}$ is probabilistic, for any event E, the following identity holds:

$$(*) : \; Prob_{C(\mathcal{A}),f:D\to R}\ (E \textit{ occurs in } \mathcal{A}_f) = Prob_{C(\mathcal{A}),T\in R^N}(\ E \textit{ occurs in } \mathcal{A}_T).$$

The above concept, notation, and identity naturally extend to more "complex" cases.[21]

---

[20] Indeed, a probabilistic oracle-calling algorithm $\mathcal{A}_{(\cdot,\ldots,\cdot)}(\cdot,\ldots,\cdot)$ is N-call if in each of its executions (i.e., for any possible choice of its inputs, oracles, and coin tosses) A makes N calls to each of its oracles.

[21] For instance, Let T be a N-long sequence whose values belong to a finite set R, and let $\mathcal{A}_{(\cdot,\cdot)}(\cdot)$ be a N-call algorithm that always queries its first oracle about a string

Let m and M be positive integers such that $m \leq M$, let σ be a string, and let $T = \{\sigma_i : i \in [1, M] - \{m\}\}$ be a sequence of strings. Then, by the notation $T_m = \sigma$ we shall denote the M-long sequence $\sigma_1, \ldots, \sigma_{m-1}, \sigma, \sigma_{m+1}, \ldots, \sigma_M$.

Let $T = (\sigma_1, \ldots, \sigma_m)$ be a sequence of strings (in Σ^r); then we say that the (Σ^r-valued sequence) $\bar{T} = (\bar{\sigma}_1, \ldots, \bar{\sigma}_M)$ is an *extension of T (over Σ^r)* if $m \leq M$ and $\bar{\sigma}_i = \sigma_i$ for $i = 1, \ldots, m$.

Let m be a non-negative integer and S a sequence of pairs of binary strings, $S = (x_1, y_1), \ldots, (x_m, y_m)$. We say that S is *extendible from Σ^{ℓ_1} to Σ^{ℓ_2}* if ℓ_1 is the length of all the x_i's, ℓ_2 is the length of all the y_i's, and, whenever $i \neq j$, we have $x_i \neq x_j$ and $y_i \neq y_j$. (The reason for insisting that $y_i \neq y_j$ whenever $x_i \neq x_j$ will become clear in Corollary 1.) If $S = (x_1, y_1), \ldots, (x_m, y_m)$ is extendible from Σ^{ℓ_1} to Σ^{ℓ_2} and the function $f : \{0,1\}^{\ell_1} \rightarrow \{0,1\}^{\ell_2}$ is such that $f(x_i) = y_i$ for $i = 1, \ldots, m$, then we say that f is an *extension of S;* in symbols, $f \in ext(S)$.

Definition: Let $\mathcal{A}$ be an algorithm calling one or more oracles. Then, in an execution of $\mathcal{A}$ where the function f was one of its oracles, we say that $\mathcal{A}$ *finds an f-collision* if $\mathcal{A}$ queries f about two different strings x and y such that $f(x) = f(y)$.

To proceed, we need to state (without proof) a "converse" of the well-known birthday paradox.

Lemma 1: $\forall$ positive integers k, $\forall$ possible inputs $z_1, z_2, \ldots$, $\forall$ $2^{k/4}$-call algorithms $\mathcal{A}_{(\cdot,\ldots,\cdot)}(\cdot, \ldots, \cdot)$, and $\forall$ possible oracles $f_1, f_2, \ldots, f_1', f_2', \ldots$,

$$Prob_{C(\mathcal{A}), f:\Sigma^{2k}\rightarrow\Sigma^k}(\mathcal{A}_{f_1,\ldots,f,f_1',\ldots}(z_1, z_2, \ldots) \textit{ finds an f-collision}) < 2^{-k/2}.$$

Corollary 1: $\forall$ positive integers k, $\forall$ positive integers $m \leq 2^{k/4}$, $\forall$ sequences $S = (x_1, y_1), \ldots, (x_m, y_m)$ extendible from Σ^{2k} to Σ^k, $\forall$ $2^{k/4}$-call algorithms $\mathcal{A}_{(\cdot,\ldots,\cdot)}(\cdot, \ldots, \cdot)$, $\forall$ possible inputs $z_1, z_2, \ldots$, and $\forall$ possible oracles $f_1, f_2, \ldots, f_1'$, $f_2', \ldots$,

$$Prob_{f\in ext(S)}(\mathcal{A}_{f_1,\ldots,f,f_1',\ldots}(z_1, z_2, \ldots) \textit{ finds an f-collision}) < 2^{-k/2}.$$

Proof of Corollary 1: Algorithm $\mathcal{A}$ has no greater a chance of finding a f-collision in the present setting than in that of Lemma 1. In fact, if it makes two queries α and β both in the set $X = \{x_1, \ldots, x_m\}$, then we are guaranteed that $f(\alpha) \neq$

in a finite set D. Then, $\mathcal{A}_{T,f_2}(x)$ denotes an execution of $\mathcal{A}$, where x is the input, f_2 the second oracle, and where the ith query of $\mathcal{A}$ to the first oracle is answered by the ith string of a sequence T. It is immediately seen that, for all possible choices of f_2, x, D, and R,

$$Prob_{f_1:D\rightarrow R, f_2}\ (E \textit{ occurs in } \mathcal{A}_{f_1,f_2}(x)) = Prob_{T\in R^N, f_2}(\ E \textit{ occurs in } \mathcal{A}_{T,f_2}(x)).$$

$f(\beta)$; on the other side, if α and β are not both in X, then the probability that $f(\alpha) = f(\beta)$ still is 2^{-k}. □

Let us now prove that $(\mathcal{P}, \mathcal{V})$ satisfies computational soundness if we choose the 4th, 5th, and 6th fundamental constants as follows:

- $c_4 =$ the smallest integer i such that $n^i > 8\ell(n) + 32$,
- $c_5 = 1/16$, and
- $c_6 = 1/16$.

(To facilitate the comprehension of what follows, we prefer to make use of the different "labels" c_5 and c_6 rather than their common numerical value 1/16.)

The proof is by contradiction. Assume that computational soundness does not hold for our choice of 4th, 5th, and 6th fundamental constants, then the following proposition holds:

$\wp 1$: *There exist an integer $n' > 1$, a n'-bit string $q' \notin \mathcal{L}$, an integer $k' > (n')^{c_4}$, and a deterministic, $2^{c_5 k'}$-call, cheating prover P' such that*

$$Prob_{f_1:\Sigma^{2k'}\to\Sigma^{k'},\, f_2:\Sigma^{k'}\to\Sigma^{k'L(n')}}(\mathcal{P}'_{f_1,f_2}(q',k') = \mathcal{C}' \wedge \mathcal{V}_{f_1,f_2}(q',k',\mathcal{C}') = YES) \geq 2^{-c_6 k'}$$

Additional notation: A function mapping $\Sigma^{2k'}$ into $\Sigma^{k'}$ will always be indicated by f_1 (with possible superscripts). Similarly, f_2 (with possible superscripts) will always denote a function mapping $\Sigma^{k'}$ into $\Sigma^{k'L(n')}$.

We shall solely focus on the non-empty outputs of a cheating prover $\mathcal{P}'$; thus, when writing $\mathcal{P}'_{f_1,f_2}(q',k') = \mathcal{C}'$, we mean that $\mathcal{P}'$ has output a non-empty string, and that this string is $\mathcal{C}'$. If this occurs, given that a cheating prover always verifies its own non-empty outputs, we must have $\mathcal{V}_{f_1,f_2}(q',k',\mathcal{C}') = YES$. Assume now that, during the verification of $\mathcal{C}'$, samplable verifier V virtually asks about a bit-location i of the virtual samplable proof; then, we write $\mathcal{C}' \ni i$.

If $\mathcal{P}'_{f_1,f_2}(q',k') = \mathcal{C}'$, to emphasize that $q' \notin \mathcal{L}$, we refer to $\mathcal{C}'$ as a *pseudo-certificate* (of $q' \in \mathcal{L}$). We further say that $\mathcal{C}'$ is *relative to* $\sigma \in \Sigma^{k'L(n')}$ if $\sigma = f_2(q'|R'_\varepsilon)$, where R'_ε is the *pseudo-root* (i.e., the k'-bit prefix) of $\mathcal{C}'$. Thus, whenever $\mathcal{P}'_{f_1,f_2}(q',k') = \mathcal{C}'$, then $\mathcal{C}'$ is relative to some string $\sigma \in \Sigma^{k'L(n')}$ sent by the second oracle to $\mathcal{P}'$ in reply to one of its queries in that execution.

Let $S = (s_1, \dots, s_i)$ be a sequence. Then ("abusing" our concatenation operator), for any value s, we let $S|s$ denote the sequence whose first i values coincide with those of S, and whose i+1st value is s.

Lemma 2: Proposition $\wp 1$ implies the following proposition

$\wp 2$: *There exist a k'-bit value R'_ε; a $\leq 2^{c_5 k'}$-long sequence S extendible from $\Sigma^{2k'}$ to $\Sigma^{k'}$; an integer $m \in [1, 2^{c_5 k'}]$; and a sequence*

$$T = \{\sigma_i \in \Sigma^{k'L(n')} : i \in [1, 2^{c_5 k'}] - \{m\}\},$$

such that

$$Prob_{f_1 \in ext(S), \sigma \in \Sigma^{k'L(n')}}(\mathcal{P}'_{f_1, T_m=\sigma}(q', k') = R'_\varepsilon \cdots) > 2^{-(c_5+c_6)k'-1}.$$

Proof of Lemma 2: According to our just established notation, proposition ℘1 implies

$$Prob_{f_1,f_2}(\mathcal{P}'_{f_1,f_2}(q', k') \neq \varepsilon) \geq 2^{-c_6 k'}.$$

Thus, because $\mathcal{P}'$ is a $2^{c_5 k'}$-oracle, and because it "verifies" all of its pseudo-certificates (i.e., each of them is relative to a string $\sigma \in \Sigma^{k'L(n')}$ obtained by $\mathcal{P}'$ in response to a query made to its second oracle), by averaging there must exist a positive integer $m \leq 2^{c_5 k'}$ such that $\mathcal{P}'$ outputs a pseudo-certificate relative to the mth reply of its second oracle with probability at least $2^{-(c_5+c_6)k'}$. Thus, denoting by $\mathcal{C}[m]$ a pseudo-certificate relative to the mth query to the second oracle, we have

$$p = Prob_{f_1,f_2}(\mathcal{P}'_{f_1,f_2}(q', k') = \mathcal{C}[m]) \geq 2^{-(c_5+c_6)k'}.$$

Let us now write $p = p_1 + p_2$ where

$$p_1 = Prob_{f_1,f_2}(\mathcal{P}'_{f_1,f_2}(q', k') = \mathcal{C}[m] \wedge \text{at least one } f_1\text{-collision})$$

and

$$p_2 = Prob_{f_1,f_2}(\mathcal{P}'_{f_1,f_2}(q', k') = \mathcal{C}[m] \wedge \text{no } f_1\text{-collisions}).$$

Now, in view of Corollary 1 and property (*) and our constraints on c_5, c_6, and k', we have $p_1 \leq 2^{-k'/4} \leq 2^{-(c_5+c_6)k'-1}$. Thus,

(3): $Prob_{f_1,f_2}(\mathcal{P}'_{f_1,f_2}(q', k') = \mathcal{C}[m] \wedge \text{no } f_1\text{-collisions}) > 2^{-(c_5+c_6)k'-1}$.

Now let us again make use of averaging for "growing" a sequence of pairs of strings, S, and for choosing the first $m-1$ entries of a sequence T as follows. We initially set S and T to be empty, and start executing algorithm $\mathcal{P}'$ on inputs q' and k' from the initial configuration, and handling oracle calls in the following manner. If $\mathcal{P}'$ queries its first oracle about a new $2k'$-bit string, x, we choose a reply y in $\Sigma^{k'}$ which (a) is different from all the second entries of the pairs currently belonging to S, and (b) "preserves Equation 4," that is, denoting by $\bar{T}$ a $2^{c_5 k'}$-long sequence with values in $\Sigma^{k'L(n')}$, such that

$$Prob_{f_1 \in ext(S|(x,y)), \bar{T} \in ext(T)}(\mathcal{P}'_{f_1,\bar{T}}(q', k') = \mathcal{C}[m] \wedge \text{no } f_1\text{-collisions}) > 2^{-(c_5+c_6)k'-1}.$$

reset $S = S|(x, y)$; feed $\mathcal{P}'$ with y, and resume the execution. If $\mathcal{P}'$ makes its jth query to the second oracle, and $j < m$, we choose a reply $\sigma_j \in \Sigma^{k'L(n')}$ so as to preserve Equation 4, that is, such that

$$Prob_{f_1 \in ext(S'), \bar{T} \in ext(T|\sigma_j)}(\mathcal{P}'_{f_1,\bar{T}}(q',k') = \mathcal{C}[m] \wedge \text{no } f_1\text{-collisions}) > 2^{-(c_5+c_6)k'-1};$$

reset $T = T|\sigma_j$; feed $\mathcal{P}'$ with y, and resume the execution. When $\mathcal{P}'$ makes its mth query to its second oracle, we stop this process, thereby having already constructed the entire desired sequence S (which is $\leq 2^{c_5 k'}$-long because $\mathcal{P}'$ is $2^{c_5 k'}$-call) and the first $m-1$ strings of T. As for the other the elements of T (i.e., from the $m+1$st on), we simply choose them so as to preserve Equation 4, that is, so that

$$Prob_{f_1 \in ext(S), \sigma \in \Sigma^{k'L(n')}}(\mathcal{P}'_{f_1,T_m=\sigma}(q',k') = \mathcal{C}[m]) \geq 2^{-(c_5+c_6)k'-1}.$$

Now notice that, given that $\mathcal{P}'$ is deterministic, $\forall\, f_1 \in ext(S)$ and $\forall\, \sigma \in \Sigma^{k'L(n')}$, the computation of $\mathcal{P}'_{f_1,T_m=\sigma}(q',k')$ is totally determined up to the mth query to the second oracle. Thus, there exists a unique k-bit string, R'_ε, such that, in any execution of $\mathcal{P}'_{f_1,T_m=\sigma}(q',k')$, the mth query to the second oracle consists of R'_ε. Thus, whenever such an execution ends in outputting a certificate relative to the mth query, R'_ε must be its k-bit prefix. □

Let now S, T and R'_ε be like in proposition $\wp 2$, and consider the following algorithm $\mathcal{A}$ calling an oracle $f_1 \in ext(S)$:

Algorithm $\mathcal{A}_{f_1}$

$A1$. For $j = 1$ to $4 \cdot 2^{\ell(n')} \cdot 2^{(c_5+c_6)k'+2}$, randomly select $\sigma_j \in \Sigma^{k'L(n')}$ and execute $\mathcal{P}'_{f_1,T_m=\sigma_j}(q',k')$.

$A2$. If in Step $A1$ algorithm $\mathcal{P}'$ has output $< 2 \cdot 2^{\ell(n')}$ pseudo-certificates (of $q' \in \mathcal{L}$) whose prefix is R'_ε, HALT without any output. Else,

$A3$. Compute BL, the set of bit-locations i such that, for some execution j,

$$\mathcal{P}'_{f_1,T_m=\sigma_j}(q',k') = (R'_\varepsilon \cdots) \ni i$$

$A4$. If, for some $i \in BL$, there is no unique bit b_i such that all questions of V about bit-location i have been consistently answered with b_i, HALT without any output. Else,

$A5$. HALT outputting the $2^{\ell(n')}$-long string τ whose ith character is b_i, if $i \in BL$, and $*$ otherwise.

Lemma 3: There exists $f'_1 \in ext(S)$ such that

$$Prob_{C(\mathcal{A})}(\mathcal{A}_{f'_1} \text{halts in Step } A5) > 1/2$$
and
$$Prob_{\sigma \in \Sigma^{k'L(n')}}(\mathcal{P}'_{f'_1,T_m=\sigma}(q',k') = R'_\varepsilon \cdots) \geq 2^{-(c_5+c_6)k'-2}.$$

Proof of Lemma 3. Say that a function $f_1 \in ext(S)$ is *lucky* if

$$Prob_{\sigma \in \Sigma^{k'L(n')}}(\mathcal{P}'_{f_1,T_m=\sigma}(q',k') = R'_\varepsilon \cdots) \geq 2^{-(c_5+c_6)k'-2}.$$

Then, a simple counting argument shows that

$$Prob_{f_1 \in ext(S)}(f_1 \text{ is lucky}) \geq 2^{-(c_5+c_6)k'-2}.$$

Now, the probability that $\mathcal{A}$ (run with an oracle $f_1 \in ext(S)$) does not halt in Step $A5$ is bounded above by the sum of (1) the probability of halting at Step $A2$ and (2) the probability of halting at Step $A4$. Now a simple application of Chernoff's bounds shows that the first probability is clearly upperbounded by $2^{-(c_5+c_6)k'-3}$. Let us now show that also the second probability is upperbounded by $2^{-(c_5+c_6)k'-3}$. To this end, notice that, whenever $\mathcal{A}_{f_1}$ halts in Step $A5$, then a f_1-collision has been found. Consider in fact, the following two conceptual steps for "locating" such a collision.

$C1$. Find a bit-location i, the leaf I containing it ($I = \lceil i/k' \rceil$), and two sibling-paths (possibly belonging to the same certificate) P_1 and P_2 between leaf I and the (alleged) root of the (alleged) $\mathcal{T}_N$, according to which the bit stored in location i is different. Then, denoting $[I] = \alpha_1 \cdots \alpha_{\log N}$, we can write P_1 and P_2 as follows:

$$P_1 = R^1_{\alpha_1 \cdots \alpha_{\log N-1}\bar{\alpha}_{\log N}}, R^1_{\alpha_1 \cdots \bar{\alpha}_{\log N-1}}, \ldots, R^1_{\bar{\alpha}_1}, R^1_\varepsilon$$

and

$$P_2 = R^2_{\alpha_1 \cdots \alpha_{\log N-1}\bar{\alpha}_{\log N}}, R^2_{\alpha_1 \cdots \bar{\alpha}_{\log N-1}}, \ldots, R^2_{\bar{\alpha}_1}, R^2_\varepsilon$$

where $R^1_{[I]} = R^1_{\alpha_1 \cdots \alpha_{\log N}}$ is the value stored in leaf I according to P_1, and $R^2_{[I]} = R^2_{\alpha_1 \cdots \alpha_{\log N}}$ is the value stored in leaf I according to P_2.
(*Comments:* First, $R^1_{[I]} \neq R^2_{[I]}$, because their $(i - ([I]-1)k')$th bits are different, since they represent their respective values stored in bit-location i. Second, $R^1_\varepsilon = R'_\varepsilon = R^2_\varepsilon$ because both sibling-path occur within a certificate or two certificates whose pseudo-root is R'_ε.)

$C2$. Find $m \in [1, \log N]$ such that

$$X = R^1_{\alpha_1 \cdots \alpha_m} | R^1_{\alpha_1 \cdots \bar{\alpha}_m} \neq R^2_{\alpha_1 \cdots \alpha_m} | R^2_{\alpha_1 \cdots \bar{\alpha}_m} = Y,$$

but $f_1(X) = f_1(Y)$.
(*Comment:* Such m must exist because of three reasons.
First, $f(R^1_{\alpha_1} | R^1_{\bar{\alpha}_1}) = R'_\varepsilon = f(R^2_{\alpha_1} | R^2_{\bar{\alpha}_1})$.
Second, $R^1_{\alpha_1 \cdots \alpha_{\log N}} | R^1_{\alpha_1 \cdots \bar{\alpha}_{\log N}} \neq R^2_{\alpha_1 \cdots \alpha_{\log N}} | R^2_{\alpha_1 \cdots \bar{\alpha}_{\log N}}$. In fact, their k'-bit prefixes
Third, $\mathcal{P}'$ verifies all its non-empty outputs, and thus has made all relevant queries to oracle f_1, including X and Y.)

Now, notice that A calls its oracle $4 \cdot 2^{\ell(n')} \cdot 2^{(c_5+c_6)k'+2}$ times. Therefore, because $c_5 = c_6 = 1/16$ and $k' > (n')^{c_4} > 8\ell(n') + 32$, $\mathcal{A}$ is a $2^{k'/4}$-call algorithm. Moreover, because sequence S is extendible from $\Sigma^{2k'}$ to $\Sigma^{k'}$, Corollary 1 implies

$$Prob_{C(\mathcal{A}), f_1 \in ext(S)}(\mathcal{A}_{f_1} \text{finds a } f_1\text{-collision}) < 2^{-k'/2} < 2^{-(c_5+c_6)k'-3}.$$

Consequently, the fraction of functions $f_1 \in ext(S)$ for which $\mathcal{A}_{f_1}$'s probability of halting in Step $A5$ exceeds $1/2$ is less than $2 \cdot 2^{-(c_5+c_6)k'-3} = 2^{-(c_5+c_6)k'-2}$. Thus there must exist a function f_1' as desired in our hypothesis. □

Definitions: Whenever $\sigma \in \Sigma^{k'L(n')}$, consider it as the concatenation of k' strings, each $L(n')$-bit long, and denote by $\sigma[i]$ the ith such segment.

For any string τ over the alphabet $\{0, 1, *\}$, define

$$P'_\tau = Prob_{\sigma \in \Sigma^{k'L(n')}}(\textstyle\bigwedge_{i=1}^{k'} SV_{\sigma[i]}(q', k', \tau) = accept \mid \mathcal{P}'_{f_1', T_m=\sigma}(q', k') = R'_\varepsilon \cdots),$$

where $SV_{\sigma_j}(q', k', \tau)$ denotes the execution of samplable verifier SV on inputs q' and k', coin tosses σ_j, and access to string τ; with the provision that SV *rejects* if it accesses a bit-location of τ storing the value $*$.

Lemma 4: There exists a string $\tilde{\tau}$, over $\{0, 1, *\}$, such that $P'_{\tilde{\tau}} > 1/2$.

Proof of Lemma 4. We shall prove our lemma by showing that algorithm $\mathcal{A}_{f_1'}$ has a positive probability of outputting a string $\tilde{\tau}$ as desired.

Let $\bar{\tau}$ be a string such that $P'_{\bar{\tau}} \leq 1/2$. Then,

$$Prob_{C(\mathcal{A})}(A_{f_1'} \text{ outputs } \bar{\tau}) < 2^{-2^{\ell(n')}}.$$

In fact, for outputting a non-empty string in Step $A5$, $\mathcal{A}_{f_1}$ should successfully compute at least $2 \cdot 2^{\ell(n')}$ pseudo-certificates with prefix R'_ε in Step $A1$. Thus, consider the first $> 2 \cdot 2^{\ell(n')}$ randomly- and independently-selected strings $\sigma \in \Sigma^{k'L(n')}$ such that

$$\mathcal{P}_{f_1', T_m=\sigma}(q', k') = (R'_\varepsilon \cdots).$$

Then, in order for $\mathcal{A}_{f_1'}$ to output $\bar{\tau}$, for each of these σ, the event

$$\bigwedge_{i=1}^{k'} SV_{\sigma[i]}(q', \bar{\tau}) = accept$$

should occur. Hence, by the definition of $P'_{\bar{\tau}}$ and the fact that its value is less than $1/2$, the probability that all these events occur is at most $2^{-2 \cdot 2^{\ell(n')}}$.

Now, because any string outputable by $\mathcal{A}_{f_1'}$ is $2^{\ell(n')}$-long, there may be at most $3^{2^{\ell(n')}}$ strings τ such that $P'_\tau \leq 1/2$. Thus,

$$Prob_{C(\mathcal{A})}(\mathcal{A}_{f_1'} = \tau \ \wedge \ P_\tau' \leq 1/2) < \sum_{j=1}^{3^{2^{\ell(n')}}} 2^{-2\cdot 2^{\ell(n')}} < 1/2.$$

But because, as shown in Lemma 3, $\mathcal{A}_{f_1'}$ halts in Step $A5$ (and thus outputs a non-empty string τ) with probability $> 1/2$, $\mathcal{A}_{f_1'}$ must output a string $\tilde{\tau}$ such that $P_{\tilde{\tau}}' > 1/2$. □

We are now ready to finish the proof of computational soundness. Define

$$P_{\tilde{\tau}} = Prob_{\sigma \in \Sigma^{k' L(n')}}(\bigwedge_{i=1}^{k'} SV_{\sigma[i]}(q', \tilde{\tau}) = accept)$$

and

$$\tilde{P}_{\tilde{\tau}} = Prob_{s \in \Sigma^{L(n')}}(SV_s(q', \tilde{\tau}) = accept).$$

Then,

$$(\tilde{P}_{\tilde{\tau}})^{k'} = P_{\tilde{\tau}} \geq P_{\tilde{\tau}}' \cdot Prob_{\sigma \in \Sigma^{k' L(n')}}(\mathcal{P}'_{f_1', T_m = \sigma}(q', k') = R_\varepsilon' \cdots) \geq 2^{-(c_5 + c_6)k' - 3}.$$

Now, because $c_5 = c_6 = 1/16$ and k' is, in particular, ≥ 8, the above inequality implies

$$\tilde{P}_{\tilde{\tau}} \geq 1/2,$$

a contradiction to the fact that $q' \notin \mathcal{L}$ and SV is a samplable verifier. □

Approximating the Random Oracles

We wish to point out, in a quick and informal way, that the just constructed CS proof-system $(\mathcal{P}, \mathcal{V})$ would maintain some key properties if one substituted its random oracles with some suitable approximations. These approximations are important not only because we are not sure that a random oracle is realizable by a physical process, but also because all physical processes that behave as random oracles may be too inconvenient to use (e.g., because they are too slow in answering their queries). Of course, deciding which approximations are satisfactory depends from the particular context at hand. Let us mention some of them.

- *Quasi-random oracles.* To begin with, one can see that for $(\mathcal{P}, \mathcal{V})$ to be a guaranteed proof-system it is not necessary that f_1 and f_2 be truly random oracles, but sufficiently random ones.
  Assume, for instance, that one may easily select oracles $f_1' : \Sigma^{2k} \to \Sigma^k$ according to a distribution F_1^k such that, for some positive constants e and d,

$$\forall k \text{ and } \forall\ 2^{dk}\text{-call algorithm } A,\ |{p_A}^{k'} - {p_A}^{k}| \leq 2^{-ek},$$

where ${p_A}^{k'}$ denotes the probability (computed selecting f_1' according to ${F_1}^k$) that $A_{f_1'}$ outputs 1, and ${p_A}^k$ the probability (computed selecting $f_1 : \Sigma^{2k} \rightarrow \Sigma^k$ at random) that A_{f_1} outputs 1. Then, such f_1' may be used instead of truly random oracles within the construction of $(\mathcal{P}, \mathcal{V})$.

- *Men in the cave.* Lipton and Rabin have, independently and in different occasions, suggested that a random oracle can be implemented by a trusted party with enough memory.

  In Rabin's suggestive language, the oracle (like respectable ones of times past) is a man living in a cave, with a coin and lots of pen-and-paper. Whenever someone queries him about a string σ, the wise and old man first consults his notes to see whether such a request was ever made before. If so, he retrieves the answer given in the past, and gives that same answer now. If not, he flips his coin the right number of times so as to construct a random string τ; gives τ as his answer; and records that, from now on, query σ should always be answered with τ.

  It would appear that such an implementation of a random oracle is useless in our application, because if our man is corruptable, it would be easy for a cheating Prover to find CS witnesses for false statements. This is not true, because there are contexts in which we can be sure that a man-in-the-cave implementation of a random oracle is trustworthy. For instance, in the context of Certified Computation (see later sections), we may wish to construct a man-in-the-cave by means of a simple probabilistic algorithm, run as a subroutine by other programs for our own use. (Jumping ahead, our oracle only "talks" to an algorithm of *ours* whose results *we* wish to test.) In this application, being the only beneficiary of a correct man-in-the-cave implementation, we have no incentive to cheat (ourselves)!
- Finally, a random oracle can be implemented by having the cryptographic random-function construction of Goldreich, Goldwasser, and Micali [17] (with suitably large security parameters) run within in a tamper-proof chip; one, that is, whose content (program and data) cannot be altered or read from the outside (without destroying it altogether). If the seeds are secret, their pseudo-random functions are provably indistinguishable from truly random ones by any observer, who access them as an oracle, that does not have sufficient resources for solving a given problem (e.g., factoring a 10,000-bit number).

  To ensure its secrecy, the seed could be selected within the chip by some "physical" method (e.g., noise diodes, Geiger counter, etc.). Alternatively, different people can each provide to the chip his own random and secret number, and the chip obtains its own seed by exclusive-oring all these inputs. This way, unless all the parties providing inputs to the chip are corrupt and reveal what their individual input was, the resulting seed will be both random and secret. (Incidentally, if the seed were known, certain unpredictability properties of the corresponding pseudo-random function would disappear, but, for all we know, the difficulty of finding CS proofs for false statements may remain sufficiently intact.)

3.2 CS Proofs with a Random String

Though in this paper we do not address the problem of explicitly constructing CS proofs with a random string, we wish to present their definition and mention their applications.

The Notion of a CS Proof-System with a Random String

In a CS proof-system with a random oracle a cheating Prover was allowed to be an algorithm which could make only a bounded number of oracle calls, but have an arbitrarily long description. Informally, this was so because finding guaranteed CS certificates for a false statement implied finding special structures in a random oracle —e.g., two different strings x and y such that $f_1(x) = f_1(y)$— discoverable only by making extraordinarily many queries. Having lots of information "built-in" in the finite-state control of an oracle-calling algorithm could not help at all in finding these structures.

In a CS proof-system with a random string, Prover and Verifier are ordinary (as opposed to oracle-calling) algorithms, sharing a short random string r. That is, whenever the security parameter is k, they share a string r that both believe to have been randomly selected among those having length k^c, where c is a positive constant.

Because string r is assumed to be universally known (at lest by all those who wish to check CS proofs of security-level k), and because a CS certificate of $\mathbf{q}$ is verifiable in our new setting given $\mathbf{q}$ and r, a good-looking certificate for a false statement is certainly discoverable by exhaustive search. Thus, the only hope to stop a malicious Prover from cheating consists of bounding the number of computational steps he can perform. But bounding a cheating Prover's running time is meaningless unless its description is also bounded. (For instance, factoring a randomly chosen k-bit integer appears to be computationally intractable when k is large, but not for an algorithm whose description is about 2^k-bit long! Indeed, the finite state control of such a Turing machine could easily encode the factorization of all k-bit integers.) For this reason, in a cryptographic CS proof-systems, a cheating prover is envisaged to be an algorithm whose running time and description, in some standard encoding, are both bounded. We actually accomplish both tasks at once, by letting cheating provers be Boolean combinatorial circuits with a bounded number of gates.

Definition: A *circuit of size* $\leq s$ is a finite function computable by at most s Boolean gates, where each gate is either a NOT-gate (with one binary input and one binary output) or an AND-gate (with two binary inputs and one binary output).

Definition: Let (P, V) be a pair of Turing machines, the second of which runs in polynomial-time. We say that such a pair (P, V) is a *CS proof-system with a random string* if there exists a sequence of 6 positive constants, $c_1, \ldots, c_6$ (refereed to as the *fundamental constants* of the system), such that the following two properties are satisfied:

1″. *Feasible Completeness.* $\forall$ integers $n > 1$, $\forall$ n-bit input $\mathbf{q} = (\bar{M}, x, t) \in \mathcal{L}$, $\forall$ unary integers k, and $\forall\ r \in \Sigma^{k^{c_1}}$,
(*i*) on inputs $\mathbf{q}, k$, and r, P halts within $(nkt)^{c_2}$ computational steps, and outputs a binary string whose length is $\leq (nk \log t)^{c_3}$, and
(*ii*) Whenever $\mathcal{C} = P(k, r, \mathbf{q})$, $V(k, r, \mathbf{q}, \mathcal{C}) = YES$.

2″. *Computational Soundness.* $\forall$ integers $n > 1$, $\forall\, n$-bit $\mathbf{q}' \notin \mathcal{L}$, $\forall\ k > n^{c_4}$, and $\forall$ (cheating) circuits P' whose size is $\leq 2^{c_5 k}$,

$$Prob_{r \in \Sigma^{k^{c_1}}}(P'(\mathbf{q}', k, r) = C' \wedge V(\mathbf{q}', k, r, C') = YES) \leq 2^{c_6 k}.$$

The Importance of CS Proofs with a Random String

CS proof-systems with a random string are important because of at least three reasons.

First, because (as we shall explain in a forthcoming paper) they are explicitly constructable under a new, but in our opinion plausible, complexity assumption.

Second, because it is much easier for two players (i.e., Prover and Verifier) or society at large to establish a common random string than a common random oracle.

Third, because the certificates for $\mathbf{q} \in \mathcal{L}$ they produce are polynomial-time recognizable.

The latter technical reason is full of notable consequences. In particular, it allows us to enlarge Blum's notion of *program checking* in new and powerful ways, and to prove that heuristics for $\mathcal{NP}$-complete problems are checkable in this broader sense. For instance, given an efficient heuristic H for graph Hamiltonicity and a specific graph G, we show how it is possible to run H twice (once on G and once of an efficiently computed graph G' for which deciding Hamiltonicity is roughly "as hard as for G") and either (1) prove that H is wrong on at least one of these two graphs, or (2) establish with a "great degree of confidence" that H's answer about graph G is correct.

These applications of CS proofs with a random string will also appear in the mentioned forthcoming paper.

3.3 Deterministic CS Proofs

Definition: Let (P, V) be a pair of Turing machines, the second of which runs in polynomial-time. We say that such a pair (P, V) is a *deterministic CS proof-system* if there exists a sequence of 5 positive constants, $c_1, \ldots, c_5$ (refereed to as the *fundamental constants* of the system), such that the following two properties are satisfied:

1'''. *Feasible Completeness.* $\forall$ integers $n > 1$, $\forall$ n-bit input $\mathbf{q} = (\bar{M}, x, t) \in \mathcal{L}$, and $\forall$ unary integers k,
(*i*) on inputs $\mathbf{q}$ and k, P halts within $(nkt)^{c_1}$ computational steps, and outputs a binary string whose length is $\leq (nk \log t)^{c_2}$, and
(*ii*) Whenever $\mathcal{C} = P(\mathbf{q}, k)$, $V(\mathbf{q}, k, \mathcal{C}) = YES$.

2'''. *Computational Soundness.* $\forall$ integers $n > 1$, $\forall n$-bit $\mathbf{q}' \notin \mathcal{L}$, $\forall$ $k > n^{c_3}$, and $\forall$ (cheating) probabilistic algorithms $P'(\cdot, \cdot)$ halting in less than $\leq 2^{c_4 k}$ steps whenever their second input is k (in unary),

$$Prob(P'(\mathbf{q}', k) = \mathcal{C}' \wedge V(\mathbf{q}', k, \mathcal{C}') = YES) \leq 2^{c_5 k}.$$

As we said, we believe that the complexity assumption needed to build deterministic CS proof-systems is much stronger than that necessary for CS proof-systems with a random string.

4 Certified Computation

In this short section we reinterpret the results obtained so far in terms of *computation* rather than proofs. More precisely, we aim at obtaining certificates ensuring that no error has occurred in a *given* execution of a *given* algorithm on a *given* input. That is, certified computation does not deal with semantic questions such as "is algorithm A correct?" Rather it addresses the following syntactic question:

Is string y what algorithm A should output on input x (no matter what A is supposed to do)?

This question is quite crucial. Consider running executing A on an input x so as to obtain a result y. How can we be sure that $y = A(x)$? In real life, such an execution occurs by means a physical computer, and all sort of errors may occur, yielding a wrong value for $A(x)$. For instance, the computer hardware may be defective. Alternatively, the hardware may function properly, but the operating system may be flawed. Alternatively yet, hardware and software may be fine, but some α rays succeed in flipping a bit together with its controls, so that the original bit value is not restored.

Also, having run A on input x so as to obtain a result y, can we convince someone else that y is the right result without having him re-do the computation himself?

Certified computation provides a way to answer these basic questions, for any algorithm and any input.

"Defining" Certified Computation. In view of our work so far, formalizing and exemplifying (at least given a random oracle) the notion of certified computation is rather straightforward, but tedious indeed. We thus think that is best to proceed at a very intuitive level.

> **Informal Definition:** A *certified-computation system* is a pair of efficient algorithms, $(\mathcal{C}, \mathcal{V})$. Given any algorithm A as input, $\mathcal{C}$ outputs an equivalent algorithm A' enjoying the following properties.
>
> 1. A' runs in essentially the same time as A does;
> 2. A' receives the same inputs applicable to A and produces the same outputs; and
> 3. for each input x, A' produces, besides the right output y, also a short and easily inspectable string, C_{Axy}, vouching that indeed $y = A(x)$ in the following sense:
>
>   If $y = A(x)$, then $\mathcal{V}(A, x, y, C_{Axy}) = YES$. Else, it is very hard to find a string σ such that $\mathcal{V}(A, x, y, \sigma) = YES$.

Of course, one may ask who verifies the correctness of the verifier (i.e., either of algorithm $\mathcal{V}$ itself or of its executions). Note, however, that such a $\mathcal{V}$ is a *unique* program, capable of verifying certificates for the correct execution of *all* other programs. It is thus meaningful to invest sufficient time in proving the correctness of this particular algorithm (e.g., by verification methods). Also, being $\mathcal{V}$ quite efficient (and running on short inputs) we may afford to execute it on very "conservative" hardware (i.e., with particular redundancy, resiliency, and so on), or even on a multiplicity of hardwares.

Constructing certified-computation systems. One possible way of constructing program certification systems essentially consists of giving a CS proof of the statement "$y = A(x)$." That is, on input x, $A' = \mathcal{C}(A)$ first runs A on input x so that, after some number T of steps, an output y is obtained. Then, A' outputs y and a CS proof that $y = A(x)$. Because the latter statement can be decided in time T, the length of its corresponding CS proof will be polynomial in $\log T$ (and, of course, $|x|$, $|y|$, and some suitable security parameter k), and so will be the time to inspect it, and the time to construct it.

Pros and cons. A certified-computation system can be considered as a *universal and guaranteed casting-out-nines procedure.* (Such a procedure is executed after a multiplication, but does does not guarantee an "arbitrarily high guarantee of correctness" for every single input.) In other words, certified computation is not a special property enjoyed by a few computational problems (such as multiplication or GCD computation.[22]), but is an *intrinsic* property of computation.

However, unless one can find a version or representation of the execution history of an algorithm that is both convenient and sufficient for our purposes,[23] our result will not be practical.

Assumptions and implementations. Certified-computation systems can be built based on a random oracle, but are particularly meaningful when constructed, under our assumption, based on a CS proofs with a random string for at least two reasons.

First, in this application, the random string needs not be agreed upon by both Prover and Verifier by virtue of some possibly difficult negotiation. When seeking reassurance that indeed $y = A(x)$, the user of a certified-computation system $(\mathcal{P}, \mathcal{V})$ controls both $\mathcal{P}$ and $\mathcal{V}$, and thus can choose their common string in a way that he believes to be genuinely random, without "asking for their consent."

---

[22] Indeed, the extended GCD algorithm may be a bit more cumbersome than the ordinary GCD onalgorithm, but more "reliable". In fact, according to Blum's point of view, by outputting not only the greatest common divisor, c, of two integers a and b, but also two integers x and y such that $ax + by = c$, the extended CGD algorithm allows one to check the validity of c quite easily. That is, by checking that indeed $ax + by = c$ and that c evenly divides both a and b, all operations much simpler and computationally less intensive than GCD computation. Note, however, that x and y consist of a quite long certificate for c's correctness (in relative terms), and that three multiplications and one addition are quite long (relative to the time required by GCD computation). By comparison, our certified-computation system may produce relatively shorter and more easily inspectable certificates, but has a probability of error.

[23] Certainly, one that does not consist of the sequence of all instantaneous Turing machine configurations.

In addition, in this application, algorithm $\mathcal{P}$ "sits on top of the user desk" and thus there is no question of it being implemented by a circuit comprising 2^k gates (nor of its making more than 2^k steps of computation), even for moderately small k. That is, if the underlying complexity assumption is true, a certified-computation system *de facto* offers the same guarantees of a probabilistic algorithm.

Conceivable applications of certified computation. Certified computation can in principle be quite useful when "contracting out" computer time. Indeed, consider an algorithm A that we believe to be correct, but is very time-consuming. Then, we can temporarily hire a super-computing company for executing A on a given input x on their computers, and agree that we will pay for their efforts if they give us back the value $y = A(x)$ together with a certificate of correct execution.

Still in principle, certified computation may facilitate the verification of certain mathematical theorems proven with the help of a computer search (as in the case of the 4-color theorem). For instance, the proof may depend on a Lemma stating that there is no sparse graph with less than 50 nodes possessing a given property, and the Lemma could be proved by means of an exhaustive search taking a few years of computing. Usually, algorithms performing an exhaustive search appear to be correct "by eye inspection". Thus, rather than (1) asking the reader of trusting that such a search has been done and has returned a negative result, or (2) asking the reader to perform such an extensive computation himself, one might publish together with the rest of the proof a compact, correct-execution certificate. If the security parameter where chosen to be —say— 1,000, then even the most skeptic reader might believe that no one has invested $2^{1,000}$ steps of computation in order to find a false certificate, nor that he has succeeded in finding one by relying on a probability of success less than $2^{-1,000}$.

5 Final Remarks

Are CS proofs really proofs? In our minds this question really goes together with an older one: do probabilistic algorithms [38] [31] really compute? There is a sense in which both answers should be NO. These negative answers, in our opinion, may stem from two different reasons: (1) a specific interpretation of the words "proof" and "computation," and (2) our mathematical tradition. The first reason is certainly true, but also "innocuous." The second is more "dangerous" and less acceptable: not because it is false that these notions break with a long past, but because the unchallenged length of a tradition should not be taken as implying that specific formalizations of fundamental intuitions are "final."

Indeed, we believe that even fundamental intuitions cannot be divorced by the large historical contexts in which they have arisen, and we expect that they will change with the changing of these contexts. Up to now, the ideas of this paper have not met much favor, exception made for the invaluable support of a trusted group of scientific friends. We hope, however, that, in the future, they will be regarded to be as natural as is, *by now*, the notion of probabilistic computation.

Truths versus proofs. According to our highest-level goal, CS proofs propose a *very appealing* relationship between proving and deciding. In the thirties, Turing, Church, and others suggested that deciding a mathematical statement consists of running a specific algorithm. Today, we are suggesting that proving a mathematical statement consists of feasibly speeding up the verification of the output of any decision algorithm. That is, proofs should guarantee to any third party an exponential speed-up in verifying that the output of any deciding process is YES; but they should not be more time-consuming than the computations whose verification they wish to facilitate. This, in our opinion, is the right relationship between proving and deciding, and one that guarantees that proving is both a useful and a distinct notion.

In order to offer this guarantee, CS proofs replace the traditional notion of truth with a *computational* one. Namely, the difference between proving true a correct statement and proving true a false one now is the difference between an easy computation and an extraordinarily hard one. Indeed, wishing to convince *any* third party (and not just a Verifier interacting with them) of the verity of a given theorem, CS Provers issue a CS certificate, a string that can be thought of as a "compressed version" of a much longer deciding computation. But the same conciseness that gives CS certificates their distinctive advantage, also causes them to "lose information" with respect to the computations they compress. It is thus possible that correct-looking certificates of false theorems exist, and whenever they do, of course, they can also be found by means of an exponential search. CS proofs guarantee, however, that essentially only an exponential search could be successful in finding misleading certificates.

In sum, the notion of a true theorem has not changed. What has changed, and in a computational-complexity direction, is the notion of what it means to *prove that a theorem is true*. Besides being very adequate in practice and very advantageous, this change also is, in our opinion, very natural. Indeed, proofs have no meaning outside efficiency: with or without modern terminology, they have always been, at least implicitly, a complexity-theoretic notion.

Let me emphasize, however, that, though very natural, the computational boundedness of a cheating prover is not, *per se*, a goal of the notion of a proof. Rather, it is the Trojan horse that lets us sneak in and achieve our desiderata. It is actually very important to establish whether some type of CS proofs exist when a cheating Prover can compute for an unbounded amount of time. [24]

We believe and hope that CS proofs maybe useful in many more applications than the ones envisaged here.

6 Acknowledgments

My most sincere thanks go to Allan Borodin, Steve Cook, Oded Goldreich, Shafi Goldwasser, Leonid Levin, and Michael Rabin for encouraging and criticizing this research, in different ways and at various stages of its development.

Thanks also to Shai Halevi, Ray Sidney and Janos Makowski (and two referees) for their comments.

References

1. S. Arora, C. Lund, R. Motwani, M. Sudan, and M. Szegedy. *Proof verification and hardness of approximation problems.* Proc. 33rd. IEEE Conference on Foundation of Computer Science, 1992, pp. 14-23.
2. S. Arora and M. Safra. *Probabilistic checking of proofs.* Proc. 33rd. IEEE Conference on Foundation of Computer Science, 1992, pp. 2-13.
3. L. Babai and L. Fortnow and L. Levin and M. Szegedy. *Checking Computation in Polylogarithmic Time.* Proc. of STOC91.
4. L. Babai and S. Moran. JCSS 1988. A preliminary version due to the first author, "Trading Group Theory for Randomness," appeared in Proc. 17th Annual Symposium on Theory of Computing, 1985, pp. 421-429.
5. M. Ben-or and S. Goldwasser and J. Kilian and A. Wigderson. *Multi Prover Interactive Proofs: How to Remove Intractability.* Proc. 20th ACM Symp. on Theory of Computing, 1988, pp. 113-131.
6. M. Blum, A. De Santis, S. Micali, and G. Persiano. *Non-Interactive Zero-Knowledge.* SIAM J. on Comp. 1991.
7. M. Blum, P. Feldman, and S. Micali. *Non-Interactive Zero-Knowledge Proof Systems and Applications.* STOC 1988.
8. M. Blum and S. Kannan. *Designing Programs that check their work.* Proc. 21st Symposium on Theory of Computing, 1989, pp. 86-97.
9. M. Blum, M. Luby, and R. Rubinfeld. *Self-Testing and Self-Correcting Programs, With Applications to Numerical Problems.* Proc. 22nd ACM Symp. on Theory of Computing, 1990, pp. 73-83.

[24] By means of a weaker construction (in particular, one not enjoying feasible completeness), we had previously observed that, via Merkle trees and the self reducibility of the permanent, a bounded prover could demonstrate membership in any $\#P$-language. Our observation in fact followed the preliminary announcement that $\#\mathcal{P}$ is efficiently accepted by a 2-Prover system, but lost any interest a few days later, with the final announcement that $\#\mathcal{P} \subset \mathcal{IP}$ [26]. We wonder if also CS proofs (at least their interactive counter-part) may be subject to the same fate.

10. M. Blum and S. Micali. *How to Generate Cryptographically-Strong Sequences of Pseudo-Random Bits.* SIAM J. on Comp. vol 13, 1984

11. G. Brassard and D. Chaum and C. Crepeau. *Minimum Disclosure Proofs of Knowledge.* J. Comput. System Sci., 37, 1988, pp. 156-189.

12. S. Cook. *The Complexity of Theorem Proving Procedures.* Proc. 3rd Annual ACM Symposium on Theory of Computing, 1971, pp. 151-158.

13. U. Feige and S. Goldwasser and L. Lovasz and S. Safra and M. Szegedi. *Approximating Clique is Almost NP-complete.* 32nd FOCS, 1991, pp. 2-12.

14. A. Fiat and A. Shamir. *How to Prove Yourselves: Practical Solutions of Identification and Signature Problems.* Proc. Crypto 86, Springer- Verlag, 263, 1987, pp.186-194.

15. L. Fortnow. *The Complexity of Perfect Zero Knowledge.* Randomness and Computation, Advances in Computer Research, ed. S. Micali, JAI Press, 1989, pp. 327-344.

16. L. Fortnow, J. Rompel, and M. Sipser. *On the Power of Multi-Prover Interactive Protocols.* Proc. 3rd Structure in Complexity Theory Conf., 1988, pp. 151-158.

17. O. Goldreich, S. Goldwasser, and S. Micali. *How To Construct Random Functions.* J. of ACM 1986

18. S. Goldwasser and S. Micali and C. Rackoff. *The Knowledge Complexity of Interactive Proof Systems.* SIAM J. Comput., 18, 1989, pp. 186-208. An earlier version of this result informally introducing the notion of a proof of knowledge appeared in Proc. 17th Annual Symposium on Theory of Computing, 1985, pp. 291-304. (Earlier yet versions include "Knowledge Complexity," submitted to the 25th Annual Symposium on the Foundations of Computer Science, 1984.)

19. S. Goldwasser, S. Micali, and R. Rivest, *A Digital Signature Scheme Secure Against Adaptive Chosen-Message Attacks*, SIAM J. Comput., Vol 17, No. 2, April 1988, pp. 281-308.

    (A preliminary version of this article appeared with the title "A paradoxical solution to the signature problem" in Proc. of 25th Annual IEEE Symposium on the Foundations of Computer Science, FL, November 1984, pp. 464-479.)

20. O. Goldreich, S. Micali, and A. Wigderson. *Proofs that Yield Nothing But Their Validity or All Languages in $\mathcal{NP}$ have Zero-Knowledge Proof Systems.* J. of the ACM, Vol 38, No. 1, July 1991, pp. 691-729.

    (A preliminary version of this paper, under the title "Proofs that yield nothing but their validity and a methodology for cryptographic protocol design," appeared in Proc. 27th Annual Symposium on Foundations of Computer Science, IEEE, New York, 1986, pp. 174-187.)

21. R. Karp. *Reducibility among combinatorial problems.* Complexity of Computer Computations, R. Miller and J. Thatcher eds., Plenum, New York, 1972, pp. 85-103.

22. J. Kilian. *A Note on Efficient Zero-Knowledge Proofs and Arguments.* Proc. 24th Ann. Symp. on Theory of Computing, Victoria, Canada, 1992.

23. R. Impagliazzo, J. Hastad, L. Levin, and M. Luby. *Pseudo-Random Generation under uniform Assumptions.* STOC 1990.

24. R. Impagliazzo, L. Levin, and M. Luby. *Pseudo-Random Generation From one-way functions.* STOC 1989
25. L. Levin. *Universal Sequential Search Problems.* Problems Inform. Transmission, Vol. 9, No. 3, 1973, pp. 265-266.
26. C. Lund and L. Fortnow and H. Karloff and N. Nisan. *Algebraic Methods for Interactive Proof Systems.* Proc. 22nd STOC, 1990.
27. R. Merkle. *A Certified Digital Signature.* Proc. Crypto 1989. Springer Verlag, 1990.
28. S. Micali. *CS Proofs.* Proc. 35th Annual Symposium on Foundations of Computer Science, 1994, pp.
(An earlier version of this paper appeared as Technical Memo MIT/LCS/TM-510. Earlier yet versions were submitted to the 25th Annual Symposium on Theory of Computing, 1993, and the 34th Annual Symposium on Foundations of Computer Science, 1993.)
29. S. Micali, Private Communication to Shafi Goldwasser, 1992.
30. A. Polishchuk and D. Spielman. *Nearly-linear Size Holographic Proofs.* Proc. STOC 1994.
31. M. Rabin. *Digitalized Signatures.* in Foundations of Secure Computation, Academic Press, 1978, pp. 155-168.
32. M. Rabin. *Digitalized Signatures as Intractable as Factorization.* MIT Laboratory for Computer Science Technical Report MIT/LCS/TR-212, Massachusetts Institute of Technology, Cambridge, MA, January 1979.
33. M. Rabin. *Probabilistic algorithms for testing primality.* J. Number Theory, Vol. 12, 1980, pp. 128-138.
34. R. Rivest, A. Shamir, and L. Adleman. *A Method for Obtaining Digital Signatures and Public-Key Cryptosystems.* Comm. ACM, Vol. 21, 1978, pp. 120-126.
35. R. Rivest. *The MD5 Message-Digest Algorithm.* Internet Activities Board, Request for Comments 1321, April 1992.
36. *Secure Hash Standard.* Federal Information Processing Standards, Publication 180, 1993.
37. A. Shamir. *IP = PSPACE.* Proc. 31st IEEE Foundation of Computer Science Conference, 1990, pp. 11-15.
38. R. Solovay and V. Strassen. *A fast Monte-Carlo test for primality.* SIAM J. Comp., Vol. 6, 1977, pp. 84-85.
39. M. Sudan. *Efficient checking of polynomials and proofs and the hardness of approximation problems.* Ph.D. Thesis, University of California at Berkeley, 1992.
40. A. Yao. *Theory and Applications of Trap-Door Functions.* Proc. 23rd IEEE on Foundations of Computer Science, 1982.

Lambek Calculus and Formal Languages

(Extended abstract)

Mati Pentus

Department of Mathematical Logic
Faculty of Mechanics and Mathematics
Moscow State University
Moscow, Russia, 119899
pentus@lpcs.math.msu.ru

Introduction

The systematic study of generating grammars was started by N. Chomsky in the 50s (cf. [10]). He defined several classes of generating grammars, which are interesting for both linguists and mathematicians, e. g. context-sensitive grammars, context-free grammars, and linear grammars. On the other hand, categorial grammars were studied by Y. Bar-Hillel, J. Lambek and others. The notion of a basic categorial grammar was introduced in [1]. In the same paper it was proved that the languages recognized by basic categorial grammars are precisely the context-free ones.

Another kind of categorial grammar was introduced by J. Lambek [15]. These grammars are based on a syntactic calculus, presently known as the Lambek calculus. Chomsky [11] conjectured that these grammars are also equivalent to context-free ones. In [12] Cohen proved that every basic categorial grammar (and, thus, every context-free grammar) is equivalent to a Lambek grammar. He also proposed a proof of the converse. However, as pointed out in [6], this proof contains an error. Buszkowski proved that some special kinds of Lambek grammars are context-free [6, 8, 9]. These grammars use weakly unidirectional types or types of order at most two.

The first result of this paper (Theorem 1) says that Lambek grammars generate only context-free languages. Thus they are equivalent to context-free grammars and also to basic categorial grammars. This fact (sometimes called *the Chomsky Conjecture*) was proved in [16] and [17].

The inteneded syntactic string models, i.e., *free semigroup models* (also called *language models* or *L-models*) for the Lambek calculus were considered in [3], [4], and [5]. The more general class of *groupoid models* has been studied in [7], [13], and [14]. In [4] W. Buszkowski established that the product-free fragment of the Lambek calculus is L-complete (i.e., complete w.r.t. free semigroup models), using the canonical model. The question of L-completeness of the full Lambek calculus remained open (cf. [2]).

The second result of this paper (Theorem 2) gives a positive answer to this question. The proof has been publised in [18] and [19].

1 Preliminaries

For any set $\mathcal{M}$ we denote by $\mathcal{M}^+$ the set of all finite non-empty strings consisting of elements of $\mathcal{M}$. The set of all subsets of $\mathcal{M}$ is denoted by $\mathbf{P}(\mathcal{M})$.

We consider the syntactic calculus introduced in [15]. The types of the Lambek calculus are built of primitive types $p_1, p_2, \ldots$ and three binary connectives $\bullet, \backslash, /$. We shall denote the set of all types by Tp. Capital letters A,B, ... range over types. Capital Greek letters range over finite (possibly empty) sequences of types. Sequents of the Lambek calculus are of the form $\Gamma \to A$, where Γ is a nonempty sequence of types.

Axioms: $A \to A$

Rules:

$$\frac{\Gamma \to A \quad \Delta \to B}{\Gamma \Delta \to A \bullet B}\ (\to\bullet) \qquad\qquad \frac{\Gamma A B \Delta \to C}{\Gamma (A \bullet B) \Delta \to C}\ (\bullet\to)$$

$$\frac{A \Pi \to B}{\Pi \to A \backslash B}\ (\to\backslash) \text{ where } \Pi \text{ is non-empty} \qquad \frac{\Pi \to A \quad \Gamma B \Delta \to C}{\Gamma \Pi (A \backslash B) \Delta \to C}\ (\backslash\to)$$

$$\frac{\Pi A \to B}{\Pi \to B / A}\ (\to/) \text{ where } \Pi \text{ is non-empty} \qquad \frac{\Pi \to A \quad \Gamma B \Delta \to C}{\Gamma (B / A) \Pi \Delta \to C}\ (/\to)$$

$$\frac{\Pi \to B \quad \Gamma B \Delta \to A}{\Gamma \Pi \Delta \to A}\ (CUT)$$

The cut-elimination theorem for this calculus is proved in [15].

2 Lambek grammars recognize context-free languages

Definition. We assume that a finite alphabet $\mathcal{T}$ and a distinguished type D are given. A *Lambek grammar* is a mapping f such that, for all $t \in \mathcal{T}$, $f(t) \subset \mathrm{Tp}$ and $f(t)$ is finite.

The *language generated by the Lambek grammar* is defined as the set of all expressions $t_1 \ldots t_n$ over the alphabet $\mathcal{T}$ for which there exists a derivable sequent $B_1 \ \ldots \ B_n \to D$ such that $B_i \in f(t_i)$ for all $i \leq n$.

Definition. We assume that two disjoint alphabets $\mathcal{T}$ and $\mathcal{W}$ are given. The elements of $\mathcal{T}$ are called *terminal symbols* and those of $\mathcal{W}$ are *auxiliary symbols*.

A *context-free rewrite rule* is of the form $X \Rightarrow e$, where X is an auxiliary symbol and e is a non-empty word in the alphabet $\mathcal{T} \cup \mathcal{W}$.

A *context-free grammar* is a finite set $\mathcal{R}$ of context-free rewrite rules, with one auxiliary symbol S designated as its *start symbol.*

By $\bar{\mathcal{G}}(\mathcal{T}, \mathcal{W}, S, \mathcal{R})$ we denote the set of all expressions over the alphabet $\mathcal{T} \cup \mathcal{W}$ that arise through some finite sequence of rewritings of the start symbol S via the rules of $\mathcal{R}$.

The *language generated by the context-free grammar* is defined as

$$\bar{\mathcal{G}}(\mathcal{T}, \mathcal{W}, S, \mathcal{R}) \cap \mathcal{T}^+.$$

Theorem 1. *For any Lambek grammar there exists a context-free grammar such that the languages generated by these grammars coincide.*

3 L-completeness of the Lambek calculus

Definition. We define *L-model* (also called *language model* or *free semigroup model*) to be a triplet $\langle \mathcal{W}^+, \circ, w \rangle$, where $\mathcal{W}$ is an arbitrary alphabet, $\circ$ denotes concatenation of words from $\mathcal{W}^+$, and w is a function $w\colon \mathrm{Tp} \to \mathbf{P}(\mathcal{W}^+)$ such that

(1) $w(A \bullet B) = w(A) \circ w(B)$;
(2) $w(A \backslash B) = \{\gamma \in \mathcal{W}^+ \mid w(A) \circ \{\gamma\} \subseteq w(B)\}$;
(3) $w(B/A) = \{\gamma \in \mathcal{W}^+ \mid \{\gamma\} \circ w(A) \subseteq w(B)\}$.

Here for any two sets $\mathcal{A} \subseteq \mathcal{W}^+$ and $\mathcal{B} \subseteq \mathcal{W}^+$ by $\mathcal{A} \circ \mathcal{B}$ we denote the set $\{\alpha \circ \beta \mid \alpha \in \mathcal{A} \text{ and } \beta \in \mathcal{B}\}$.

Definition. A sequent $A_1 \ldots A_n \to B$ is *true* in a model $\langle \mathcal{W}^+, \circ, w \rangle$ iff

$$w(A_1) \circ \ldots \circ w(A_n) \subseteq w(B).$$

Theorem 2. *A sequent is derivable in the Lambek calculus if and only if it is true in every L-model.*

Theorem 3. *A sequent is derivable in the Lambek calculus if and only if it is true in every L-model over an alphabet $\mathcal{W}$ consisting of two symbols.*

References

1. Y. Bar-Hillel, C. Gaifman, and E. Shamir. On categorial and phrase-structure grammars. *Bull. Res. Council Israel Sect. F*, 9F:1–16, 1960.
2. J. van Benthem. *Language in Action: Categories, Lambdas and Dynamic Logic.* North-Holland, Amsterdam, (Studies in Logic 130), 1991.
3. W. Buszkowski. Undecidability of some logical extensions of Ajdukiewicz-Lambek calculus. *Studia Logica*, 37:59–64, 1978.
4. W. Buszkowski. Compatibility of categorial grammar with an associated category system. *Zeitschrift für mathematische Logik und Grundlagen der Mathematik*, 28:229–238, 1982.
5. W. Buszkowski. Some decision problems in the theory of syntactic categories. *Zeitschrift für mathematische Logik und Grundlagen der Mathematik*, 28:539–548, 1982.
6. W. Buszkowski. The equivalence of unidirectional Lambek categorial grammars and context-free grammars. *Zeitschrift für mathematische Logik und Grundlagen der Mathematik*, 31:369–384, 1985.
7. W. Buszkowski. Completeness Results for Lambek Syntactic Calculus. *Zeitschrift für mathematische Logik und Grundlagen der Mathematik*, 32:13–28, 1986.
8. W. Buszkowski. Generative power of categorial grammars. In R.T. Oehrle, E. Bach, and D. Wheeler, editors, *Categorial Grammars and Natural Language Structures*, pages 69–94, Reidel, Dordrecht, 1988.
9. W. Buszkowski. On generative capacity of the Lambek calculus. In J. van Eijck, editor, *Logics in AI*, pages 139–152, Springer, Berlin, 1991.

10. N. Chomsky. *Syntactic Structures.* Mouton, The Hague, 1957.
11. N. Chomsky. Formal properties of grammars. In R.D. Luce et al., editors, *Handbook of Mathematical Psychology*, vol. 2, pages 323–418, Wiley, New York, 1963.
12. J.M. Cohen. The equivalence of two concepts of categorial grammar. *Information and Control*, 10:475–484, 1967.
13. K. Došen. A Completeness Theorem for the Lambek Calculus of Syntactic Categories. *Zeitschrift für mathematische Logik und Grundlagen der Mathematik*, 31:235–241, 1985.
14. K. Došen. *A brief survey of frames for the Lambek calculus.* Bericht 5-90, Konstanzer Berichte zur Logik und Wissenschaftstheorie, Universität Konstanz, 1990.
15. J. Lambek. The mathematics of sentence structure. *American Mathematical Monthly*, 65(3):154–170, 1958.
16. M. Pentus. *Lambek grammars are context free.* Preprint No. 8 of the Department of Math. Logic, Steklov Math. Institute, Series Logic and Computer Science, Moscow, 1992.
17. M. Pentus. Lambek grammars are context free. *Proceedings of the 8th Annual IEEE Symposium on Logic in Computer Science*, 429–433, 1993.
18. M. Pentus. Lambek calculus is L-complete. ILLC Prepublication Series LP–93–14, Institute for Logic, Language and Computation, University of Amsterdam, 1993.
19. M. Pentus. Models for the Lambek calculus. *Annals of Pure and Applied Logic*, 75:179–213, 1995.

Zil'ber's Trichotomy and o-minimal Structures

(Extended abstract) *

Ya'acov Peterzil

Department of Mathematics
University of Haifa
Haifa, Israel
kobi@mathcs2.haifa.ac.il

A first-order structure $\mathcal{M}$ is called *a geometric structure* (see [2]) if it has the following properties:
(i) $acl(-)$ satisfies the Exchange principle.
Namely, given $a, b, \bar{c}$ from M, if $a \in acl(b, \bar{c}) \setminus acl(\bar{c})$ then $b \in acl(a, \bar{c})$.
(ii) For any formula $\varphi(\bar{x}, \bar{y})$ there is $n \in \mathbb{N}$ such that for any $\bar{b}$ in M, $\varphi(\bar{x}, \bar{b})$ has either less than n solutions in M or infinitely many.

O-minimal and strongly minimal structures are geometric structures. The field of p-adics and pseudo-finite fields are geometric structures as well.

Given a geometric structure $\mathcal{M}$ one can assign a dimension to definable sets in a natural way which in all the field cases mentioned above is just the algebro-geometric dimension of the Zariski closure. *A curve* is any definable 1-dimensional subset of M^2 and a definable (or interpretable) family $\mathcal{F}$ of curves is called *normal* if any two curves from $\mathcal{F}$ which are given by different parameters intersect at most finitely many times. If $\mathcal{F}$ is normal its dimension is taken to be the dimension of the parameter set.

Given a geometric structure, one and only one of the following holds.
Z1. Every interpretable normal family of curves $\mathcal{F}$ is of dimension at most 1 and for all but finitely many curves $\mathcal{C} \in \mathcal{F}$, for all but finitely many points $\langle a, b\rangle \in \mathcal{C}$, either $\dim(\mathcal{C} \cap (\{a\} \times M)) = 1$ or $\dim(\mathcal{C} \cap (M \times \{b\})) = 1$.
Z2. Every interpretable normal family of curves is of dimension at most 1, but Z1 does not hold.
Z3. There is an interpretable normal family of curves of dimension greater than 1.

In the early 1980's Boris Zil'ber (see [6]), in his analysis of $\aleph_1$-categorical structures, suggested that the above trichotomy corresponds to the interpretability (or the lack of which) of certain algebraic structures in $\mathcal{M}$. He called Z2 and Z3 the module-like and field-like cases, respectively, and conjectured that if a strongly minimal structure satisfies Z3 then it can interpret a field. We formulate this correspondence as follows:

Definition 1 A class $\mathcal{K}$ of geometric structures is said to satisfy the Zil'ber Principle, ZP, if for every $\mathcal{M} \in \mathcal{K}$,

* This abstract discusses a joint work of the author and Sergei Starchenko. For the proof of the main theorem see [4]. An expanded version of this abstract has appeared in [5].

(i) $\mathcal{M}$ satisfies Z1 if and only if no group is interpretable in $\mathcal{M}$.
(ii) $\mathcal{M}$ satisfies Z2 if and only if all definable sets arise from an interpretable vector space (or more generally a module) structure.
(iii) $\mathcal{M}$ satisfies Z3 if and only if a field can be interpreted in $\mathcal{M}$.

Hrushovski (see [1]) disproved Zil'ber's conjecture by constructing a strongly minimal structure satisfying Z3, with no group interpretable in it. Later, Hrushovski and Zil'ber showed that a more restrictive class of strongly minimal structures called Zariski geometries (see [3]) satisfies the Zil'ber Principle.

The Trichotomy Theorem below implies that the class of o-minimal structures satisfies the Zil'ber Principle (with the exception that that in the Z2 case it is only locally, around each nontrivial point, that the definable sets arise from a vector space). The theorem seems to reduce in many cases model theoretic questions on abstract o-minimal structures to algebraic and differential questions on o-minimal expansions of real closed fields.

We call a point a in an o-minimal structure *nontrivial* if there is an open interval I containing a and a definable function $F(x, y)$ on $I \times I$ which is continuous and strictly monotone in each variable.

The Trichotomy Theorem[4] *Let $\mathcal{M}$ be an o-minimal, ω^+-saturated structure. Given $a \in M$, one and only one of the following hold:*
T1. *a is trivial;*
T2. *The structure that $\mathcal{M}$ induces on some convex neighborhood of a is an ordered vector space over an ordered division ring;*
T3. *The structure that $\mathcal{M}$ induces on some open interval around a is an o-minimal expansion of a real closed field.*

References

1. E. Hrushovski, *A new strongly minimal set*, Annals of Pure and Applied Logic 62 (1993), pp. 147-166.
2. —— and A. Pillay, *Groups definable in local fields and pseudo-finite fields*, Israel Journal of Mathematics 85 (1994), pp. 203-262.
3. —— and B. Zil'ber, *Zariski Geometries*, Bulletin (New Series) of the AMS 28, #2, April 1993.
4. Y. Peterzil and S. Starchenko *A trichotomy theorem for o-minimal structures*, preprint.
5. ————, *Geometry, calculus and Zil'ber's conjecture*, Bulletin of Symbolic Logic.
6. B. Zil'ber, *Structural properties of models of $\aleph_1$-categorical theories*, in: R. Barcon Marcus et al. eds., Logic Methodology and Philosophy of Science VII (North Holland, Amsterdam, 1986) pp. 115-128.

The Higher Infinite in Proof Theory

Michael Rathjen*

School of Mathematics
University of Leeds
Leeds, LS2 9JT, England
rathjen@amsta.leeds.ac.uk

1 Introduction

The higher infinite usually refers to the lofty reaches of Cantor's paradise, notably to the realm of large cardinals whose existence cannot be proved in the established formalisation of Cantorian set theory, i.e. Zermelo-Fraenkel set theory with the axiom of choice. Proof theory, on the other hand, is commonly associated with the manipulation of syntactic objects, that is finite objects par excellence. However, finitary proof theory became already infinitary in the 1950's when Schütte re-obtained Gentzen's ordinal analysis for number theory in a particular transparent way through the use of an infinitary proof system with the so-called ω-rule (cf. [49]). Nowadays one even finds vestiges of large cardinals in ordinal-theoretic proof theory. Large cardinals have worked their way down through generalized recursion (in the shape of recursively large ordinals) to proof theory wherein they appear in the definition procedures of so-called *collapsing functions* which give rise to ordinal representation systems. The surprising use of ordinal representation systems employing "names" for large cardinals in current proof-theoretic ordinal analyses is the main theme of this paper.

The exposition here diverges from the presentation given at the conference in two regards. Firstly, the talk began with a broad introduction, explaining the current rationale and goals of ordinal-theoretic proof theory, which take the place of the original Hilbert Program. Since this part of the talk is now incorporated in the first two sections of the BSL-paper [43] there is no point in reproducing it here. Secondly, we shall omit those parts of the talk concerned with infinitary proof systems of ramified set theory as they can also be found in [43] and even more detailed in [40]. Thirdly, thanks to the aforementioned omissions, the advantage of present paper over the talk is to allow for a much more detailed account of the actual information furnished by ordinal analyses and the role of large cardinal hypotheses in devising ordinal representation systems.

2 Observations on ordinal analyses

How are ordinals connected with formal systems? Well, this question is way more difficult to answer than "How are vector spaces measured by cardinals?"

* Research supported by a Heisenberg-Fellowship of the Deutsche Forschungsgemeinschaft.

Since the answer is crucial to the branch of logic reported on in this paper, we shall gather together some wellknown and some not so wellknown facts. In doing so, we also aim at averting certain misconceptions about ordinal-theoretic proof theory.[2]

Furthermore, using results of [15], we characterize the provably recursive functions of theories for which an ordinal analysis has been given.

Definition 2.1 For a set X and and a binary relation $\prec$ on X, let $\mathrm{LO}(X, \prec)$ abbreviate that $\prec$ linearly orders the elements of X and that for all u, v, whenever $u \prec v$, then $u, v \in X$.

A *linear ordering* is a pair $\langle X, \prec\rangle$ satisfying $\mathrm{LO}(X, \prec)$.

Let T be a framework for formalizing a certain part of mathematics. T should be a true theory which contains a modicum of arithmetic.

Let A be a subset of $\mathbb{N}$ ordered by $\prec$ such that A and $\prec$ are both definable in the language of T. If the language of T allows for quantification over subsets of $\mathbb{N}$, like that of second order arithmetic or set theory, *well-foundedness* of $\langle A, \prec\rangle$ will be formally expressed by

$$\mathrm{WF}(A, \prec) := \forall X \subseteq \mathbb{N}\, [\forall u \in A (\forall v \prec u\, v \in X \rightarrow u \in X) \rightarrow \forall u \in A\, u \in X.] \tag{1}$$

If, however, the language of T does not provide for quantification over arbitrary subsets of $\mathbb{N}$, like that of Peano arithmetic, we shall assume that it contains a new unary predicate $\mathbf{U}$. $\mathbf{U}$ acts like a free set variable, in that no special properties of it will ever be assumed. We will then resort to the following formalization of well-foundedness:

$$\mathrm{WF}(A, \prec) := \forall u \in A (\forall v \prec u\, \mathbf{U}(v) \rightarrow \mathbf{U}(u)) \rightarrow \forall u \in A\, \mathbf{U}(u), \tag{2}$$

where $\forall v \prec u \ldots$ is short for $\forall v (v \prec u \rightarrow \ldots)$.

We also set

$$\mathrm{WO}(A, \prec) := \mathrm{LO}(A, \prec) \wedge \mathrm{WF}(A, \prec). \tag{3}$$

If $\langle A, \prec\rangle$ is well-founded, we use $|\prec|$ to signify its set-theoretic order-type. For $a \in A$, the ordering $\prec\restriction a$ is the restriction of $\prec$ to $\{x \in A : x \prec a\}$.

The ordering $\langle A, \prec\rangle$ is said to be *provably well-founded in* T if

$$T \vdash \mathrm{WO}(A, \prec). \tag{4}$$

The *proof-theoretic ordinal* $|T|$ of T is often defined as follows:

$$|T| := = \sup \{\alpha : \alpha \text{ provably recursive in } T\} \tag{5}$$

where an ordinal α is said to be provably recursive in T if there is a recursive well–ordering $\langle A, \prec\rangle$ with order–type α such that

$$T \vdash \mathrm{WO}(A, \prec)$$

with A and $\prec$ being provably recursive in T.

The calibration of $|T|$ is then called *ordinal analysis of* T.

[2] The present section is complementary to [43], §2.

The above definition of $|T|$ has the advantage of being mathematically precise.[3] But as to the activity named "ordinal analysis" it is left completely open what constitutes such an analysis. One often encounters this kind of sloppy talk of ordinals in proof theory. Among the uninitiated it might give the impression that the calibration of $|T|$ is akin to computing numerical invariants in other branches of mathematics, i.e. computing dimensions of vector spaces. This likening is not completely mistaken, but what is most problematic about it is that ordinals are not as easily bestowed upon us as natural numbers are. Before one can go about determining the proof-theoretic ordinal of T, one needs to be furnished with representations of ordinals. Not surprisingly, a great deal of ordinally informative proof theory has been concerned with developing and comparing particular ordinal representation systems. Moreover, to obtain the reductions of classical (non-constructive) theories to constructive ones (as related, for instance, in [12], [43],§2) it appears to be pivotal to work with very special and well-structured ordinal representation systems.

But before attempting to delineate the type of ordinal representation systems that are actually used in ordinal analyses, it should be mentioned that, in general, $|T|$ has several equivalent characterizations; though some of these hinge upon the mathematical strength of T.

Proposition 2.2 *(i) Suppose that for every elementary well-ordering $\langle A, \prec\rangle$, whenever $T \vdash \mathrm{WO}(A, \prec)$, then*

$$T \vdash \forall u\,[\,A(u) \to (\forall v \prec u P(v)) \to P(u)] \to \forall u\,[A(u) \to P(u)]$$

holds for all provably recursive predicates P of T. Then

$$\begin{aligned} |T| &= \sup\{\alpha : \ \alpha \textit{ is provably elementary in } T\} \qquad (6)\\ &= \sup\{\alpha : \ \alpha \textit{ is provably recursive in } T\}. \end{aligned}$$

Moreover, if $T \vdash$ WO$(A, \prec)$ and $A, \prec$ *are provably recursive in T, then one can find an elementary well-ordering $\langle B, <\rangle$ and a recursive function f such that $T \vdash$* WO$(B, <)$, *f is provably recursive in T, and T proves that f supplies an order isomorphism between $\langle B, <\rangle$ and $\langle A, \prec\rangle$.*

(ii) If T proves comparability of well-orderings, then

$$|T| = \sup\{\alpha : \ \alpha \textit{ is provably arithmetic in } T\}. \qquad (7)$$

(iii) If T proves comprehension for analytic sets of integers, i.e. lightface Σ^1_1 sets of integers, then

$$|T| = \sup\{\alpha : \ \alpha \textit{ is provably analytic in } T\}. \qquad (8)$$

[3] It even rules out some of the pathological candidates of the "dreary list" in [23], p. 334.

Proof: Probably folklore. But the only reference I know is [37]. For (ii) and (iii) see [37], Theorem 1.2 and Corollary 1.3. (i) follows by refining the proof of [37], 1.2.(ii). □

Examples for (i) are the theories $\mathbf{I\Sigma_1}$, $\mathbf{WKL_0}$ and **PA**. Examples for (ii),(iii) are $\mathbf{ATR_0}$ and $\Pi^1_1 - \mathbf{CA_0}$, respectively.

Definition 2.3 *Elementary recursive arithmetic,* **ERA**, is a weak system of number theory, in a language with $0, 1, +, \times, E$ (exponentiation), $<$, whose axioms are:

1. the usual recursion axioms for $+, \times, E, <$.
2. induction on Δ_0-formulae with free variables.

ERA is referred to as elementary recursive arithmetic since its provably recursive functions are exactly the Kalmar *elementary functions,* i.e. the class of functions which contains the successor, projection, zero, addition, multiplication, and modified subtraction functions and is closed under composition and bounded sums and products (cf. [46]).

The next definition garners some features (similar to [15]) that ordinal representation systems used in proof theory always have, and collectively calls them *"elementary ordinal representation system"*. One reason for singling out this notion is that it leads to an elegant characterization of the provably recursive functions of theories equipped with transfinite induction principles for such ordinal representation systems. Furthermore, though only based on empirical facts about ordinal representation systems surfacing in proof theory, this definition can also be viewed as a first (naive) step towards answering the question: "What is a natural well-ordering?"

Definition 2.4 An *elementary ordinal representation system* (EORS) for a limit ordinal λ is a structure $\langle A, \lhd, n \mapsto \lambda_n, +, \times, x \mapsto \omega^x \rangle$ such that:

(i) A is an elementary subset of $\mathbb{N}$.
(ii) $\lhd$ is an elementary well-ordering of A.
(iii) $|\lhd| = \lambda$.
(iv) Provably in **ERA**, $\lhd \restriction \lambda_n$ is a proper initial segment of $\lhd$ for each n, and $\bigcup_n \lhd \restriction \lambda_n = \lhd$. In particular, $\mathbf{ERA} \vdash \forall y\, \lambda_y \in A \wedge \forall x \in A \exists y\, [x \lhd \lambda_y]$.
(v) $\mathbf{ERA} \vdash \mathrm{LO}(A, \lhd)$
(vi) $+, \times$ are binary and $x \mapsto \omega^x$ is unary. They are elementary functions on elementary initial segments of A. They correspond to ordinal addition, multiplication and exponentiation to base ω, respectively. The initial segments of A on which they are defined are maximal.
$n \mapsto \lambda_n$ is an elementary function.
(vii) $\langle A, \lhd, +, \times, \omega^x \rangle$ satisfies "all the usual algebraic properties" of an initial segment of ordinals. In addition, these properties of $\langle A, \lhd, +, \times, \omega^x \rangle$ can be proved in **ERA**.

(viii) Let $\bar{n}$ denote the nth element in the ordering of A. Then the correspondence $n \leftrightarrow \bar{n}$ is elementary.

(ix) Let $\alpha = \omega^{\beta_1} + \cdots \omega^{\beta_k}, \beta_1 \geq \cdots \geq \beta_k$ (Cantor normal form). Then the correspondence $\alpha \leftrightarrow \langle \beta_1, \ldots, \beta_k \rangle$ is elementary.

Elements of A will often be referred to as *ordinals*, and denoted $\alpha, \beta, \ldots$.

In a sense the preceding definition manages to characterize natural well-orderings of order-type ε_0 as any two such well-orderings arising from EORSs are recursively isomorphic (mainly due to 2.4(ix)). Of course, this cannot be expected to hold for larger order-types.

As for the computational complexity of EORSs involved in ordinal analyses, it appears that they are even Δ_0-representable (cf. [52]). Be this as it may, ordinal analysts never expected that the peculiarities of "real" ordinal representation systems, including their naturalness, could be fathomed via complexity theory.[4] Sommer has addressed the issue at great length in [52, 53]. Here are his conclusions:

Observation 2.5 Synopsis of discussion in [52]

- *It is an empirical fact that with regard to complexity measures considered in complexity theory the ordinal representation systems emerging in proof theory are of low computational complexity and their basic properties are provable in weak fragments of arithmetic.*
  The latter includes that computations on ordinals in actual proof-theoretic ordinal analyses can also be handled in such weak theories.
- *The complexity of ordinal representation systems involved in proof-theoretic ordinal analyses cannot be described in terms of the complexity of the representations of these ordinals, but only in terms of the difficulty in recognizing the well-foundedness of these representations.*

We continue to gather information about ordinal analyses.

Definition 2.6 Suppose $\mathrm{LO}(A, \lhd)$ and $F(u)$ is a formula. Then $\mathrm{TI}_{\langle A, \lhd \rangle}(F)$ is the formula

$$\forall n{\in}A\, [\forall x \lhd n F(x) \rightarrow F(n)] \rightarrow \forall n{\in}A\, F(n). \tag{9}$$

$\mathrm{TI}(A, \lhd)$ is the schema consisting of $\mathrm{TI}_{\langle A, \lhd \rangle}(F)$ for all F.

Given a linear ordering $\langle A, \lhd \rangle$ and $\alpha{\in}A$ let $A_\alpha = \{\beta{\in}A : \beta \lhd \alpha\}$ and $\lhd_\alpha$ be the restriction of $\lhd$ to A_α.

In what follows, quantifiers and variables are supposed to range over the natural numbers. When n denotes a natural number, $\bar{n}$ is the canonical name in the language under consideration which denotes that number.

[4] Though at times they got carried away pointing out the computational complexity of their orderings as if it were their decisive feature.

Observation 2.7 *Every ordinal analysis of a classical (intuitionistic) theory* **T** *that has ever appeared in the literature provides an EORS* $\langle A, \lhd, \ldots \rangle$ *such that* **T** *and* **PA** $+ \bigcup_{\alpha \in A} \mathrm{TI}(A_{\bar{\alpha}}, \lhd_{\bar{\alpha}})$ (**HA** $+ \bigcup_{\alpha \in A} \mathrm{TI}(A_{\bar{\alpha}}, \lhd_{\bar{\alpha}})$) *prove the same arithmetic sentences.*

Moreover, regardless of the underlying logic, **T** *and* **HA** $+ \bigcup_{\alpha \in A} \mathrm{TI}(A_{\bar{\alpha}}, \lhd_{\bar{\alpha}})$ *prove the same* Π^0_2 *statements.*

Proof: **PA** $+ \bigcup_{\alpha \in A} \mathrm{TI}(A_{\bar{\alpha}}, \lhd_{\bar{\alpha}})$ can be interpreted in **HA** $+ \bigcup_{\alpha \in A} \mathrm{TI}(A_{\bar{\alpha}}, \lhd_{\bar{\alpha}})$ via the Gödel–Gentzen $\neg\neg$-translation. Since the theorems of **HA** $+ \bigcup_{\alpha \in A} \mathrm{TI}(A_{\bar{\alpha}}, \lhd_{\bar{\alpha}})$ are closed under the Markov-rule for primitive recursive predicates (using, for instance, Friedman's A-translation), it follows that the theories prove the the same Π^0_2 propositions, hence **HA** $+ \bigcup_{\alpha \in A} \mathrm{TI}(A_{\bar{\alpha}}, \lhd_{\bar{\alpha}})$ proves the same Π^0_2 sentences as **T**. □

The latter result can be considerably improved.

Definition 2.8 For each $\alpha \in A$, $\mathrm{ERWF}(\lhd, \bar{\alpha})$ is the schema

$$\forall \mathbf{x} \exists y [f(\mathbf{x}, y) \underline{\lhd} f(\mathbf{x}, y+1) \vee f(\mathbf{x}, y) \notin A \vee \bar{\alpha} \underline{\lhd} f(\mathbf{x}, y)]$$

for each (definition of an) elementary function f.

$\mathrm{ERWF}(\lhd)$ is the schema

$$\forall \mathbf{x} \exists y [f(\mathbf{x}, y) \underline{\lhd} f(\mathbf{x}, y+1) \vee f(\mathbf{x}, y) \notin A]$$

for each elementary function f.

The schemata $\mathrm{PRWF}(\lhd, \bar{\alpha})$ and $\mathrm{PRWF}(\lhd)$ are defined identically, except that f ranges over the primitive recursive functions.

Definition 2.9 $\mathbf{DRA}_{\langle A, \lhd \rangle}$ (*Descent Recursive Arithmetic*) is the theory whose axioms are **ERA** $+ \bigcup_{\alpha \in A} \mathrm{ERWF}(\lhd, \bar{\alpha})$.

$\mathbf{DRA}(\lhd^+)$ is the theory whose axioms are **ERA** $+ \mathrm{ERWF}(\lhd)$.

The difference is that $\mathbf{DRA}(\lhd)$ asserts only the non-existence of elementary infinitely descending sequences below each $\alpha \in A$, where α is given at the meta-level.

Combined with 2.7 the latter result leads to a neat characterization of the provably recursive functions of **T** due to the following observation:

Proposition 2.10 *The provably recursive functions of* $\mathbf{DRA}_{\langle A, \lhd \rangle}$ *are all functions* f *of the form*

$$f(\mathbf{m}) = g(\mathbf{m}, \textit{least } n.h(\mathbf{m}, n) \underline{\lhd} h(\mathbf{m}, n+1)) \qquad (10)$$

where g *and* h *are elementary functions and* **ERA** $\vdash \forall \mathbf{xy}\, h(\mathbf{x}, y) \in A_{\bar{\alpha}}$ *for some* $\alpha \in A$.

The above class of recursive functions will be referred to as the *descent recursive functions over A*.

Proposition 2.11 (Friedman, Sheard [15, 4.4])
$\mathbf{DRA}_{\langle A,\lhd\rangle}$ *and* $\mathbf{PA} + \bigcup_{\alpha\in A} \mathrm{TI}(A_{\bar{\alpha}}, \lhd_{\bar{\alpha}})$ *prove the same* Π^0_2 *sentences.*

From 2.7 and 2.11 we get:

Observation 2.12 *Suppose an ordinal analysis of the formal system T has been attained using an EORS $\langle A, \lhd, \ldots\rangle$. Then the provably recursive functions of T are the descent recursive functions over A.*

We shall list some complimentary results.

Definition 2.13 If T is a theory, the 1-*consistency of* T is the schema

$$\forall u[Pr_T(\ulcorner F(\dot{u})\urcorner) \rightarrow F(u)]$$

for Σ^0_1 formulae $F(u)$ with one free variable u.

Theorem 2.14 (Friedman and Sheard [15, 4.5]) *The following are equivalent over* **PRA***:*

(i) 1-consistency of $\mathbf{PA} + \bigcup_{\alpha\in A} \mathrm{TI}(A_{\bar{\alpha}}, \lhd_{\bar{\alpha}})$
(ii) $\mathrm{PRWF}(\lhd^+)$
(ii) $\mathrm{ERWF}(\lhd^+)$.

Observation 2.15 *Again, let T be a theory for which an ordinal analysis has been carried out via $\langle A, \lhd\rangle$. Then the following are equivalent over* **PRA***:*

(i) 1-consistency of T
(ii) $\mathrm{PRWF}(\lhd^+)$
(ii) $\mathrm{ERWF}(\lhd^+)$.

The ordinal representation systems used in ordinal analyses are distinguished by another property. Suppose **T** successfully underwent an ordinal analysis by employing an EORS $\langle A, \lhd, \ldots\rangle$. Further, assume $T \vdash \mathrm{WO}(B, \prec)$ for some elementary (or recursive) well-ordering $\langle B, \prec\rangle$. Then a question suggesting itself is whether it is possible to determine an initial segment $\lhd_\alpha$ of $\lhd$ and T-provably recursive function f such that

$$\mathbf{T} \vdash \mathbf{f} : \mathbf{B} \xrightarrow{1-1} \mathbf{A}_{\bar{\alpha}} \wedge \forall \mathbf{x}, \mathbf{y} {\in} \mathbf{B}[\mathbf{x} \prec \mathbf{y} \leftrightarrow \mathbf{f}(\mathbf{x}) \lhd_{\bar{\alpha}} \mathbf{f}(\mathbf{y})]\,? \qquad (11)$$

The content of (11) is that $\langle A, \lhd\rangle$ provides a universal measure for the provable well-orderings of T in that each such well-ordering is T-recursively embedded in an initial segment of $\lhd$.

In the case of **PA** a positive answer to (11) can be obtained from Gentzen's proof of $|\mathbf{PA}| \leq \varepsilon_0$ (cf. [54, 13.4]). Fortunately, this is not the only example. The proof of the following result requires a refined analysis of infinitary derivations.

Observation 2.16 *In practice, that is to say when a reduction as in 2.7 has been attained, the answer to question (11) is "YES".*[5]

[5] *Ordinal analyses providing reductions as in 2.7 have also been distinguished in [34], where they were christened "profound".*

A caveat is in order here. Taken in isolation, property (11) does not guarantee a meaningful ordinal notation system as the pathological example (iv) of [23], p. 334 demonstrates.

The preceding pointed out some markings of EORSs found in proof theory. Another feature that we deem more important than the ones mentioned hitherto is their versatileness in establishing equivalences between classical nonconstructive theories and intuitionistic constructive theories (cf. [43]) based on radically different ontologies. Thus far we have only given a rather unsatisfying and imprecise answer to the question: "What is so particular about the ordinal representation system used in ordinal analyses?" In connection with this question, it has been suggested (cf. [23], [11]) that it is important to address the broader question of "What is a natural well-ordering?" A criterion for naturalness put forward in [23] is uniqueness up to recursive isomorphism. Furthermore, in [23], Kreisel seems to seek naturalness in algebraic characterizations of ordered structures. Feferman, in [10], discerns the properties of completeness, repleteness, relative categoricity and preservation of these under iteration of the critical process as significant features of systems of natural representation. Girard [17] appears to propose dilators to capture the abstract notion of a notation system for ordinals.

However, in the ensuing sections we shall not be particularly heedful of these suggestions and rather try to reflect on the main question from new angles.

3 Large cardinals and ordinal representation systems I

3.1 A brief history of ordinal representation systems up till the early 1980s

Several natural well-orderings that later came to be used in proof theory had arisen in a purely set-theoretic context. The Cantor normal form of ordinals with exponentiation to the base ω provides an ordinal representation system for ε_0. Veblen's work [55], whose main tools are the operations of derivation and transfinite iteration applied to continuous increasing functions of ordinals, distinguished several ordinals (e.g. Γ_0) which Feferman and Schütte then employed in their investigations on predicativity.

Still from a set-theoretic stance, Bachmann [4] utilized Veblen's methods for building hierarchies of normal functions and added the new procedure of *diagonalization*. A hierarchy of normal functions $\{\varphi_\alpha\}_{\alpha\in B}$ is defined by simultaneously defining the indexing set B such that with each limit $\alpha\in B$ is associated a fundamental sequence $\langle\alpha[\xi]:\xi<\tau_\alpha\rangle$ of ordinals $\alpha[\xi]\in B$ of length τ_α with $\alpha[\xi]<\alpha$. Depending on the type of τ_α the function φ is defined from previously defined functions by one of the procedures. Bachmann's novel idea was the systematic use of uncountable ordinals in the indexing set to keep track of the functions defined by diagonalization.

When in the sixties important proof-theoretic ordinals were located in Bachmann's system, it became the standard source of notations for ordinals required in proof theory. Bachmann's hierarchy was extended by Pfeiffer [32] and Isles [20]. By the end of the 1960s the conceptually straightforward Bachmann method had been pushed as far as it could be. Unfortunately, the dependence of the construction on fundamental sequences for each limit indexing ordinals, with certain additional "dove-tailing" properties, adds enormous complexity to the very definition of the φ_α and severely hampers their applicability in ordinal analyses.

At the end of the 1960s the definitions of ordinal representation systems were so contaminated by details that future progress of ordinal-theoretic proof theory was at stake. Fortunately, around 1970, this impasse was overcome by Feferman who, in unpublished work, made conceptional improvements in the Bachmann approach. In contrast to the definition of Bachmann-style hierarchies, Feferman's definition does not require simultaneous assignment of fundamental sequences to limit ordinals. The definition of the φ_α's is uniform for all α since it does not hinge on a previous assignment of cofinality type τ_α to α.

The new approach was carried out and pushed further by Aczel, Weyhrauch, Bridge and Buchholz (cf. [11]) in the early 1970s. Considerable conceptual improvements and extensions of ordinal representation systems in the late 1970s and early 1980s are due to Buchholz, Jäger, Pohlers and Schütte (cf. [33]).

In this section we shall exhibit three ordinal representation systems which featured in ordinal analyses of extensions of Kripke-Platek set theory from around 1980 on, the first one being an epitome of the finale of the history reported above. Their respective definition procedures make use of weakly inaccessible, weakly Mahlo and weakly compact cardinals. Our objective is to show how large cardinal assumptions are actually employed for devising ordinal representation systems, also with the intention to rectify certain opinions held about ordinal representation systems. Such systems are by no means cooked up or impenetrable. As a rule, they utilize and extend wellknown set-theoretic hierarchies, for instance Mahlo's π- and ρ-number hierarchies [24].

3.2 Ordinal functions based on a weakly inaccessible cardinal

KPi is a set theory which originates from Kripke-Platek set theory and in addition has an axiom which says that any set is contained in an admissible set. Thus the standard models of **KPi** in **L** are the segments $\mathbf{L}_\kappa$ with κ recursively inaccessible. The ordinal analysis for **KPi** (cf. [21]) used an EORS built from ordinal functions which had originally been defined with the help of a weakly inaccessible cardinal. In this subsection we expound on the development of this particular EORS with an eye towards the role of cardinals therein.

Let

$$\mathbf{I} := \text{"first weakly inaccessible cardinal"} \tag{12}$$

and let

$$(\alpha \mapsto \Omega_\alpha)_{\alpha<\mathbf{I}} \tag{13}$$

be a function that enumerates the cardinals below **I**. Further let

$$\Re^{\mathbf{I}} := \{\mathbf{I}\} \cup \{\Omega_{\xi+1} : \xi < \mathbf{I}\}. \tag{14}$$

Variables κ, π will range over $\Re^{\mathbf{I}}$.

Definition 3.1 An ordinal representation system for the analysis of **KPi** can be derived from the following functions and Skolem hulls of ordinals defined by recursion on α:

$$C^{\mathbf{I}}(\alpha, \beta) = \begin{cases} \textit{closure of } \beta \cup \{0, \mathbf{I}\} \\ \textit{under:} \\ +, (\xi \mapsto \omega^{\xi}) \\ (\xi \mapsto \Omega_{\xi})_{\xi < \mathbf{I}} \\ (\xi\pi \longmapsto \psi^{\xi}(\pi))_{\xi < \alpha} \end{cases} \tag{15}$$

$$\psi^{\alpha}(\pi) \simeq \min\{\rho < \pi : C^{\mathbf{I}}(\alpha, \rho) \cap \pi = \rho \wedge \pi \in C^{\mathbf{I}}(\alpha, \rho)\}. \tag{16}$$

Note that if $\rho = \psi^{\alpha}(\pi)$, then $\psi^{\alpha}(\pi) < \pi$ and $[\rho, \pi) \cap C^{\mathbf{I}}(\alpha, \rho) = \emptyset$, thus the order-type of the ordinals below π which belong to the Skolem hull $C^{\mathbf{I}}(\alpha, \rho)$ is ρ . In more pictorial terms, ρ is the αth *collapse* of π.

Lemma 3.2 *If $\pi \in C^{\mathbf{I}}(\alpha, \pi)$, then $\psi^{\alpha}(\pi)$ is defined; in particular $\psi^{\alpha}(\pi) < \pi$.*

Proof: Note first that for a limit ordinal λ,

$$C^{\mathbf{I}}(\alpha, \lambda) = \bigcup_{\xi < \lambda} C^{\mathbf{I}}(\alpha, \xi)$$

since the right hand side is easily shown to be closed under the clauses that define $C^{\mathbf{I}}(\alpha, \lambda)$. Thus we can pick $\omega \leq \eta < \pi$ such that $\pi \in C^{\mathbf{I}}(\alpha, \eta)$. Now define

$$\begin{aligned} \eta_0 &= \sup C^{\mathbf{I}}(\alpha, \eta) \cap \pi \\ \eta_{n+1} &= \sup C^{\mathbf{I}}(\alpha, \eta_n) \cap \pi \\ \eta^* &= \sup_{n<\omega} \eta_n. \end{aligned} \tag{17}$$

Since the cardinality of $C^{\mathbf{I}}(\alpha, \eta)$ is the same as that of η and therefore less than π, the regularity of π implies that $\eta_0 < \pi$. By repetition of this argument one obtains $\eta_n < \pi$, and consequently $\eta^* < \pi$. The definition of η^* then ensures

$$C^{\mathbf{I}}(\alpha, \eta^*) \cap \pi = \bigcup_n C^{\mathbf{I}}(\alpha, \eta_n) \cap \pi = \eta^* < \pi.$$

Therefore, $\psi^{\alpha}(\pi) < \pi$. □

Let $\varepsilon_{\mathbf{I}+1}$ be the least ordinal $\alpha > \mathbf{I}$ such that $\omega^{\alpha} = \alpha$. The next definition singles out a subset $\mathcal{T}(\mathbf{I})$ of $C^{\mathbf{I}}(\varepsilon_{\mathbf{I}+1}, 0)$ which gives rise to an ordinal representation system, i.e., there is an elementary ordinal representation system $\langle \mathcal{OR}, \lhd, \hat{\Re}, \hat{\psi}, \ldots \rangle$, so that

$$\langle \mathcal{T}(\mathbf{I}), <, \Re, \psi, \ldots \rangle \cong \langle \mathcal{OR}, \lhd, \hat{\Re}, \hat{\psi}, \ldots \rangle. \tag{18}$$

"..." is supposed to indicate that more structure carries over to the ordinal representation system.

Definition 3.3 $\mathcal{T}(\mathbf{I})$ is defined inductively as follows:

1. $0, \mathbf{I} \in \mathcal{T}(\mathbf{I})$.
2. If $\alpha_1, \ldots, \alpha_n \in \mathcal{T}(\mathbf{I})$ and $\omega^{\alpha_1} + \cdots + \omega^{\alpha_n} > \alpha_1 \geq \ldots \geq \alpha_n$, then $\omega^{\alpha_1} + \cdots + \omega^{\alpha_n} \in \mathcal{T}(\mathbf{I})$.
3. If $\alpha \in \mathcal{T}(\mathbf{I})$, $0 < \alpha < \mathbf{I}$ and $\alpha < \Omega_\alpha$, then $\Omega_\alpha \in \mathcal{T}(\mathbf{I})$.
4. If $\alpha, \pi \in \mathcal{T}(\mathbf{I})$, $\pi \in C^{\mathbf{I}}(\alpha, \pi)$ and $\alpha \in C^{\mathbf{I}}(\alpha, \psi^\alpha(\pi))$, then $\psi^\alpha(\pi) \in \mathcal{T}(\mathbf{I})$.

The side conditions in 3.3.2, 3.3.3 are easily explained by the desire to have unique representations in $\mathcal{T}(\mathbf{I})$. The requirement $\alpha \in C^{\mathbf{I}}(\alpha, \psi^\alpha(\pi))$ in 3.3.4 also serves the purpose of unique representations (and more) but is probably a bit harder to explain. The idea here is that from $\psi^\alpha(\pi)$ one should be able to retrieve the stage (namely α) where it was generated. This is reflected by $\alpha \in C^{\mathbf{I}}(\alpha, \psi^\alpha(\pi))$.

It can be shown that the foregoing definition of $\mathcal{T}(\mathbf{I})$ is deterministic, that is to say every ordinal in $\mathcal{T}(\mathbf{I})$ is generated by the inductive clauses of 3.3 in exactly one way. As a result, every $\gamma \in \mathcal{T}(\mathbf{I})$ has a unique representation in terms of symbols for $0, \mathbf{I}$ and function symbols for $+, (\alpha \mapsto \Omega_\alpha), (\alpha, \pi \mapsto \psi^\alpha(\pi))$. Thus, by taking some primitive recursive (injective) coding function $\lceil \cdots \rceil$ on finite sequences of natural numbers, we can code $\mathcal{T}(\mathbf{I})$ as a set of natural numbers as follows:

$$\ell(\alpha) = \begin{cases} \lceil 0, 0 \rceil & \text{if } \alpha = 0 \\ \lceil 1, 0 \rceil & \text{if } \alpha = \mathbf{I} \\ \lceil 2, \ell(\alpha_1), \cdots, \ell(\alpha_n) \rceil & \text{if } \alpha = \omega^{\alpha_1} + \cdots + \omega^{\alpha_n} \\ \lceil 3, \ell(\beta) \rceil & \text{if } \alpha = \Omega_\beta \\ \lceil 4, \ell(\beta), \ell(\pi) \rceil & \text{if } \alpha = \psi^\beta(\pi), \end{cases}$$

where the distinction by cases refers to the unique representation of 3.3. With the aid of ℓ, the ordinal representation system of (18) can be defined by letting $\mathcal{OR}$ be the image of ℓ and setting $\lhd := \{(\ell(\gamma), \ell(\delta)) : \gamma < \delta \wedge \delta, \gamma \in \mathcal{T}(\mathbf{I})\}$ etc. However, for a proof that this definition of $\langle \mathcal{OR}, \lhd, \Re, \hat{\psi}, \ldots \rangle$ in point of fact furnishes an elementary ordinal representation system, we have to refer to the literature (cf. [6, 7, 42]).

3.3 Ordinal functions based on a weakly Mahlo cardinal

In a paper from 1911 Mahlo [24] investigated two hierarchies of regular cardinals. In view of its early appearance this work is astounding for its refinement and its audacity in venturing into the higher infinite. Mahlo called the cardinals considered in the first hierarchy π_α-*numbers*. In modern terminology they are spelled out as follows:

κ is 0-*weakly inaccessible* iff κ is regular;

κ is $(\alpha+1)$-*weakly inaccessible* iff κ is a regular limit of α-weakly inaccessibles

κ is λ-*weakly inaccessible* iff κ is α-weakly inaccessible for every $\alpha < \lambda$

for limit ordinals λ. This hierarchy could be extended through diagonalization, by taking next the cardinals κ such that κ is κ-weakly inaccessible and after that choosing regular limits of the previous kind etc.

Mahlo also discerned a second hierarchy which is generated by a principle superior to taking regular fixed-points. Its starting point is the class of ρ_0-numbers which later came to be called *weakly Mahlo cardinals*. Weakly Mahlo cardinals are larger than any of those that can be obtained by the above processes from below. Remarkably, Gaifman [16] showed that in a mathematical precise sense a weakly Mahlo cardinal is the least upper bound of diagonalizing the regular fixed-point operation from below.

Here we shall define an extension of Mahlo's π-hierarchy by using ordinals above a weakly Mahlo to keep track of diagonalization.

The resulting EORS of [35] has been used in [36] to give an ordinal analysis of **KPM**. **KPM** is an extension of **KPi** by a schema stating that for every Σ_1-definable (class) function there exists an admissible set closed under this function. Its canonical models are the sets $\mathbf{L}_\mu$ with μ recursively Mahlo.

Let

$$\mathbf{M} := \text{first weakly Mahlo cardinal} \tag{19}$$

and set

$$\Re^{\mathbf{M}} := \{\pi < \mathbf{M} : \pi \text{ regular}, \pi > \omega\}. \tag{20}$$

Variables κ, π will range over $\Re^{\mathbf{M}}$.

Definition 3.4 An ordinal representation system for the analysis of **KPM** can be derived from the following functions and Skolem hulls of ordinals, defined by recursion on α:

$$C^{\mathbf{M}}(\alpha,\beta) = \begin{cases} \textit{closure of } \beta \cup \{0,\mathbf{M}\} \\ \textit{under:} \\ +, (\xi \longmapsto \omega^{\xi}) \\ (\xi\delta \mapsto \chi^{\xi}(\delta))_{\xi<\alpha} \\ (\xi\pi \longmapsto \psi^{\xi}(\pi))_{\xi<\alpha} \end{cases} \tag{21}$$

$$\chi^{\alpha}(\delta) \simeq \delta^{th} \text{ regular } \pi < \mathbf{M} \text{ s.t. } C^{\mathbf{M}}(\alpha,\pi) \cap \mathbf{M} = \pi \tag{22}$$

$$\psi^{\alpha}(\pi) \simeq \min\{\rho < \pi :\ C^{\mathbf{M}}(\alpha,\rho) \cap \pi = \rho \ \wedge\ \pi \in C^{\mathbf{M}}(\alpha,\rho)\}. \tag{23}$$

Lemma 3.5 *For all* α,

$$\chi^{\alpha} : \mathbf{M} \to \mathbf{M}$$

i.e. χ^{α} *is a total function on* $\mathbf{M}$.

Proof: Set

$$X_{\alpha} := \{\rho < \mathbf{M} : C^{\mathbf{M}}(\alpha,\rho) \cap \mathbf{M} = \rho\}.$$

We want to show that X_α is closed and unbounded in $\mathbf{M}$. As $\mathbf{M}$ is weakly Mahlo the latter will imply that X_α contains $\mathbf{M}$-many regular cardinals, ensuring that χ^{α} is total on $\mathbf{M}$.

Unboundedness: Given $\eta < \mathbf{M}$, define

$$\begin{aligned}\eta_0 &= \sup(C^{\mathbf{M}}(\alpha, \eta+1) \cap \mathbf{M})\\ \eta_{n+1} &= \sup(C^{\mathbf{M}}(\alpha, \eta_n) \cap \mathbf{M})\\ \eta^* &= \sup_n \eta_n.\end{aligned}$$

One easily verifies $C^{\mathbf{M}}(\alpha, \eta^*) \cap \mathbf{M} = \eta^*$. Hence, $\eta < \eta^*$ and $\eta^* \in X_\alpha$.

Closedness: If $X_\alpha \cap \lambda$ is unbounded in a limit $\lambda < \mathbf{M}$, then

$$C^{\mathbf{M}}(\alpha, \lambda) = \bigcup_{\xi \in X_\alpha \cap \lambda} C^{\mathbf{M}}(\alpha, \xi),$$

whence

$$C^{\mathbf{M}}(\alpha, \lambda) \cap \mathbf{M} = \sup\{\xi : \xi \in X_\alpha \cap \lambda\} = \lambda,$$

verifying $\lambda \in X_\alpha$. □

For a comparison with Mahlo's π_α numbers let $\mathbf{I}_\alpha$ be the function that enumerates, monotonically, the α-weakly inaccessibles. Neglecting finitely many exceptions, the function $\mathbf{I}_\alpha$ enumerates Mahlo's π_α numbers.

Proposition 3.6 *Set* $\Lambda^* :=$ *least* $\pi.\, \mathbf{I}_\pi(0) = \pi$.

(i) $\forall \alpha < \Lambda^* \, \forall \xi < \mathbf{M} \; \mathbf{I}_\alpha(\xi) = \chi^\alpha(\xi)$
(ii) $\Lambda^* = \chi^{\mathbf{M}}(0)$.
(iii) For all $\alpha < \mathbf{M}$, *a sufficient condition for* $\mathbf{I}_\alpha(\xi) = \chi^\alpha(\xi)$ *to hold is* $\omega^{\alpha+1} \leq \xi < \mathbf{M}$.
(iv) The diagonal set $\{\kappa < \mathbf{M} : \kappa$ *is* κ*-weakly inaccessible*$\}$ *is enumerated by the function* $(\alpha \mapsto \chi^{\mathbf{M}}(\alpha))_{\alpha < \mathbf{M}}$.

Ever higher levels of diagonalizations are obtained by the functions χ^{M^M}, $\chi^{M^{M^M}}$, etc.

The preceding gives rise to an EORS $\mathcal{T}(\mathbf{M})$ (similarly as sketched for $\mathcal{T}(\mathbf{I})$) which is essentially order isomorphic to $C^{\mathbf{M}}(\varepsilon_{\mathbf{M}+1}, 0)$. This EORS exactly captures the strength of **KPM**.

3.4 Ordinal functions based on a weakly compact cardinal

Here we shall venture much further, assuming the existence of a weakly compact cardinal. The original impetus was to find an ordinal representation system strong enough for the ordinal analysis of **KP** + Π_3-Reflection (cf. [42]). By Π_3-*Reflection* we mean the schema

$$\phi \rightarrow \exists z\, [\text{“}z \text{ transitive”} \wedge z \neq \emptyset \wedge \phi^z]$$

where ϕ is a set-theoretic Π_3-formula and ϕ^z is the result of restricting all quantifiers to z.

A limit ordinal κ is said to be Π_3-*reflecting* if $\mathbf{L}_\kappa \models \Pi_3$-Reflection.

The connection of weak compactness with Π_3-Reflection was established by Richter and Aczel [45]. The first step to evince this analogy consists in an interesting characterization of the notion of weak compactness (or Π^1_1-Indescribability) in terms of higher type operations.

Let $F : {}^{\kappa}\kappa \to {}^{\kappa}\kappa$. F is *κ-bounded* if for every $f : \kappa \to \kappa$ and $\xi < \kappa$, the value $F(f)(\xi)$ is determined by less than κ values of f, i.e.

$$\forall f \in {}^{\kappa}\kappa\, \exists \gamma < \kappa\, \forall g \in {}^{\kappa}\kappa\, [g \restriction \gamma = f \restriction \gamma \to F(f)(\xi) = F(g)(\xi)].$$

$0 < \alpha < \kappa$ is a *witness for* F if for every $f : \kappa \to \kappa$,

$$f''\alpha \subseteq \alpha \to F(f)''\alpha \subseteq \alpha.$$

Definition 3.7 $\kappa > 0$ is *2-regular* if every κ-bounded $F : {}^{\kappa}\kappa \to {}^{\kappa}\kappa$ has a witness.

Theorem 3.8 ([45], Theorem 1.14) **(ZFC)** *κ is 2-regular iff κ is weakly compact.*

2-regularity has a straightforward analogue in terms of recursion theory on ordinals (cf. [5, 19]). Let κ be an admissible ordinal. A partial function $f \subseteq \kappa \times \kappa$ is said to be *partial κ-recursive* if its graph is *κ-recursively enumerable*, i.e. Σ_1-definable over L_κ (where L_κ denotes the κth level of Gödel's constructible hierarchy). The partial κ-recursive functions can be parametrized by a κ-recursively enumerable predicate of three arguments, with indices from the ordinals $< \kappa$ (cf. [5], V.4.6 or [47], VII,1.9). In the following definition we write $\{\xi\}_\kappa$ to denote the κ-recursive partial function with index ξ, and write $\{\xi\}_\kappa : \kappa \to \kappa$ to mean that $\{\xi\}_\kappa$ is total on κ.

Definition 3.9 Let κ be an admissible ordinal and $\xi < \kappa$. *$\{\xi\}_\kappa$ maps κ-recursive functions to κ-recursive functions* if

$$\forall \beta < \kappa \left[\{\beta\}_\kappa : \kappa \to \kappa \to \{\{\xi\}_\kappa(\beta)\}_\kappa : \kappa \to \kappa\right].$$

Suppose $\{\xi\}_\kappa$ maps κ-recursive functions to κ-recursive functions. An admissible $\pi < \kappa$ is a *witness for* ξ if $\xi < \pi$ and $\{\xi\}_\pi$ maps π-recursive functions to π-recursive functions.

An admissible κ is *2-admissible* if every $\xi < \kappa$ such that $\{\xi\}_\kappa$ maps κ-recursive functions to κ-recursive functions has a witness.

The next result gives the final link for the analogy.

Theorem 3.10 ([45], Theorem 1.16) *κ is 2-admissible iff κ is Π_3-reflecting.*

Turning back to the main objective of this subsection, we recall Mahlo's second method of generating large cardinals, the ρ-numbers (cf. [24, 25, 26, 16]).

Definition 3.11 Mahlo formulated his ρ numbers by using an operation which is now known as *Mahlo's operation:*

$$M(X) = \{\alpha \in X : X \cap \alpha \text{ is stationary in } \alpha\}.$$

The ρ_α-numbers are obtained by iterating this process:

$$\begin{gathered}\kappa \text{ is } 0\textit{-weakly Mahlo} \text{ iff } \kappa \text{ is regular;}\\ \kappa \text{ is } (\alpha+1)\textit{-weakly Mahlo} \text{ iff } \{\tau<\kappa : \tau \text{ is } \alpha\text{-weakly Mahlo}\} \text{ is stationary in } \kappa\\ \kappa \text{ is } \lambda\textit{-weakly Mahlo} \text{ iff } \kappa \text{ is } \alpha\text{-weakly Mahlo for every } \alpha<\lambda\end{gathered}$$

for limit ordinals λ.

Proceeding similarly as with Mahlo's first hierarchy, we shall locate the ρ-number in a hierarchy based on the first weakly compact cardinal. Let

$$\mathbf{K} := \text{first weakly compact cardinal.} \tag{24}$$

Definition 3.12 defined by recursion on α:

$$C^{\mathbf{K}}(\alpha,\beta) = \begin{cases} \textit{closure of } \beta \cup \{0, \mathbf{K}\} \\ \textit{under:} \\ +, (\xi \longmapsto \omega^{\xi}) \\ (\xi\delta \mapsto \Xi^{\xi}(\delta))_{\xi<\alpha} \\ (\xi\sigma\pi \longmapsto \Psi^{\xi}_{\sigma}(\pi))_{\sigma\leq\xi<\alpha} \end{cases} \tag{25}$$

$$\mathbf{M}^0 = \{\rho < \mathbf{K} : C^{\mathbf{K}}(0,\rho) \cap \mathbf{K} = \rho\} \tag{26}$$

and for $\alpha > 0$:

$$\mathbf{M}^{\alpha} = \left\{\pi < \mathbf{K} : \begin{array}{c} C^{\mathbf{K}}(\alpha,\pi)\cap\mathbf{K} = \pi \ \wedge \ \ \pi \text{ regular } \wedge \\ (\forall\xi \in C^{\mathbf{K}}(\alpha,\pi)\cap\alpha)(\mathbf{M}^{\xi} \text{ is stationary in } \pi)\end{array}\right\} \tag{27}$$

$$\Xi^{\alpha}(\delta) \simeq \delta^{th} \text{ element of } \mathbf{M}^{\alpha} \tag{28}$$

$$\Psi^{\alpha}_{\beta}(\pi) \simeq \min\{\rho \in \mathbf{M}^{\beta}\cap\pi : \ C^{\mathbf{K}}(\alpha,\rho)\cap\pi = \rho \wedge \pi \in C^{\mathbf{K}}(\alpha,\rho)\} \tag{29}$$

providing $\beta \leq \alpha$ and π is regular and $\omega < \pi < \mathbf{K}$.

The sets $\mathbf{M}^{\alpha}$ are related to Mahlo's hierarchy as follows:

$$\begin{aligned} \mathbf{M}^0 &= \varepsilon\text{-numbers below } \mathbf{K} \\ \mathbf{M}^1 &= \text{regular cardinals} > \omega \text{ below } \mathbf{K} \\ \mathbf{M}^2 &= \text{weakly Mahlo cardinals below } \mathbf{K} \\ \mathbf{M}^3 &= 2\text{-weakly Mahlo cardinals below } \mathbf{K} \\ &\vdots \\ \mathbf{M}^{\alpha} &= \alpha\text{-weakly Mahlo cardinals below } \mathbf{K} \\ &\vdots \\ \mathbf{M}^{\mathbf{K}} &= \{\kappa < \mathbf{K} : \kappa \text{ is } \kappa\text{-weakly Mahlo}\} \end{aligned} \tag{30}$$

where $\omega \leq \alpha < \text{least}\rho.\rho \in \mathbf{M}^{\mathbf{K}}$.

Let $V = \bigcup_{\alpha} V_{\alpha}$ be the cumulative hierarchy of sets, i.e. $V_0 = \emptyset$, $V_{\alpha+1} = \{X : X \subseteq V_{\alpha}\}$ and $V_{\lambda} = \bigcup_{\xi<\lambda} V_{\xi}$ for limit ordinals λ.

Theorem 3.13 *For all $\alpha < \varepsilon_{\mathbf{K}+1}$, $\mathbf{M}^\alpha$ is stationary in $\mathbf{K}$ and hence $\Xi^\alpha(\delta)$ is defined for all $\delta < \mathbf{K}$.*

Proof: Each ordinal $\mathbf{K} < \beta < \varepsilon_{\mathbf{K}+1}$ has a unique representation of the form $\beta = \omega^{\beta_1} + \cdots + \omega^{\beta_n}$ with $\beta > \beta_1 \geq \cdots \geq \beta_n$ and $n > 0$, denoted $\beta =_{NF} \omega^{\beta_1} + \cdots + \omega^{\beta_n}$. Due to uniqueness, we can define an injective mapping

$$f : \varepsilon_{\mathbf{K}+1} \longrightarrow L_{\mathbf{K}}$$

by letting[6]

$$f(\beta) = \begin{cases} \beta & \text{if } \beta < \mathbf{K} \\ \{1\} & \text{if } \beta = \mathbf{K} \\ \langle 2, f(\beta_1), \ldots, f(\beta_n)\rangle & \text{if } \beta =_{NF} \omega^{\beta_1} + \cdots + \omega^{\beta_n} \text{ and } \mathbf{K} < \beta. \end{cases}$$

Putting

$$f(\alpha) \lhd f(\beta) :\Longleftrightarrow \alpha < \beta,$$

$\lhd$ defines a well-ordering on a subset of $L_{\mathbf{K}}$ of order type $\varepsilon_{\mathbf{K}+1}$.

To show the Theorem, we proceed by induction on α, or, equivalently, by induction on $\lhd$.

For any set E that is closed and unbounded in $\mathbf{K}$, we have to verify that $\mathbf{M}^\alpha \cap E \neq \emptyset$. Using the induction hypothesis, for all $\beta < \alpha$, $\mathbf{M}^\beta$ is stationary in $\mathbf{K}$. Define

$$U_1 := \{f(\alpha)\},\ \ U_2 := \{\langle x, y\rangle : x \lhd y\},\ \text{and } U_3 := \bigcup_{\beta<\alpha} (\mathbf{M}^\beta \times \{f(\beta)\}).$$

In what follows,

- $\mathbf{fun}(G)$ abbreviates that G is a function;
- $\mathbf{dom}(G)$, $\mathbf{ran}(G)$ denote the domain and the range of G, respectively.
- $\mathbf{pow}(a)$ denotes the powerset of a;
- $\mathbf{club}(X)$ says that X is a closed and unbounded class.
- $G''x$ is the set $\{G(y) : y \in x\}$.

The following sentences are satisfied in the structure $\langle V_{\mathbf{K}}, \in, U_1, U_2, U_3, E\rangle$:

(1) $\forall G \forall \delta[\mathbf{fun}(G) \wedge \mathbf{dom}(G) = \delta \wedge \mathbf{ran}(G) \subseteq On \ \rightarrow\ \exists \gamma (G''\delta \subseteq \gamma)]$
(2) $\forall a \exists b \exists \beta \exists g[b = \mathbf{pow}(a) \wedge \mathbf{fun}(g) \wedge \mathbf{dom}(g) = b \ \wedge\ \mathbf{ran}(g) = \beta \wedge\ g \text{ injective}]$
(3) $U_1 \neq \emptyset\ \wedge\ \forall \gamma \exists \delta[\gamma < \delta \wedge \delta \in E]$
(4) $\forall X \forall s \forall t[t \in U_1 \wedge \langle s, t\rangle \in U_2 \wedge \mathbf{club}(X) \ \rightarrow\ \{y : \langle y, s\rangle \in U_3\} \cap X \neq \emptyset]$

Employing the Π^1_1–indescribability of $\mathbf{K}$, there exists $\pi < \mathbf{K}$ such that the structure

$$\langle V_\pi, \in, U_1 \cap \pi, U_2 \cap \pi, U_3 \cap \pi, E \cap \pi\rangle$$

satisfies:

[6] $\langle x, y\rangle := \{\{x\}, \{x, y\}\}$; $\langle x_1, \ldots, x_{n+1}\rangle := \langle\langle x_1, \ldots, x_n\rangle, x_{n+1}\rangle$ for $n > 2$.

(a) $\forall G \forall \delta[\mathbf{fun}(G) \wedge \mathbf{dom}(G) = \delta \wedge \mathbf{ran}(G) \subseteq On \ \rightarrow \ \exists \gamma(G''\delta \subseteq \gamma)]$

(b) $\forall a \exists b \exists \beta \exists g[b = \mathbf{pow}(a) \wedge \mathbf{fun}(g) \wedge \mathbf{dom}(g) = b \wedge \mathbf{ran}(g) = \beta \wedge$ g injective]

(c) $U_1 \cap \pi \neq \emptyset \ \wedge \ \forall \gamma \exists \delta(\gamma < \delta \ \wedge \ \delta \in E \cap \pi)$

(d) $\forall X \forall s \forall t[t \in U_1 \cap \pi \wedge (\langle s, t \rangle) \in U_2 \cap \pi \wedge \mathbf{club}(X) \ \rightarrow \{y : \langle y, s \rangle \in U_3 \cap \pi\} \cap X \neq \emptyset]$

By virtue of (a), observing that $\forall G$ is second order, and (b), π must be inaccessible. Due to (c), $f(\alpha) \in V_\pi$ and E is unbounded in π; whence $\pi \in E$. (d) ensures that

$$(*) \ \ (\forall \beta < \alpha)[f(\beta) \in V_\pi \rightarrow \mathbf{M}^\beta \text{ stationary in } \pi].$$

Next, we want to verify

$$(+) \ \ (\forall \eta \in C^{\mathbf{K}}(\alpha, \pi))[f(\eta) \in V_\pi].$$

Set $X := \{\eta \in C^{\mathbf{K}}(\alpha, \pi) : f(\eta) \in V_\pi\}$. Clearly, $\pi \cup \{0, \mathbf{K}\} \subseteq X$.

If $\eta =_{NF} \omega^{\eta_1} + \cdots + \omega^{\eta_n}$ and $\eta_1, \ldots, \eta_n \in X$, then $\eta \in X$ since π is closed under $+$ and $\zeta \mapsto \omega^\zeta$ and V_π is closed under $\langle \cdot, \cdot \rangle$.

If $\beta \in X \cap \alpha$, then, according to $(*)$, $\mathbf{M}^\beta$ is stationary in π, yielding

$$\Xi^\beta(\delta) = f(\Xi^\beta(\delta)) < \pi \text{ for all } \delta \in X \cap \mathbf{K}.$$

If $\kappa, \xi, \delta \in X$ und $\xi \leq \delta < \alpha$, then $f(\kappa) = \kappa < \pi$ and therefore $\Psi^\xi_\kappa(\delta) < \pi$. So it turns out that X enjoys all the closure properties defining $C^{\mathbf{K}}(\alpha, \pi)$. This verifies $(+)$. Using $(*)$ and $(+)$, we obtain

$$(\forall \beta \in C^{\mathbf{K}}(\alpha, \pi) \cap \alpha)[\mathbf{M}^\beta \text{ is stationary in } \pi].$$

Whence, $\pi \in \mathbf{M}^\alpha \cap E$. □

The desired EORS, which encapsulates the strength of $\mathbf{KP} + \Pi_3$-Reflection, is essentially isomorphic to $\langle C^{\mathbf{K}}(\varepsilon_{\mathbf{K}+1}, 0), < \rangle$.

4 Recursively large ordinals and ordinal representation systems

The previous section gave ample examples of how large cardinal hypotheses enter the definition procedures of collapsing functions. The latter are then employed in the shape of terms to "name" a countable set of ordinals, and when one succeeds in establishing recursion relations for the ordering between those terms, the set of terms gives rise to an ordinal representation system. It has long been suggested (cf. [11], p. 436) that, instead, one should be able to interpret the collapsing functions as operating directly on the recursively large counterparts of those cardinals. For example, taking such an approach in Definition 3.1 would consist in letting

$$\mathbf{I} := \text{first recursively inaccessible ordinal}$$

and conceiving of $\alpha \mapsto \Omega_\alpha$ as enumerating the admissible ordinals and their limits. The difficulties with this approach arise with the proof of Lemma 3.2. One wants to show that for any admissible π satisfying $\pi \in C^{\mathbf{I}}(\alpha, \pi)$, one has $\psi_\pi(\alpha) < \pi$. In the cardinal setting this comes down to a simple cardinality argument. To get a similar result for an admissible π one would have to work solely with π-recursive operations. How this can be accomplished is far from being clear as the definition of $C^{\mathbf{I}}(\alpha, \rho)$ for $\rho < \pi$ usually refers to higher admissibles than just π. Notwithstanding that, the admissible approach is workable as was shown in [39, 41, 48]. A key idea therein is that the higher admissibles which figure in the definition of $\psi_\pi(\alpha)$ can be mimicked via names within the structure $\mathbf{L}_\pi$ in a π-recursive manner.

The drawback of the admissible approach is that it involves quite horrendous definition procedures and computations, which when taken as the first approach are at the limit of human tolerance.

On the other hand, the admissible approach provides a natural semantics for the terms in the EORSs. Recalling the notion of *good Σ_1-definition* from admissible set theory (see [5], II.5.13), given a set theory T, we say that an ordinal α has a *good Σ_1-definition in T* if there is a Σ_1-formula $\phi(u)$ such that

$$\mathbf{L}_{\mathbf{I}} \models \phi[\alpha] \text{ and } T \vdash \exists! x \phi(x).$$

In case of **KP** it turns out that all the ordinals of the corresponding EORS possess a good Σ_1-definition in **KP** (cf. [38]). As for **KPi**, the admissible approach canonically associates with each ordinal $\alpha \in \mathcal{T}(\mathbf{I}) \cap \mathbf{I}$ a good Σ_1-definition in **KPi**. However, via this interpretation $\mathcal{T}(\mathbf{I}) \cap \mathbf{I}$ only forms a proper subset of the **KPi**-definable ordinals. Therefore, to illuminate the nature of the ordinals in $\mathcal{T}(\mathbf{I})$, it would be desirable to find another property which distinguishes them within the **KPi**-definable ordinals.

In the above **KPi** just served the purpose of an example for a general phenomenon. The same considerations apply to **KPM** etc.

5 Large Cardinals and ordinal representation systems II

This section is devoted to the strongest large cardinal notions that have been used in developing ordinal representation systems. These cardinals exhibit strong indescribability properties which bear some resemblance to supercompact cardinals. The resulting ordinal representation systems have been put to use in ordinal analyses of the subsystems of second order arithmetic based on Π^1_n-Comprehension for $n \geq 2$. When drawing connections to ordinal recursion theory, these cardinals should be viewed as cardinal analogues of stable and n-stable ordinals.(cf. [19])

Proofs for all results in this section are in [44].

To begin with we recall some definitions from ordinal recursion theory.

Definition 5.1 An ordinal κ is said to be stable if $\mathbf{L}_\kappa \prec_1 \mathbf{L}$, i.e. $\mathbf{L}_\kappa$ is a Σ_1-elementary substructure of $\mathbf{L}$.

Let $\rho > \kappa$. κ is ρ-stable if $\mathbf{L}_\kappa \prec_1 \mathbf{L}_\rho$.

Another rendering of stability comes in terms of ordinal recursion theory (cf. [19], VIII.5.1):

κ is stable iff κ is closed under all ∞-partial recursive ordinal functions.

Likewise,

κ is ρ-stable iff κ is closed under all (∞, ρ)-partial recursive functions.

The connection of the system of Π^1_2-Comprehension ($\Pi^1_2 - \mathbf{CA}$ hereafter) with set theory comes through the fact that $\mathbf{KP} + \Sigma_1$-Separation is a conservative extension of $\Pi^1_2 - \mathbf{CA} + \mathbf{BI}$, where **BI** is the so-called principle of *Bar Induction.*

Σ_n-separation is the schema of axioms

$$\exists z(z = \{x \in a : \phi(x)\})$$

for all set-theoretic Σ_n-formulae ϕ.

BI is the schema

$$\forall X\left(\mathrm{WO}(<_X) \wedge \forall n\left[\forall m <_X n\, \Phi(m) \to \Phi(n)\right] \to \forall n \Phi(n)\right)$$

for all formulae Φ of the language of second order arithmetic, where

$$m <_X n := 2^m \cdot 3^n \in X.$$

Assuming *Infinity* to be among the axioms of **KP**, the precise relationship is as follows:

Theorem 5.2 $\mathbf{KP} + \Sigma_1$*-Separation and* $(\Pi^1_2 - \mathbf{CA}) + \mathbf{BI}$ *prove the same sentences of second order arithmetic.*

The ordinals κ such that $\mathbf{L}_\kappa \models \mathbf{KP} + \Sigma_1$-*Separation* are familiar from ordinal recursion theory. They are called *nonprojectible* (cf. [5]) and are exactly those ordinals $\kappa > \omega$ such that κ is a limit of (smaller) κ-stable ordinals.

Stronger comprehension is linked to set theories as follows:

Proposition 5.3 *Let $n > 0$.*

$$\mathbf{KP} + \Sigma_n\textit{-Collection} + \Sigma_n\textit{-Separation}$$

and

$$(\Pi^1_{n+1} - \mathbf{CA}) + (\mathbf{\Sigma^1_{n+1}} - \mathbf{AC}) + \mathbf{BI}$$

prove the same sentences of second order arithmetic.

To characterize the standard models of $\mathbf{KP} + \Sigma_n$-Collection + Σ_n-Separation, we introduce the notion of *n-stability.*

Definition 5.4 An ordinal κ is said to be *n-stable* if $\mathbf{L}_\kappa \prec_n \mathbf{L}$, i.e. $\mathbf{L}_\kappa$ is a Σ_n-elementary substructure of $\mathbf{L}$.

For $\rho > \kappa$, we say that κ is n-ρ-stable if $\mathbf{L}_\kappa \prec_n \mathbf{L}_\rho$.

n-stability can be reduced to stability in terms of relativized stability.

Let $A \subseteq \mathbf{L}$ be a class. κ is *stable in* A if $\langle \mathbf{L}_\kappa; A_\kappa \rangle \prec_1 \langle \mathbf{L}; A \rangle$, where $A_\kappa = \mathbf{L}_\kappa \cap A$

Let S_1 be the class of stable ordinals, and for $n > 0$, let S_{n+1} be the class of ordinals stable in S_n.

Proposition 5.5 (**ZFC**) *κ is $n+1$-stable iff κ is stable in S_n.*

Similar to the connection between Σ_1-Separation and nonprojectability one has:

Proposition 5.6 *The following are equivalent for limit ordinals κ:*

(i) $\mathbf{L}_\kappa \models \Sigma_n$*-Collection* + Σ_n*-Separation.*
(ii) For every $a \in \mathbf{L}_\kappa$ there exists $M \in \mathbf{L}_\kappa$ such that $a \subseteq M$ and $M \prec_n \mathbf{L}_\kappa$.

The next definition introduces what we consider to be the cardinal analogue of stability.

Definition 5.7 Let $\eta > 0$. A cardinal κ is *η-shrewd* if for all $P \subseteq V_\kappa$ and every set-theoretic formula $\phi(v_0, v_1)$, whenever

$$V_{\kappa+\eta} \models \phi[P, \kappa],$$

then there exist $0 < \kappa_0, \eta_0 < \kappa$ such that

$$V_{\kappa_0+\eta_0} \models \phi[P \cap V_{\kappa_0}, \kappa_0].$$

κ is *shrewd* if κ is η-shrewd for every $\eta > 0$.

Corollary 5.8 *If κ is δ-shrewd and $0 < \eta < \delta$, then κ is also η-shrewd.*

Apparently, the notion of shrewdness has not been put into the dictionary of large cardinals. There are some similarities between the notions of η-shrewdness and η-indescribability (see [8], Ch.9, §4). However, the notions are quite different in other aspects. For instance, it is impossible, for any κ, that κ is κ-indescribable. Therefore, if κ is η-indescribable and $\rho < \eta$, it does not necessarily follow that κ is also ρ-indescribable (see [8], 9.4.6). Another difference is that if π is measurable, then for every β, the set $\{\kappa < \pi : \kappa \text{ is } \beta\text{-indescribable}\}$ is stationary in π whereas there need not be any $\pi + 2$-shrewd cardinals below π.

A negative reason for calling the above cardinals *shrewd* is a shortage of names for cardinals. A positive reason is the following: If there is a shrewd cardinal κ in the universe, then, loosely speaking, for any notion of large cardinal N which does not make reference to the totality of all ordinals, whenever there exists an N-cardinal then the least such is below κ. So, for instance, if there are measurable and shrewd cardinals in the universe, then the least measurable is smaller than any of the shrewd cardinals.

A way of evincing the analogy between shrewdness and stability more closely consists in relating shrewdness to power recursion with search over the set-theoretic universe. Power recursion has been studied by Moschovakis [28] and Moss [29]. Central examples of power recursive functions (not requiring search) are $\alpha \mapsto V_\alpha$ and $\alpha \mapsto \aleph_\alpha$. However, limitations of space prevent us from going into details.

As suggested by 5.5, we shall also consider a notion of shrewdness with regard to a given class.

Let $\mathcal{L}_{set}$ denote the language of set theory. Let $\mathbf{U}$ be a fresh unary predicate symbol. Given a language $\mathcal{L}$ let $\mathcal{L}(\mathbf{U})$ denote its extension by $\mathbf{U}$.

If $\mathcal{A}$ is a class, we denote by $\langle V_\alpha; \mathcal{A}\rangle$ the structure $\langle V_\alpha; \in; \mathcal{A} \cap V_\alpha\rangle$. For an $\mathcal{L}_{set}(\mathbf{U})$-sentence ϕ, let the meaning of "$\langle V_\alpha; \mathcal{A}\rangle \models \phi$" be determined by interpreting $\mathbf{U}(t)$ as $t \in \mathcal{A} \cap V_\alpha$.

Definition 5.9 Let $\mathcal{A}$ be a class. Let $\eta > 0$. A cardinal κ is *$\mathcal{A}$-η-shrewd* if for all $P \subseteq V_\kappa$ and every formula $\phi(v_0, v_1)$ of $\mathcal{L}_{set}(\mathbf{U})$, whenever

$$\langle V_{\kappa+\eta}; \mathcal{A}\rangle \models \phi[P, \kappa],$$

then there exist $0 < \kappa_0, \eta_0 < \kappa$ such that

$$\langle V_{\kappa_0+\eta_0}; \mathcal{A}\rangle \models \phi[P \cap V_{\kappa_0}, \kappa_0].$$

κ is *$\mathcal{A}$-shrewd* if κ is $\mathcal{A}$-η-shrewd for every $\eta > 0$.

Corollary 5.10 *If κ is $\mathcal{A}$-δ-shrewd and $0 < \eta < \delta$, then κ is $\mathcal{A}$-η-shrewd.*

To situate the notion of shrewdness with regard to consistency strength in the usual hierarchy of large cardinals, we recall the notion of a subtle cardinal.

Definition 5.11 A cardinal κ is said to be *subtle* if for any sequence $\langle S_\alpha : \alpha < \kappa\rangle$ such that $S_\alpha \subseteq \alpha$ and C closed and unbounded in κ, there are $\beta < \delta$ both in C satisfying

$$S_\delta \cap \beta = S_\beta.$$

Since subtle cardinals are not covered in many of the standard texts dealing with large cardinals, we mention the following facts (see [22], §20):

Remark 5.12 *Let $\kappa(\omega)$ denote the first ω-Erdös cardinal.*

(i) $\{\pi < \kappa(\omega) : \pi$ is subtle$\}$ is stationary in $\kappa(\omega)$.
(ii) "Subtlety" relativises to $\mathbf{L}$, i.e. if π is subtle, then $\mathbf{L} \models$ "π is subtle".

Lemma 5.13 *Assume that π is a subtle cardinal and that $\mathcal{A} \subseteq V_\pi$. Then for every $B \subseteq \pi$ closed and unbounded in π there exists $\kappa \in B$ such that*

$$\langle V_\pi; \mathcal{A}\rangle \models \text{"}\kappa \textit{ is } \mathcal{A}\textit{-shrewd"}.$$

There are similarities between the cardinal notions of shrewdness and supercompactness. To bring out this analogy, we introduce two new cardinal notions. The first of them embodies considerable consistency strength.

Definition 5.14 Let $\mathcal{A}$ be a class. Assume $\eta > 0$. κ is *strongly $\mathcal{A}$-η-reducible* if for every $P \subseteq V_{\kappa+\eta}$ there exist $0 < \kappa_0, \eta_0 < \kappa$ and $Q \subseteq V_{\kappa_0+\eta_0}$ and an elementary embedding i such that $Q \cap V_{\kappa_0} = P \cap V_{\kappa_0}$ and

$$i : \langle V_{\kappa_0+\eta_0}; \in; \mathcal{A}; Q\rangle \longrightarrow \langle V_{\kappa+\eta}; \in; \mathcal{A}; P\rangle$$

with critical point κ_0 and $i(\kappa_0) = \kappa$.

κ is *strongly $\mathcal{A}$-reducible* if κ is strongly $\mathcal{A}$-η-reducible for all $\eta > 0$.

κ is *strongly η-reducible* if κ is strongly V-η-reducible. κ is *strongly reducible* if κ is strongly η-reducible for all $\eta > 0$.

Using elementary equivalence ($\equiv$) of structures instead of elementary embeddability one arrives at the following notion:

Definition 5.15 Let $\mathcal{A}$ be a class. If $\eta > 0$, κ is *$\mathcal{A}$-η-reducible* if for every $P \subseteq V_{\kappa+\eta}$ there exist $0 < \kappa_0, \eta_0 < \kappa$ and $Q \subseteq V_{\kappa_0+\eta_0}$ such that

$$\langle V_{\kappa_0+\eta_0}; \in; \kappa_0; \mathcal{A}; Q; x\rangle_{x\in V_{\kappa_0}} \equiv \langle V_{\kappa+\eta}; \in; \kappa; \mathcal{A}; P; x\rangle_{x\in V_{\kappa_0}}. \tag{31}$$

κ is *$\mathcal{A}$-reducible* if κ is $\mathcal{A}$-η-reducible for every η. κ is η-reducible if κ is V-η-reducible. κ is *reducible* if κ is η-reducible for every η.

Note that $Q \cap V_{\kappa_0} = P \cap V_{\kappa_0}$ springs from (31).

To make the foregoing definition resemble more closely the definition of strong reducibility, notice that in the situation of (31) there exists a *partial* embedding p from $V_{\kappa_0+\eta_0}$ into $V_{\kappa+\eta}$ satisfying $p \restriction V_{\kappa_0+\eta_0} = \mathrm{id} \restriction V_{\kappa_0+\eta_0}$ and $p(\kappa_0) = \kappa$. Moreover, p can be canonically extended so as to being defined on all elements of $V_{\kappa_0+\eta_0}$ which are definable in the structure $\langle V_{\kappa_0+\eta_0}; \in; \mathfrak{V}; \kappa_0; \mathcal{A}; Q; x\rangle_{x\in V_{\kappa_0}}$.

We will use

$$p : \langle V_{\kappa_0+\eta_0}; \in; \mathcal{A}; Q\rangle \underset{\equiv}{\longrightarrow} \langle V_{\kappa+\eta}; \in; \mathcal{A}; P\rangle$$

as a shorthand for conveying the foregoing situation.

The aspired analogy between shrewdness and strong reducibility resides in the fact that (weak) reducibility is closely related to shrewdness.

Proposition 5.16 *If κ is $\mathcal{A}$-ρ-shrewd and $0 < \eta < \rho$, then κ is $\mathcal{A}$-η-reducible.*

The circle of analogies will be completed by the next proposition, which also shows that the notion of a strongly reducible cardinal is equivalent to *supercompactness.*

Definition 5.17 κ is *δ-supercompact* if there is a transitive class M and an elementary embedding

$$j : V \longrightarrow M$$

such that $\mathrm{crit}(j) = \kappa$ and $\delta < j(\kappa)$, and ${}^{\delta}M \subseteq M$.

κ is *supercompact* if κ is δ-supercompact for every $\delta \geq \kappa$.

Proposition 5.18 *κ is strongly reducible iff κ is supercompact.*

A similar equivalence can be shown for $\mathcal{A}$-supercompact cardinals (cf. [51], 6.7).

Proposition 5.19 *κ is strongly $\mathcal{A}$-reducible iff κ is $\mathcal{A}$-supercompact.*

Sufficiently strong ordinal representation systems for the analyses of the systems (Π^1_n − **CA**) utilize the notion of $\mathcal{A}$-reducibility for classes $\mathcal{A}$ which depend on the given n. The pertaining collapsing functions are obtained from inverses of partial elementary embeddings as explained in 5.15. The details will appear in [44].

6 Large sets in constructive set theory

Ideally, one wants to have mathematical results which allow one to state how it is that large cardinals come to be utilized in proof-theoretic ordinal analyses. Something that suggests more than merely an analogue. One idea pursued here is, that one should study the same notion of largeness in different settings. To give an example, we start off with a definition.

Definition 6.1 A non-empty set A is *regular* if A is transitive, and for every $a \in A$ and set $R \subseteq a \times A$ if $\forall x \in a\, \exists y\, (\langle x, y\rangle \in R)$, then there is a set $b \in A$ such that

$$\forall x \in a\, \exists y \in b\, (\langle x, y\rangle \in R) \;\wedge\; \forall y \in b\, \exists x \in a\, (\langle x, y\rangle \in R).$$

In particular, if $R : a \to A$ is a function, then the image of R is an element of A.

In the context of **ZFC** we have that V_κ is regular iff κ is a regular cardinal. The analogy between admissible sets and regular sets is drawn by restricting the class of relations (or functions) to the A-recursive ones. In contradistinction to the latter approach we suggest a study of regularity such that the only changes being made take place in the surrounding environment.[7]

The particular environment will be Aczel's constructive set theory, **CZF**. As for the main question raised above, we have no conclusive answers, but the results presented here might give some new insights. Proofs will be published elsewhere.

This section deals with large cardinal properties in the context of intuitionistic set theories. Since in intuitionistic set theory $\in$ is not a linear ordering on ordinals the notion of a cardinal does not play a central role. Consequently, one talks about *"large set properties"* instead of *"large cardinal properties"*. Friedman and Ščedrov [14] studied large set properties in the context of **IZF**. When stating these properties one has to proceed rather carefully. Classical equivalences of cardinal notion might no longer prevail in the intuitionistic setting , and one therefore wants to choose a rendering which intuitionistically retains the most strength. On the other hand certain notions have to be avoided so as not

[7] Feferman [13] is in a similar vein, but undertakes a different approach.

to imply excluded third. To give an example, cardinal notions like measurability, supercompactness and hugeness have to be expressed in terms of elementary embeddings rather than ultrafilters.

The axioms of **IZF** are Extensionality, Pairing, Union, as usual, and the following:

Infinity $\exists x \forall u[u \in x \leftrightarrow (\emptyset \in x \vee \exists v \in x(u = v \cup \{v\}))]$
Set Induction $\forall x[\forall y \in x\phi(y) \rightarrow \phi(x)] \rightarrow \forall x\phi(x)$
Separation $\forall a \exists b \forall x[x \in b \leftrightarrow x \in a \wedge \phi(x)]$
Collection $\forall a[\forall x \in a \exists y \phi(x, y) \rightarrow \exists b \forall x \in a \exists y \in b\, \phi(x, y)]$
Powerset $\forall x \exists y \forall z(z \in y \leftrightarrow z \subseteq x)$

for all set-theoretic formulae ϕ.

Regarding proof-theoretic strength, the upshot of [14] is that the equiconsistency of **ZF** and **IZF** propagates to extensions with large set axioms. The proof employs a $\neg\neg$-interpretation.

Theorem 6.2 (Friedman and Ščedrov, [14]) *If* **LSA** *is a large set axiom pertaining to any of the large cardinal axioms asserting the existence of an inaccessible, Mahlo, measurable, supercompact or n-huge cardinal, then:*

$$\mathbf{IZF} + \mathbf{LSA} \textit{ and } \mathbf{ZF} + \mathbf{LSA} \textit{ are equiconsistent.}$$

To be of interest, the latter systems should not imply excluded third. This follows from the next theorem.

Theorem 6.3 (Friedman and Ščedrov, [14]) *With* **LSA** *as above, the theory* **IZF** + **LSA** *has the disjunction property and the number existence property. Moreover,* **IZF** + **LSA** *is equiconsistent with* **IZF** + **LSA** + *Church's thesis.*

For our purpose the foregoing results appear to be disappointing since large set assumptions retain their consistency strength on the basis of **IZF**. The situation changes radically when we exchange **IZF** for **CZF**. The latter theory is due to Aczel (cf. [1, 2, 3]) and extends Myhill's constructive set theory **CST** (cf. [30]) which grew out of endeavours to discover a (simple) formalism that relates to Bishop's constructive mathematics as **ZFC** relates to classical Cantorian mathematics. The novel ideas were to replace Powerset by the (classically equivalent) Exponentiation Axiom and to discard full Comprehension while retaining full Collection. Aczel extended **CST** to **CZF** and corroborated the constructiveness of the latter theory by interpreting it in Martin-Löf's intuitionistic type theory (cf. [27]).

6.1 The System CZF

In this subsection we will summarize the language and axioms for Aczel's constructive set theory or **CZF**. The language of **CZF** is the first order language of ZF whose only non-logical symbol is $\in$. The logic of **CZF** is intuitionistic first order logic with equality. Its non-logical axioms comprise *Extensionality, Pairing, Union* in their usual forms, and *Infinity* and *Set Induction* as stated for **IZF**. **CZF** has additionally axiom schemata which we will now proceed to summarize.

Restricted Separation

$$\forall a \exists b \forall x[x \in b \leftrightarrow x \in a \wedge \phi(x)]$$

for all *restricted* formulae ϕ. A set-theoretic formula is *restricted* if it is constructed from prime formulae using $\neg, \wedge, \vee, \rightarrow, \forall x{\in}y$, and $\exists x{\in}y$ only.

Strong Collection

$$\forall a\big[\forall x \in a \exists y \phi(x,y) \rightarrow \\ \exists b\,[\forall x \in a\, \exists y \in b\, \phi(x,y) \wedge \forall y \in b\, \exists x \in a\, \phi(x,y)]\big]$$

for all formulae ϕ.

Subset Collection

$$\forall a \forall b \exists c \forall u\,\big[\forall x \in a\, \exists y \in b\, \phi(x,y,u) \rightarrow \\ \exists d \in c\,[\forall x \in a\, \exists y \in d\, \phi(x,y,u) \wedge \forall y \in d\, \exists x \in a\, \phi(x,y,u)]\big]$$

for all formulae ϕ.

The mathematically important axiom of *Dependent Choices* (**DC**) could be included among the axioms of **CZF** without changing any essential properties of **CZF**, including its interpretation in type theory.

The Subset Collection schema easily qualifies for the most intricate axiom of **CZF**. To explain this axiom in different terms, we introduce the notion of *fullness*.

Definition 6.4 For sets A, B let ${}^A B$ be the class of all functions with domain A and with range contained in B.

Let $\mathbf{mv}({}^A B)$ be the class of all sets $R \subseteq A \times B$ satisfying $\forall u{\in}A\, \exists v{\in}B\, \langle u, v\rangle{\in}R$. A set C is said to be *full* in $\mathbf{mv}({}^A B)$ if $C \subseteq \mathbf{mv}({}^A B)$ and

$$\forall R{\in}\mathbf{mv}({}^A B)\, \exists S{\in}C\, S \subseteq R.$$

Additional axioms we shall consider are:

Exponentiation: $\forall x \forall y \exists z\, z = {}^x y$.

Fullness: $\forall x \forall y \exists z$ "z full in $\mathbf{mv}({}^x y)$".

Proposition 6.5 *Let* $\mathbf{CZF}^-$ *be* **CZF** *without* Subset Collection.

(i) $\mathbf{CZF}^- \vdash$ Subset Collection $\leftrightarrow$ Fullness.
(ii) $\mathbf{CZF} \vdash$ Exponentiation.

Let **TND** be the principle of excluded third, i.e. the schema consisting of all formulae of the form $A \vee \neg A$.

The first central fact to be noted about **CZF** is:

Proposition 6.6 **CZF + TND = ZF**.

Proof: Note that classically Collection implies Separation. Powerset follows classically from Exponentiation. □

To stay in the world of **CZF** one has to keep away from any principles that imply **TND**. Moreover, it is fair to say that **CZF** is such an interesting theory owing to the non-derivability of Powerset and Separation. Therefore one ought to avoid any principles which imply Powerset or Separation.

In what follows we shall investigate largeness notions corresponding to inaccessibility, Mahloness and weak compactness. Bowing to the demands of brevity, we content ourselves with listing the definitions and results.

6.2 Inaccessibility

Let $\mathbf{Reg}(A)$ be the statement that A is a regular set (cf. (6.1)). The next axiom comprises that the universe is a union of regular sets.

Regular Extension Axiom (REA)

$$\forall x \exists y[x \subseteq y \wedge \mathbf{Reg}(y)]$$

Definition 6.7 A set I is said to be *inaccessible* if $\mathbf{Reg}(I)$ and I is a model of **CZF + REA** in a strong sense, i.e. the structure $\langle I, \in \restriction (I \times I)\rangle$ is a model of Pairing, Union, Infinity, restricted Separation, and **REA** and the following holds:

$$(\mathbf{A}) \quad \forall A, B{\in}I\, \exists C{\in}I \text{ "}C \text{ is full in } \mathbf{mv}({}^{A}B)\text{"}$$

Due to $\mathbf{Reg}(I)$ and $(\mathbf{A})$, $\langle I, \in \restriction (I \times I)\rangle$ is also a model of strong collection and subset collection.

Corollary 6.8 *The following theories are the same theories, i.e. they prove the same formulae:*

(i) $\mathbf{CZF} + \exists I\, \mathbf{inac(I)} + \mathbf{TND}$
(ii) $\mathbf{ZF} + \exists \mathbf{I}\, \mathbf{inac(I)}$

They are equiconsistent with $\mathbf{ZFC} + \exists\kappa$ *"κ inaccessible cardinal"*

Theorem 6.9

$$\mathbf{CZF} + \forall x \exists I\, [x{\in}I \;\wedge\; \text{"}I \textit{ inaccessible}\text{"}]$$

can be interpreted in

$$\mathbf{KP} + \forall\alpha\exists\kappa\, [\alpha{\in}\kappa \;\wedge\; \text{"}\kappa \textit{ recursively inaccessible}\text{"}].$$

The interpretation preserves validity of Π_2-sentences. The theories have the same proof-theoretic strength.

6.3 Mahloness

Definition 6.10 A set M is said to be *Mahlo* if it is inaccessible and for each set $R \subseteq M \times M$, whenever

$$\forall x{\in}M\,\exists y{\in}M\,\langle x,y\rangle{\in}R,$$

then for every $u{\in}M$ there exists an inaccessible $I{\in}M$ with $u{\in}I$ and:

$$\forall x{\in}I\,\exists y{\in}I\,\langle x,y\rangle{\in}R.$$

Definition 6.11 Let A, α be sets. A is *α-inaccessible* iff A is inaccessible and for all $\beta{\in}\alpha$:

$$\forall a{\in}A\,\exists B{\in}A[a{\in}B \wedge \text{“}B \text{ is } \beta\text{-inaccessible]}.$$

Proposition 6.12 (CZF) *If M is Mahlo then M is M-inaccessible.*

Corollary 6.13 **CZF** + $\exists M$ *“M Mahlo”* + **TND** *and* **ZF** + $\exists$**M** *“M Mahlo” are the same theories.*

They are equiconsistent with **ZFC** + $\exists\pi$ *“π Mahlo cardinal”.*

Theorem 6.14

$$\mathbf{CZF} + \forall x \exists M\,[x{\in}M \;\wedge\; \text{“}M \textit{ Mahlo”}]$$

can be interpreted in

$$\mathbf{KP} + \forall\alpha\exists\kappa\,[\alpha{\in}\kappa \;\wedge\; \text{“}\kappa \textit{ recursively Mahlo ordinal”}].$$

The interpretation preserves validity of Π_2-sentences. The theories have the same proof-theoretic strength.

6.4 Weak compactness

Theorem 3.8 suggests 2-regularity as the natural rendering of weak compactness in **CZF**. However, due to the absence of the axiom of choice in **CZF**, we prefer to introduce a slightly different notion.

Definition 6.15 Recall that $\mathbf{mv}({}^{A}B) = \{R \subseteq A \times B : \forall u{\in}A\,\exists v{\in}B\,\langle u,v\rangle{\in}R\}$.

An inaccessible set K is called *2-strong* if the following holds true for all sets S:

$$\begin{aligned}&\forall r{\in}\mathbf{mv}({}^{K}K)\forall u{\in}K\,\exists x{\in}K\,\exists v{\in}K[x \subseteq R \wedge \langle x,u,v\rangle{\in}S] \rightarrow \\ &\quad \exists I{\in}K(\mathbf{inac(I)} \wedge \forall \mathbf{R}{\in}\mathbf{mv}({}^{\mathbf{I}}\mathbf{I})\forall \mathbf{u}{\in}\mathbf{I}\,\exists \mathbf{x}{\in}\mathbf{I}\,\exists \mathbf{v}{\in}\mathbf{I}[\mathbf{x} \subseteq \mathbf{R} \wedge \langle \mathbf{x},\mathbf{u},\mathbf{v}\rangle{\in}\mathbf{S}]).\end{aligned}$$

Corollary 6.16 (CZF) *If K is 2-strong, then for any formula ϕ,*

$$\begin{aligned}&\forall R{\in}\mathbf{mv}({}^{K}K)\forall u{\in}K\,\exists x{\in}K\,\exists v{\in}K[x \subseteq R \wedge \phi(x,u,v)] \rightarrow \\ &\quad \exists I{\in}K(\mathbf{inac(I)} \wedge \forall \mathbf{r}{\in}\mathbf{mv}({}^{\mathbf{I}}\mathbf{I})\forall \mathbf{u}{\in}\mathbf{I}\,\exists \mathbf{x}{\in}\mathbf{I}\,\exists \mathbf{v}{\in}\mathbf{I}[\mathbf{x} \subseteq \mathbf{R} \wedge \phi(\mathbf{x},\mathbf{u},\mathbf{v})])].\end{aligned}$$

Lemma 6.17 (**ZFC**) *For all ordinals κ, V_κ is 2-strong iff κ is weakly compact.*

Definition 6.18 Let α, C be sets. C is *α-Mahlo* if C is inaccessible and for all $\beta\in\alpha$:

$$\forall R\in\mathbf{mv}({}^{C}C)\,\exists B\in C\left[\text{“}B\text{ is }\beta\text{-Mahlo”}\ \wedge\ \forall x\in B\,\exists y\in B\langle x,y\rangle\in R\right].$$

Proposition 6.19 (**CZF**) *If C is 2-strong, then C is C-Mahlo.*

Theorem 6.20

$$\mathbf{CZF}+\forall x\exists K\,[x\in K\ \wedge\ \text{“}K\text{ 2-strong”}]$$

can be interpreted in

$$\mathbf{KP}+\forall\alpha\exists\kappa\,[\alpha\in\kappa\ \wedge\ \text{“}\kappa\ \Pi_3\text{-reflecting”}].$$

The interpretation preserves validity of Π_2-sentences. The theories have the same proof-theoretic strength.

References

[1] P. Aczel: *The Type Theoretic Interpretation of Constructive Set Theory*, in: MacIntyre, A. Pacholski, L., and Paris, J. (eds.), *Logic Colloquium '77*, North–Holland, Amsterdam 1978.

[2] P. Aczel: *The Type Theoretic Interpretation of Constructive Set Theory: Choice Principles*, in: Troelstra, A. S., van Dalen, D. (eds), *The L.E.J. Brouwer Centenary Symposium*, North–Holland, Amsterdam 1982.

[3] P. Aczel: *The Type Theoretic Interpretation of Constructive Set Theory: Inductive Definitions*, in: Marcus, R. B. et al. (eds), *Logic, Methodology, and Philosopy of Science VII*, North–Holland, Amsterdam 1986.

[4] H. Bachmann: *Die Normalfunktionen und das Problem der ausgezeichneten Folgen von Ordinalzahlen*, Vierteljahresschrift Naturforsch. Ges. Zürich 95 (1950) 115–147.

[5] J. Barwise: *Admissible Sets and Structures* (Springer, Berlin 1975).

[6] W. Buchholz: *A new system of proof–theoretic ordinal functions*, Ann. Pure Appl. Math. Logic **32** (1986) 195–207.

[7] W. Buchholz, K. Schütte: *Proof theory of impredicative subsystems of analysis* (Bibliopolis, Naples, 1988).

[8] F. Drake: *Set Theory: An introduction to large cardinals* (Amsterdam, North Holland, 1974)

[9] S. Feferman: *Systems of predicative analysis*, Journal of Symbolic Logic 29 (1964) 1–30.

[10] S. Feferman: *Systems of predicative analysis II. Representations of ordinals*, Journal of Symbolic Logic 33 (1968) 193–220.

[11] S. Feferman: *Proof theory: a personal report*, in: G. Takeuti, Proof Theory, 2nd edition (North-Holland, Amsterdam, 1987) 445–485.

[12] S. Feferman: *Hilbert's program relativized: Proof-theoretical and foundational reductions*, JSL 53 (1988) 364–384.

[13] S. Feferman: *Gödel's program for new axioms: Why, where, how and what?* to appear in: Gödel '96 conference, Brno. 23 pages.
[14] H. Friedman and S. Ščedrov: *Large sets in intuitionistic set theory*, Annals of Pure and Applied Logic 27 (1984) 1–24.
[15] H. Friedman and S. Sheard: *Elementary descent recursion and proof theory*, Annals of Pure and Applied Logic 71 (1995) 1–45.
[16] H. Gaifman: *A generalization of Mahlo's method for obtaining large cardinal numbers*, Israel Journal of Mathematics 5 (1967) 188–200.
[17] J.-.Y. Girard: *A survey of Π^1_2-logic. Part I: Dilators*, Annals of Mathematical Logic 21 (1981) 75–219.
[18] E. Griffor and M. Rathjen: *The strength of some Martin–Löf type theories.* Archive for Mathematical Logic 33 (1994) 347–385.
[19] P.G. Hinman: *Recursion-theoretic hierarchies* (Springer, Berlin, 1978).
[20] D. Isles: *Regular ordinals and normal forms*, in: A. Kino, J. Myhill, R.E. Vesley (eds.): Intuitionism and proof theory (North-Holland, Amsterdam, 1968) 288–300.
[21] G. Jäger and W. Pohlers: *Eine beweistheoretische Untersuchung von* $\mathbf{\Delta^1_2}$- $\mathbf{CA+BI}$ *und verwandter Systeme*, Sitzungsberichte der Bayerischen Akademie der Wissenschaften, Mathematisch–Naturwissenschaftliche Klasse (1982).
[22] A. Kanamori, M. Magidor: *The evolution of large cardinal axioms in set theory.* In: G. H. Müller, D.S. Scott (eds.) *Higher Set Theory.* Lecture Notes in Mathematics 669 (Springer, Berlin, 1978) 99-275.
[23] G. Kreisel: *A survey of proof theory*, Journal of Symbolic Logic 33 (1968) 321–388.
[24] P. Mahlo: *Über lineare transfinite Mengen*, Berichte über die Verhandlungen der Königlich Sächsischen Gesellschaft der Wissenschaften zu Leipzig, Mathematisch-Physische Klasse, 63 (1911) 187–225.
[25] P. Mahlo: *Zur Theorie und Anwendung der ρ_0-Zahlen*, ibid. 64 (1912) 108–112.
[26] P. Mahlo: *Zur Theorie und Anwendung der ρ_0-Zahlen*, ibid. 65 (1913) 268–282.
[27] P. Martin-Löf: *Intuitionistic Type Theory*, (Bibliopolis, Naples 1984).
[28] Y.N. Moschovakis: *Recursion in the universe of sets*, mimeographed note, 1976.
[29] L. Moss: *Power set recursion*, Annals of Pure and Applied Logic 71 (1995) 247–306.
[30] J. Myhill: *Constructive Set Theory*, JSL 40 (1975) 347–382.
[31] D. Normann: *Set recursion*, in: Fenstad et al. (eds.): *Generalized recursion theory II* (North-Holland, Amsterdam, 1978) 303–320.
[32] H. Pfeiffer: *Ausgezeichnete Folgen für gewisse Abschnitte der zweiten und weiterer Zahlklassen* (Dissertation, Hannover, 1964).
[33] W. Pohlers: *Contributions of the Schütte school in Munich to proof theory*, in: G. Takeuti, *Proof Theory*, 2nd edition (North-Holland, Amsterdam, 1987) 406–431.
[34] W. Pohlers: *A short course in ordinal analysis*, in: P. Aczel, H. Simmons, S. Wainer (eds.): *Proof Theory* (Cambridge University Press, Cambridge, 1992) 27–78. .
[35] M. Rathjen: *Ordinal notations based on a weakly Mahlo cardinal*, Archive for Mathematical Logic 29 (1990) 249–263.
[36] M. Rathjen: *Proof-Theoretic Analysis of KPM*, Arch. Math. Logic 30 (1991) 377–403.
[37] M. Rathjen: *The role of parameters in bar rule and bar induction*, Journal of Symbolic Logic 56 (1991) 715–730.
[38] M. Rathjen: *Fragments of Kripke-Platek set theory with infinity*, in: P. Aczel, H. Simmons, S. Wainer (eds.): *Proof Theory* (Cambridge University Press, Cambridge, 1992) 251–273.

[39] M. Rathjen: *How to develop proof–theoretic ordinal functions on the basis of admissible sets.* Mathematical Quarterly 39 (1993) 47–54.
[40] M. Rathjen: *Admissible proof theory and beyond.* In: Logic, Methodology and Philosophy of Science IX (D. Prawitz, B. Skyrms and D. Westerstahl, eds.), Elsevier Science B.V. (1994) 123–147.
[41] M. Rathjen: *Collapsing functions based on recursively large ordinals: A well–ordering proof for KPM.* Archive for Mathematical Logic 33 (1994) 35–55.
[42] M. Rathjen: *Proof theory of reflection.* Annals of Pure and Applied Logic 68 (1994) 181–224.
[43] M. Rathjen: *Recent advances in ordinal analysis: Π^1_2-CA and related systems.* Bulletin of Symbolic Logic 1 (1995) 468–485.
[44] M. Rathjen: *An ordinal analysis of Π^1_2 comprehension and related systems,* preprint.
[45] Richter, W. and Aczel, P.: *Inductive definitions and reflecting properties of admissible ordinals.* In: J.E. Fenstad, Hinman (eds.) Generalized Recursion Theory (North Holland, Amsterdam, 1973) 301-381.
[46] H.E. Rose: *Subrecursion: functions and hierarchies.* (Clarendon Press, Oxford, 1984).
[47] G.E. Sacks: *Higher recursion theory* (Springer, Berlin, 1990).
[48] A. Schlüter: *Provability in set theories with reflection*, preprint, 1995.
[49] K. Schütte: *Beweistheorie* (Springer, Berlin, 1960).
[50] H. Schwichtenberg: *Proof theory: Some applications of cut-elimination.* In: J. Barwise (ed.): *Handbook of Mathematical Logic* (North Holland, Amsterdam, 1977) 867–895.
[51] R.M. Solovay, W.N. Reinhardt, A. Kanamori: *Strong Axioms of Infinity and Elementary Embeddings*, Annals of Mathematical Logic 13 (1978) 73–116.
[52] R. Sommer: *Ordinal arithmetic in $I\Delta_0$.* In: P. Clote and J. Krajicek (eds.): *Arithmetic, proof theory, and computational complexity* (Clarendon Press, Oxford, 1993) 320–363.
[53] R. Sommer: *Ordinal functions in fragments of arithmetic*, Preprint (1992) 28 pages.
[54] G. Takeuti: *Proof theory*, second edition (North Holland, Amsterdam, 1987).
[55] O. Veblen: *Continous increasing functions of finite and transfinite ordinals,* Trans. Amer. Math. Soc. 9 (1908) 280–292.

There May Be No Nowhere Dense Ultrafilter

Saharon Shelah *

Institute of Mathematics
The Hebrew University of Jerusalem
91904 Jerusalem, Israel
shelah@math.huji.ac.il
and
Department of Mathematics
Rutgers University
New Brunswick, NJ 08854, USA

Abstract. We show the consistency of ZFC + "there is no NWD-ultrafilter on ω", which means: for every non-principal ultrafilter $\mathcal{D}$ on the set of natural numbers, there is a function f from the set of natural numbers to the reals, such that for every nowhere dense set A of reals, $\{n : f(n) \in A\} \notin \mathcal{D}$. This answers a question of van Douwen, which was put in more general context by Baumgartner.

0 Introduction

We prove here the consistency of "there is no NWD-ultrafilter on ω" (non-principal, of course). This answers a question of van Douwen [vD81] which appears as question 31 of [B6]. Baumgartner [B6] considers the question which he dealt more generally with J-ultrafilter where

Definition 1. 1. An ultrafilter $\mathcal{D}$, say on ω, is called a J-ultrafilter where J is an ideal on some set X (to which all singletons belong, to avoid trivialities) *if* for every function $f : \omega \longrightarrow X$ for some $A \in \mathcal{D}$ we have $f''(A) \in J$.

2. The NWD–ultrafilters are the J-ultrafilters for

$$J = \{B \subseteq \mathcal{Q} : B \text{ is nowhere dense}\}$$

($\mathcal{Q}$ is the set of all rationals; we will use an equivalent version, see 22).

This is also relevant for the consistency of "every (non-trivial) c.c.c. σ-centered forcing notion adds a Cohen real", see [Sh:F151].

The most natural approach to a proof of the consistency of "there is no NWD–ultrafilter" was to generalize the proof of CON(there is no P-point) (see [Sh:b, VI, §4] or [Sh:f, VI, §4]), but I (and probably others) have not seen how.

We use an idea taken from [Sh 407], which is to replace the given maximal ideal I on ω by a quotient; moreover, we allow ourselves to change the quotient.

* The research partially supported by "Basic Research Foundation" of the Israel Academy of Sciences and Humanities. Publication 594.

In fact, the forcing here is simpler than the one in [Sh 407]. A related work is Goldstern, Shelah [GoSh 388].

We similarly may consider the consistency of "no α–ultrafilter" for limit $\alpha < \omega_1$ (see [B6] for definition and discussion of α–ultrafilters). This question and the problems of preservation of ultrafilters and distinguishing existence properties of ultrafilters will be dealt with in a subsequent work [Sh:F187].

In §3 we note that any ultrafilter with property M (see Definition 25) is an NWD–ultrafilter, hence it is consistent that there is no ultrafilter (on ω) with property M.

I would like to thank James Baumgartner for arousing my interest in the questions on NWD–ultrafilters and α-ultrafilters and Benedikt on asking about the property M as well as Shmuel Lifches for corrections, the participants of my seminar in logic in Madison Spring'96 for hearing it, and Andrzej Rosłanowski for corrections and introducing the improvements from the lecture to the paper.

1 The basic forcing

In Definition 3 below we define the forcing notion $\mathbb{Q}^1_{I,h}$ which will be the one used in the proof of the main result 24. The other forcing notion defined below, $\mathbb{Q}^2_{I,h}$, is a relative of $\mathbb{Q}^1_{I,h}$. Various properties are much easier to check for $\mathbb{Q}^2_{I,h}$, but unfortunately it does not do the job. The reader interested in the main result of the paper only may concentrate on $\mathbb{Q}^1_{I,h}$.

Definition 2. Let I be an ideal on ω containing the family $[\omega]^{<\omega}$ of finite subsets of ω.

1. We say that an equivalence relation E is an I*-equivalence relation* if:
   (a) $\mathrm{dom}(E) \subseteq \omega$,
   (b) $\omega \setminus \mathrm{dom}(E) \in I$,
   (c) each E-equivalence class is in I.
2. For I-equivalence relations E_1, E_2 we write $E_1 \leq E_2$ *if*
   (i) $\mathrm{dom}(E_2) \subseteq \mathrm{dom}(E_1)$,
   (ii) $E_1 \restriction \mathrm{dom}(E_2)$ refines E_2,
   (iii) $\mathrm{dom}(E_2)$ is the union of a family of E_1-equivalence classes.

Definition 3. Let I be an ideal on ω to which all finite subsets of ω belong and let $h : \omega \longrightarrow \omega$ be a non-decreasing function. Let $\ell \in \{1, 2\}$. We define a forcing notion $\mathbb{Q}^\ell_{I,h}$ (if $h(n) = n$ we may omit it) intended to add $\langle y^n_i : i < h(n), n < \omega \rangle$, $y^n_i \in \{-1, 1\}$. We use x^n_i as variables.

1. $p \in \mathbb{Q}^\ell_{I,h}$ if and only if $p = (H, E, A) = (H^p, E^p, A^p)$ and
   (a) E is an I–equivalence relation on $\mathrm{dom}(E) \subseteq \omega$,
   (b) $A = \{n \in \mathrm{dom}(E) : n = \min(n/E)\}$,
   (c) if $\ell = 1$, *then* H is a function with range $\subseteq \{-1, 1\}$ and domain

$$B^p_1 = \{x^n_i : i < h(n) \text{ and } n \in \omega \setminus \mathrm{dom}(E) \textit{ or } \\ n \in \mathrm{dom}(E) \text{ and } i \in [h(\min(n/E)), h(n))\},$$

(d) if $\ell = 2$, *then*

(α) H is a function on $\mathrm{dom}(H) = B_2^p \cup B_3^p$, where

$$B_2^p = \{x_i^m : m \in \omega,\ A^p \cap (m+1) = \emptyset,\ i < h(m)\} \qquad \text{and}$$
$$B_3^p = \{x_i^m : m \in \mathrm{dom}(E^p) \setminus A^p \text{ or } m \notin \mathrm{dom}(E^p) \text{ but } A^p \cap m \neq \emptyset,\ i < h(m)\},$$

(β) for $x_i^m \in B_3^p$, $H(x_i^m)$ is a function of the variables $\{x_j^n : (n,j) \in w_p(m,i)\}$ to $\{-1,1\}$, where

$$w_p(m) = w_p(m,i) = \{(\ell, j) : \ell \in A^p \cap m \text{ and } j < h(\ell)\},$$

for $n \in A^p$ we stipulate $H^p(x_i^n) = x_i^n$ and

(γ) $H \restriction B_2^p$ is a function to $\{-1,1\}$.

(e) if $\ell = 2$ and $x_i^n \in B_3^p$, $n^* = \min(n/E^p) < n$ and $y_i^m \in \{-1,1\}$ for $m \in A^p \cap n^*$, $i < h(m)$ and $z_j^n \in \{-1,1\}$ for $j < h(n^*)$ *then* for some $y_j^{n^*} \in \{-1,1\}$ for $j < h(n^*)$ we have

$$j < h(n^*) \quad \Rightarrow \quad z_j^n = (H^p(x_j^n))(\ldots, y_i^m, \ldots)_{(m,i) \in w_p(n,j)}.$$

When it can not cause any confusion, or we mean "for both $\ell = 1$ and $\ell = 2$", we omit the superscript ℓ.

2. Defining functions like $H(x_i^m)$, $x_i^m \in B_3^p$ (when $\ell = 2$), we may allow to use dummy variables. In particular, if $H^p(x_i^m)$ is $-1, 1$ we identify it with constant functions with this value.
3. We say that a function $f : \{x_i^n : i < h(n),\ n < \omega\} \longrightarrow \{-1,1\}$ satisfies a condition $p \in \mathbb{Q}_{I,h}^\ell$ if:

   (a) $f(x_i^n) = H^p(x_i^n)$ *when* $x_i^n \in B_1^p$ and $\ell = 1$, or $x_i^n \in B_2^p$ and $\ell = 2$,

   (b) $f(x_i^n) = H^p(x_i^n)(\ldots, f(x_j^m), \ldots)_{(m,j) \in w_p(n,i)}$ *when* $\ell = 2$ and $x_i^n \in B_3^p$,

   (c) $f(x_i^n) = f(x_i^{\min(n/E^p)})$ *when* $\ell = 1$, $n \in \mathrm{dom}(E^p)$ and $i < h(\min(n/E^p))$.
4. The partial order $\leq = \leq_{\mathbb{Q}_{I,h}^\ell}$ is defined by $p \leq q$ if and only if:

   (α) $E^p \leq E^q$,

   (β) every function $f : \{x_i^n : i < h(n),\ n < \omega)\} \longrightarrow \{-1,1\}$ satisfying q satisfies p.

Proposition 4. *$(\mathbb{Q}_{I,h}^\ell, \leq_{\mathbb{Q}_{I,h}^\ell})$ is a partial order.* ■

Remark. We may reformulate the definition of the partial orders $\leq_{\mathbb{Q}_{I,h}^\ell}$, making them perhaps more direct. Thus, in particular, if $p, q \in \mathbb{Q}_{I,h}^1$ then $p \leq_{\mathbb{Q}_{I,h}^1} q$ if and only if the demand (α) of 3(4) holds and

(β) for each $x_i^n \in B_1^q$:*

(i) *if $x_i^n \in B_1^p$ then $H^q(x_i^n) = H^p(x_i^n)$,*

(ii) *if $n \in \mathrm{dom}(E^p) \setminus \mathrm{dom}(E^q)$, $i < h(\min(n/E^p))$ then $H^q(x_i^n) = H^q(x_i^{\min(n/E^p)})$,*

(iii) *if* $n \in \mathrm{dom}(E^q) \setminus A^p$, $\min(n/E^p) > \min(n/E^q)$ *and* $h(\min(n/E^q)) \leq i < h(\min(n/E^p))$ *then* $H^q(x_i^n) = H^q(x_i^{\min(n/E^p)})$.

The corresponding reformulation for the forcing notion $\mathbb{Q}^2_{I,h}$ *is more complicated, but it should be clear too.*

One may wonder why we have h in the definition of $\mathbb{Q}^\ell_{I,h}$ and we do not fix that e.g. $h(n) = n$. This is to be able to describe nicely what is the forcing notion $\mathbb{Q}^\ell_{I,h}$ below a condition p like. The point is that $\mathbb{Q}^\ell_{I,h}\restriction\{q : q \geq p\}$ is like $\mathbb{Q}^\ell_{I,h}$ but we replace I by its quotient and we change the function h. More precisely:

Proposition 5. *If* $p \in \mathbb{Q}^\ell_{I,h}$ *and* $A^p = \{n_k : k < \omega\}$, $n_k < n_{k+1}$, $h^* : \omega \longrightarrow \omega$ *is* $h^*(k) = h(n_k)$ *and* $I^* = \{B \subseteq \omega : \bigcup_{k\in B} (n_k/E) \in I\}$ then $\mathbb{Q}^\ell_{I,h} \restriction \{q : p \leq_{\mathbb{Q}^\ell_{I,h}} q\}$ *is isomorphic to* $\mathbb{Q}^\ell_{I^*,h^*}$.

PROOF Natural. ■

Definition 6. We define a $\mathbb{Q}_{I,h}$–name $\underset{\sim}{\bar\eta} = \langle \underset{\sim}{\eta}_n : n < \omega\rangle$ by:
$\underset{\sim}{\eta}_n$ is a sequence of length $h(n)$ of members of $\{-1, 1\}$ such that

$$\underset{\sim}{\eta}_n[G_{\mathbb{Q}_{I,h}}](i) = 1 \quad \Leftrightarrow \quad (\exists p \in G_{\mathbb{Q}_{I,h}})(H^p(x_i^n) = 1).$$

[Note that in both cases $\ell = 1$ and $\ell = 2$, if $H^p(x_i^n) = 1$, $x_i^n \in \mathrm{dom}(H^p)$ and $q \geq p$ then $H^q(x_i^n) = 1$; remember 3(2).]

Proposition 7.

1. *If* $n < \omega$, $A^p \cap (n+1) = \emptyset$ then $p \Vdash$ "$\underset{\sim}{\eta}_n = \langle H^p(x_i^n) : i < h(n)\rangle$".
2. *For each* $n < \omega$ *the set* $\{p \in \mathbb{Q}_{I,h} : A^p \cap (n+1) = \emptyset\}$ *is dense in* $\mathbb{Q}_{I,h}$.
3. *If* $p \in \mathbb{Q}_{I,h}$ *and* $a \subseteq A^p$ *is finite or at least* $\bigcup_{n\in a} (n/E^p) \in I$, *and*

$$f : \{x_i^n : i < h(n) \text{ and } n \in a\} \longrightarrow \{-1, 1\},$$

then *for some unique* q *which we denote by* $p^{[f]}$, *we have:*
(a) $p \leq q \in \mathbb{Q}_{I,h}$,
(b) $E^q = E^p \restriction \bigcup\{n/E^p : n \in A \setminus a\}$,
(c) *for* $n \in a$, $i < h(n)$ *we have* $H^q(x_i^n)$ *is* $f(x_i^n)$.

PROOF Straight. ■

Definition 8.

1. $p \leq_n q$ (in $\mathbb{Q}_{I,h}$) *if* $p \leq q$ and:

$$k \in A^p \ \&\ |A^p \cap k| < n \quad \Rightarrow \quad k \in A^q.$$

2. $p \leq_n^* q$ *if* $p \leq q$ and:

$$k \in A^p \ \&\ |A^p \cap k| < n \quad \Rightarrow \quad k \in A^q \ \&\ k/E^p = k/E^q.$$

3. $p \leq_n^{\otimes} q$ *if* $p \leq_{n+1} q$ and:

$$n > 0 \quad \Rightarrow \quad p \leq_n^* q \qquad \text{and} \qquad \mathrm{dom}(E^q) = \mathrm{dom}(E^p).$$

4. For a finite set $\mathbf{u} \subseteq \omega$ we let $\mathrm{var}(\mathbf{u}) \stackrel{\mathrm{def}}{=} \{x_i^n : i < h(n),\ n \in \mathbf{u}\}$.

Proposition 9.

1. *If* $p \leq q$, $\mathbf{u}$ *is a finite initial segment of* A^p *and* $A^q \cap \mathbf{u} = \emptyset$, then *for some unique* $f : \{x_i^n : i < h(n)$ *and* $n \in \mathbf{u}\} \longrightarrow \{-1, 1\}$ *we have* $p \leq p^{[f]} \leq q$ *(where* $p^{[f]}$ *is from 7(3)).*
2. *If* $p \in \mathbb{Q}_{I,h}^{\ell}$ *and* $\mathbf{u}$ *is a finite initial segment of* A^p then
   $(*)_1$ $f \in {}^{\mathrm{var}(\mathbf{u})}\{-1, 1\}$ *implies* $p \leq p^{[f]}$ *and*
   $$p^{[f]} \Vdash \text{“} (\forall n \in \mathbf{u})(\forall i < h(n))(\underset{\sim}{\eta}_n(i) = f(x_i^n)) \text{”},$$
   $(*)_2$ *the set* $\{p^{[f]} : f \in {}^{\mathrm{var}(\mathbf{u})}\{-1, 1\}\}$ *is predense above* p *(in* $\mathbb{Q}_{I,h}^{\ell}$*).*
3. $\leq_n$ *is a partial order on* $\mathbb{Q}_{I,h}^{\ell}$, *and* $p \leq_{n+1} q \ \Rightarrow\ p \leq_n q$. *Similarly for* $<_n^*$ *and* $<_n^{\otimes}$. *Also*
   $$p \leq_n^{\otimes} q \quad \Rightarrow \quad p \leq_n^* q \quad \Rightarrow \quad p \leq_n q \quad \Rightarrow \quad p \leq q.$$
4. *If* $p \in \mathbb{Q}_{I,h}^{\ell}$, $\mathbf{u}$ *is a finite initial segment of* A^p, $|\mathbf{u}| = n$ *and*
   $$f : \{x_i^n : i < h(n) \text{ and } n \in \mathbf{u}\} \longrightarrow \{-1, 1\} \quad \text{and} \quad p^{[f]} \leq q \in \mathbb{Q}_{I,h}^{\ell},$$
   then *for some* $r \in \mathbb{Q}_{I,h}^{\ell}$ *we have* $p \leq_n^* r \leq q$, $r^{[f]} = q$.
5. *If* $p \in \mathbb{Q}_{I,h}^{2}$, $\mathbf{u}$ *is a finite initial segment of* A^p, $|\mathbf{u}| = n + 1$ *and*
   $$f : \{x_i^n : i < h(n) \text{ and } n \in \mathbf{u}\} \longrightarrow \{-1, 1\} \quad \text{and} \quad p^{[f]} \leq q,$$
   then *for some* $r \in \mathbb{Q}_{I,h}^{2}$ *we have* $p <_n^{\otimes} r \leq q$ *and* $r^{[f]} = q$.

PROOF
1) Define $f : \{x_i^n : i < h(n) \text{ and } n \in \mathbf{u}\} \longrightarrow \{-1, 1\}$ by:

$$f(x_i^n) \text{ is the constant value of } H^q(x_i^n)$$

(it is a constant function by 3(1)(e), 3(1)(f(γ))).
2) By 7 and 9(1).
3) Check.
4) First let us define the required condition r in the case $\ell = 1$. So we let

$$\mathrm{dom}(E^r) = \bigcup_{n \in \mathbf{u}} (n/E^p) \cup \mathrm{dom}(E^q),$$
$$E^r = \big\{(n_1, n_2) : n_1\, E^q\, n_2 \text{ or for some } n \in \mathbf{u} \text{ we have: } \{n_1, n_2\} \subseteq (n/E^p)\big\},$$
$$A^r = \mathbf{u} \cup A^q$$

(note that if $n_1 \, E^q \, n_2$ then $n_1 \notin \mathbf{u}$). Next, for $x_i^n \in B_1^r$ (where B_1^r is given by 3(1)(e)) we define

$$H^r(x_i^n) = \begin{cases} H^q(x_i^n) & \text{if } n \notin \bigcup_{k\in\mathbf{u}} k/E^p \text{ and } x_i^n \in \mathrm{dom}(H^q), \\ H^p(x_i^n) & \text{if } n \in \bigcup_{k\in\mathbf{u}} k/E^p \text{ and } x_i^n \in \mathrm{dom}(H^p). \end{cases}$$

It should be clear that $r = (H^r, E^r, A^r) \in \mathbb{Q}^1_{I,h}$ is as required.

If $\ell = 2$ then we define r in a similar manner, but we have to be more careful defining the function H^r. Thus E^r and A^r are defined as above, B_2^r, B_3^r and $w_r(m,i)$ for $x_i^m \in B_3^r$ are given by 3(1)(f). Note that $B_2^r = B_2^p$ and $B_3^r \subseteq B_3^p$. Next we define:

if $x_i^m \in B_2^r$ then $H^r(x_i^m) = H^p(x_i^m)$,
if $x_i^m \in B_3^r$, $m \cap A^r \subseteq \mathbf{u}$ then $H^r(x_i^m) = H^p(x_i^m)$,
if $x_i^m \in B_3^r$ and $\min(\mathrm{dom}(E^q)) < m$ then

$$H^r(x_i^m)(\ldots, x_j^k, \ldots)_{(k,j)\in w_r(m,i)} =$$
$$H^p(x_i^m)(x_j^k, H^q(x_{j'}^{k'})(\ldots, x_{j''}^{k''}, \ldots)_{(k'',j'')\in w_q(k',j')})_{\substack{(k,j)\in w_r(m,i)\\ (k',j')\in w_p(m,i)\setminus w_r(m,i)}}.$$

Note that if $(k',j') \in w_p(m,i) \setminus w_r(m,i)$, $x_i^m \in B_3^r$ then $k' \in A^p \setminus (\mathbf{u} \cup A^q)$ and $w_q(k',j') \subseteq w_r(m,i)$.

5) Like the proof of (4). Let $n^* = \max(\mathbf{u})$. Put $\mathrm{dom}(E^r) = \mathrm{dom}(E^p)$ and declare that $n_1 \, E^r \, n_2$ if one of the following occurs:

(a) for some $n \in \mathbf{u} \setminus \{n^*\}$ we have $\{n_1, n_2\} \subseteq (n/E^p)$, *or*
(b) $n_1 \, E^q \, n_2$ (so $n \in \mathbf{u} \Rightarrow \neg n \, E^p \, n_1$), *or*
(c) $\{n_1, n_2\} \subseteq B$, where

$$B \stackrel{\mathrm{def}}{=} n^*/E^p \cup \bigcup\{m/E^p : m \in \mathrm{dom}(E^p) \setminus \mathrm{dom}(E^q),\ \min(m/E^p) > n^*\}.$$

We let $A^r = \mathbf{u} \cup A^q$ (in fact A^r is defined from E^r). Finally the function H^r is defined exactly in the same manner as in (4) above (for $\ell = 2$). ■

Corollary 10. *If $p \in \mathbb{Q}^\ell_{I,h}$, $n < \omega$ and $\underset{\sim}{\tau}$ is a $\mathbb{Q}^\ell_{I,h}$-name of an ordinal,* then *there are $\mathbf{u}, q$ and $\bar\alpha = \langle \alpha_f : f \in {}^{\mathrm{var}(\mathbf{u})}\{-1,1\}\rangle$ such that:*

(a) $p \le_n^* q \in \mathbb{Q}^\ell_{I,h}$,
(b) $\mathbf{u} = \{\ell \in A^p : |\ell \cap A^p| < n\}$,
(c) *for $f \in {}^{\mathrm{var}(\mathbf{u})}\{-1,1\}$ we have $q^{[f]} \Vdash$ "$\underset{\sim}{\tau} = \alpha_f$",*
(d) $q \Vdash$ "$\underset{\sim}{\tau} \in \{\alpha_f : f \in {}^{\mathrm{var}(\mathbf{u})}\{-1,1\}\}$" *(which is a finite set).*

PROOF Let $k = \prod_{\ell\in\mathbf{u}} 2^{h(\ell)}$. Let $\{f_\ell : \ell < k\}$ enumerate ${}^{\mathrm{var}(\mathbf{u})}\{-1,1\}$. By induction on $\ell \le k$ define r_ℓ, α_{f_ℓ} such that:

$$r_0 = p, \quad r_\ell \le_n^* r_{\ell+1} \in \mathbb{Q}^\ell_{I,h}, \quad r_{\ell+1}^{[f_\ell]} \Vdash_{\mathbb{Q}^\ell_{I,h}} \text{"}\underset{\sim}{\tau} = \alpha_{f_\ell}\text{"}.$$

The induction step is by 9(4). Now $q = r_k$ and $\langle \alpha_f : f \in {}^{\mathrm{var}(\mathbf{u})}\{-1,1\}\rangle$ are as required. ■

Corollary 11. *If $\ell = 2$ then in 10(a) we may require $p \leq^{\otimes}_{n} q \in \mathbb{Q}^{\ell}_{I,h}$.*

PROOF Similar: just use 9(5) instead of 9(4). ■

Definition 12. Let I be an ideal on ω containing $[\omega]^{<\omega}$ and let E be an I–equivalence relation.

1. We define a game $GM_I(E)$ between two players. The game lasts ω moves. In the nth move the first player chooses an I-equivalence relation E^1_n such that
$$E^1_0 = E, \qquad [n > 0 \quad \Rightarrow \quad E^2_{n-1} \leq E^1_n],$$
and the second player chooses an I-equivalence relation E^2_n such that $E^1_n \leq E^2_n$. In the end, the second player wins if
$$\bigcup \{\operatorname{dom}(E^2_n) \setminus \operatorname{dom}(E^1_{n+1}) : n \in \omega\} \in I$$
(otherwise the first player wins).
2. For a countable elementary submodel N of $(\mathcal{H}(\chi), \in, <^*)$ such that $I, E \in N$ we define a game $GM^N_I(E)$ in a similar manner as $GM_I(E)$, but we demand additionally that the relations played by both players are from N (i.e. $E^1_n, E^2_n \in N$ for $n \in \omega$).

Proposition 13.

1. *Assume that I is a maximal (non-principal) ideal on ω and E is an I–equivalence relation. Then the game $GM_I(E)$ is not determined. Moreover, for each countable $N \prec (\mathcal{H}(\chi), \in, <^*)$ such that $I, E \in N$ the game $GM^N_I(E)$ is not determined.*
2. *For the conclusion of (1) it is enough to assume that $\mathcal{P}(\omega)/I \models$ ccc.*

PROOF 1) As each player can imitate the other's strategy.
2) Easy, too, and will not be used in this paper. ■

Proposition 14.

1. *Let $p \in \mathbb{Q}^{\ell}_{I,h}$. Suppose that the first player has no winning strategy in $GM_I(E^p)$.* Then *in the following game Player I has no winning strategy: in the nth move,*
   Player I chooses a $\mathbb{Q}^{\ell}_{I,h}$-name $\underset{\sim}{\tau}_n$ of an ordinal and
   Player II chooses $p_n, \mathbf{u}_n, w_n$ such that: w_n is a set of $\leq \prod_{\ell \in \mathbf{u}_n} 2^{h(\ell)}$ ordinals, $p \leq p_n \leq^*_n p_{n+1}$, $p_n \leq_{n+1} p_{n+1}$, $\mathbf{u}_n$ a finite initial segment of A^{p_n} with n elements and $p_n \Vdash$ "$\underset{\sim}{\tau}_n \in w_n$", moreover
$$f \in {}^{\operatorname{var}(\mathbf{u}_n)}\{-1, 1\} \quad \Rightarrow \quad p^{[f]}_n \text{ forces a value to } \underset{\sim}{\tau}_n.$$

In the end, the second player wins if for some $q \geq p$ we have

$$q \Vdash \text{“}(\forall n \in \omega)(\underset{\sim}{\tau}_n \in w_n)\text{”}.$$

We can let Player II choose $k_n < \omega$ and demand $|\mathbf{u}_n| \leq k_n$, and in the end Player II wins if $\liminf\langle k_n : n < \omega\rangle < \omega$ or there is q as above.

2. *Let $p \in \mathbb{Q}^{\ell}_{I,h}$ and let N be a countable elementary submodel of $(\mathcal{H}(\chi), \in, <^*)$ such that $p, I, h \in N$. If the first player has no winning strategy in $GM^N_I(E^p)$ then Player I has no winning strategy in the game like above but with restriction that $\underset{\sim}{\tau}_n, p_n \in N$.*

PROOF 1) As in [Sh 407, 1.11, p.436].
Let $\mathbf{St}_p$ be a strategy for Player I in the game from 14. We shall define a strategy $\mathbf{St}$ for the first player in $GM_I(E^p)$ during which the first player, on a side, plays a play of the game from 14, using $\mathbf{St}_p$, with $\langle p_\ell : \ell < \omega\rangle$ and he also chooses $\langle q_\ell : \ell < \omega\rangle$.

Then, as $\mathbf{St}$ cannot be a winning strategy in $GM_I(E)$, in some play in which the first player uses his strategy $\mathbf{St}$ he loses, and then $\langle p_\ell : \ell < \omega\rangle$ will have an upper bound as required.

In the nth move (so $E^1_\ell, E^2_\ell, q_\ell, p_\ell, \mathbf{u}_\ell, w_\ell$ for $\ell < n$ are defined), the first player in addition to choosing E^1_n chooses $q_n, p_n, \mathbf{u}_n$, such that:

(a) $p = p_{-1} \leq q_0 = p_0$, $p_n \in \mathbb{Q}^{\ell}_{I,h}$, $q_n \in \mathbb{Q}^{\ell}_{I,h}$,
(b) $p_n \leq^*_n p_{n+1} \in \mathbb{Q}^{\ell}_{I,h}$,
(c) $\mathbf{u}_0$ is $\emptyset$,
(d) $\mathbf{u}_{n+1} = \mathbf{u}_n \cup \{\min(A^{q_{n+1}} \setminus \mathbf{u}_n)\}$, so $|\mathbf{u}_{n+1}| = n+1$,
(e) $E^1_0 = E^p$, $E^1_{n+1} = E^{p_n} \restriction (\mathrm{dom}(E^{p_n}) \setminus \bigcup_{i \in \mathbf{u}_n} i/E^{p_n})$,
(f) q_n is defined as follows:
 (f_0) if $n = 0$ then $E^{q_n} = E^2_0$,
 (f_1) if $n > 0$ then $\mathrm{dom}(E^{q_n}) = \mathrm{dom}(E^{p_{n-1}})$ and $x\, E^{q_n}\, y$ if and only if
  either $x\, E^2_n\, y$,
  or for some $k \in \mathbf{u}_{n-1}$ we have $x, y \in k/E^{p_{n-1}}$,
  or $x, y \in (\mathrm{dom}(E^1_n) \setminus \mathrm{dom}(E^2_n)) \cup \min(\mathrm{dom}(E^2_n))/E^2_n$,
 (f_2) H^{q_n} is such that $p_{n-1} \leq q_n$,
(g) $p_n \leq^*_n q_{n+1} \leq^*_{n+1} p_{n+1}$, $p_n \leq_{n+1} q_{n+1}$ (so $p_n \leq_{n+1} p_{n+1}$),
(h) if $f \in {}^{\mathrm{var}(\mathbf{u}_n)}\{-1,1\}$ then $p^{[f]}_n$ forces a value to $\underset{\sim}{\tau}_n$.

In the first move, when $n = 0$, the first player plays $E^1_0 = E^p$ (as the rules of the game require, according to (e)). The second player answers choosing an I-equivalence relation $E^2_0 \geq E^1_0$. Now, on a side, Player I starts to play the game of 14 using his strategy $\mathbf{St}_p$. The strategy says him to play a name $\underset{\sim}{\tau}_0$ of an ordinal. He defines q_0 by (f) (so $q \in \mathbb{Q}^{\ell}_{I,h}$ is a condition stronger than p and such that $E^{q_0} = E^2_0$) and chooses a condition $p_0 \geq q_0$ deciding the value of the name $\underset{\sim}{\tau}_0$, say $p_0 \Vdash \underset{\sim}{\tau}_0 = \alpha$. He pretends that the second player answered (in the game of 14) by: p_0, $\mathbf{u}_0 = \emptyset$, $w_0 = \{\alpha\}$. Next, in the play of $GM_I(E^p)$, he plays $E^1_1 = E^{p_0}$ as declared in (e).

Now suppose that we are at the $(n+1)^{\text{th}}$ stage of the play of $GM_I(E^p)$, the first player has played E^1_{n+1} already and on a side he has played the play of the game 14 as defined by (a)–(h) and $\mathbf{St}_p$ (so in particular he has defined a condition p_n and $E^1_{n+1} = E^{p_n}\restriction(\mathrm{dom}(E^{p_n}) \setminus \bigcup_{i\in\mathbf{u}_n} i/E^{p_n})$ and $\mathbf{u}_n$ is the set of the first n elements of A^{p_n}). The second player plays an I–equivalence relation $E^2_{n+1} \geq E^1_{n+1}$. Now the first player chooses (on a side, pretending to play in the game of 14): a name $\underset{\sim}{r}_{n+1}$ given by the strategy $\mathbf{St}_p$, a condition $q_{n+1} \in \mathbb{Q}^\ell_{I,h}$ determined by (f) (check that (g) is satisfied), $\mathbf{u}_{n+1}$ as in (d) and a condition $p_{n+1} \in \mathbb{Q}^\ell_{I,h}$ satisfying (g), (h) (the last exists by 10). Note that, by (g) and 9, the condition p_{n+1} determines a suitable set w_{n+1}. Thus, Player I pretends that his opponent in the game of 14 played $p_{n+1}, \mathbf{u}_{n+1}, w_{n+1}$ and he passes to the actual game $GM_I(E^p)$. Here he plays E^1_{n+2} defined by (e).

The strategy $\mathbf{St}$ described above cannot be the winning one. Consequently, there is a play in $GM_I(E^p)$ in which Player I uses $\mathbf{St}$, but he looses. During the play he constructed a sequence $\langle(p_n, \mathbf{u}_n, w_n) : n \in \omega\rangle$ of legal moves of Player II in the game of 14 against the strategy $\mathbf{St}_p$. Let $E^q = \lim_{n<\omega} E^{p_n}$ (i.e. $\mathrm{dom}(E^q) = \bigcap_{n<\omega} \mathrm{dom}(E^{p_n})$, $x\, E^q\, y$ if and only if for every large enough n, $x\, E^{p_n}\, y$) and let $H^q(x^m_i)$ will be $H^{p_n}(x^m_i)$ for any large enough n (it is eventually constant). It follows from the demand (g) that E^q-equivalence classes are in I. Moreover, $\mathrm{dom}(E^1_{n+1}) \setminus \mathrm{dom}(E^2_{n+1}) \subseteq k/E^q$, where k is the $(n+1)^{\text{th}}$ member of A^q. Therefore

$$\omega \setminus \mathrm{dom}(E^q) = \omega \setminus \bigcap_{n\in\omega} \mathrm{dom}(E^{p_n}) \subseteq$$
$$\omega \setminus \mathrm{dom}(E^{p_0}) \cup \bigcup\{\mathrm{dom}(E^2_n) \setminus \mathrm{dom}(E^1_{n+1}) : n \in \omega\} \in I$$

(remember, Player I lost in $GM_I(E^p)$). Now it should be clear that $q \in \mathbb{Q}^\ell_{I,h}$ and it is stronger than every p_n (even $p_n \leq^*_n q$). Hence Player II wins the corresponding play of 14, showing that $\mathbf{St}_p$ is not a winning strategy.

2) The same proof. ■

Proposition 15. *If in 14 we assume $\ell = 2$ and demand $p_n \leq^\otimes_n p_{n+1}$ instead $p_n \leq^*_n p_{n+1}$* then *Player II has a winning strategy.*

PROOF Using 11, the second player can find suitable conditions p_n (in the game of 14) such that $p_n \leq^\otimes_{n+1} p_{n+1}$. But note that the partial orders $\leq^\otimes_n$ have the fusion property, so the sequence $\langle p_n : n < \omega\rangle$ will have an upper bound in $\mathbb{Q}^2_{I,h}$. ■

Remark. We could have used $<^\otimes_n$ also in [Sh 407].

Definition 16 see [Sh:f, VI, 2.12, A-F].

1. A forcing notion $\mathbb{P}$ has the PP-property if:
   ($\otimes^{PP}$) *for every* $\eta \in {}^{\omega}\omega$ from $\mathbf{V}^{\mathbb{P}}$ and a strictly increasing $x \in {}^{\omega}\omega \cap \mathbf{V}$ *there is* a closed subtree $T \subseteq {}^{<\omega}\omega$ such that:
   (α) $\eta \in \lim(T)$, i.e. $(\forall n < \omega)(\eta \restriction n \in T)$,
   (β) $T \cap {}^{n}\omega$ is finite for each $n < \omega$,
   (γ) for arbitrarily large n there are k, and $n < i(0) < j(0) < i(1) < j(1) < \ldots < i(k) < j(k) < \omega$ and for each $\ell \leq k$, there are $m(\ell) < \omega$ and $\eta^{\ell,0}, \ldots, \eta^{\ell,m(\ell)} \in T \cap {}^{j(\ell)}\omega$ such that $j(\ell) > x(i(\ell) + m(\ell))$ and
   $$(\forall \nu \in T \cap {}^{j(k)}\omega)(\exists \ell \leq k)(\exists m \leq m(\ell))(\eta^{\ell,m} \trianglelefteq \nu).$$
2. We say that a forcing notion $\mathbb{P}$ has the strong PP-property if
   ($\oplus^{sPP}$) for every function $g : \omega \longrightarrow \mathbf{V}$ from $\mathbf{V}^{\mathbb{P}}$ there exist a set $B \in [\omega]^{\aleph_0} \cap \mathbf{V}$ and a sequence $\langle w_n : n \in B \rangle \in \mathbf{V}$ such that for each $n \in B$
   $$|w_n| \leq n \qquad \text{and} \qquad g(n) \in w_n.$$

Remark. Of course, if a proper forcing notion has the strong PP-property then it has the PP-property.

Conclusion 17 *Assume that for each* $p \in \mathbb{Q}^{\ell}_{I,h}$ *and for each countable* $N \prec (\mathcal{H}(\chi), \in, <^*)$ *such that* $p, I, h \in N$*, the first player has no winning strategy in* $GM^N_I(E^p)$ *(e.g. if* I *is a maximal ideal). Then*

(*) $\mathbb{Q}^{\ell}_{I,h}$ *is proper,* α*-proper, strongly* α*-proper for every* $\alpha < \omega_1$*, is* ${}^{\omega}\omega$*-bounding and it has the PP-property, even the strong PP-property.* ■

By [Sh:f, VI, 2.12] we know

Theorem 18. *Suppose that* $\langle \mathbb{P}_i, \underset{\sim}{\mathbb{Q}}_j : j < \alpha, i \leq \alpha \rangle$ *is a countable support iteration such that*

$$\Vdash_{\mathbb{P}_j} \text{“} \underset{\sim}{\mathbb{Q}}_j \textit{ is proper and has the PP-property”}.$$

Then $\mathbb{P}_\alpha$ *has the PP-property.* ■

2 NWD ultrafilters

A subset A of the set $\mathcal{Q}$ of rationals is *nowhere dense* (NWD) if its closure (in $\mathcal{Q}$) has empty interior. Remember that the rationals are equipped with the order topology and both "closure" and "interior" refer to this topology. Of course, as $\mathcal{Q}$ is dense in the real line, we may consider these operations on the real line and get the same notion of nowhere dense sets. For technical reasons, in forcing considerations we prefer to work with ${}^{\omega}2$ instead of the real line. So naturally we want to replace rationals by ${}^{<\omega}2$. But what are nowhere dense subsets of

${}^{<\omega}2$ then? (One may worry about the way we "embed" ${}^{<\omega}2$ into ${}^{\omega}2$.) Note that we have a natural lexicographical ordering $<_{\ell x}$ of ${}^{<\omega}2$:

$\eta <_{\ell x} \nu$ if and only if
either there is $\ell < \omega$ such that $\eta\restriction\ell = \nu\restriction\ell$ and $\eta(\ell) < \nu(\ell)$
or $\eta^\frown\langle 1\rangle \trianglelefteq \nu$
or $\nu^\frown\langle 0\rangle \trianglelefteq \eta$.

Clearly $({}^{<\omega}2, <_{\ell x})$ is a linear dense order without end-points (and consequently it is order–isomorphic to the rationals). Now, we may talk about nowhere dense subsets of ${}^{<\omega}2$ looking at this ordering only, but we may relate this notion to the topology of ${}^{\omega}2$ as well.

Proposition 19. *For a set $A \subseteq {}^{<\omega}2$ the following conditions are equivalent:*

1. *A is nowhere dense,*
2. $(\forall\eta \in {}^{<\omega}2)(\exists\nu \in {}^{<\omega}2)[\eta \trianglelefteq \nu \ \&\ (\forall\rho \in {}^{<\omega}2)(\nu \trianglelefteq \rho \Rightarrow \rho \notin A)]$,
3. *the set*
$$A^* \stackrel{\mathrm{def}}{=} \{\eta \in {}^{\omega}2 : (\forall n \in \omega)(\exists\nu \in A)(\eta\restriction n \trianglelefteq \nu)\}$$
is nowhere dense (in the product topology of ${}^{\omega}2$),
4. *there is a sequence $\langle\eta_n : n < \omega\rangle$ such that for each $n < \omega$*
(i)$_n$ $\eta_n : [n, \ell_n) \longrightarrow 2$ *for some $\ell_n > n$ and*
(ii)$_n$ $(\forall\rho \in A)(\eta_n \not\subseteq \rho)$,
5. *there is a sequence $\langle\eta_n : n < \omega\rangle$ such that for each $n < \omega$ condition* **(i)**$_n$ *(see above) holds and*
(ii)$^*_n$ $(\forall\nu \in {}^{n}2)(\{\rho \in {}^{<\omega}2 : \nu \cup \eta_n \trianglelefteq \rho\} \cap A = \emptyset)$,
6. *there are $B \in [\omega]^{\aleph_0}$ and $\langle\eta_n : n \in B\rangle$ such that for each $n \in B$ the conditions* **(i)**$_n$, **(ii)**$_n$ *above are satisfied.*

PROOF 1. $\Rightarrow$ 2. Suppose $A \subseteq {}^{<\omega}2$ is nowhere dense but for some sequence $\eta \in {}^{<\omega}2$, for every $\nu \in {}^{<\omega}2$ extending η there is $\rho \in A$ such that $\nu \trianglelefteq \rho$. Look at the interval $(\eta^\frown\langle 0\rangle, \eta^\frown\langle 1\rangle)_{<_{\ell x}}$ (of $({}^{<\omega}2, <_{\ell x})$). We claim that A is dense in this interval. Why? Suppose

$$\eta^\frown\langle 0\rangle \leq_{\ell x} \eta^*_0 <_{\ell x} \eta^*_1 \leq_{\ell x} \eta^\frown\langle 1\rangle.$$

Assume $\ell g(\eta^*_0) \leq \ell g(\eta^*_1)$. Take $\nu \stackrel{\mathrm{def}}{=} \eta^{*\frown}_1\langle 0\rangle$. By the definition of the order $<_{\ell x}$ we have then

$$\eta^*_0 <_{\ell x} \nu^\frown\langle 0\rangle <_{\ell x} \nu^\frown\langle 1\rangle <_{\ell x} \eta^*_1 \qquad \text{and} \qquad \eta \triangleleft \nu.$$

By our assumption we find $\rho \in A$ such that $\nu^\frown\langle 0, 1\rangle \trianglelefteq \rho$. Then

$$\nu^\frown\langle 0\rangle <_{\ell x} \rho <_{\ell x} \nu^\frown\langle 1\rangle \qquad \text{and hence} \qquad \rho \in (\eta^*_0, \eta^*_1)_{<_{\ell x}}.$$

Similarly if $\ell g(\eta^*_1) \leq \ell g(\eta^*_0)$.

2. $\Rightarrow$ 3. Should be clear if you remember that sets

$$[\nu] \stackrel{\mathrm{def}}{=} \{\eta \in {}^{\omega}2 : \nu \triangleleft \eta\} \qquad \text{(for } \nu \in {}^{<\omega}2)$$

constitute the basis of the topology of ${}^{\omega}2$.

3. $\Rightarrow$ 4. Suppose A^* is nowhere dense in ${}^{\omega}2$. Let $n < \omega$. Considering all elements of 2^n build (e.g. inductively) a function $\eta_n^* : [n, \ell_n^*) \longrightarrow 2$ such that $n < \ell_n^*$ and

$$(\forall \nu \in 2^n)([\nu^\frown \eta_n^*] \cap A^* = \emptyset).$$

This means that for each $\nu \in 2^n$ the set $\{\rho \in A : \nu^\frown \eta_n^* \trianglelefteq \rho\}$ is finite (otherwise use König lemma to construct an element of A^* in $[\nu^\frown \eta_n^*]$). Taking sufficiently large $\ell_n > \ell_n^*$ and extending η_n^* to η_n with domain $[n, \ell_n)$ we get that $(\forall \rho \in A)(\eta_n \not\subseteq \rho)$ (as required).

4. $\Rightarrow$ 5. $\Rightarrow$ 6. Read the conditions.

6. $\Rightarrow$ 1. Let B, $\langle \eta_n : n \in B\rangle$ be as in 6. Suppose $\nu_0, \nu_1 \in {}^{<\omega}2$, $\nu_0 <_{\ell x} \nu_1$. Assume $\ell g(\nu_0) \leq \ell g(\nu_1) = m$. Take any $n \in B \setminus (m+1)$ and let $\nu = \nu_1{}^\frown \underbrace{\langle 0, \ldots, 0\rangle}_{n-m}{}^\frown \eta_n$. We know that no element of A extends ν. But this implies that the interval $(\nu^\frown\langle 0\rangle, \nu^\frown\langle 1\rangle)_{<_{\ell x}}$ is disjoint from A (and is contained in the interval $(\nu_0, \nu_1)_{<_{\ell x}}$). Similarly if $\ell g(\nu_1) \leq \ell g(\nu_0)$. ■

Lemma 20. *Let $n, k^* < \omega$. Assume that $\bar{\nu}^k = \langle \nu_i^k : n \leq i < i_k\rangle$ for $k < k^* < \omega$, $n \leq i_k < \omega$, $\nu_i^k \in \bigcup_{j \geq i} {}^{[i,j)}2$ and $w_k \subseteq [n, i_k)$, $|w_k| \geq k^*$ and:*

$$\textit{if } k < k^*,\ m_1 < m_2 \textit{ are in } w_k \textit{ then } \max\operatorname{dom}(\nu_{m_1}^k) < m_2.$$

Lastly let

$$i(*) = \max\{\sup\operatorname{dom}(\nu_i^k) + 1 : k < k^* \textit{ and } i \in (n, i_k)\}.$$

Then *we can find $\rho \in {}^{[n,i(*))}2$ such that:*

$$(\forall k < k^*)(\exists i \in w_k)(\nu_i^k \subseteq \rho).$$

Proof By induction on k^* (for all possible other parameters). For $k^* = 0, 1$ it is trivial.
Let $n_k^0 = \min(w_k)$ and $n_k^1 = \min(w_k \setminus (n_k^0 + 1))$. Let $\ell < k^*$ be with minimal n_ℓ^1. Apply the induction hypothesis with n_ℓ^1, $\bar{\nu}^k = \langle \bar{\nu}_i^k : n_\ell^1 \leq i < i_k\rangle$ for $k < k^*, k \neq \ell$ and $\langle w_k \setminus n_\ell^1 : k < k^*, k \neq \ell\rangle$ here standing for n, $\bar{\nu}^k$ for $k < k^*$, $\langle w_k : k < k^*\rangle$ there and get $\rho_1 \in {}^{[n_\ell^1, i(*))}2$. Note that $w_k \setminus n_\ell^1 \supseteq w_k \setminus n_k^1$ has at least $|w_k| - 1$ elements. Let $\rho \in {}^{[n,i(*))}2$ be such that $\rho_1 \subseteq \rho$ and $\bar{\nu}_{n_\ell^0}^\ell \subseteq \rho$. ■

Proposition 21. *Assume that $\mathbb{R}$ is a proper forcing notion with the PP-property. Then*

($\oplus^{\mathrm{nwd}}$) *for every nowhere dense set $A \subseteq {}^{<\omega}2$ in $\mathbf{V}^{\mathbb{R}}$ there is a nowhere dense set $A^* \subseteq {}^{<\omega}2$ in $\mathbf{V}$ such that $A \subseteq A^*$.*

PROOF Let $A \in \mathbf{V}^{\mathbb{R}}$ be a nowhere dense subset of ${}^{<\omega}2$. Thus, in $\mathbf{V}^{\mathbb{R}}$, we can, for each $n < \omega$, choose $\nu_n \in \bigcup_{\ell \geq n} {}^{[n,\ell)}2$ such that:

$$(\forall \nu \in {}^{n}2)(\forall \rho \in {}^{<\omega}2)(\nu^\frown\nu_n \trianglelefteq \rho \ \Rightarrow\ \rho \notin A).$$

So $\langle \nu_n : n < \omega \rangle \in \mathbf{V}^{\mathbb{R}}$ is well defined. Next for each n we choose an integer $\ell_n \in (n, \omega)$, a sequence $\eta_n \in {}^{[n,\ell_n)}2$ and a set $w_n \subseteq [n, \ell_n)$ such that:

- $|w_n| > n$,
- $(\forall m \in w_n)(\nu_m \subseteq \eta_n)$, so in particular $(\forall m \in w_n)(\max \mathrm{dom}(\nu_m) < \ell_n)$, and
- for any $m_1 < m_2$ from w_n we have $\max\mathrm{dom}(\nu_{m_1}) < m_2$.

So $\bar{w} = \langle w_n : n < \omega \rangle$, $\bar{\eta} = \langle \eta_n : n < \omega \rangle \in \mathbf{V}^{\mathbb{R}}$ are well defined.

Since $\mathbb{R}$ has the PP-property it is ${}^{\omega}\omega$-bounding, and hence there is a strictly increasing $x \in {}^{\omega}\omega \cap \mathbf{V}$ such that $(\forall n \in \omega)(\ell_n < x(n))$. Applying the PP-property of $\mathbb{R}$ to x and the function $n \mapsto (\eta_n, w_n)$ we can find $\langle \langle V^n_\ell : \ell \leq k_n \rangle : n < \omega \rangle$ in $\mathbf{V}$ and $\langle \langle (i_\ell(n), j_\ell(n)) : \ell \leq k_n \rangle : n < \omega \rangle$ in $\mathbf{V}$ such that:

(a) $i_0(n) < j_0(n) < i_1(n) < j_1(n) < \ldots < i_{k_n}(n) < j_{k_n}(n)$,
(b) $j_{k_n}(n) < i_0(n+1)$ for $n < \omega$,
(c) $x(i_\ell(n)) < j_\ell(n)$,
(d) $V^n_\ell \subseteq \{(\eta, w) : \eta \in {}^{[i_\ell(n), j_\ell(n))}2$ and $w \subseteq [i_\ell(n), j_\ell(n))$, $|w| > i_\ell(n)\}$ for $\ell \leq k_n$, $n < \omega$,
(e) $|V^n_\ell| \leq i_\ell(n)$,
(f) for every $n < \omega$, for some $\ell \leq k_n$ and $(\eta, w) \in V^n_\ell$ we have $w = w_{i_\ell(n)}$, $\eta_{i_\ell(n)} \subseteq \eta$.

[Note that $i_\ell(n)$ corresponds to $i(\ell) + m(\ell)$ in definition 16(1), so we do not have $m_\ell(n)$ here.] Working in $\mathbf{V}$, by 20, for each $n < \omega$, $\ell \leq k_n$ there is $\rho^n_\ell \in {}^{[i_\ell(n), j_\ell(n))}2$ such that:

$$(\forall (\eta, w) \in V^n_\ell)(\exists m_1, m_2 \in w)(m_2 = \min(w \setminus (m_1 + 1))\ \&\ \eta \restriction [m_1, m_2) \subseteq \rho^n_\ell).$$

Let $\rho_n \in {}^{[i_0(n), i_0(n+1))}2$ be such that $\ell \leq k_n \ \Rightarrow\ \rho^n_\ell \subseteq \rho_n$. As we have worked in $\mathbf{V}$, $\langle \rho_n : n < \omega \rangle \in \mathbf{V}$. Let

$$A^* = \{\rho \in {}^{<\omega}2 : \neg(\exists n \in \omega)(\rho_n \subseteq \rho)\}.$$

Clearly $A^* \in \mathbf{V}$ is as required. ■

Let us recall definition 1 reformulating it slightly for technical purposes. (Of course, the two definitions are equivalent; see the discussion at the beginning of this section.)

Definition 22. We say that a non-principal ultrafilter $\mathcal{D}$ on ω is an NWD-ultrafilter *if* for any sequence $\langle \eta_n : n < \omega \rangle \subseteq {}^{<\omega}2$ for some $A \in \mathcal{D}$ the set $\{\eta_n : n \in A\}$ is nowhere dense in ${}^{<\omega}2$.

Lemma 23. *Let $\mathcal{D}$ be a non-principal ultrafilter on ω and I be the dual ideal (and $h : \omega \longrightarrow \omega$ non-decreasing $\lim_{n\to\infty} h(n) = \infty$). Then:*

1. *in $\mathbf{V}^{\mathbb{Q}^1_{I,h}}$ we cannot extend $\mathcal{D}$ to an NWD-ultrafilter.*
2. *If $\underset{\sim}{\mathbb{Q}}$ is a $\mathbb{Q}^1_{I,h}$-name of a proper forcing notion with the PP-property,* then *also in $\mathbf{V}^{\mathbb{Q}^1_{I,h} * \underset{\sim}{\mathbb{Q}}}$ we cannot extend $\mathcal{D}$ to an NWD-ultrafilter.*

PROOF 1) Let $\bar{\underset{\sim}{\eta}} = \langle \underset{\sim}{\eta}_n : n < \omega \rangle$ be the name defined in 6, but now we interpret the value -1 as 0. So $\Vdash$"$\underset{\sim}{\eta}_n \in {}^{h(n)}2$" (for each $n < \omega$). Clearly it is enough to show that

$$(*) \qquad \Vdash_{\mathbb{Q}^1_{I,h}} \text{" if } X \subseteq \omega \text{ and the set } \{\underset{\sim}{\eta}_n : n \in X\} \text{ is nowhere dense then there is } Y \in \mathcal{D} \text{ disjoint from } X\text{"}.$$

So suppose that $\underset{\sim}{\tau}$ is a $\mathbb{Q}^1_{I,h}$-name for a subset of ω and a condition $p^* \in \mathbb{Q}^1_{I,h}$ forces that $\{\underset{\sim}{\eta}_n : n \in \underset{\sim}{\tau}\}$ is nowhere dense.
By 19, for some $\mathbb{Q}^1_{I,h}$-names $\bar{\underset{\sim}{\nu}} = \langle \underset{\sim}{\nu}_m : m < \omega \rangle$ we have

$$p^* \Vdash \text{"}\underset{\sim}{\nu}_m \in \bigcup_{\ell \geq m} {}^{[m,\ell)}2 \text{ and for every } m < \omega \text{ for no } n \in \underset{\sim}{\tau} \text{ we have } \underset{\sim}{\nu}_m \subseteq \underset{\sim}{\eta}_n\text{"}.$$

By 14 (or actually by its proof) without loss of generality:

> for every $n \in A^{p^*}$, for some $k_n \in (n, \min(A^{p^*} \setminus (n+1)))$, for every $f : \{x^m_j : m \in A^{p^*} \cap (n+1) \text{ and } j < h(m)\} \longrightarrow \{-1, 1\}$, the condition $p^{*[f]}$ forces a value to $\underset{\sim}{\tau} \cap k_n$, and $\underset{\sim}{\tau} \cap k_n \setminus n \neq \emptyset$.

[Why? Give a strategy to Player I in the game there for p^* trying to force the needed information, so for some such play Player II wins and replaces p^* by q from there.]
Again by 14 we may assume that

> for every $f : \{x^m_j : j < h(m) \text{ and } m \in A^{p^*} \cap (n+1)\} \longrightarrow \{-1, 1\}$, $n \in A^{p^*}$, for some $\bar{\nu}^f$ we have
>
> $$p^{*[f]} \Vdash \text{"}\bar{\nu}^f \text{ is an initial segment of } \bar{\underset{\sim}{\nu}} \text{ and } \ell g(\bar{\nu}^f) = n + 1 \text{"}.$$

For $n \in A^{p^*}$ and $f : \{x^m_j : j < h(m) \text{ and } m \in A^{p^*} \cap (n+1)\} \longrightarrow \{-1, 1\}$ and $k \in A^{p^*} \setminus (n+1)$ let:

(a) $f^{[k,p^*]}$ be the function with domain $\{x^m_j : j < h(m) \text{ and } m \in A^{p^*} \cap (k+1)\}$ extending f that is constantly 1 on $\mathrm{dom}(f^{[k,p^*]}) \setminus \mathrm{dom}(f)$,

(b) $\bar{\rho}^f$ be an ω-sequence $\langle \rho^f_\ell : \ell < \omega \rangle$ such that for each $k \in A^{p^*} \setminus (n+1)$ we have $\bar{\rho}^f \restriction (k+1) = \bar{\nu}^{f^{[k,p^*]}} \restriction (k+1)$.

Now, for every $n \in A^{p^*}$, we can find $\rho^*_n \in {}^{<\omega}2$ such that for every function

$$f : \{x^m_j : j < h(m) \text{ and } m \in A^{p^*} \cap (n+1)\} \longrightarrow \{-1, 1\}$$

for some $\ell(f) \in (h(n), \omega)$ we have $\rho^f_{\ell(f)} \subseteq \rho^*_n$ (so $\ell(f) < \ell g(\rho^*_n)$).
[Why? Let $\{f_j : j < j(*)\}$ list the possible f's, and we chose by induction on $j \le j(*)$, $\rho^j \in {}^{<\omega}2$ such that $\rho^j \lhd \rho^{j+1}$, and ρ^{j+1} satisfies the requirement on f_j, e.g. $\rho_0 = \langle\underbrace{0, \ldots, 0}_{h(n)}\rangle$, $\rho^{j+1} = \rho^j \frown \rho^{f_j}_{\ell g(\rho^j)}$].

Now choose by induction on $\zeta < \omega$, $n_\zeta \in A^{p^*}$ such that $n_\zeta < n_{\zeta+1}$, and $\ell g(\rho^*_{n_\zeta}) < h(n_{\zeta+1})$. Without loss of generality $\bigcup\limits_{\zeta<\omega} (n_\zeta / E^{p^*}) \in I$. Then

$$\begin{array}{ll} \text{either} & \bigcup\{n/E^{p^*} : n \in A^{p^*} \text{ and } (\exists \zeta < \omega)(n_{2\zeta} < n < n_{2\zeta+1})\} \in \mathcal{D} \\ \text{or} & \bigcup\{n/E^{p^*} : n \in A^{p^*} \text{ and } (\exists \zeta < \omega)(n_{2\zeta+1} < n < n_{2\zeta+2})\} \in \mathcal{D}, \end{array}$$

so by renaming the latter holds. (Again, it suffices that the ideal I is such that the quotient algebra $\mathcal{P}(\omega)/I$ satisfies the c.c.c.) Lastly we define a condition $r \in \mathbb{Q}^1_{I,h}$:

$$\mathrm{dom}(E^r) = \bigcup_{\zeta<\omega} n_{2\zeta}/E^{p^*} \cup \bigcup\{n/E^{p^*} : n \in A^{p^*} \text{ and } (\exists \zeta < \omega)(n_{2\zeta+1} < n < n_{2\zeta+2})\},$$

$$n_{2\zeta}/E^r = (n_{2\zeta}/E^{p^*}) \cup \bigcup\{m/E^{p^*} : m \in A^{p^*} \cap (n_{2\zeta+1}, n_{2\zeta+2})\}$$

(note that this defines correctly an I–equivalence relation E^r), $A^r = \{n_{2\zeta} : \zeta < \omega\}$. The function H^r is defined by cases (interpreting the value 0 as -1, where appears):

$$\begin{array}{lll} H^r(x^m_j) = H^{p^*}(x^m_j) & \text{if} & m \in (\omega \setminus \mathrm{dom}(E^{p^*})) \text{ and } j < h(m), \\ H^r(x^m_j) = H^{p^*}(x^m_j) & \text{if} & m \in \mathrm{dom}(E^{p^*}) \text{ and } j \in [h(\min(m/E^{p^*})), h(m)) \\ H^r(x^m_j) = 1 & \text{if} & m \in \mathrm{dom}(E^{p^*}) \text{ and } \min(m/E^{p^*}) \in (n_{2\zeta}, n_{2\zeta+1}] \\ & & \text{and } j < h(\min(m/E^{p^*})) \\ H^r(x^m_j) = \rho^*_{n_{2\zeta}}(j) & \text{if} & m \in \mathrm{dom}(E^{p^*}) \text{ and } \min(m/E^{p^*}) \in (n_{2\zeta+1}, n_{2\zeta+2}) \\ & & \text{and } j \in \mathrm{dom}(\rho^*_{n_{2\zeta}}) \text{ and } j \ge h(n_{2\zeta}) \\ H^r(x^m_j) = 1 & & \text{otherwise (but } x^m_j \in \mathrm{dom}(H^r)). \end{array}$$

Now check that $p^* \le r \in \mathbb{Q}^1_{I,h}$ and for each $n \in \mathrm{dom}(E^r) \setminus \bigcup\limits_{\zeta<\omega} n_{2\zeta}/E^{p^*}$:

$$r \Vdash \text{“ } \underset{\sim}{\eta}_n \text{ violates the property of } \underset{\sim}{\bar{\nu}} \text{ and hence } n \notin \underset{\sim}{\tau}\text{”}.$$

As $\mathrm{dom}(E^r) \setminus \bigcup\limits_{\zeta<\omega} n_{2\zeta}/E^{p^*} \in \mathcal{D}$ we have finished.

2) Should be clear by (*) of the proof of 23(1) and 21.
However we will give an alternative proof of 23(2). We start as in the proof of 23(1): suppose some $(p^*, \underset{\sim}{r}^*) \in \mathbb{Q}^1_{I,h} * \underset{\sim}{\mathbb{Q}}$ forces “$\underset{\sim}{F}$ is an NWD-ultrafilter on ω

extending $\mathcal{D}$". As $\Vdash$"$\underset{\sim}{\eta}_n[G_{\mathbb{Q}^1_{I,h}}] \in {}^{h(n)}2$", for some $(\mathbb{Q}^1_{I,h} * \underset{\sim}{\mathbb{Q}})$-name $\underset{\sim}{\tau}$ for a subset of ω

$$(p^*, \underset{\sim}{r}^*) \Vdash \text{" } \underset{\sim}{\tau} \in \underset{\sim}{F} \text{ and } (\forall \eta \in {}^{<\omega}2)(\exists \nu \in {}^{<\omega}2)(\eta \trianglelefteq \nu \ \& \ (\forall n \in \underset{\sim}{\tau})(\neg \nu \trianglelefteq \underset{\sim}{\eta}_n)) \text{ "}.$$

So for some $\mathbb{Q}^1_{I,h} * \underset{\sim}{\mathbb{Q}}$-name $\bar{\underset{\sim}{\nu}} = \langle \underset{\sim}{\nu}_n : n < \omega \rangle$

$$(p^*, \underset{\sim}{r}^*) \Vdash \text{" } \underset{\sim}{\nu}_\ell \in \bigcup_{j \in [\ell, \omega)} {}^{[\ell, j)}2 \text{ and for no } n \in \underset{\sim}{\tau} \text{ we have } \underset{\sim}{\nu}_\ell \subseteq \underset{\sim}{\eta}_n\text{"}.$$

So for some $\mathbb{Q}^1_{I,h} * \underset{\sim}{\mathbb{Q}}$–names $\underset{\sim}{d}_\ell$, $\underset{\sim}{w}_\ell$

$$(p^*, \underset{\sim}{r}^*) \Vdash \text{" } \omega > \underset{\sim}{d}_\ell > \ell, \ \underset{\sim}{w}_\ell \subseteq [\ell, \underset{\sim}{d}_\ell), \ |\underset{\sim}{w}_\ell| > (4 \cdot \textstyle\prod_{s \leq n} h(s))! \text{ and}$$
$$[m_1 < m_2 \text{ in } \underset{\sim}{w}_\ell \quad \Rightarrow \quad \max \operatorname{dom}(\underset{\sim}{\nu}_{m_1}) < m_2]\text{"}.$$

Let $p^* \in G_{\mathbb{Q}^1_{I,h}} \subseteq \mathbb{Q}^1_{I,h}$ and $G_{\mathbb{Q}^1_{I,h}}$ generic over $\mathbf{V}$. Now in $\mathbf{V}[G_{\mathbb{Q}^1_{I,h}}]$, the forcing notion $\underset{\sim}{\mathbb{Q}}[G_{\mathbb{Q}^1_{I,h}}]$ is ${}^\omega\omega$-bounding (this follows from the PP-property) and also $\mathbb{Q}^1_{I,h}$ is ${}^\omega\omega$-bounding.
Hence for some $r' \in \underset{\sim}{\mathbb{Q}}[G_{\mathbb{Q}^1_{I,h}}]$ and strictly increasing $x \in {}^\omega\omega \cap \mathbf{V}$ we have:

$$r' \Vdash_{\underset{\sim}{\mathbb{Q}}[G_{\mathbb{Q}^1_{I,h}}]} \text{" } \underset{\sim}{d}_n < x(n) \text{ and } m \in \underset{\sim}{w}_n \ \Rightarrow \ \operatorname{dom}(\underset{\sim}{\nu}_m) \subseteq [0, x(n))\text{"}.$$

In $\mathbf{V}[G_{\mathbb{Q}^1_{I,h}}]$, by the property of $\mathbb{Q}$, there are r^{**}, $r' \leq r^{**} \in \underset{\sim}{\mathbb{Q}}[G_{\mathbb{Q}^1_{I,h}}]$ and a sequence $\langle \langle (i_\ell(n), j_\ell(n)) : \ell \leq k_n \rangle : n < \omega \rangle$ such that

$$i_0(n) < j_0(n) < i_1(n) < j_1(n) < \ldots < j_{k_n}(n) < i_\ell(n+1), j_\ell(n) > x(i_\ell(n))$$

and there are $\bar{\nu}^*_{n,\ell,t} = \langle \nu^*_{n,\ell,t,j} : j \in [i_\ell(n), j_\ell(n)) \rangle$ for $t < i_\ell(n), \ell \leq k_n$ and $\bar{w}^*_{n,\ell,t} = \langle w^*_{n,\ell,t,j} : j \in [i_\ell(n), i_{\ell+1}(n))$ for $t < i_\ell(n), \ell \leq k_n \rangle$ such that

$$r^{**} \Vdash_{\underset{\sim}{\mathbb{Q}}} \text{"}\langle \underset{\sim}{\nu}_{i_\ell(n)+j} : j \in [i_\ell(n), j_\ell(n)) \rangle \text{ is } \bar{\nu}^*_{n,\ell,t} \text{ and}$$
$$\langle \underset{\sim}{w}_{i_\ell(n)+j} : j \in [i_\ell(n) : j_\ell(n)) \rangle \text{ is } \bar{w}^*_{n,\ell,t} \text{ for some } t < i_\ell(n)\text{"}.$$

Back in $\mathbf{V}$ we have a $\mathbb{Q}^1_{I,h}$-name $\underset{\sim}{r}^{**}$ and $\langle \langle (\underset{\sim}{i}_\ell(n), \underset{\sim}{j}_\ell(n)) : \ell \leq \underset{\sim}{k}_n \rangle : n < \omega \rangle$ and $\langle \langle \bar{\underset{\sim}{\nu}}^*_{n,\ell,t} : t < i_\ell(n) \rangle : \ell < \underset{\sim}{k}_n, n < \omega \rangle$ and $\langle \langle \bar{w}^*_{n,\ell,t} : t < i_\ell(n) \rangle : \ell < \underset{\sim}{k}_n, n < \omega \rangle$ are forced (by p^*) to be as above.

By 14, increasing p^*, we get

for every $f : \{x^n_i : i < h(m), m \in A^{p^*} \cap (n+1)\} \longrightarrow \{-1, 1\}$, $n \in A^{p^*}$, the condition $p^{*[f]}$ forces a value to

$$\langle \langle (\underset{\sim}{i}_\ell(m), \underset{\sim}{j}_\ell(m)) : \ell \leq \underset{\sim}{k}_m \rangle : m \leq n \rangle,$$
$$\langle \bar{\underset{\sim}{\nu}}^*_{n,\ell,t} : t < \underset{\sim}{i}_\ell(n), \ell \leq \underset{\sim}{k}_n \rangle,$$
$$\langle \bar{\underset{\sim}{w}}^*_{m,\ell,t} : t < \underset{\sim}{i}_\ell(n), \ell < \underset{\sim}{k}_n \rangle$$

moreover, without loss of generality

$$n \in A^{p^*} \quad \Rightarrow j_{\underset{\sim}{k}_n}(n) < \min(A^{p^*} \setminus (n+1)).$$

Now by 20, without loss of generality for each $n \in A^{p^*}$ we can find a function ρ_n from $[n, \min(A^{p^*} \setminus (n+1))]$ to $\{-1,1\}$ such that:

if $f : \{x_i^m : i < h(m), m \in A^{p^*} \cap (n+1)\} \longrightarrow \{-1,1\}$, $n \in A^{p^*}$
then $(p^{*[f]}, \underset{\sim}{r}^{**})$ forces that ρ_n extends some $\underset{\sim}{\nu}_\ell$.

Now we continue as in the proof of 23(1). ■

3 The consistency proof

Theorem 24. *Assume CH and* $\diamondsuit_{\{\gamma<\omega_2 : \mathrm{cf}(\gamma)=\omega_1\}}$.
Then there is an $\aleph_2$*–cc proper forcing notion* $\mathbb{P}$ *of cardinality* $\aleph_2$ *such that*

$$\Vdash_{\mathbb{P}} \text{" there are no NWD–ultrafilters on } \omega \text{ "}.$$

PROOF Define a countable support iteration $\langle \mathbb{P}_i, \underset{\sim}{\mathbb{Q}}_j : i \le \omega_2, j < \omega_2 \rangle$ of proper forcing notions and sequences $\langle \underset{\sim}{\mathcal{D}}_i : i < \omega_2 \rangle$ and $\langle \underset{\sim}{\bar{\eta}}^i : i < \omega_2 \rangle$ such that for each $i < \omega_2$:

1. $\underset{\sim}{\mathcal{D}}_i$ is a $\mathbb{P}_i$–name for a non–principal ultrafilter on ω,
2. $\underset{\sim}{\mathbb{Q}}_i$ is a $\mathbb{P}_i$–name for a proper forcing notion of size $\aleph_1$ with the PP–property,
3. $\underset{\sim}{\bar{\eta}}^i$ is a $\mathbb{P}_i * \underset{\sim}{\mathbb{Q}}_i$–name for a function from ω to ${}^{<\omega}2$,
4. $\Vdash_{\mathbb{P}_i * \underset{\sim}{\mathbb{Q}}_i}$ "if $X \subseteq \omega$ and the set $\{\underset{\sim}{\eta}^i_n : n \in X\} \subseteq {}^{<\omega}2$ is nowhere dense then there is $Y \in \underset{\sim}{\mathcal{D}}_i$ disjoint from X",
5. if $\underset{\sim}{\mathcal{D}}$ is a $\mathbb{P}_{\omega_2}$–name for an ultrafilter on ω then the set
$$\{i < \omega_2 : \ \mathrm{cf}(i) = \omega_1 \quad \& \quad \underset{\sim}{\mathcal{D}}_i = \underset{\sim}{\mathcal{D}} \restriction \mathcal{P}(\omega)^{\mathbf{V}^{\mathbb{P}_i}}\}$$
is stationary.

Let us first argue that if we succeed with the construction then, in $\mathbf{V}^{\mathbb{P}_{\omega_2}}$, we will have

$$2^{\aleph_0} = \aleph_2 \quad + \quad \text{"there is no NWD-ultrafilter on } \omega\text{"}.$$

Why? As each $\underset{\sim}{\mathbb{Q}}_i$ is (a name) for a proper forcing notion of size $\aleph_1$, the limit $\mathbb{P}_{\omega_2}$ is a proper forcing notion with a dense subset of size $\aleph_2$ and satisfying the $\aleph_2$–cc. Since $\mathbb{P}_{\omega_2}$ is proper, each subset of ω (in $\mathbf{V}^{\mathbb{P}_{\omega_2}}$) has a canonical countable name (i.e. a name which is a sequence of countable antichains; every condition in the nth antichain decides if the integer n is in the set or not; of course we do not require that the antichains are maximal). Hence $\Vdash_{\mathbb{P}_{\omega_2}} 2^{\aleph_0} \le \aleph_2$ (remember that we have assumed $\mathbf{V} \models$CH). Moreover, by 18 + 21 we know that $\mathbb{P}_{\omega_2}$ satisfies $(\oplus^{\text{nwd}})$ of 21, i.e.

$$\Vdash_{\mathbb{P}_{\omega_2}} \text{"each nowhere dense subset of } {}^{<\omega}2 \text{ can be covered by a nowhere dense subset of } {}^{<\omega}2 \text{ from } \mathbf{V}\text{"}.$$

Now suppose that $\underset{\sim}{\mathcal{D}}$ is a $\mathbb{P}_{\omega_2}$–name for an ultrafilter on ω. By the fifth requirement, we find $i < \omega_2$ such that $\underset{\sim}{\mathcal{D}}_i = \underset{\sim}{\mathcal{D}} \restriction \mathcal{P}(\omega)^{\mathbf{V}^{\mathbb{P}_i}}$ (and $\mathrm{cf}(i) = \omega_1$). Since $\mathbb{P}_{\omega_2}$ satisfies $(\oplus^{\text{nwd}})$, we have

$\Vdash_{\mathbb{P}_{\omega_2}}$ "if $X \subseteq \omega$ and the set $\{\eta^i_n : n \in X\} \subseteq {}^{<\omega}2$ is nowhere dense then there is an element of $\mathcal{D}\restriction\mathcal{P}(\omega)^{\mathbf{V}^{\mathbb{P}_i}}$ disjoint from X"

[Why? Cover $\{\eta^i_n : n \in X\}$ by a nowhere dense set $A \subseteq {}^{<\omega}2$ from $\mathbf{V}$ and look at the set $X_0 = \{n \in \omega : \eta^i_n \in A\}$. Clearly $X_0 \in \mathbf{V}^{\mathbb{P}_i * \mathbb{Q}_i}$ and $X \subseteq X_0$. Applying the fourth clause to X_0 we find $Y \in \mathcal{D}_i = \mathcal{D}\restriction\mathcal{P}(\omega)^{\mathbf{V}^{\mathbb{P}_i}}$ such that $Y \cap X_0 = \emptyset$. Then $Y \cap X = \emptyset$ too.]
But this means that, in $\mathbf{V}^{\mathbb{P}_{\omega_2}}$, the function $\bar{\eta}^i$ exemplifies that $\mathcal{D}$ is not an NWD ultrafilter (remember $\mathcal{D}\restriction\mathcal{P}(\omega)^{\mathbf{V}^{\mathbb{P}_i}} \subseteq \mathcal{D}$). Moreover, as CH implies the existence of NWD-ultrafilters, we conclude that actually $\Vdash_{\mathbb{P}_{\omega_2}} 2^{\aleph_0} = \aleph_2$.

Let us describe how one can carry out the construction. Each $\mathbb{Q}_i$ will be $\mathbb{Q}^1_{I_i,h}$ for some increasing function $h \in {}^\omega\omega$ (e.g. $h(n) = n$) and a ($\mathbb{P}_i$–name for a) maximal non–principal ideal I_i on ω. By 22, 17 we know that $\mathbb{Q}^1_{I_i,h}$ satisfies the demands (2)–(4) for the ultrafilter $\mathcal{D}_i$ dual to I_i and the function $\bar{\eta}^i$ as in the proof of 22. Thus, what we have to do is to say what are the names $\mathcal{D}_i$. To choose them we will use the assumption of $\diamondsuit_{\{\gamma<\omega_2:\mathrm{cf}(\gamma)=\omega_1\}}$. In the process of building the iteration we choose an enumeration $\langle (p_i, \tau_i) : i < \omega_2\rangle$ of all pairs (p, τ) such that p is a condition in $\mathbb{P}_{\omega_2}$ (in its standard dense subset of size $\aleph_2$) and τ is a canonical (countable) $\mathbb{P}_{\omega_2}$–name for a subset of ω. We require that $p_i \in \mathbb{P}_i$ and τ_i is a $\mathbb{P}_i$–name (of course, it is done by a classical bookkeeping argument). Note that each subset of $\mathcal{P}(\omega)$ from $\mathbf{V}^{\mathbb{P}_{\omega_2}}$ has a name which may be interpreted as a subset X of ω_2: if $i \in X$ then p_i forces that τ_i is in our set. Now we may describe how we choose the names $\mathcal{D}_i$. By $\diamondsuit_{\{\gamma<\omega_2:\mathrm{cf}(\gamma)=\omega_1\}}$ we have a sequence $\langle X_i : i < \omega_2\ \&\ \mathrm{cf}(i) = \omega_1\rangle$ such that

(i) $X_i \subseteq i$ for each $i \in \omega_2$, $\mathrm{cf}(i) = \omega_1$,
(ii) if $X \subseteq \omega_2$ then the set

$$\{i \in \omega_2 : \mathrm{cf}(i) = \omega_1\ \&\ X_i = X \cap i\}$$

is stationary.

Arriving at stage $i < \omega_2$, $\mathrm{cf}(i) = \omega_1$ we look at the set X_i. We ask if it codes a $\mathbb{P}_i$–name for an ultrafilter on ω (i.e. we look at $\{(p_\alpha, \tau_\alpha) : \alpha \in X_i\}$ which may be interpreted as a $\mathbb{P}_i$–name for a subset of $\mathcal{P}(\omega)$). If yes, then we take this name as $\mathcal{D}_i$. In all remaining cases we take whatever we wish, we may even not define the name $\bar{\eta}^i$ (note: this leaves us a lot of freedom and one may use this to get some additional properties of the final model). So why we may be sure that the fifth requirement is satisfied? Suppose that we have a $\mathbb{P}_{\omega_2}$–name for an ultrafilter on ω. This name can be thought of as a subset X of ω_2. If $i < \omega_2$ is sufficiently closed then $X \cap i$ is a $\mathbb{P}_i$–name for an ultrafilter on ω which is the restriction of $\mathcal{D}$ to $\mathbf{V}^{\mathbb{P}_i}$. So we have a club $C \subseteq \omega_2$ such that for each $i \in C$, if $\mathrm{cf}(i) = \omega_1$ the $X \cap i$ is of this type. By **(ii)** the set

$$S \stackrel{\text{def}}{=} \{i < \omega_2 : i \in C\ \&\ \mathrm{cf}(i) = \omega_1\ \&\ X_i = X \cap i\}$$

is stationary. But easily, for each $i \in S$, the name $\underset{\sim}{\mathcal{D}}_i$ has been chosen in such a way that $\underset{\sim}{\mathcal{D}}_i = \underset{\sim}{\mathcal{D}} \restriction \mathcal{P}(\omega)$, so we are done. ■

We note that this implies that there is also no ultrafilter with property M. This was asked by Benedikt in [Bn].

Definition 25. A non-principal ultrafilter $\mathcal{D}$ on ω has the M-property (or property M) if:

> if for some real $\varepsilon > 0$, for $n < \omega$ we have a tree $T_n \subseteq {}^{<\omega}2$ such that $\mu(\lim(T_n)) \geq \varepsilon$
> then $(\exists A \in \mathcal{D})(\bigcap_{n \in A} \lim(T_n) \neq \emptyset)$

(where μ stands for the Lebesgue measure on ${}^{\omega}2$).

Proposition 26. *If a non-principal ultrafilter $\mathcal{D}$ on ω is not NWD, then $\mathcal{D}$ does not have the property M.*

PROOF Let

$$S^\varepsilon_\ell = \{T \cap {}^{\ell \geq}2 : T \subseteq {}^{<\omega}2,\ T \text{ a tree not containing a cone},\ \mu(\lim(T)) > \varepsilon\}$$

(note that S^ε_ℓ is a set of trees not a set of nodes) and let $S^\varepsilon = \bigcup_\ell S^\varepsilon_\ell$.

Now let $t_1 \prec t_2$ if: $t_1 \in S^\varepsilon_{\ell_1}$, $t_2 \in S^\varepsilon_{\ell_2}$, $\ell_1 < \ell_2$ and $t_1 = t_2 \cap {}^{\ell_1 \geq}2$. So S^ε is a tree with ω levels, each level is finite. As $\mathcal{D}$ is not NWD, we can find $\eta^\varepsilon_n \in \lim(S^\varepsilon)$ for $n < \omega$ such that:

$$\text{if } A \in \mathcal{D} \text{ then } \{\eta^\varepsilon_n : n \in A\} \text{ is somewhere dense.}$$

Now let $T^\varepsilon_n \subseteq {}^{<\omega}2$ be a tree such that $\langle T^\varepsilon_n \cap {}^{\ell \geq}2 : \ell < \omega \rangle = \eta^\varepsilon_n$. We claim that:

$$\langle T^\varepsilon_n : n < \omega \rangle \text{ exemplifies } \mathcal{D} \text{ does not have the } M\text{-property.}$$

Clearly T^ε_n is a tree of the right type, in particular

$$\mu(\lim(T^\varepsilon_n)) = \inf\{|T^\varepsilon_n \cap {}^{\ell}2|/2^\ell : \ell < \omega\} \geq \varepsilon.$$

So assume $A \in \mathcal{D}$ and we are going to prove that $\bigcap_{n \in A} \lim(T^\varepsilon_n)$ is empty. We know that $\{\eta^\varepsilon_n : n \in A\}$ is somewhere dense, so there is $\ell^* < \omega$ and $t^* \in S^\varepsilon_{\ell^*}$ such that:

$$\ell^* < \ell < \omega \ \&\ t^* \prec t \in S^\varepsilon_\ell \quad \Rightarrow \quad (\exists n \in A)(t \lhd \eta^\varepsilon_n).$$

Now $\frac{|t^* \cap {}^{\ell^*}2|}{2^{\ell^*}}$ is $> \varepsilon$ (so S^ε_ℓ was defined). So we choose $\ell > \ell^*$, such that:

> if $\nu \in {}^{\ell}2$, $\nu \restriction \ell^* \in t^*$
> then $t'_\nu = \{\rho \in {}^{\ell}2 : \rho \restriction \ell^* \in t^* \text{ and } \rho \neq \nu\} \in S^\varepsilon_\ell$,

hence there is $n = n_\nu \in A$ such that t'_ν appears in η^ε_n. Now clearly

$$\bigcap_{n\in A} \lim(T^\varepsilon_n) \supseteq \bigcap_{\substack{\nu\in {}^{\ell}2\\ \nu\restriction\ell^*\in t^*}} \lim(T^\varepsilon_{n_\nu})$$
$$\supseteq \{\eta \in {}^{<\omega}2 : \eta \restriction \ell \in \bigcap\{t'_\nu : \nu \in {}^{\rho}2, \nu \restriction \ell \in t^*\}\} = \emptyset,$$

finishing the proof. ■

Conclusion 27 *In the universe $\mathbf{V}^{\mathbb{P}_{\omega_2}}$ from 24, there is no (non-principal) ultrafilter (on ω) with property M.* ■

Concluding Remarks 28 We may consider some variants of $\mathbb{Q}^2_{I,h}$.

In definition 3 we have $\mathrm{dom}(H^p)$ is as in 3(1) but: $H^p \restriction B^p_1$ gives constants (not functions) and for $x^m_i \in B^p_3 \setminus B^p_1$, letting $n = \min(m/E^p)$ the function $H^p(x^m_i)$ depends just on $\{x^n_j : j \le i\}$. Moreover, it is such that changing the value of x^n_i changes the value, so $H^p(x^m_i) = x^n_i \times f^p_{x^m_i}(x^n_0, \dots, x^n_{i-1})$. Call this $\mathbb{Q}^3_{I,h}$.

A second variant is when we demand the functions $f^p_{x^m_i}(x^n_0, \dots, x^n_{i-1})$ to be constant, call it $\mathbb{Q}^4_{I,h}$.

Both have the properties proved $\mathbb{Q}^2_{I,h}$. In particular, in the end of the proof of 9(5), we should change: $H^r(x^m_i)$ is defined exactly as in the proof of 9(4) except that when $i < h(n^*)$, $k = \min(m/E^p)$, $k \notin \mathrm{dom}(E^q)$, $k \notin \mathbf{u}$ (so m, k, n^* are E^r–equivalent) we let $H^r(x^k_i) = H^q(x^m_i) \times f(x^{n^*}_i) \times x^{n^*}_i$ (the first two are constant), so $H^r(x^m_i)$ is computed as before using this value.

References

[B6] James E. Baumgartner. Ultrafilters on ω. *The Journal of Symbolic Logic*, **60**:624–639, 1995.

[Bn] Michael Benedikt. Ultrafilters Which Extend Measure. *The Journal of Symbolic Logic*, to appear.

[GoSh 388] Martin Goldstern and Saharon Shelah. Ramsey ultrafilters and the reaping number—Con($\mathfrak{r} < \mathfrak{u}$). *Annals of Pure and Applied Logic*, **49**:121–142, 1990.

[Sh:F187] Saharon Shelah. More consistency results related to ultrafilters.

[Sh:f] Saharon Shelah. *Proper and improper forcing*, accepted. Springer.

[Sh:b] Saharon Shelah. *Proper forcing*, volume 940 of *Lecture Notes in Mathematics*. Springer-Verlag, Berlin-New York, xxix+496 pp, 1982.

[Sh 407] Saharon Shelah. CON($\mathfrak{u} > \mathfrak{i}$). *Archive for Mathematical Logic*, **31**:433–443, 1992.

[Sh:F151] Shelah, Saharon. For Olek Blaszczyk: Complete σ-centered Boolean Algebra not adding Cohen reals.

[vD81] Eric K. van Douwen. Remote points. *Dissertationes Mathematicae*, **188**:3–45, 1981.

Towards Recursive Model Theory*

Alexei P. Stolboushkin**

Research Division
Fourth Dimension Software
555 Twin Dolphin Dr., 500, Redwood City, CA 94065-2102
aps@4ds.com

Abstract. We argue that the models that are relevant to computer science are recursive and that Recursive Model Theory deserves being studied systematically, with at least the same vigor as Finite Model Theory has been. We study the status of some fundamental theorems from the classical model theory in this context and establish failure of several of them, including (generalized) Completeness, Compactness, Beth's Definability, Craig's Interpolation, and Lyndon's Lemma.

1 Introduction

Classical Model Theory deals with *all models*. If, for whatever reason, the class of models is restricted, this may potentially change model-theoretical laws that we take for granted. Take, for instance, the central for logic notion of *truth*. There may be sentences that are uniformly true in all the models of a certain class, but refutable in models not in the considered class. Hence, restricting the class of models may expand the class of true sentences. Conversely, the class of satisfiable sentences may shrink. If this actually happens, the equivalence between Model Theory and Proof Theory implied by Gödel's Completeness Theorem discontinues to hold, although in a specific situation a remedy can possibly be found by changing the axiomatization.

Recently, the Model Theory of finite models (those with finite universes) has been intensively investigated. The main motivation for Finite Model Theory has been the fact that, in several computer science applications, notably in databases, the models are often finite, and many issues in the theory of databases can be studied in the context of Finite Model Theory. Surprisingly or not, Finite Model Theory looks very much different from its classical counterpart.

The author is generally interested in Logic in Computer Science, and while finite models often are relevant to Computer Science, without question, not all the models that show up in CS applications are finite. Even in databases which have long been the Finite Model Theory refuge, infinite models not only show up, but actually move towards the central stage. In other CS playgrounds, say, in verification, finite models have never had any noticeable fraction of the market.

* Abstract of Plenary Talk given at Logic Colloquium'95, Haifa, Israel.

** This work has been partially supported by NSF Grant CCR 9403809. A part of this work was done at the UCLA Department of Mathematics.

Having said this, it is on the other hand the author's conviction that considering arbitrary models is irrelevant, as far as Computer Science is concerned. By way of example, suppose we knew that a constraint was going to be universally satisfied for all the databases over countable domains. Would it be a sufficient evidence to altogether eliminate this constraint from the list of constraints? My answer is "yes", because all the databases that a computer can handle are countable. Moreover, they are *recursive*, or *constructive*.

Of course, computer scientists do play with non-recursive structures. For example, a series of papers has appeared recently that deal with and establish interesting results about constraint databases over o-minimal and quasi o-minimal domains—a diverse family that includes several recursive structures, but at the same time includes such monstrous structures as the reals—see [PVV95, BDLW96, ST96]). The strength of these results is in providing uniform techniques for a wide class of domains, recursive and non-recursive alike, however, at the end it should not be forgotten that reals can not be represented inside a computer. In my view, "databases over the reals" can be used as a figure of speech, but should not be confused with real databases that are organized over strings, integers, and other *recursive* domains.

Generally, recursive models have been studied for decades. But the question, *"what Logic would look like if we had recursive models only?"*, has never been, to the best of the author's knowledge, systematically addressed before. In this paper, we consider several central results of Classical Model Theory:

- Gödel's Completeness Theorem
- Compactness Theorem
- Craig's Interpolation Theorem
- Beth's Definability Theorem
- Lyndon's Lemma on Monotonicity

from this viewpoint, and show that they all fail over recursive models—the situation remarkably similar to that with finite models. This indicates that Recursive Model Theory deserves being studied systematically, with at least the same vigor as the theory of finite models has been.

I wish to point the reader's attention to another paper in Recursive Model Theory [HH96] where Harel and Hirst (independently of this paper) show failure of Compactness and Completeness, and also prove several other results, including a very nice asymptotic law for recursive models.

Finally, a few words about the techniques used in this paper. We start off by pulling together several results about nonstandard models of Peano Arithmetic that jointly imply finite axiomatizability of the standard model $\mathbb{N}$ of arithmetic in the class of recursive models. This result is used as instrumental in refuting several of the aforementioned classical results in the context of recursive models. Of them the only one that requires a technically involved argument is Lyndon's Lemma. Generally, we further develop the technique introduced in [Sto95] for refuting Lyndon's Lemma over finite models. This requires substantial changes, but at the end, the proof turns out to be even a bit simpler.

2 Preliminaries

2.1 Recursive models

A model is called *recursive* iff it is isomorphic to a model over the natural numbers (or an initial segment thereof) with all the signature functions and predicates defined by recursive functions.

For example, every finite model is recursive. There are, however, infinite recursive models, some of them well known: the standard model $\mathbb{N}$ of Peano Arithmetic, or rational numbers with equality, $+$ and $\times$. Recursive models should not, however, be confused with models with decidable theories, as these two notions are independent. Indeed, the theory of real numbers with $+, \times, 0, 1$ is decidable, but the model is not recursive by virtue of its uncountability; on the other hand, the aforementioned standard model of arithmetic, while recursive, is undecidable.

2.2 Tennenbaum's Theorem and generalizations

The goal of this subsection is to pull together several results jointly implying finite axiomatizability of $\mathbb{N}$, in the class of recursive models. As is well known, Tennenbaum [Ten59] showed that the standard model of arithmetic is the only recursive model of Peano Arithmetic PA. However, PA is not finitely axiomatizable. Second however, McAloon [McA82] extended the result by showing that $\mathbb{N}$ continues to be the only recursive model of $I\Delta_0$, and, by implication, of $I\Sigma_n$, for any n. IX denotes a restricted version of PA with the induction axiom applied to X-formulas only.

Now, although the question of finite axiomatizability of $I\Delta_0$ continues to be open, it is known that $I\Sigma_n$, for $n \geqslant 1$, is finitely axiomatizable—see, e.g., Exercise 10.4 in [Kay91]. Hence:

Theorem 1. *The standard model $\mathbb{N}$ of Peano Arithmetic is finitely axiomatizable in the class of recursive models.*

Fix a finite axiomatization $\mathcal{A}$ of $\mathbb{N}$—for example, a finite axiomatization of $I\Sigma_1$.

2.3 Monotone formulas

The classical result by Roger Lyndon [Lyn59] is that any first order formula, monotone in a certain predicate, is equivalent to some formula of the same signature that is positive in the predicate. Apart from the fact that the lemma is used in proofs of Lyndon Interpolation Theorem, it reflects a very interesting logical phenomenon and is hence extremely interesting by itself.

By definition, a formula is *positive* in a predicate symbol iff, syntactically, every occurrence of this predicate symbol in the formula is under an even number of negations.

Monotonicity of a formula in a predicate is, on the other hand, a semantical property. Namely, a formula $\varphi(P, \mathbf{x})$ with free variables $\mathbf{x}$ is *monotone* in a predicate symbol P iff for every model M, tuple $\mathbf{a}$, and every extension $Q \supseteq P^M$,

$$M \models \varphi(P^M, \mathbf{a}) \text{ implies } M \models \varphi(Q, \mathbf{a}).$$

The "every model" refers to every model of the class we consider. The standard definition means "every possible model", but now we are interested in recursive models only.

All the same, a formula positive in a predicate symbol is necessarily monotone in this symbol. In the other direction, over all models Lyndon's Lemma says that every monotone formula is equivalent to a positive formula. To show that Lyndon's Lemma fails over recursive models we will need to construct a formula monotone in a predicate symbol *over all recursive models,* but not equivalent, over *recursive models,* to any positive formula.

3 Classical Theorems in Recursive Models

The goal of this section is to look at several important classical results about first order logic and to clarify their status in the recursive model theory.

To begin with, several classical theorems are syntactical by nature, and their status does not depend on a particular class of models being considered. A good example is the Prenex Normal Form.

Several other results, however, directly relate to the class of models, and it makes a perfect sense to inquire as to their status. Surprisingly or not, we establish here failure of several such results.

Theorem 2 No Gödel's Completeness. *In recursive models, first order truth is not recursively axiomatizable.*

PROOF: By Theorem 1, $\mathbb{N}$ is finitely axiomatizable. If the truth over recursive models were recursively axiomatizable, then so would be arithmetical truth. But, by Gödel's Incompleteness Theorem, it is not. *Q.E.D.*

Theorem 3 No Compactness. *Compactness Theorem fails over recursive models.*

In other words, there exists a set of first order sentences whose every finite subset has a recursive model, while the set itself does not have a recursive model.

PROOF: Recall that $\mathcal{A}$ is the finite axiomatization from Theorem 1. Consider the arithmetical signature *plus* an additional constant symbol ∞. Let ε_k, for $k \in \{0, 1, 2, \ldots\}$, be the following sentence of this signature:

$$\mathcal{A} \wedge \infty > k.$$

The $\infty > k$ is an obvious abbreviation. Then the set

$$\Xi = \{\varepsilon_k\}_{k=0}^{\infty}$$

does not have a recursive model. However, every finite subset of Ξ is satisfied in the standard model when ∞ is interpreted as a large enough number (greater than all the ks in the sentences). *Q.E.D.*

Let us now formulate Craig's Interpolation Theorem and Beth's Definability Theorem that we are going to refute next:

Theorem 4 Craig's Interpolation Theorem. *Let φ, ψ be two sentences, such that $\varphi \models \psi$. There exists a sentence δ, such that:*

1. $\varphi \models \delta$
2. $\delta \models \psi$, *and*
3. *any relation, or function symbol, occurring in δ (except possibly for equality) does also occur in both φ, and ψ*

This sentence δ is often called an interpolant of φ, ψ.

Consider a signature Ω, and an extra predicate symbol $R \notin \Omega$. Let $\varphi(R)$ be a sentence over the signature $\Omega \bigcup \{R\}$. We say that $\varphi(R)$ ***defines the predicate R*** *implicitly* iff for any model $\mathcal{M}$ of the signature Ω there exists a unique interpretation $R^{\mathcal{M}}$ of the predicate R, such that the sentence $\varphi(R)$ is valid in $\mathcal{M}$ under this interpretation of R.

Theorem 5 Beth's Definability Theorem. *Let Ω be a signature, $R \notin \Omega$ be a new predicate symbol of an arity k, and let a sentence $\varphi(R)$ over the signature $\Omega \bigcup \{R\}$ define the predicate R implicitly. There exists a formula $\psi(x_1, x_2, \ldots, x_k)$ over the signature Ω such that*

$$\varphi(R) \models \forall x_1 \forall x_2 \ldots \forall x_k \left(R(x_1, x_2, \ldots, x_k) \longleftrightarrow \psi(x_1, x_2, \ldots, x_k)\right).$$

In other words, over all models implicit definability does not increase the power of first order logic in defining predicates. It will also be useful to have in mind the following "modern" proof of Beth's Definability Theorem: essentially, it demonstrates the well known fact that for any reasonable class of models, and any reasonable logic, the interpolation property implies the definability property.
PROOF OF THEOREM 5: Since $\varphi(R)$ defines R implicitly, by definition, we have (for the signature $\Omega \bigcup \{R, R'\}$, where R' is yet another new predicate symbol of the same arity k):

$$\varphi(R) \wedge \varphi(R') \models \forall x_1 \forall x_2 \ldots \forall x_k (R(x_1, x_2, \ldots, x_k) \longrightarrow R'(x_1, x_2, \ldots, x_k)).$$

Let us enrich the signature with the additional constants (0-ary function symbols) $c_1, c_2, \ldots, c_k$. Now,

$$\varphi(R) \wedge \varphi(R') \models R(c_1, c_2, \ldots, c_k) \longrightarrow R'(c_1, c_2, \ldots, c_k), \text{ or, in other words,}$$

$$\varphi(R) \wedge R(c_1, c_2, \ldots, c_k) \models \varphi(R') \longrightarrow R'(c_1, c_2, \ldots, c_k).$$

By Craig's Interpolation Theorem, there exists an interpolant $\psi(c_1, c_2, \ldots, c_k)$ over the signature $\Omega \bigcup \{c_1, c_2, \ldots, c_k\}$ (but *without* R, R'), such that:

1. $\varphi(R) \wedge R(c_1, c_2, \ldots, c_k) \models \psi(c_1, c_2, \ldots, c_k)$, and
2. $\psi(c_1, c_2, \ldots, c_k) \models \varphi(R') \longrightarrow R'(c_1, c_2, \ldots, c_k)$

Now, the sentence 1 refers to the models of the signature $\Omega \bigcup \{R, c_1, c_2, \ldots, c_k\}$, while 2 does to ones of $\Omega \bigcup \{R', c_1, c_2, \ldots, c_k\}$. As any model of the former signature is also a model of the latter, if we interpret R' as R, we finally have:

- $\varphi(R) \wedge R(c_1, c_2, \ldots, c_k) \models \psi(c_1, c_2, \ldots, c_k)$, and
- $\psi(c_1, c_2, \ldots, c_k) \models \varphi(R) \longrightarrow R(c_1, c_2, \ldots, c_k)$

in the signature $\Omega \bigcup \{R, c_1, c_2, \ldots, c_k)$, or

- $\varphi(R) \models R(c_1, c_2, \ldots, c_k) \longrightarrow \psi(c_1, c_2, \ldots, c_k)$, and
- $\varphi(R) \models \psi(c_1, c_2, \ldots, c_k) \longrightarrow R(c_1, c_2, \ldots, c_k)$

Combining these two assertions, we have:

$$\varphi(R) \models R(c_1, c_2, \ldots, c_k) \longleftrightarrow \psi(c_1, c_2, \ldots, c_k), \text{ or, further,}$$

$$\varphi(R) \models \forall x_1 \forall x_2 \ldots \forall x_k (R(x_1, x_2, \ldots, x_k) \longleftrightarrow \psi(x_1, x_2, \ldots, x_k)),$$

for the signature $\Omega \bigcup \{R\}$. *Q.E.D.*

The situation changes dramatically over all recursive models:

Theorem 6 No Beth's Definability. *Beth's Definability Theorem fails for recursive models.*

Specifically, there exist a signature Ω, and a sentence $\varphi(P)$ in the signature $\Omega \bigcup \{P\}$, where P is a unary predicate symbol not in Ω, such that:

1. *in every recursive model M of the signature Ω, $\varphi(P)$ is satisfied for one and only one unary predicate P_M substituted for P*
2. *there exists a recursive model M of the signature Ω such that for any formula $\psi(x)$ in the signature Ω with one free variable x:*

$$M \not\models \varphi(\{a \in |M| : \psi(m)\})$$

PROOF: The set of all true arithmetical sentences is implicitly definable in arithmetic (see, e.g., [Rog67]). Let Ω be the arithmetical signature $+, \times, <, 0, 1$, and let $\mathcal{T}(P)$ implicitly defines the set of numbers of true arithmetical sentences. This set, however, is not *explicitly* definable in arithmetic [Rog67].

To finish the proof, define the required $\varphi(P)$ as follows:

$$(\mathcal{A} \wedge \mathcal{T}(P)) \vee (\neg \mathcal{A} \wedge (\forall x)(P(x) \longrightarrow \textbf{false})),$$

where, again, $\mathcal{A}$ is the axiomatization from Theorem 1.

Observe that $\varphi(P)$ defines the set of numbers of true arithmetical sentences in the standard model, and an empty set otherwise. *Q.E.D.*

Now, let us refute Craig's Interpolation Theorem.

Theorem 7 No Craig's Interpolation. *Craig's Interpolation fails over recursive models.*

In other words, there exist two sentences φ, ψ such that $\varphi \models \psi$ over all recursive models, but no interpolant sentence δ exists such that:

1. *$\varphi \models \delta$ over all recursive models*
2. *$\delta \models \psi$ over all recursive models, and*
3. *any relation, or function symbol, occurring in δ (except possibly for equality) does also occur in both φ, and ψ*

PROOF: A closer look at our proof of Theorem 5 reveals that, if Craig's Interpolation Theorem (Theorem 4) held over recursive models, we would be able to prove Beth's Definability Theorem over recursive models as well. But we already know that the latter fails for recursive models. *Q.E.D.*

4 Failure of Lyndon's Lemma

Finally, we are going to refute Lyndon's Lemma. Our method of refutation builds on the technique developed in [Sto95] for finite models. There, the author introduced Positive Pebble Games and proved that they capture "preservation" under positive formulas. Independently, similar games (together with one direction of proof of the Theorem 9) were introduced in [McC95].[3] Let us recall the definitions from [Sto95].

4.1 Positive preservation and pebble games

Definition 8 positive n-preservation. Let A, B be two models of a certain relational signature $\Omega = \Delta \bigcup \Theta$, Δ and Θ be disjoint, and let $\mathbf{a} = a_1, a_2, \ldots, a_k$, $\mathbf{b} = b_1, b_2, \ldots, b_k$ be elements of $|A|$, $|B|$, respectively.

We say that $(A, \mathbf{a})$ is *Θ-positively n-preservable by* $(B, \mathbf{b})$, iff any first order Θ-positive formula $\varphi(x_1, x_2, \ldots, x_k)$ of quantifier depth n with k free variables $x_1, x_2, \ldots, x_k$ that is true in A under $x_1, x_2, \ldots, x_k := a_1, a_2, \ldots, a_k$, is also true in B under $x_1, x_2, \ldots, x_k := b_1, b_2, \ldots, b_k$.

If $k = 0$, we will simply say that A is Θ-positively n-preservable by B.

Let $\Omega = \Delta \bigcup \Theta$ be a relational signature, Δ and Θ be disjoint, and let A, B be two models of the signature. Let $\mathbf{a} = a_1, a_2, \ldots, a_k$, $\mathbf{b} = b_1, b_2, \ldots, b_k$ be elements of $|A|$, $|B|$, respectively. Consider the following game of two players on the pair $\langle (A, \mathbf{a}), (B, \mathbf{b}) \rangle$, called n-positive pebble game, for $n \geq 0$.

Each player initially has n pebbles, numbered $1, 2, \ldots, n$. In the first step of the game, the first player, whom we will call *Spoiler*, chooses a model among A, B, and places his pebble with number 1 onto some element of the model.

[3] An anonymous referee has suggested that the idea of "positive pebble games" might have been known to certain model theorists in some form.

Then the second player, whom we will call *Duplicator*, takes the other model, and places his pebble with number 1 onto some element of this model.

After $(r-1)$-th step of the game, for $1 < r \leq n$, each player retains the pebbles numbered $r, r+1, \ldots, n$, while the pebbles numbered $1, 2, \ldots, r-1$ are somehow placed onto elements of the models A, B. Then in the r-th step of the game, again, the first player chooses a model among A, B, and places his pebble r onto some element of the model. Then the second player takes the other model, and places his pebble r onto some element of this model.

The n-positive pebble game ends after its n-th step. After the game ends, each of the models A, B has pebbles numbered $1, 2, \ldots, n$ placed somehow onto its elements. Let $a_{k+1}, a_{k+2}, \ldots, a_{k+n}$ be the elements of A covered with, respectively, pebbles numbered $1, 2, \ldots, n$, and let $b_{k+1}, b_{k+2}, \ldots, b_{k+n}$ be the elements of B covered with, respectively, pebbles numbered $1, 2, \ldots, n$.

Let A' be the substructure of A generated with $a_1, a_2, \ldots, a_{k+n}$, and let B' be the substructure of B generated with $b_1, b_2, \ldots, b_{k+n}$.

By definition, the second player, Duplicator, wins the game, if the mapping $a_i \mapsto b_i$, for $i = 1, 2, \ldots, k+n$, is correctly defined and this mapping is:

1. an isomorphism of Δ-reducts of A' and B';
2. a homomorphism of Θ-reduct of A' onto Θ-reduct of B'.

Otherwise the first player, Spoiler, wins the game.

Theorem 9. *Let A, B be two models of a finite relational signature $\Omega = \Delta \bigcup \Theta$ (with or without constants), Δ and Θ be disjoint, and let $\mathbf{a} = a_1, a_2, \ldots, a_k$, $\mathbf{b} = b_1, b_2, \ldots, b_k$ be elements of $|A|$, $|B|$, respectively.*

$(A, \mathbf{a})$ is Θ-positively n-preservable by $(B, \mathbf{b})$, iff Duplicator has a winning strategy in the n-positive pebble game on the pair $\langle (A, \mathbf{a}), (B, \mathbf{b}) \rangle$.

4.2 Rotated grids

The goal of this subsection is to define the class of recursive models that is going to be instrumental in refuting Lyndon's Lemma. These structures will be called *Rotated Grids of Finite Height,* or simply *Rotated Grids.*

First, let's attempt an informal explanation. Imagine a grid on points (i, j) on the integer surface, with two unary functions—North and West—connecting (i, j) with $(i, j+1)$ and $(i-1, j)$, respectively. In the grid structure, there is going to be a binary relation H which is the union of the graphs of the two functions.

Now, rotate this grid 45 degrees clockwise, draw two parallel horizontal lines, and cut off everything above the upper, and below the lower line. This constitutes an informal definition of a rotated grid.

Additionally, the second binary relation, $<$, is defined in a rotated grid as the transitive closure of H. If $<$ is a proper subset of the transitive closure of H, the result is called a rotated *under*grid.

On the side, in the rotated (under)grids we will have the standard model of arithmetic (with $+, \times, 0, 1$ and $\leqslant$—distinguished from $<$), whose elements—natural numbers—are distinguished from the pairs by a unary predicate N.

These two parts of the models are going to be related by a binary relation $\equiv$ that connects all the points of a horizontal level in an (under)grid to the "number" of this level. There are finitely many levels.

Formally, the non-arithmetical part of an (under)grid of k levels is defined as the collection of elements:

$$\{(i,j) \mid i = 1, \ldots, k;\ j \in \mathbb{Z}\},$$

with the binary predicate H defined as follows:

$$((i,j),(l,m)) \in H \ \textbf{iff} \ l = i + 1 \wedge (j = m \vee j + 1 = m).$$

Notice that, for technical convenience, the first coordinate of a tuple numbers the level—this is different from the informal explanation attempted above, although the structure continues to be precisely the same.

The relation $\equiv$ connecting the grid part with $\mathbb{N}$ is defined as follows:

$$(i,j) \equiv i.$$

The intention of this relation $\equiv$ is to make the finiteness of the height of grids axiomatizable.

Finally, a rotated undergrid of height k is called *canonical* iff adding the pair $((0,0),(k,0))$ to its relation H makes it a rotated grid.

Theorem 10. *For any n, a rotated grid of height 2^{n+2} is positively n-preservable by its canonical undergrid.*

PROOF: Due to Theorem 9, it suffices to demonstrate a winning strategy for the second player in the positive n-game played in the rotated grid and its canonical undergrid.

First, in the arithmetical part, the games is played by replicating all the moves, while in the non-arithmetical part (in the grids), the moves will at least preserve the levels. Because of the definition of the relation $\equiv$, specifics of the pebble positions within the arithmetical part become irrelevant to winning the game, and we can henceforth concentrate on the grid parts only.

On the grids, Duplicator starts from drawing an imaginary line cutting the two grids into two parts. This line is drawn through the points $(0, -2^{n+1})$ and $(2^{n+2}, 2^{n+1})$. The part of the grid on or at the left from the line will be called left, and the other part will be called right.

Generally, any move of Spoiler in the right part of either of the grids is answered by replicating. In the left part, when Spoiler pebbles an element (i,j) in the grid, Duplicator answers by pebbling $(i, j+1)$ in the undergrid. Conversely, when Spoiler pebbles (i,j) in the undergrid, duplicator answers $(i, j-1)$ in the grid. At the step k of the game, Duplicator moves the imaginary line 2^{n+1-k} steps left or right, as follows:

- if the last move was in the left part, move the line to the right
- if the last move was in the right part, move the line to the left

("left" and "right" here refer to changing the second coordinate).

It is easy to see by induction, that the distance from the imaginary line to the closest pebble is, after the kth step, at least 2^{n-k}. Hence, between right and left pebbles, the relation H is impossible.

Observe, that although the line moves throughout the game, a pebble once attributed to the left (right) part, continues to be in this part. Therefore, separately, the positions of the left (right) pebbles in the two structures are isomorphic. To show that Duplicator wins the game, it then suffices to show that, if in the grid, ℓ, r are two pebbled elements from the left and right parts, respectively, and $r < \ell$,[4] then in the undergrid, the corresponding pebbled elements r', ℓ' are in the same relation $r' < \ell'$.

Indeed, let $\ell = (h_\ell, v_\ell), r = (h_r, v_r)$. $r < \ell$ implies $h_r < h_l$ and $v_r \leq v_\ell$. By definition, $\ell' = (h_\ell, v_\ell{+}1), r' = (h_r, v_r)$. Taking into account that $(h_\ell, v_r{+}h_\ell{-}h_r)$ is, by definition, in the right part, we can conclude that $v_\ell + 1 < v_r + h_\ell - h_r$. Hence, $r' < \ell'$. *Q.E.D.*

4.3 Monotone separation of rotated grids

A binary relation $<$ defined in a set U is called *cyclic,* iff there exist elements $e_1, e_2, \ldots, e_n \in U, n > 0$, such that $e_1 < e_2 < \cdots < e_n < e_1$. $<$ is called *acyclic* iff it is not cyclic.

An acyclic relation $<$ in a set U is called *downward infinite* iff there exist infinitely many elements $\{e_i\}_{i=-\infty}^{0}$, such that $\cdots < e_{-2} < e_{-1} < e_0$. An acyclic $<$ is called *downward finite* iff it is not downward infinite.

To begin with, let's axiomatize a class of models of the rotated grids' signature where H is an acyclic downward finite relation in $\overline{N}$ (not all such models though). This class will include rotated (under)grids. In what follows, $\mathcal{A}_N$ is the axiom system $\mathcal{A}$ from Theorem 1, specialized to the predicate N. Consider the following list of axioms:

$$\mathcal{A}_N \tag{1}$$

$$\forall x \notin N \exists! n \in N (x \equiv n) \tag{2}$$

$$\forall xy \notin N \forall nm \in N (H(x,y) \wedge x \equiv n \wedge y \equiv m \longrightarrow m = n+1) \tag{3}$$

Obviously, in the recursive models of this axiom system, H is acyclic and downward finite. For the rest of the section, consider only recursive models of this axiom system.

The following is a variation of the technique used in [Sto95].

In the grid structure, a is called an *immediate predecessor* of b, iff $H(a,b)$, and a *predecessor* of b, iff a is an immediate predecessor of either b, or one of its predecessors. Similarly, we can define the notion of $<$*-predecessor*. The following formula $\beta(x)$ with one free variable x asserts that if $b \leq x$, a is an immediate predecessor of b, and $c \leq b$, then $c < x$ and $c < b$:

$$\neg N(x) \wedge \forall yzu \notin N (H(z,y) \wedge u \leq z \wedge y \leq x \longrightarrow z < x \wedge z < y).$$

---

[4] Notice that, by the definition of the imaginary line, $\ell < r$ is impossible.

The following formula $\alpha(x)$ asserts the property β for x and for all its $<$-predecessors:

$$\forall y \notin N(y \leq x \longrightarrow \beta(y)).$$

In the above formulas, $u \leq v$ abbreviates $u = v \vee u < v$. Note that it is different from the arithmetical comparison $\leqslant$ defined in N. An important property implied by α is observed in the following:

Lemma 11. *Assume both the relations H and $<$ are acyclic and downward finite.*

If, for an element e, $\alpha(e)$, then the relation $<$ restricted to the set of H-predecessors of e together with e itself, extends the transitive closure of H on this set.

PROOF: Clearly, $<$ extends H. By induction, assume that if $(a, b) \in TC(H)$ and the difference in the levels of a and b is smaller than k, then the lemma holds.

Take $(a, b) \in TC(H)$ with the level difference k. Take an element c such that $(a, c) \in TC(H), H(c, b)$. Since, by induction, $\beta(b)$ and $a < c$, we have $a < b$.
Q.E.D.

Informally, our axioms for $<$ are organized into two groups—positive and negative—whose disjunction constitutes the axiomatization. The **positive part** is very simple, it asserts that $<$ is transitive and contains the transitive closure of H in N:

$$\forall xy \notin N(H(x, y) \longrightarrow x < y) \tag{4}$$

$$\forall xyz \notin N(x < y \wedge y < z \longrightarrow x < z) \tag{5}$$

For example, the rotated grids are going to be accepted because of this positive part alone.

The **negative part** accepts models where the relation $<$ is "weird". One example is when the relation disagrees with the level numbers, but, more importantly, when the following situation occurs:

- all the immediate predecessors of a certain a satisfy α
- a certain $b < a$
- for no immediate predecessor c of a, $c \leq b$

In this situation, it is guaranteed that b is not a predecessor of a (in the sense of H), still it is smaller than a, hence, the model is not an undergrid! Here are the **negative axioms**:

$$\exists xy \notin N \exists nm \in N(x \equiv n \wedge y \equiv m \wedge m \leqslant n \wedge x < y) \tag{6}$$

$$\exists xy \notin N(x < y \wedge \forall u \notin N(H(u, y) \longrightarrow \alpha(u)) \wedge \neg \exists z \notin N(H(z, y) \wedge z \leq x)) \tag{7}$$

The formal proof consists of a few lemmas:

Lemma 12. *The conjunction of the negative axioms is monotone in $<$.*

PROOF: If Axiom 6 is satisfied once, it is going to be satisfied in any extension of $<$. So the interesting case is when Axiom 6 is not satisfied for $<$, but Axiom 7 is satisfied, say, b, a witness the axiom. Extend $<$ arbitrarily. Again, assume Axiom 6 is not satisfied in the extension, hence $<$ is downward finite.

Call an element e *sound* (with respect to the current definition of $<$) iff all its predecessors are sound, and for any element f, $f < e$ iff f is a predecessor of e. Observe that all the elements at the level 1 are sound in the extension, and that soundness of an element implies the property α.

Observe that by Lemma 11, $<$ extends the transitive closure of H on the set of H-predecessors of a (because α holds for all the immediate predecessors of a). The same remains true in the extension. Therefore, if b, a do not witness Axiom 7 anymore, among the predecessors of a there is going to be an element a' which is not sound, but its all predecessors are sound. Hence, there exists a b' such that $b' < a'$, but $b' \not< c$ for any immediate predecessor c of a'.

These $b' < a'$ will now witness Axiom 7 in the extension.

Clearly, our proof relies on the downward finiteness of H.

Q.E.D.

Lemma 13. *If $<$ strictly extends the transitive closure of H, one of the negative axioms is satisfied.*

PROOF: Again, Axiom 6 is the trivial case. If the extension, however, does not satisfy Axiom 6, there is going to be an element which is not sound with all its predecessors being sound, which finishes the proof as in the case of Lemma 12.

Q.E.D.

Finally, our axiomatization is going to be:

$$\text{Axiom } 1 \wedge \cdots \wedge \text{Axiom } 3 \wedge$$

$$(\text{Axiom } 4 \wedge \text{Axiom } 5 \vee \text{Axiom } 6 \wedge \text{Axiom } 7)$$

Clearly, this axiomatization accepts every rotated grid, but does not accept any rotated undergrid. Combining this with Lemmas 12, 13, and Theorem 10, we finally have:

Theorem 14. *There exist a signature Ω and a sentence μ in this signature such that:*

- *μ is monotone in a predicate symbol $P \in \Omega$ over the class of recursive models of the signature Ω*
- *mu is not equivalent, over the class of recursive models of the signature, to any formula π in the signature Ω positive in the predicate symbol P*

5 Conclusion

There are many natural problems related to Recursive Model Theory whose status needs to be resolved. One fascinating open problem that we want to specifically attract public attention to is, whether an analogue of Gurevich-Shelah Theorem holds for recursive models. Recall that, although Lyndon's Lemma fails over finite models, the fixpoints of positive and monotone formulas over finite models were shown to be the same (see [GS86]). Likewise, the fixpoints continue to be the same in the standard model of arithmetic (and in some even more powerful recursive structures), even though the status of Lyndon's Lemma there continues to be open (see [Dou87]).

The Beth definability theorem as used in this paper sometimes is called the "weak Beth definability theorem", while the "Beth definability theorem" is used to refer to the version of the theorem in which "one and only one" is replaced to "at most one". Clearly, this relaxed version is refuted by this paper as well.

Acknowledgment

Firstly, I wish to thank Yiannis Moschovakis for long, and also extremely helpful, discussions that we have had. I am also grateful to Julia Knight, Marion Srebrny, Michael Taitslin, and Jouko Väänänen for their useful comments.

References

[BDLW96] M. Benedikt, G. Dong, L. Libkin, and L. Wong. Relational expressive power of constraint query languages. In *Proc. 15th ACM Symp. on Principles of Database Systems*, pages 5–16, 1996.

[Dou87] R.L. Dougherty. Monotone but not positive subsets of the Cantor space. *Journal of Symbolic Logic*, 52(3):817–818, 1987.

[GS86] Y. Gurevich and S. Shelah. Fixed-point extensions of first-order logic. *Annals of Pure and Applied Logic*, 32:265–280, 1986.

[HH96] D. Harel and T. Hirst. More about recursive structures: Descriptive complexity and zero-one laws. In E. Clarke, editor, *Proc. 11th IEEE Symp. on Logic in Computer Science*. IEEE Computer Society Press, Los Alamitos, CA, 1996. to appear.

[Kay91] R. Kaye. *Models of Peano Arithmetic*. Clarendon Press, Oxford, England, 1991.

[Lyn59] R. Lyndon. Properties preserved under homomorphism. *Pacific J. of Mathematics*, 9(1):143–154, 1959.

[McA82] K. McAloon. On the complexity of models of arithmetic. *Journal of Symbolic Logic*, 47:403–415, 1982.

[McC95] G.L. McColm. The dimension of the negation of transitive closure. *Journal of Symbolic Logic*, 60(2):392–414, 1995.

[PVV95] J. Paradaens, J. Van den Bussche, and D. Van Gucht. First-order queries on finite structures over reals. In *Proc. 10th IEEE Symp. on Logic in Computer Science*, pages 79–87. IEEE Computer Society Press, 1995.

[Rog67] H. Rogers Jr. *Theory of recursive functions and effective computability.* McGraw-Hill, 1967.

[ST96] A.P. Stolboushkin and M.A. Taitslin. Linear vs. order constraint queries over rational databases. In *Proc. 15th ACM Symp. on Principles of Database Systems*, pages 17–27, 1996.

[Sto95] A.P. Stolboushkin. Finitely monotone properties. In *Proc. 10th IEEE Symp. on Logic in Computer Science*, pages 324–330, Los Alamitos, CA, 1995. IEEE Computer Society Press.

[Ten59] S. Tennenbaum. Non-Archimedean models for arithmetic. *Notices of the AMS*, 6(3):270, 1959.

Accessible Segments of the Fast Growing Hierarchy *

Stanley Scott Wainer

School of Mathematics
University of Leeds
Leeds, LS2 9JT, England
pmt6ssw@sun.leeds.ac.uk

Abstract. We examine two ways of "bootstrapping" segments of the fast growing hierarchy by autonomous generation. One method closes off at ε_0 with the provably recursive functionals of arithmetic, whereas the other exhausts the provably recursive functions of $\Pi_1^1 - CA_0$.

1 Introduction.

Autonomously generated hierarchies are constructed according to the principle: *proceed to level α if α is already coded or recognized at some earlier level.* The question addressed here is : how might this principle be applied in a subrecursive (rate of growth) context, and what is the effect? Clearly it will depend upon the way in which we choose to subrecursively code or recognize countable ordinals. Typically this could be done by means of some "natural" scale or hierarchy $\alpha \mapsto f_\alpha$ such that the growth-rate of function f_α reflects its rank α. However this is a delicate matter since (in contrast with generalized recursion theory, where ordinal comparison is a fundamental property) one cannot expect to compare and compute ordinals by sub-recursive functions without reference to some prior given notation system. Either one simply accepts this situation and does the best one can (and we will, later, in section 4); or alternatively one could shift the goalposts and try to reconsider the problem in a more amenable, generalized setting. One suitable place to look is the theory of type two recursive functionals where, as noted by Kleene (1958) and in stark contrast with the recursive functions, a good notation-free hierarchy already exists - by classifying total recursive functionals according to the ordinal heights of their trees of unsecured sequences (see also Wainer (1995)). This opens the possibility of coding and comparing recursive ordinals in terms of the majorization relationship between certain descent-recursive functionals which represent them.

* This paper presents newer material not surveyed in the author's conference lecture (much of which can already be found in ref.10), but expanding the same overall theme.

2 Ordinal Presentations and Scales.

By a *presentation* of a countable ordinal $\alpha > 0$ is meant a nested sequence $\alpha[0] \subset \alpha[1] \subset \alpha[2] \subset \ldots$ of finite subsets such that $\alpha = \bigcup_n \alpha[n]$. In addition we will require that (i) $0 \in \alpha[0]$, (ii) if $\beta + 1 \in \alpha[n]$ then $\beta \in \alpha[n]$, and also (iii) if $\beta \in \alpha[n]$ and $\beta + 1 < \alpha$ then $\beta + 1 \in \alpha[n+1]$.

Obviously a presentation of α induces a sub-presentation of each $\beta < \alpha$ by $\beta[n] = \alpha[n] \cap \beta$. It also induces for each n, the *n-predecessor* function $P_n(\beta) = \max \beta[n]$. Thus for each non-zero $\beta \leq \alpha$,

$$\beta[n] = \{0, \ldots, P_n^3(\beta), P_n^2(\beta), P_n(\beta)\} \ .$$

This minimal apparatus is all that's needed to construct "majorization hierarchies", or "scales" of increasing functions $\{f_\beta : \beta \leq \alpha\}$. For an alternative general approach, see Buchholz, Cichon and Weiermann (1994).

Definition 1. *Given a fixed presentation for ordinal α, we shall call any sequence of increasing functions $\{f_\beta : \beta \leq \alpha\}$ an α-scale if it satisfies the majorization condition:*

$$\gamma \in \beta[n] \Rightarrow f_\gamma(n) < f_\beta(n).$$

Note.
There is a crucial dependence on the chosen ordinal presentation : for given any strictly increasing function g we can define a presentation for ω by $\omega[n] = \{0, 1, 2, \ldots, g(n) - 1\}$, so that any ω-scale must satisfy $f_\omega(n) \geq g(n)$. Thus "natural" hierarchies must depend upon "natural" choices of presentations.

Theorem 2. *Let $\Phi : (N \to N) \to (N \to N)$ be any operator taking increasing functions to increasing functions, which is monotone in the sense that for all increasing h_0, h_1 and all $k \in N$,*

$$\forall m \geq k \ (h_0(m) < h_1(m)) \Rightarrow \forall m \geq k \ (\Phi(h_0)(m) < \Phi(h_1)(m)).$$

Let f_0 be any increasing function such that $f_0(n) < \Phi(f_0)(n)$ for every n. Let α be equipped with a fixed presentation as above and for $0 < \beta \leq \alpha$ define

$$f_\beta(n) = \Phi(f_{P_n(\beta)})(n).$$

Then $\{f_\beta : \beta \leq \alpha\}$ is an α-scale.

Proof. By induction on β.

To see that each f_β is increasing note that for $m > n$,

$$P_n(\beta) \in \beta[n+1] = P_{n+1}(\beta)[n+1] \cup \{P_{n+1}(\beta)\} \subset P_{n+1}(\beta)[m] \cup \{P_{n+1}(\beta)\}$$

so by the induction hypothesis, $\forall m > n(f_{P_n(\beta)}(m) \leq f_{P_{n+1}(\beta)}(m))$. Therefore by the monotonicity of Φ we have for each n,

$$f_\beta(n) = \Phi(f_{P_n(\beta)})(n) \leq \Phi(f_{P_n(\beta)})(n+1) \leq \Phi(f_{P_{n+1}(\beta)})(n+1) = f_\beta(n+1).$$

Secondly, if $\gamma \in \beta[n]$ then $\gamma \in P_n(\beta)[n] \cup \{P_n(\beta)\}$, so by the induction hypothesis $f_\gamma(n) \leq f_{P_n(\beta)}(n)$. If $P_n(\beta) \neq 0$ then $P_n^2(\beta) \in P_n(\beta)[n]$, so again by the induction hypothesis and the monotonicity of Φ,

$$f_\gamma(n) \leq f_{P_n(\beta)}(n) = \Phi(f_{P_n^2(\beta)})(n) < \Phi(f_{P_n(\beta)})(n) = f_\beta(n).$$

If $P_n(\beta) = 0$ then $\gamma = 0$ and $f_0(n) < \Phi(f_0)(n) = f_\beta(n)$ by the assumption concerning f_0.

Example : G Scales.
Choose $\Phi_1(f)(n) = f(n) + 1$ in 2.2, to obtain the "Slow Growing" scale of functions $\{G_\beta : \beta \leq \alpha\}$ where

$$G_0(n) = 0\ , \qquad G_\beta(n) = G_{P_n(\beta)}(n) + 1.$$

This is the *minimum* scale with respect to a chosen presentation, since $G_\beta(n) = \text{card } \beta[n]$.

Example : H Scales.
Choose $\Phi_2(f)(n) = f(n+1)$ in 2.2, to obtain the "Hardy" scale $\{H_\beta : \beta \leq \alpha\}$ where

$$H_0(n) = n\ , \qquad H_\beta(n) = H_{P_n(\beta)}(n+1).$$

Example : F Scales.
Choose $\Phi_3(f)(n) = f^{n+1}(n)$ in 2.2, to obtain the "Fast Growing" scale of functions $\{F_\beta : \beta \leq \alpha\}$ where

$$F_0(n) = n + 1, \qquad F_\beta(n) = F^{n+1}_{P_n(\beta)}(n).$$

Recall that f^k denotes the k-times iterate of f.
Remark. In addition to providing well known classifications of the provably recursive functions of arithmetical theories, the F-hierarchy has a more general recursion-theoretic significance in that it is the subrecursive analogue of the classical jump hierarchy. For an obvious formulation of a subrecursive (bounded) jump operator is $f \mapsto sj(f)$ where

$$\begin{aligned} sj(f)(e, x, y) &= \{e\}^f(x) && \text{if defined in } y \text{ steps} \\ &= 0 && \text{otherwise} \end{aligned}$$

and if f has at least exponential growth then $sj(f)$ has the same elementary-recursive degree as $f \oplus \Phi_3(f)$. Thus up to elementary degree, $F_{2+\beta}$ is the β-th iterate of the subrecursive jump. See Heaton and Wainer (1996) and, for a detailed exposition, Heaton (1997).

The Bootstrapping Problem.
The problem addressed here is to make some sense of the following principle of autonomous inductive generation along the F-hierarchy:

$$\begin{cases} \tau_0 &= 0 \\ \tau_{n+1} &= \text{the ordinal "recognized" by } F_{\tau_n}. \end{cases}$$

Thus (i) what could "recognition" mean ? (ii) is τ_{n+1} uniquely defined ? and (iii) what is the closure ordinal $\tau = \sup \tau_n$? We give two solutions.

3 Recognition by Descent Functionals.

The predecessor functions P_n associated with a given presentation of an ordinal α allow a natural reformulation of Friedman's descent functional:

$$f \mapsto \text{ least } k : P_{f(k-1)} P_{f(k-2)} \cdots P_{f(1)} P_{f(0)}(\alpha) = 0$$

which "witnesses" the well-foundedness of α (see Friedman and Sheard (1995)). In our present context it will be convenient to adopt the more restrictive formulation where f is replaced by its successive iterates thus:

$$D_\alpha(f)(n) = \text{ least } k : P_{f^{k-1}(n)} P_{f^{k-2}(n)} \cdots P_{f(n)} P_n(\alpha) = 0.$$

Not only this, but we shall even go a stage further and use, instead of D_α, the functional

$$H_\alpha(f)(n) = f^k(n) \text{ where } k = D_\alpha(f)(n).$$

Note that this is just a functional version of the H-scale defined in section 2 since it satisfies the same recursive definition:

$$H_0(f)(n) = n\ , \qquad H_\beta(f)(n) = H_{P_n(\beta)}(f)(f(n))$$

but with the successor function replaced by an arbitrary (strictly increasing) function f. Similarly, a functional version of the F-scale is

$$F_0(f)(n) = f(n)\ , \qquad F_\beta(f)(n) = F_{P_n(\beta)}(f)^{n+1}(n).$$

The reason for using H rather than D is that it connects neatly with the fast-growing F.

Definition 3. *Given presentations for α and β, define a presentation for $\alpha + \beta$ by:*

$$\alpha + \beta[n] = \alpha[n] \cup \{\alpha + \gamma : \gamma \in \beta[n]\}$$

and a presentation for ω^β as follows:

$$\omega^\beta[n] = \{\omega^{\gamma_1}.m_1 + \omega^{\gamma_2}.m_2 + \cdots + \omega^{\gamma_{k-1}}.m_{k-1} + \omega^{\gamma_k}.m_k : m_i \leq n\} \cup \{0\}$$

where $\beta[n] = \{\gamma_k, \gamma_{k-1}, \ldots, \gamma_2, \gamma_1\}_<$, i.e. $\gamma_i = P_n^i(\beta)$ for each $i = 1, \ldots, k$ and k is the least such that $\gamma_k = 0$.

Lemma 4. *Given fixed presentations for α and β, and with the induced presentations for $\alpha + \beta$ and ω^β defined as above, we have for all f,*

(i) $H_{\alpha+\beta}(f) = H_\alpha(f) \circ H_\beta(f)$

(ii) $H_{\omega^\beta}(f) = F_\beta(f)$.

Proof. By inductions on β.

Firstly, if $\beta = 0$ we have for all $n \in N$,

$$H_{\alpha+0}(f)(n) = H_\alpha(f)(n) = H_\alpha(f)(H_0(f)(n)),$$

and if $\beta > 0$ then since $P_n(\alpha + \beta) = \alpha + P_n(\beta)$, we have

$$H_{\alpha+\beta}(f)(n) = H_{\alpha+P_n(\beta)}(f)(f(n))$$

and so by the induction hypothesis,

$$H_{\alpha+\beta}(f)(n) = H_\alpha(f)(H_{P_n(\beta)}(f)(f(n))) = H_\alpha(f)(H_\beta(f)(n)).$$

Secondly, if $\beta = 0$ then $P_n(\omega^\beta) = 0$ so

$$H_{\omega^0}(f)(n) = H_0(f)(f(n)) = f(n) = F_0(f)(n).$$

If $\beta > 0$ let $\beta[n] = \{\gamma_k, \gamma_{k-1}, \ldots, \gamma_2, \gamma_1\}_<$ where $\gamma_i = P_n^i(\beta)$ for each $i = 1, \ldots, k$ and k is minimal such that $\gamma_k = 0$. Then by definition of $\omega^\beta[n]$,

$$P_n(\omega^\beta) = \omega^{\gamma_1}.n + \omega^{\gamma_2}.n + \ldots + \omega^{\gamma_{k-1}}.n + \omega^{\gamma_k}.n$$

so by the induction hypothesis and repeated applications of part (i) we have

$$H_{\omega^\gamma.n}(f) = F_\gamma(f)^n$$

and hence

$$\begin{aligned}
H_{\omega^\beta}(f)(n) &= F_{\gamma_1}(f)^n \circ F_{\gamma_2}(f)^n \circ \cdots \circ F_{\gamma_{k-1}}(f)^n \circ F_{\gamma_k}(f)^n \circ f(n) \\
&= F_{\gamma_1}(f)^n \circ F_{\gamma_2}(f)^n \circ \cdots \circ F_{\gamma_{k-1}}(f)^n \circ F_{\gamma_k}(f)^{n+1}(n) \\
&= F_{\gamma_1}(f)^n \circ F_{\gamma_2}(f)^n \circ \cdots \circ F_{\gamma_{k-1}}(f)^n \circ F_{\gamma_{k-1}}(f)(n) \\
&= F_{\gamma_1}(f)^n \circ F_{\gamma_2}(f)^n \circ \cdots \circ F_{\gamma_{k-1}}(f)^{n+1}(n) \\
&= F_{\gamma_1}(f)^n \circ F_{\gamma_2}(f)^n \circ \cdots \circ F_{\gamma_{k-2}}(f)^{n+1}(n) \\
&= \ldots\ldots\ldots\ldots \\
&= F_{\gamma_1}(f)^n \circ F_{\gamma_2}(f)^{n+1}(n) \\
&= F_{\gamma_1}(f)^n \circ F_{\gamma_1}(f)(n) \\
&= F_{\gamma_1}(f)^{n+1}(n) \text{ where } \gamma_1 = P_n(\beta) \\
&= F_\beta(f)(n).
\end{aligned}$$

The next result shows how the ordering between (set-theoretic) ordinals is reflected by the majorization relation between descent functionals.

Lemma 5. *Let $\{\alpha[n] : n \in N\}$ and $\{\alpha'[[n]] : n \in N\}$ be any two chosen presentations of ordinals α and α' where $\alpha \leq \alpha'$. Let g be any strictly increasing function such that for every $n, \alpha[n] \subset \alpha'[[g(n)]]$. Then*

$$H_\alpha(g) \leq H'_{\alpha'}(g) \circ g$$

where $H_\alpha(g), H'_{\alpha'}(g)$ denote the H-scales determined by the respective presentations for α and α'.

Proof. The existence of such functions g with the stated properties is immediate, provided $\alpha \leq \alpha'$, though of course g may not be recursive (it depends on how "effective" the given presentations of α and α' are, and they needn't even be recursive ordinals).

We first show, by induction on γ, that for all n,

$$\gamma \in \alpha[n] \Rightarrow H_\gamma(g)(n) \leq H'_\gamma(g)(g(n)).$$

This holds trivially if $\gamma = 0$, and otherwise we have by the induction hypothesis, if $\gamma \in \alpha[n]$,

$$H_\gamma(g)(n) = H_{P_n(\gamma)}(g)(g(n)) \leq H'_{P_n(\gamma)}(g)(g^2(n)).$$

But $P_n(\gamma) \in \gamma[n] \subset \gamma[[g(n)]]$ so letting β denote the maximum element of $\gamma[[g(n)]]$ we have $P_n(\gamma) \leq \beta$ and hence

$$H_\gamma(g)(n) \leq H'_{P_n(\gamma)}(g)(g^2(n)) \leq H'_\beta(g)(g^2(n)) = H'_\gamma(g)(g(n)).$$

Now applying this with $\gamma = P_n(\alpha) \in \alpha'[[g(n)]]$ we obtain

$$H_\alpha(g)(n) = H_{P_n(\alpha)}(g)(g(n)) \leq H'_{P_n(\alpha)}(g)(g^2(n)) \leq H'_{\alpha'}(g)(g(n))$$

as required.

Theorem 6. *Let H_α and $H'_{\alpha'}$ be the descent functionals determined by arbitrary but fixed presentations of their respective ordinals α and α'. Then if α' is a limit we have*

$$\alpha < \alpha' \Leftrightarrow \exists g \ \exists k \ \forall f \ (g <_k f \rightarrow H_\alpha(f) \circ f <_k H'_{\alpha'}(f))$$

where g and f range over (strictly) increasing functions on N, $k \in N$, and $g <_k f$ means that $g(n) < f(n)$ for all $n \geq k$.

Proof. Suppose $\alpha < \alpha'$ where α' is a limit. Choose k and g so that, with the notation of Lemma 3.3, $\alpha + 2 \in \alpha'[[k]]$ and for every $n, \alpha[n] \subset \alpha'[[g(n)]]$. Then if $g <_k f$ we also have $\alpha[n] \subset \alpha'[[f(n)]]$ for $n \geq k$, so by the Lemma we obtain, for each $n \geq k$,

$$H_\alpha(f)(f(n)) \leq H'_\alpha(f)(f^2(n)) = H'_{\alpha+2}(f)(n) < H'_{\alpha'}(f)(n)$$

the final inequality coming about because of 2.2.

Conversely if $\alpha' \leq \alpha$ then, given any g and k we can certainly find a strictly increasing f such that $g <_k f$ and $\alpha'[[n]] \subset \alpha[f(n)]$ for every n. But then by Lemma 3.3 with α, α' reversed, we have

$$H'_{\alpha'}(f) \leq H_\alpha(f) \circ f.$$

This completes the proof.

Bootstrapping the Fast Growing Functionals.

Define a sequence of ordinal presentations τ_n as follows by induction on n,

$$\begin{cases} \tau_0 &= 0 \\ \tau_{n+1} &= \text{that ordinal } \beta \text{such that for some presentation of it, } F_{\tau_n} = H_\beta. \end{cases}$$

By 3.2, an immediate candidate for β is ω^{τ_n} with the "standard" presentation described earlier. Its uniqueness is ensured by 3.3. Thus $\tau_{n+1} = \omega^{\tau_n}$.

Theorem 7. $\tau_{n+1} = \omega^{\omega^{\cdot^{\cdot^{\cdot^{\omega}}}}}$ *with exponential stack of height n, and hence* $\sup \tau_n = \varepsilon_0$.

The proof theory of Peano Arithmetic shows that each F_{τ_n} is a provably recursive functional of PA in the sense that $F_{\tau_n}(f)$ is a provably recursive function of $PA + \forall x(f(x) < f(x+1))$; and conversely every provably recursive functional of PA is dominated by some F_{τ_n}.

4 Recognition by Minimum Scales.

In this section we choose as our notion of "ordinal recognition" the following: a fast-growing function F recognizes α if $F = G_\alpha$, the topmost element of the minimum (slow growing) α-scale $\{G_\beta : \beta \leq \alpha\}$.

As already noted, these function-hierarchies depend crucially upon the chosen presentation of α. In contrast with the results of section 3 which concern type-2 *functionals*, we cannot expect the majorization relation between *functions* to preserve the ordering between set-theoretic ordinals without restricting our attention to a fixed initial segment α with a chosen presentation, for only then do we have

$$\beta_1 < \beta_2 \leq \alpha \;\Leftrightarrow\; \exists k\ (G_{\beta_1} <_k G_{\beta_2}).$$

This "intensionality" is the major stumbling block in subrecursion theory. Why should one ordinal presentation be chosen in preference to another? - presumably because it is in some sense more "natural", but although we seem able to agree about the naturalness of some presentations when we see them, no clear definition of the concept is at hand.

All we do know, immediately from the foregoing, is that given arbitrary ordinal presentations $\{\alpha[n] : n \in N\}$ and $\{\alpha'[[n]] : n \in N\}$ where $\alpha \leq \alpha'$, there must be a function $g : N \to N$ such that for every $n, \alpha[n] \subset \alpha'[[g(n)]]$ and hence

$$\forall \beta \leq \alpha\ \exists k\ (G_\beta \leq_k G'_\beta \circ g \leq_k G'_{\alpha'} \circ g)$$

where G, G' are the slow growing hierarchies determined by the respective presentations. Now experience does seem to show that "natural" presentations possess a certain subrecursive stability in the sense that the function g above will usually turn out to be elementary or primitive recursive, so their respective hierarchies will be subrecursively comparable.

With this in mind we proceed to construct a particular ordinal presentation which will turn out to be just large enough for our purposes. The construction is based on generalizations of the fast growing hierarchy, operating on the higher "tree-ordinal" classes $\Omega_0, \Omega_1, \Omega_2, \ldots$ given by the following iterated inductive definition.

Definition 8. *$\Omega_0 = N$ and for $k > 0$, Ω_k is generated by the inductive clauses*

(i) $0 \in \Omega_k$,

(ii) *if $\alpha \in \Omega_k$ then $\alpha + 1 := \alpha \cup \{\alpha\} \in \Omega_k$,*

(iii) *if $\alpha : \Omega_i \to \Omega_k$ for some $i < k$ then $\alpha \in \Omega_k$.*

Notation.
Henceforth, lower case greek letters denote tree ordinals, not set-theoretic ordinals, although of course each tree ordinal α has a set-theoretic ordinal height, denoted $|\alpha|$. For each countable ordinal there are many different tree ordinals $\alpha \in \Omega_1$ with that same height. We shall usually reserve λ to denote "limit" tree ordinals arising by clause (iii) above, and call i its "type". Furthermore, we sometimes write, more suggestively, $\lambda = \sup_{\Omega_i} \lambda_\zeta$ where λ_ζ denotes the value of the function λ at argument $\zeta \in \Omega_i$. The sub-tree partial ordering $\prec$ is defined as the transitive closure of (i) $\alpha \prec \alpha + 1$ and (ii) $\lambda_\zeta \prec \lambda$ for every $\zeta \in \Omega_i$.

Definition 9. *For $\alpha \in \Omega_1$ and $n \in N$ define $\alpha[n]$ as follows:*

$$0[n] = \emptyset, \qquad \alpha + 1[n] = \alpha[n] \cup \{\alpha\}, \qquad \lambda[n] = \lambda_n[n].$$

Then call α structured if it satisfies the following conditions

(i) $\forall n\ (\alpha[n] \subset \alpha[n+1])$

(ii) $\forall \beta \prec \alpha\ \exists n\ (\beta \in \alpha[n])$

(iii) $\forall \beta\ \forall n\ (\beta \in \alpha[n] \wedge \beta + 1 \prec \alpha \to \beta + 1 \in \alpha[n+1])$.

Thus if α is structured, $\{\beta : \beta \prec \alpha\}$ is well-ordered by $\prec$ and a presentation of $|\alpha|$ is induced by $|\alpha|[n] = \{|\beta| : \beta \in \alpha[n]\}$.

There are various ways to ensure the structuredness of $\alpha \in \Omega_1$. One sufficient condition (see e.g. Wainer (1990), Kadota (1993)) is

$$\forall \lambda \preceq \alpha\ \forall n\ (\lambda_n \in \lambda[n+1]).$$

The question then is how to construct large structured tree ordinals in Ω_1. This leads us into the realm of "collapsing functions" as used in proof theoretical ordinal analysis, but there is a fairly obvious procedure to follow in our present context. Since the fast growing F-hierarchy uses countable ordinal presentations as indices to name large numbers, a straight generalization of it should enable us to name large tree ordinals in Ω_k, using indices from Ω_{k+1}. The named elements of Ω_k may then be used to index large elements of Ω_{k-1}, then large elements of Ω_{k-2} etcetera until we finish up naming large elements of Ω_1.

Definition 10. *For each $k \in N$ define the function*

$$\varphi^{(k)} : \Omega_{k+1} \times \Omega_k \to \Omega_k$$

by recursion over Ω_{k+1} as follows:

$$\begin{aligned} \varphi^{(k)}(0,\beta) &= \beta + 1 \\ \varphi^{(k)}(\alpha+1,\beta) &= \varphi^{(k)}(\alpha,-)^{1+\beta}(\beta) \\ \varphi^{(k)}(\lambda,\beta) &= \sup_{\Omega_i} \varphi^{(k)}(\lambda_\zeta,\beta) \quad \textit{if type } (\lambda) = i < k \\ \varphi^{(k)}(\lambda,\beta) &= \varphi^{(k)}(\lambda_\beta,\beta) \quad \textit{if type } (\lambda) = k. \end{aligned}$$

Note.
At level $k = 0$ we have, for all structured $\alpha \in \Omega_1$ and all $n \in N$,

$$\varphi^{(0)}(\alpha, n) = F_\alpha(n)$$

where F_α is defined with respect to the presentation induced by α.

Definition 11. *For each k let $\omega_k \in \Omega_{k+1}$ be the identity on Ω_k. Define $\tau = \sup \tau_k \in \Omega_1$ where $\tau_0 = 0$ and for each $k > 0$,*

$$\tau_k = \varphi^{(1)}(\varphi^{(2)}(\ldots\varphi^{(k-1)}(\varphi^{(k)}(0,\omega_{k-1}),\omega_{k-2})\ldots,\omega_1),\omega_0).$$

By Kadota (1993), τ is structured.
By Wainer (1989), $G_{\tau_{k+1}} = F_{\tau_k}$.
This establishes our second bootstrapping principle.

Theorem 12. *For each $k \in N$,*

$$\tau_{k+1} = \textit{ that } \beta \prec \tau \textit{ such that } F_{\tau_k} = G_\beta.$$

With some careful technical modification one can carry through the ordinal analysis of the theories ID_k by Buchholz (1987) using, instead of his collapsing functions $D : \Omega_{k+1} \to \Omega_1$, the functions

$$\alpha \mapsto \varphi^{(1)}(\varphi^{(2)}(\ldots\varphi^{(k)}(\alpha,\omega_{k-1})\ldots,\omega_1),\omega_0).$$

His analysis computes the proof theoretic ordinal of ID_k as $D(\varepsilon_{\omega_k+1})$ which, in terms of φ, is equivalent to substituting $\alpha := \varphi^{(k+1)}(\varphi^{(k+2)}(0,\omega_{k+1}),\omega_k)$ in the above expression. This gives τ_{k+2} precisely. Thus $|\tau|$ is the proof-theoretic ordinal of $ID_{<\omega}$ and, again by Buchholz's analysis, the functions elementary-recursive in the sequence $\{F_\beta : \beta \prec \tau\}$ are exactly those provably recursive in the theory $\Pi^1_1 - CA_0$.

References.

(1) W. Buchholz, "An independence result for $(\Pi_1^1 - CA) + (BI)$", Annals of Pure and Applied Logic Vol. 23, 1987, 131-155.

(2) W. Buchholz, E.A. Cichon and A. Weiermann, "A uniform approach to fundamental sequences and subrecursive hierarchies", Math. Logic Quarterly Vol. 40, 1994, 273-286.

(3) H. Friedman and M. Sheard, " Elementary descent recursion and proof theory", Annals of Pure and Applied Logic Vol. 71, 1995, 1-45.

(4) A.J. Heaton and S.S. Wainer, "Axioms for subrecursion theories", in S.B. Cooper, T.A. Slaman, S.S. Wainer (Eds.) Computability, Enumerability, Unsolvability, LMS Lecture Notes 224, Cambridge Univ. Press 1996, 123-138.

(5) A.J. Heaton, "A jump operator for subrecursion theory", in preparation.

(6) N. Kadota, "On Wainer's notation for a minimal subrecursive inaccessible ordinal", Math. Logic Quarterly Vol. 39,1993, 217-227.

(7) S.C. Kleene, "Extension of an effectively generated class of functions by enumeration", Colloquium Mathematicum Vol. 6, 1958, 67-78.

(8) S.S. Wainer, "Slow growing versus fast growing", Journal of Symbolic Logic Vol. 54, 1989, 608-614.

(9) S.S. Wainer, "Hierarchies of provably computable functions", in P. Petkov (Ed.) Mathematical Logic, Plenum Press 1990, 211-220.

(10) S.S. Wainer, "The hierarchy of terminating recursive programs over N", in D. Leivant (Ed.), Logic and Computational Complexity, Springer Lecture Notes in Computer Science 960, 1995, 281-299.

Author Index

Lightning Source UK Ltd.
Milton Keynes UK
UKOW03n0800210317
297142UK00014B/212/P